Human Plasma Lipoproteins

CLINICAL BIOCHEMISTRY

PRINCIPLES · METHODS

APPLICATIONS

3

Series Editors
H.Ch.Curtius
M.Roth

Walter de Gruyter
Berlin · New York
1989

Human Plasma Lipoproteins

Editors

J.C. Fruchart

J. Shepherd

Walter de Gruyter
Berlin · New York
1989

Editors

Professor Jean-Charles Fruchart
Service de Recherche sur les Lipoprotéins
et l'Athérosclérose (SERLIA)
Institut Pasteur
1, rue du Professeur Albert-Calmette
F-59019 Lille Cedex
France

Professor James Shepherd, Ph. D.
University of Glasgow
Department of Pathological Biochemistry
Royal Infirmary
GB-Glasgow G4 OSF
Great Britain

Series Editors

Dr. Hans-Christoph Curtius
Head, Division of Clinical Chemsitry
Department of Pediatrics
University of Zurich
Steinwiesstrasse 75
CH-8032 Zurich
Switzerland

Dr. Marc Roth
Director, Central Laboratory of
Clinical Chemistry
Cantonal Hospital
University of Geneva
CH-1211 Geneva 4
Switzerland

Publishers

Walter de Gruyter & Co.
Genthiner Straße 13
D-1000 Berlin 30
Fed. Republic of Germany
Telephone (030) 260 05-0
Telefax (030) 260 05-251 · Telex 184 027

Walter de Gruyter, Inc.
200 Saw Mill River Road
Hawthorne, N.Y. 10532
USA
Telephone (914) 747-0110
Telefax (914) 747-1326 · Telex 64 66 77

Library of Congress Cataloging-in-Publication Data

Human plasma lipoproteins / editors, J. C. Fruchart, J. Shepherd.
(Clinical biochemistry ; 3)
Includes index.
ISBN 0-89925-549-3 (U.S.)
1. Blood lipoproteins — Pathophysiology. 2. Blood lipoproteins — Metabolism — Disorders. 3. Blood lipoproteins. I. Fruchart, J. C. (Jean-Charles), 1945– II. Shepherd, J. (James), 1944–.
III. Series: Clinical biochemistry (Berlin, Germany) ; 3.
[DNLM: 1. Hyperlipoproteinemias — complications. 2. Lipoproteins — blood. 3. Lipoproteins — metabolism. QU 85 H918]
RC632.L54H86 1989 616.3'997—dc20 89-1443

Deutsche Bibliothek Cataloging-in-Publication Data

Human plasma lipoproteins / eds.: J. C. Fruchart ; J. Shepherd. — Berlin ; New York : de Gruyter, 1989
(Clinical biochemistry ; 3)
ISBN 3-11-010734-1
NE: Fruchart, Jean-Charles [Hrsg.]; GT

Printed in Germany.
Typesetting: Arthur Collignon GmbH, Berlin. — Printing: Gerike GmbH, Berlin. — Binding: Dieter Mikolai, Berlin. — Cover design: Hansbernd Lindemann, Berlin.

Preface

Lipoprotein disorders are among the most common metabolic derangements encountered in industrialised countries. Figures for the United States of America, for example, lead us to the belief that more than half of the adult population of that country harbor plasma cholesterol levels which are high enough to produce a significantly increased risk of coronary artery disease. In Europe, the situation is, if anything, worse. It is therefore not surprising that ischaemic heart disease is the commonest cause of death in the western world.

Can anything be done to stem the tide of this epidemic? Epidemiological surveys give some cause for optimism in this regard and intervention studies have confirmed that a downward trend in heart disease mortality can be achieved by attacking those factors which are linked to coronary risk. Elevated blood lipoprotein levels fall into that category.

The present volume focuses on the structure and metabolism of human plasma lipoproteins and outlines those defects which may give rise to hyperlipidemia and increased risk atherosclerosis. Although genetic hyperlipidemias are examined in some detail, particular attention is given to the secondary derangements of lipid metabolism which commonly follow from dietary indiscretion, or over-indulgence in alcohol, or are linked to hepatic, renal, or endocrine disease. Where treatment of the primary condition fails to restore normolipemia, the reader is offered information on lipid lowering diet or drug therapy.

This volume is intended primarily for the general physician or chemical pathologist who is seeking a broad background to the aetiology of hyperlipoproteinemia, but will also be useful as a reference volume for scientists and clinicians involved in lipoprotein research, or for the occasional student who wishes to delve more deeply into a rapidly developing and fascinating area of investigative medicine.

July 1989

Jean-Charles Fruchart
James Shepherd

Contributors

Peter Alaupovic, Ph. D.
Head, Lipoprotein and Atherosclerosis Research Program
Oklahoma Medical Research Foundation
825 NE 13th Street
Oklahoma City, OK 73104
U.S.A.

Jean-Louis Beaumont
Professor of Medicine
Inserm U. 32
Hôpital Henri Mondor
F-94010 Créteil
France

Violette Beaumont, M. D.
Inserm U. 32
Hôpital Henri Mondor
F-94010 Créteil
France

Ulricke Beisiegel, Dr.
Medizinische Kern- und Poliklinik
Universitätskrankenhaus Eppendorf
Martinistraße 52
D-2000 Hamburg 20
West-Germany

Piers R. Blackett, M. D.
Assistant Professor of Pediatrics
Department of Pediatrics
Health Sciences Center
University of Oklahoma
P.O. Box 26901
Oklahoma City, Oklahoma 73190
U.S.A.

Richard Deckelbaum
Associate Professor of Pediatrics
Biochemical Genetics and Metabolism
The Rockefeller University
1230 York Avenue
New York, N.Y. 10021-6399
U.S.A.

Philippe Douste-Blazy
Professor
Unité de Prévention des Facteurs de Risque de l'Athérosclérose
Service de Cardiologie
Pavillon Baudot
Chu Purpan
Place du Dr. Baylac
F-31059 Toulouse Cédex
France

Shlomo Eisenberg, M. D.
Professor of Medicine
Lipid Research Laboratory
Department of Medicine B
Hadassah University Hospital
Ein-Kerem
Jerusalem
Israel

Jean-Charles Fruchart
Professor of Biochemistry
Head, Laboratory of Lipid and Lipoprotein Studies
Institute Pasteur
University and Inserm U. 325
1 rue du Professeur Calmette
F-59019 Lille, Cédex
France

Henry N. Ginsberg, M. D.
Associate Professor of Medicine
Department of Medicine
Columbia University
College of Physicians and Surgeons
630 West 168th Street
New York, N.Y. 10032
U.S.A.

Ira J. Goldberg, M. D.
Assistant Professor of Medicine
Department of Medicine
Columbia University
College of Physicians and Surgeons
630 West 168th Street
New York, N.Y. 10032
U.S.A.

Antonio M. Gotto, Jr., M. D., D. Phil.
Department of Medicine
Baylor College of Medicine
Internal Medicine Service
The Methodist Hospital
Houston, TX 77030
U.S.A.

John R. Guyton, M. D.
Associate Professor of Medicine
Department of Medicine
Baylor College of Medicine
6565 Fannin MS A-601
Houston, TX 77030
U.S.A.

Hans U. Kloer, Dr.
Professor of Medicine
Laboratory of Lipid and Lipoprotein
Studies OMRF
825 Northeast 13th Street
Oklahoma City, OK 73140
U.S.A.

Gerhard Max Kostner, Dr.
Professor
Institute for Medical Biochemistry
Karl-Franzens-Universität
Harrachgasse 21
A-8010 Graz
Austria

Peter Laggner, Dr. phil.
Professor
Institut für
Röntgenfeinstrukturforschung
der Österreichischen Akademie
der Wissenschaften
und der Forschungsgesellschaft
Joanneum
Steyrergasse 17
A-8010 Graz
Austria

Aapo Lehtonen, M. D.
Department of Medicine
Turku City Hospital
SF-20700 Turku
Finland

Charles S. Lieber, M. D.
Professor of Medicine and Pathology
Mount Sinai School of Medicine
of the City University of New York
Director, Alcohol Research and
Treatment Center
Veterans Administration Medical Center
(151/G)
130 West Kingsbridge Road
Bronx, New York 10468
U.S.A.

Santica Marcovina, Dr.
Director, Department of
Immunochemistry and Lipoproteins
Scientific Institute San Raffaele
Via Olgettina 60
I-20132 Milano
Italy

Rubino C. Mordasini, Dr., M. D.
Heart Center Clinic
Beau-Site
CH-3000 Bern
Switzerland

Paul John Nestel, M. D., FRACP
Professor of Clinical Medicine
Chief, CSIRO, Division of Human Nutrition
The Flinders University of South Australia
Kintore Avenue
Adelaide SA 5000
Australia

Christopher J. Packard, BSc, Ph. D., MRCPath
Department of Pathological Biochemistry
Royal Infirmary
University of Glasgow
GB-Glasgow G4 OSF
Great Britain

Jean-Pierre Pignon, Dr.
Department of Medical Statistics
Institute Gustave-Roussy
39, rue Camille Desmoulins
F-94805 Villejuif, Cédex
France

Pascal Puchois, Dr.
Laboratory of Lipid and Lipoprotein Studies
Pasteur Institute
1, rue du Professeur Calmette
F-59019 Lille, Cédex
France

Walter F. Riesen, Dr. Dr.
Professor of Clinical Chemistry
Lipid Laboratory
University of Bern
Tiefenau-Hospital
CH-3004 Bern
Switzerland

Dominique Robert, Dr., M. D.
Chef, Service Cantonal d'Hémodialyse
Dialysis unit La Providence
CH-2000 Neuchâtel
Switzerland

Wolfgang Johann Schneider, Ph. D.
Professor of Biochemistry
Lipid and Lipoprotein Research Group
Department of Biochemistry
The University of Alberta
328 Heritage Medical Research Centre
Edmonton, Alberta T6G 2S2
Canada

James Shepherd, BSc, MB ChB, Ph. D., MRCPath, MRCP (Glasgow)
Professor of Pathological Biochemistry
Honorary Consultant
Department of Pathological Biochemistry
Royal Infirmary
University of Glasgow
GB-Glasgow G4 OSF
Great Britain

Matti J. Tikkanen, M. D.
Associate Professor of Medicine
Third Department of Medicine
Central University Hospital
SF-00290 Helsinki
Finland

Abbreviations

ABL	abetalipoproteinemia
ACAT	acyl-CoA:cholesterol acyl transferase
AIH	autoimmune hyperlipidemia
apo, apo-Lp	apolipoprotein(s)
ATP	adenosine triphosphate
CAPD	continuus ambulatory peritoneal dialysis
CE	cholesterylester
CETP	cholesteryl ester transfer protein
CHD	coronary heart disease
CMR	chylomicrons remnants
CPPT	coronary primary prevention trial
CYM	chylomicrons
CV	coefficient of variation
DEAE	diethylaminoethyl
DHA	docosahexanoic acid
DSC	differential scanning calorimetry
EC	endothelial cells
EDTA	ethylenediamine tetraacetic acid
EGF	epidermal growth factor
ELISA	enzyme linked immunosorbent assay
EPA	eicosapentaenoic acid
ESR	electron spin resonance
FC	free cholesterol
FCR	fractional catabolic rate
FFA	free fatty acids
FH	familial hypercholesterolemia
GSH	glutathione
HDL	high density lipoproteins
HL	hepatic lipase
HMG CoA reductase	3-hydroxy-3 methylglutaryl coenzyme A reductase
HPLC	high performance liquid chromatography
HTG	hypertriglyceridemia
HTGL	hepatic triglyceride lipase
HTG-VLDL	hypertriglyceridemic VLDL
IDDM	insulin dependent diabetes mellitus
IDL	intermediate density lipoproteins
IG	immunoglobulin
IGT	impaired glucose tolerance
ISA	intrinsic sympathomimetic activity
LCAT	lecithin:cholesterol acyl transferase
LCT	long-chain fatty acid triglycerides

LDL	low density lipoproteins
LMWH	low molecular weight heparin
LP	human plasma lipoproteins
LPL	lipoprotein lipase
LRC CPPT	lipid research clinics coronary primary prevention trial
LTP	lipid transfer proteins
M_r	relative molecular mass
MCT	medium-chain fatty acid triglycerides
MEOS	microsomal enzyme oxidising system
M-HDL	modified HDL (lipid transfer protein modified HDL)
M-LDL	modified LDL (lipid transfer protein modified LDL)
MRFIT	multiple risk factor intervention trial
NIDDM	noninsulin dependent diabetes mellitus
NMR	nuclear magnetic resonance
N-VLDL	normal VLDL
PDGF	platelet derived growth factor
PHLA	post heparin lipolytic activity
PL	phospholipid
PUFA	polyunsaturated fatty acids
S_f	Svedberg flotation coefficient
SDS-PAGE	sodium dodecyl sulfate polyacrylamide gel electrophoresis
SMC	smooth muscle cells
T_3	triiodothyronine
TG	triglycerides
TMS	trimethylsilyl
TNM	tetranitromethane
TSH	thyroid stimulating hormone
Ty	thyroxine
VLDL	very low density lipoproteins
WHHL rabbits	watanabe heritable hyperlipidemic rabbits

Contents

Chapter 1
Lipoproteins in Health and Disease: Lipoprotein Nomenclature and the Classification of Hyperlipoproteinemia

J. Shepherd and J.-C. Fruchart

1 Introduction

Lipids play a central role in human metabolism, participating in such diverse functions as the maintenance of cellular integrity, the storage of energy, the provision of metabolic intermediates and the transmission and transduction of signals. Although we have the capacity to synthesise them in abundance, most of us consume enough or more than enough in our diet to satisfy our daily requirements. The average Western diet provides approximately 40% of our total energy needs in the form of triglyceride. Each day we ingest 100 – 150 g of this lipid along with 0.5 – 1.0 g of cholesterol and require to transmit them from their sites of absorption in the gut to the liver and interstitial tissue for utilisation or storage. Mobilisation of such hydrophobic substances through the aqueous environment of the plasma necessitated the evolution of a system for the production of stable dispersions of fat in water. This was achieved by surrounding the insoluble globules of dietary cholesteryl ester and triglyceride with a shell of so-called amphipathic proteins and phospholipids, hydrophilic on one face and hydrophobic on the other. The lipid-protein complexes or chylomicrons generated in this process are common to all vertebrate species. But, although responsible for the bulk of lipid transport through the bloodstream, chylomicrons are not the sole vehicles involved in this activity. In the fasting state, when intestinal lipoprotein production is minimal (although not completely suspended), smaller, triglyceride-rich particles continue to be elaborated and secreted by the liver (see Chapter 3). Intravascular remodelling results in their conversion to species of reduced size and lipid content which express a wide spectrum of physical properties. This characteristic polydispersity among the plasma lipoproteins has complicated not only our efforts to unravel their structure and function but has also confounded our efforts to classify them in a systematic fashion. Notwithstanding, several attempts have been made to devise operational definitions for them based on the methods employed in their fractionation. This approach, though not entirely satisfactory, at least is practical and, because of the inherent complexities of the lipoprotein spectrum within the circulation, is likely to see continuing application for the foreseeable future.

2 Lipoproteins in Healthy Subjects: Nomenclature

Human plasma lipoproteins may be categorised according to their

(1) buoyant density
(2) electrophoretic mobility
(3) size
(4) apolipoprotein content

and each approach has its own merits and detractions. Table 1 outlines the relationships which exist between these classification procedures.

2.1 Lipoprotein classification on the basis of buoyant density

The ultracentrifuge was first used, albeit inadvertently, by McFarlane (1) to examine the lipoprotein spectrum in normal and pathological serum in 1935. His report, later confirmed by others (2, 3), indicated that a labile protein (the "x protein") whose shape was influenced by the presence of salt and by sample dilution, cosedimented proximal to albumin in the analytical cell. Its nature remained a mystery until 14 years later when Gofman and his colleagues (4) hit upon the revolutionary idea that this "protein" might in fact represent a component within their system which was accumulating with time on the slowly sedimenting albumin boundary. That is, the "x protein" was buoyant at the density of albumin; and further experimentation showed that if the solution background density within the analytical cell was raised to substantially greater than 1.03 kg/l the "x protein" actually floated – the concept of ultracentrifugal flotation of lipoproteins was born. From these beginnings, Gofman and his associates (5) proceeded to characterise the spectral distribution of the human plasma lipoproteins using analytical ultracentrifugation (Fig. 1). Their detection of absorption minima at solvent densities of 1.006, 1.019, 1.063, 1.125 and 1.210 kg/l laid the foundation for the classification of lipoproteins into species of very low density (VLDL, $\varrho < 1.006$ kg/l), intermediate density (IDL, $\varrho = 1.006-1.019$ kg/l), low density (LDL, $\varrho = 1.019-1.063$ kg/l) and high density (HDL_2, $\varrho = 1.063-1.125$ kg/l; HDL_3, $\varrho = 1.125-1.210$ kg/l) and

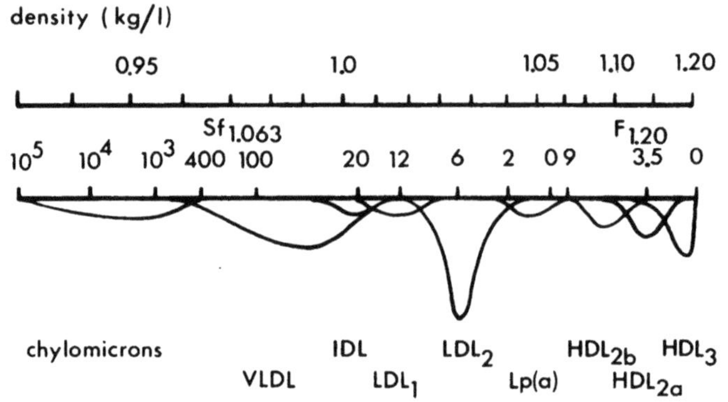

Fig. 1. Analytical ultracentrifugation of plasma lipoproteins. (Adapted from ref. 7.)

led to the development of the sequential preparative flotation procedure (6) which is widely employed today to isolate the major plasma lipoprotein fractions using a salt-generated stepped increase in background density. The facility of this approach has ensured its continued use even although it is known (8 – 10) that some lipoproteins are disrupted or modified during ultracentrifugation. Table 1 describes some of the properties of the five major plasma lipoprotein density classes and relates their densities to flotation rate in the analytical ultracentrifuge. At a background density of 1.063 kg/l and a temperature of 26 °C, chylomicrons display flotation (S_f) rates of greater than 400, while the equivalent values for VLDL, IDL and LDL are 20 – 400, 12 – 20 and 0 – 12 respectively. When the salt density is raised to 1.20 kg/l, HDL_2 and HDL_3 exhibit flotation rates of 3.5 – 9 and 0 – 3.5. It should be noted that the particle distributions encapsulated within each of these flotation intervals are polydisperse.

2.2 Lipoprotein classification by electrophoretic mobility

The lipoprotein fractions in the plasma carry an electrical charge and consequently will migrate in an electric field, permitting their resolution into four major classes (Table 1, Fig. 2). Chylomicrons become trapped in the interstices of the support medium and therefore remain at the origin of the electrophoretogram. VLDL migrates with the α_2 (or pre-β) globulin fraction, LDL with the β-globulins, and HDL with the α_1-globulins. Thus, these operational terms, based on the independent

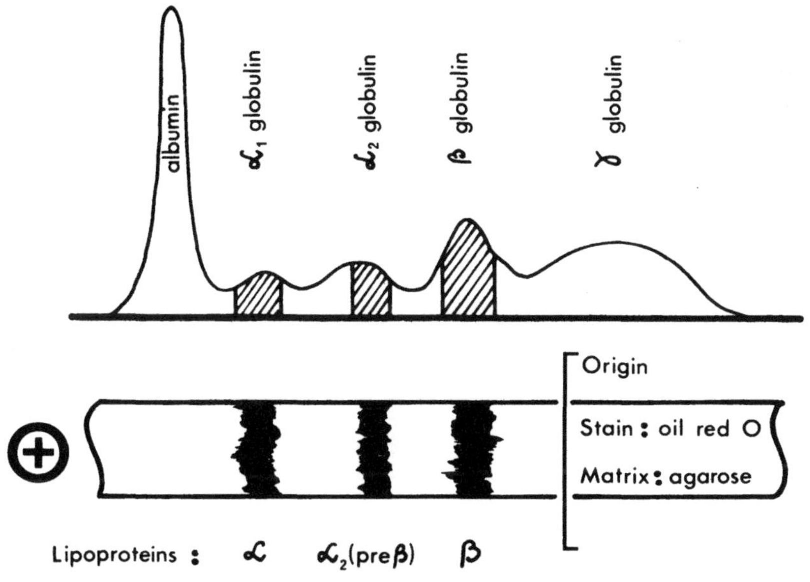

Fig. 2. Electrophoresis of serum lipoproteins. Serum lipoproteins are separable into three major classes (α, pre-β and β) on the basis of their electrophoretic mobility in a variety of matrices including paper, agarose, cellulose acetate and polyacrylamide. Agarose is most commonly used and is considered to give the best results.

Table 1. Operational classification of the human plasma lipoproteins

Lipoprotein	Buoyant density (kg/l)	Flotation rate at 26 °C and at a background density of 1.063 kg/l	Flotation rate at 26 °C and at a background density of 1.20 kg/l	Mean particle mass (10^6 daltons)	Electrophoretic mobility	Apolipoprotein content % A	B	C	D	E	Mean diameter (nm)
Chylomicrons	<0.95	>400	–	1000	origin	1	40	47	trace	12	100–1000
Very low density lipoproteins	0.95–1.006	20–400	–	4.5–100	α_2 (pre-β)	trace	54	38	trace	8	30–70
Intermediate density lipoproteins	1.006–1.019	12–20	–	4.0	α_2–β	–					25
Low density lipoproteins	1.019–1.063	0–12	–	3.0	β	1.4	90	5	trace	3.5	20
High density lipoprotein Subfraction 2	1.063–1.125	–	0–3.5	0.4	α_1	82	3.7	8.9	2.9	2.1	10
High density lipoprotein Subfraction 3	1.125–1.21	–	3.5–9	0.2	α_1	95	Nil	2.8	2.2	0.4	8

properties of density and electrophoretic mobility, are often used interchangeably although they are in fact not synonymous. A variety of support media, including paper (11, 12), agarose (13, 14), starch gel (15, 16), agar (17), and more recently polyacrylamide (18, 19) have been employed for lipoprotein electrophoresis. The latter, whose use requires that the lipoproteins be stained prior to electrophoresis, is often helpful for special analyses but is less appropriate for routine service since the lipoprotein species which accumulate in the plasma in some dyslipidemias may migrate anomalously and create interpretational difficulties. Finally, it is important to note that lipoprotein electrophoresis should never be used by itself to differentiate normal from abnormal lipid transport (20). The procedure actually visualises the ester bonds on triglyceride and cholesteryl esters. Phospholipids and free cholesterol are not detected. Consequently, the electrophoretogram of plasma obtained from an individual with, for example, biliary cirrhosis, who may have tremendous increments of free cholesterol and phospholipid in his plasma often with the accumulation of an abnormal lipoprotein (lipoprotein x — see Chapter 2) may not look particularly unusual even although his cholesterol concentrations may be substantially in excess of 25 mmol/l. For diagnostic purposes therefore, lipoprotein electrophoretograms must be coupled with more quantitative methods such as measurement of plasma cholesterol and triglyceride. The use of the electrophoretogram alone in this circumstance is to be deprecated.

2.3 Lipoprotein classification on the basis of particle size

The electrophoretic mobilities and hydrated densities of the plasma lipoproteins form the basis of their operational classification and have been exploited to fractionate the particles for both diagnostic and preparative purposes. A third colligative property of the lipoproteins, their size, has not yet been widely used to effect their separation even although three decades have passed since molecular sieve chromatography was first recommended for this purpose (21). In the interim, the technique has been adopted in a number of laboratories to characterise circulating lipoproteins from various species (22, 23, 24). A six percent agarose gel is the most appropriate commercially available medium for separating the major lipoproteins in the plasma. Chylomicrons and VLDL are excluded from the gel and appear in the void volume while LDL and HDL penetrate and are retarded by the matrix mesh (Fig. 3). Selection of gels with larger pore size (e. g. 2% agarose) permits resolution of the particles within VLDL into two peaks, one at the void volume of the column bed and the other within the included volume of the gel. In this system, LDL and HDL elute near the total column volume and are incompletely resolved from each other. Recently, a simple and rapid method has been developed for lipoprotein analysis using high performance gel permeation chromatography (25) which permits direct measurement of the cholesterol, phospholipid or triglyceride content of each lipoprotein class present in small amounts (10–20 µl) of untreated serum. While this approach is not appropriate for routine lipoprotein analysis, its high resolving power and direct applicability to the examination of untreated serum samples obviates the potential detrimental effects of preliminary ultracentrifugation.

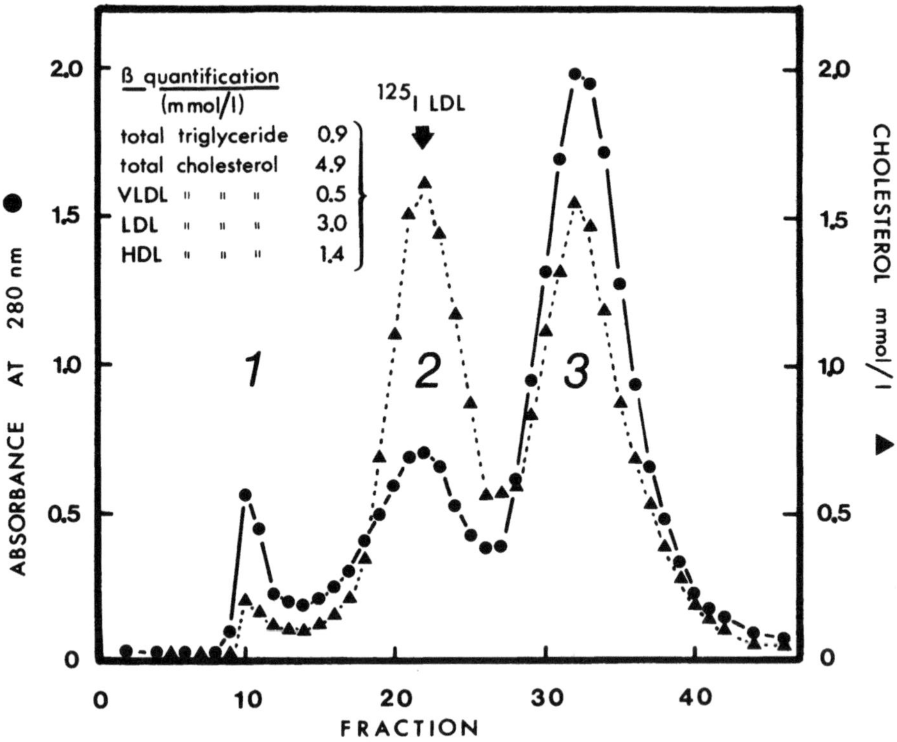

Fig. 3. Agarose gel filtration of plasma lipoproteins. A5m (6%) agarose gel profile of normal plasma lipoproteins. Total plasma lipoproteins were isolated from a control subject by ultracentrifugation and subjected, with a ^{125}I-labelled LDL marker, to gel filtration. The cholesterol content and absorption of each fraction at 280 nm were measured. Peaks 1, 2 and 3 correspond to VLDL, LDL and HDL respectively. The subject's plasma lipid values are given. The procedure (24) is adapted from Rudel et al. (22).

Later in the chapter we shall touch briefly upon the potential usefulness of molecular sieve chromatography in the identification of the common hyperlipoproteinemias.

2.4 Lipoprotein classification by apoprotein content

The apolipoprotein components of the plasma lipoproteins are their major antigenic determinants and confer upon them their metabolic properties. It was therefore not surprising that in the mid 1960's they became the focus of research in the field, and soon thereafter began to be used as a means of classifying the lipoprotein species in human plasma. Credit for developing this approach must go to Alaupovic and his colleagues (26, 27) who coined the ABC nomenclature system which pertains today. According to their concept, the plasma lipoproteins can be viewed as a system of discrete species or families distinguishable from each other on the basis of their apolipoprotein composition. Initially, three families were identified (28); LP-A, containing the "protein" A, and associated with HDL; LP-B, the most abundant fraction in human plasma and the major component of LDL; and LP-C, found primarily in the VLDL density interval, and to a lesser extent in HDL. However, it soon became clear that the "proteins" which formed the basis of the system were not single polypeptides. Apolipoprotein A, for example, turned out to be two distinct proteins (29, 30) to which were appended for identification purposes the Roman numerals AI and AII. By the same token, the apolipoprotein C family is now known to contain three members, CI, CII and CIII. One additional complication has arisen recently. Chemical and physical studies of these proteins have shown that several of them exist in a number of polymorphic forms to which have been designated Arabic numerals. Thus, apolipoprotein AI exists as a number of variants AI-1, AI-2, AI-3 etc. The chemical origins of these variants and the present status of this system of nomenclature is addressed in Chapter 2.

3 Lipoproteins in Disease: Classification of the Hyperlipoproteinemias

In 1967 Fredrickson and his colleagues (31) pioneered a new approach to the description of abnormalities of plasma lipid transport which relied heavily on the use of lipoprotein electrophoresis on paper. Publication of their concepts cleared the way for investigators to probe the primary defects responsible for the development of dyslipoproteinemias. As a result, a number of molecular abnormalities (see Chapters 2 and 3) linked to disorders of lipid transport have come to light. Eventually, this information will permit a radical revision of our classification of the hyperlipoproteinemias. At present however, such a change would be premature and of limited benefit to the lipid clinician. Fredrickson's approach still remains the cornerstone for the practical management of patients with plasma lipid disorders. Technical developments have now relegated lipoprotein electrophoresis to a supportive role in diagnosis. Individual lipoprotein families are currently measured in terms of their cholesterol concentration, except for chylomicrons which continue to be estimated qualitatively (see Chapter 4 for details). The procedure (32), called lipoprotein quantification or β-quantification (Fig. 4), involves direct assessment of total plasma cholesterol and triglyceride. A sample of the same plasma is then subjected to ultracentrifugation in order to separate buoyant VLDL from denser LDL and HDL. The VLDL at the top of the tube and the infranatant LDL, HDL and residual plasma proteins are harvested separately by tube slicing and their cholesterol contents measured. Subsequent quantitation of the HDL cholesterol concentration following precipitation of the LDL in the infranate by the addition of heparin and manganese chloride (33, 34) then permits calculation of the LDL cholesterol content of the sample. As an alternative for those laboratories which do not have access to an ultracentrifuge, it is possible to estimate LDL cholesterol (in mmol/l) by using the following formula (35):

$$\text{LDL chol} = \text{total chol} - \text{HDL chol} - \frac{\text{total Tg}}{2.2}$$

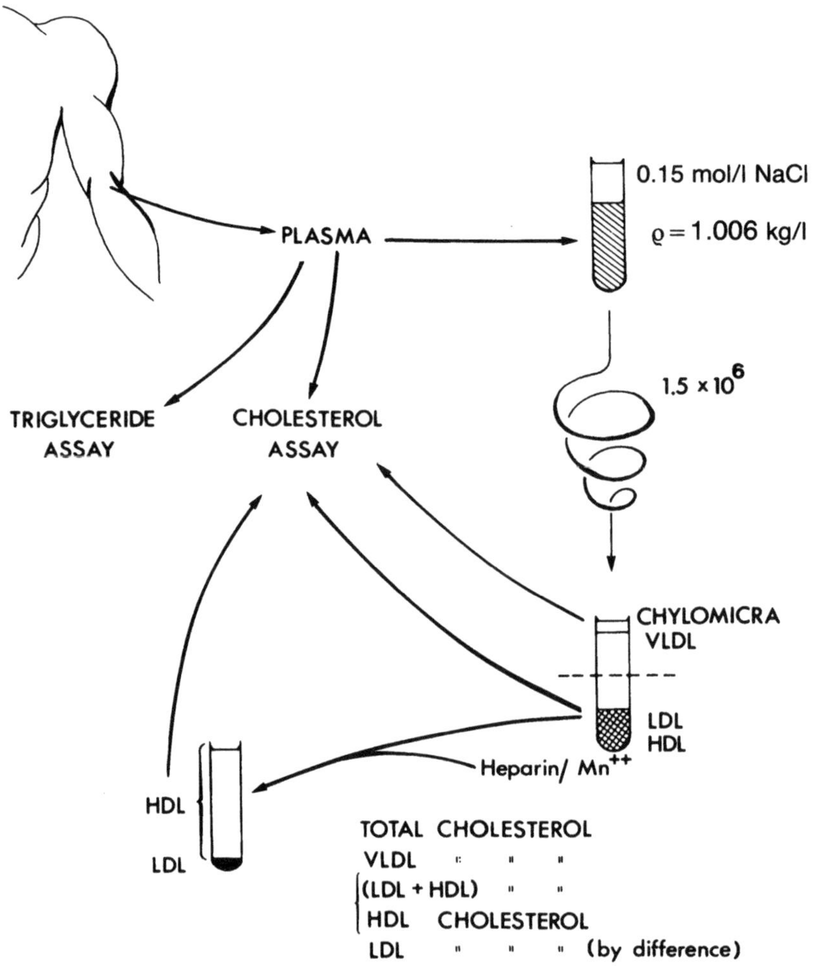

Fig. 4. Lipoprotein quantitation in anglehead rotors. Lipoprotein quantification is carried out as follows:

a) Total plasma cholesterol is measured directly.
b) VLDL ($\varrho < 1.006$ kg/l) is separated from the denser classes, LDL and HDL, by overnight centrifugation and a cholesterol assay is made on the isolated VLDL fraction.
c) LDL is precipitated from the $\varrho > 1.006$ kg/l infranate by addition of heparin and manganese chloride (92 mmol/l, ref. 34) and the cholesterol content of the HDL which remains in the supernatant is measured.
d) Finally, the LDL cholesterol is calculated by subtracting the value obtained for HDL cholesterol from the total cholesterol content of the $\varrho > 1.006$ kg/l fraction.

This procedure is adapted from the Lipid Research Clinics Manual of Laboratory Operations (32).

Table 2. Hyperlipoproteinemia: Phenotype and Genotype

Phenotype	Synonym	Clinical presentation	Laboratory findings	Associated genetic disorder
Type 1	Primary hyperchylomicronemia Burger-Grutz disease	Childhood disease Eruptive papular xanthomatosis Hepatosplenomegaly Pancreatitis Lipemia retinalis	Triglyceride ↑↑ Cholesterol ↑ Fasting chylomicronemia Post heparin lipase deficiency	Familial lipoprotein-lipase deficiency
Type IIa	Hypercholesterolemia	Present in childhood but usually manifest in adult life by premature vascular disease. Tendinous xanthomatosis and family history may be present	Triglyceride → Cholesterol ↑↑ LDL increased	Polygenic hyper-cholesterolemia Familial hyper-cholesterolemia LDL receptor deficiency
Type IIb	Combined hyperlipidemia	Present in childhood but usually manifest in adult life by premature vascular disease. Tendinous xanthomatosis and family history may be present	Triglyceride ↑ Cholesterol ↑↑ LDL and VLDL increased	Familial combined hyperlipidemia Familial hyper-cholesterolemia
Type III	Familial dysbetalipoproteinemia	Planar palmar xanthomata	Triglyceride ↑ Cholesterol ↑	Apolipoprotein E_2 homozygosity
	Remnant removal disease	Tuberous and tendinous xanthomata	Chylomicron and VLDL remnants (and IDL) increased	
	Floating beta disease	Premature central and peripheral vascular disease		
Type IV	Endogenous hypertriglyceridemia	Premature vascular disease. Occasional eruptive xanthomatosis – hepatosplenomegaly and pancreatitis	Triglyceride ↑↑ Cholesterol → VLDL increased	Familial hyper-triglyceridemia Famial combined hyper-lipidemia
Type V	Mixed hyperlipoproteinemia	Eruptive papular xanthomatosis Hepatosplenomegaly Pancreatitis Lipemia retinalis	Triglyceride ↑↑ Cholesterol ↑ Fasting chylomicronemia VLDL increased LDL and HDL decreased	Apolipoprotein CII deficiency Familial combined hyperlipidemia Familial hyper-triglyceridemia

This approach requires the direct measurement of HDL cholesterol in the plasma following heparin/manganese precipitation of VLDL and is based on three assumptions, viz.:

(1) that the sample is free of chylomicrons;
(2) that essentially all of the plasma triglycerides are in VLDL;
(3) that the molar ratio of triglyceride to cholesterol in the VLDL is 2.2.

These assumptions are not valid at plasma triglyceride concentrations greater than 3.5 mmol/l, when ultracentrifugal analysis becomes mandatory.

Table 2 outlines the phenotypic classification system developed by Fredrickson and his colleagues in 1967 (31) and endorsed by Beaumont et al. (36) in the WHO Bulletin in 1970. Each of the phenotypes may be primary, or secondary to other disease states, the more common of which – diabetes mellitus, alcoholism, obstructive liver disease, renal disease, hypothyroidism, and autoimmunity – are examined in some detail later in this volume. Here we shall concentrate on the clinical and biochemical features of the primary condition and shall attempt (Table 2) to relate them, where possible, to the molecular defects known to affect lipoprotein processing (Chapter 3).

3.1 Type I hyperlipoproteinemia

This rare autosomal recessive abnormality becomes manifest in childhood as an intolerance of dietary fat (37, 38). The affected child usually presents with a widespread cutaneous maculopapular rash, often accompanied by abdominal colic which may progress to full-blown pancreatitis. Fundal inspection reveals the presence of lipemia in the retinal vessels, reflecting the role of hyperchylomicronemia in the pathogenesis of the disease. Plasma obtained from affected individuals is very turbid and separates into a creamy upper layer with a clear infranatant upon standing for 24 hours at 4 °C. Triglyceride levels may reach 50 mmol/l or more (upper reference value = 2.3 mmol/l) and analytical ultracentrifugation (38) or gel filtration (24) of the lipoproteins (Fig. 5) clearly shows that there is an accumulation of large triglyceride-rich particles of the size range and density of chylomicrons. Plasma HDL and LDL are low, and the levels of their associated apolipoproteins, Apo AI, AII and B are reduced by about half (39). Most cases of Type I hyperlipoproteinemia derive from defective expression of the plasma enzyme lipoprotein lipase (40) which normally is responsible for the clearance of triglyceride-rich particles from the circulation. However, a few arise (Table 2) as a result of congenital absence (41) of the Apo CII activator of the enzyme (see Chapter 3). The profile of symptoms in such individuals is usually milder and in general the condition presents later in life. Apo CII deficient patients show a rapid and remarkable resolution of their hypertriglyceridemia upon infusion of small amounts of this protein or even of normal plasma (41), but long term therapy of this kind would be impracticable. Instead, treatment of all Type I hyperlipoproteinemic subjects is directed at reducing intestinal production of chylomicrons by severely limiting dietary fat intake, long chain fatty acids being replaced by medium chain triglyceride with less than 14 carbon atoms in the fatty acid moiety. The latter do not result in

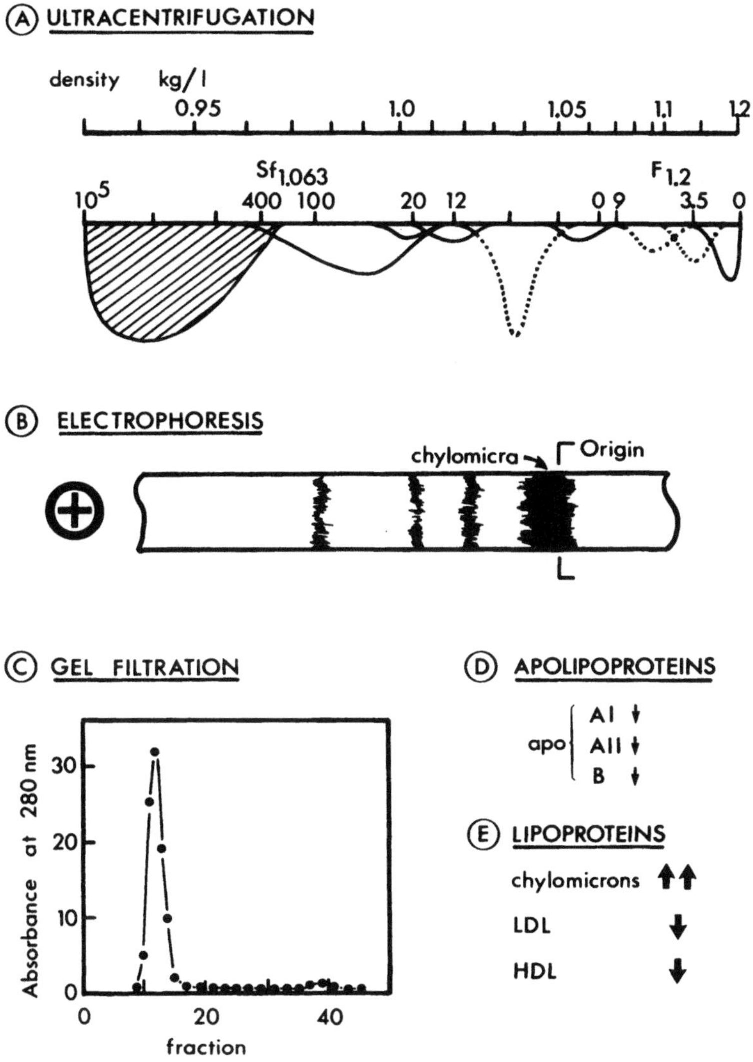

Fig. 5. Type I hyperlipoproteinemia.
Type I hyperlipoproteinemia or primary hyperchylomicronemia results from defective plasma clearance of intestinally derived chylomicron particles which, as a result, persists in the plasma even in the post absorptive state. The presence of these large, triglyceride-rich particles in the sample is detectable by ultracentrifugation (A), electrophoresis (B) or gel filtration (C). Concomitant reductions in plasma LDL and HDL produce a fall in the plasma levels of the major apoprotein constituents of these lipoprotein classes (apolipoproteins AI, AII and B). This subject's plasma lipoprotein profile was

	mmol/l
Total triglyceride	15.25
Total cholesterol	5.0
VLDL cholesterol	3.1
LDL cholesterol	1.1
HDL cholesterol	0.8

the generation of triglyceride-rich particles in the intestine, but are transported, albumin-bound, in the portal blood to the liver. Type I hyperlipoproteinemia does not seem to predispose to early vascular disease.

3.2 Type II hyperlipoproteinemia

Type II hyperlipoproteinemia is common, affects both sexes equally, and is expressed at all ages. Its clinical importance derives from its association with premature vascular disease. In a recent survey of 3820 apparently healthy individuals in the city of Glasgow (Table 3), 16.1% of males and 9.9% of females aged less than 40 years had plasma cholesterol levels above the action limit for their age (6.5 mmol/l).

Table 3. Distribution of plasma cholesterol levels in a Glasgow population ($n = 3820$)

	Males		Females	
	aged 25–40 years ($n = 667$)	aged >40 years ($n = 1205$)	aged 25–40 years ($n = 654$)	aged >40 years ($n = 1294$)
Cholesterol >7.5 mmol/l	4.2%	6.4%	2.8%	13.2%
Cholesterol >7.0 mmol/l	8.7%	16.2%	4.5%	21.5%
Cholesterol >6.5 mmol/l	16.1%	28.4%	9.9%	34.7%
Cholesterol >6.0 mmol/l	25.4%	45.6%	20.1%	51.7%

Most of these subjects probably suffer from a polygenic inability to metabolise efficently their normal dietary cholesterol load (Table 2); and dietary intervention achieved a favorable clinical response in the majority of cases. In a few instances, however, the patient will harbor the monogenic autosomal dominant condition known as familial hypercholesterolemia (42). The key features of this condition are premature coronary atherosclerosis, tendinous xanthomatosis and a family history of ischemic heart disease. Affected individuals, whose plasma cholesterol levels are usually significantly elevated, rarely show an adequate response to dietary restriction alone. Heterozygosity for the condition which appears with a frequency of 1 in 500 of the population, is associated with LDL cholesterol values which are about twice normal while the rarer, more severely affected homozygotes (1 in a million) may present with levels which reach 20 mmol/l or more. Studies initiated in Dallas, Texas, by J. L. Goldstein and M. S. Brown over 15 years ago have uncovered the mechanisms responsible for the disease (43). They showed that, normally, cells contain on their plasma membranes receptors that bind LDL with high affinity and initiate a sequence of events leading to the endocytic uptake and degradation of the lipoprotein with release of its core cholesterol for the cell's use. Several lines of evidence support the premise that a defect in the activity of this pathway might be responsible for the high LDL levels which characterise familial hypercholester-

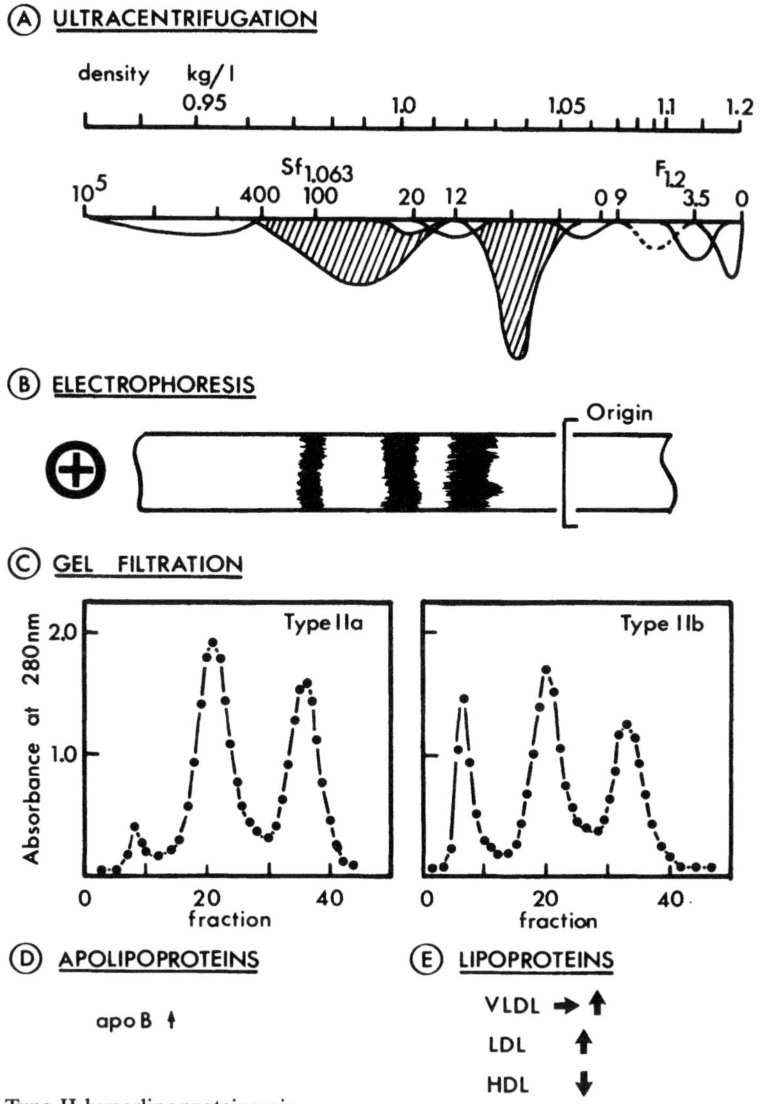

Fig. 6. Type II hyperlipoproteinemia.
Type II hyperlipoproteinemia is associated with a plasma accumulation of LDL (Type IIa) or VLDL and LDL (Type IIb). This results in an increase in plasma apolipoprotein B and commonly leads to a reduction in circulating HDL. The plasma lipoprotein profiles of the samples used in this illustration were

	Type IIa mmol/l	Type IIb mmol/l
Total triglyceride	1.2	3.7
Total cholesterol	11.2	10.3
VLDL cholesterol	0.6	0.7
LDL cholesterol	9.2	8.2
HDL cholesterol	1.3	1.4

The notations of the ultracentrifugation spectrum, the electrophoretogram and the gel filtration profiles are given in Figures 1, 2 and 3 respectively.

olemia. First, early kinetic studies (44, 45) clearly showed that affected individuals fail to catabolise LDL normally; secondly, evidence from cell culture experiments (46), later confirmed *in vivo* (47), indicated that the activity of the receptor pathway was instrumental in regulating environmental (i. e. plasma) LDL levels; and thirdly, the best response to therapy (48, 49) was seen in those heterozygous patients given drug therapy specifically designed to stimulate cellular LDL receptor activity – homozygotes with ineffective receptors, did not respond. The sustained investigations of the Dallas group have not only confirmed the involvement of the LDL receptor in the pathogenesis of familial hypercholesterolemia but have begun to unravel the molecular bases of the condition (see Chapter 5 for details). Available evidence indicates that a variety of mutations may affect the LDL receptor gene (50 – 52), leading to an absolute defect in receptor production or to the generation of polymorphic forms defective in their ability to initiate and sustain endocytic assimilation of LDL. It is therefore not surprising that expression of the disease and the progress of its clinical course shows such variation from patient to patient.

Diagnosis of Type II hyperlipoproteinemia is based on the detection of hypercholesterolemia in an individual due to a selective increase in LDL alone (Type IIa) or in LDL and VLDL (Type IIb). Ultracentrifugation, lipoprotein electrophoresis and gel exclusion chromatography (Fig. 6) all provide useful confirmatory data, and apolipoprotein B measurements usually show that the plasma levels of this protein are approximately twice normal (39). Laboratory diagnosis of familial hypercholesterolemia on the basis of the assay of LDL receptor expression on isolated lymphocytes or cultured fibroblasts is much less straightforward (53 – 55). The difficulty arises not with the homozygotes, but among heterozygotes whose cell receptor assays may show substantial overlap with normal. Recent refinements in the assay procedure (56) may help to resolve this diagnostic problem.

3.3 Type III hyperlipoproteinemia

Normally, the triglyceride-rich lipoproteins, chylomicrons and VLDL are readily distinguishable from LDL by ultracentrifugation, gel filtration or electrophoresis. However, one in five thousand individuals accumulates in his plasma a lipoprotein fraction with physicochemical properties which lie between those of VLDL and LDL. This fraction contains particles with the electrophoretic mobility and high cholesterol content of LDL, the flotation characteristics of VLDL, and a size intermediate between the two (Fig. 7). The metabolic defect responsible for the condition is not yet clear (57), although two independent factors seem to combine together to produce it. First, the primary sequence of amino acids in apolipoprotein E (58 – 60) is defective in dysbetalipoproteinemic patients (see Chapter 2). Though insignificant in terms of its circulating plasma mass, this protein plays a key role in the clearance of triglyceride-rich particle remnants from the plasma. The amino acid sequence defect reduces the rate of elimination of the remnants from the circulation (61) and results in their accumulation in the bloodstream. It therefore follows that apolipoprotein E levels in the plasma would rise, and values of approximately 2 – 3 times normal (39) are common (Fig. 7). However, the E protein defect is not solely responsible for the hyperlipidemia (31, 62, 63) since it occurs in

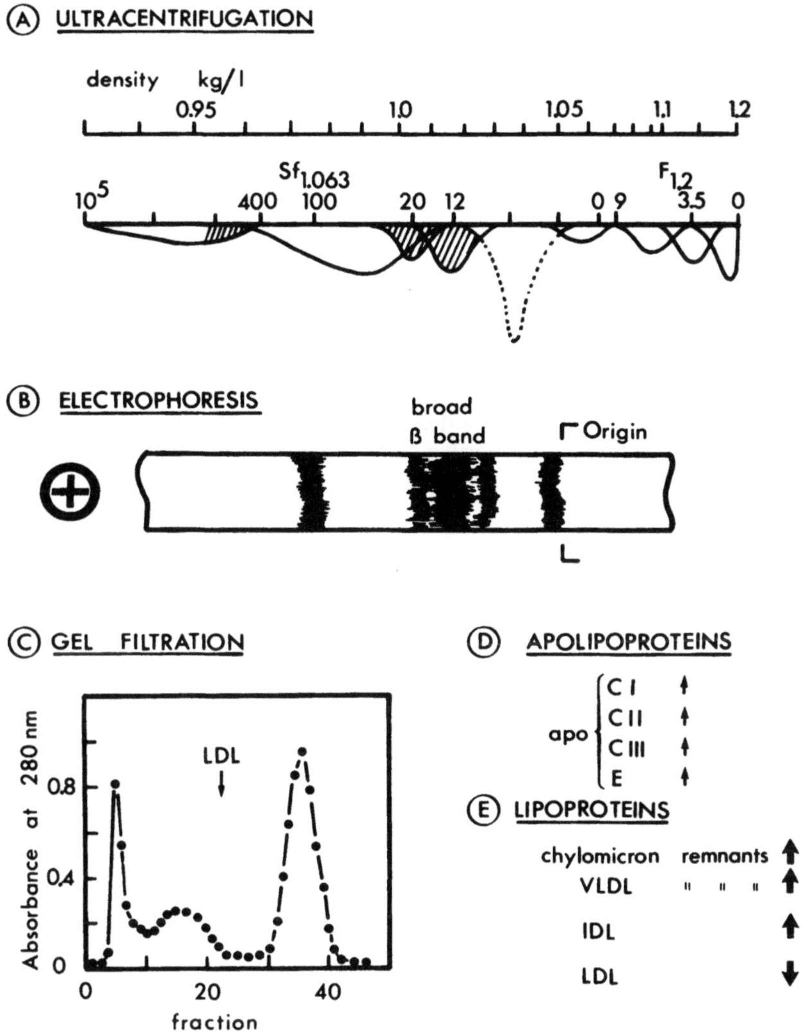

Fig. 7. Type III hyperlipoproteinemia.
A defect in the conversion cascade from VLDL to LDL results in the accumulation of remnants of VLDL and chylomicrons in the plasma. These retain the flotation characteristics of their parent particles but adopt the electrophoretic mobility of LDL. The distinction between β and pre-β lipoproteins on the electrophoretogram is therefore obscured by the presence of a "broad β band", due to the presence of VLDL remnants or IDL in the plasma. These particles are smaller than LDL on gel filtration, and are rich in apolipoproteins C and E, which accumulate in the circulation. LDL levels are usually low in the plasma of Type III hyperlipoproteinemic patients. The lipoprotein profile for this subject was

	mmol/l
Total triglyceride	3.6
Total cholesterol	10.4
VLDL cholesterol	4.3
LDL cholesterol	3.7
HDL cholesterol	2.4

about one percent of the population and yet, as noted above, only about one in five thousand develops Type III hyperlipoproteinemia. Consequently, a second factor must be implicated in its pathogenesis. What this triggering influence may be is not yet clear although the intriguing observation that Type III hyperlipoproteinemia is almost exclusively a disease of adult life (31, 57) suggests that it may be developmental. Interestingly, although the level of LDL in the plasma of Type III hyperlipoproteinemic patients is reduced rather than elevated, they are predisposed to premature atherosclerosis (57) which has been reported to have a predeliction for peripheral vessels. This observation certainly is in accord with the hypothesis of Zilversmit (64) that cholesterol-rich chylomicron remnant particles are major contributors to atherosclerosis in humans; and in fact it is of interest to note that these remnants are the only naturally occurring lipoprotein particles capable of initiating foam cell formation in cultured macrophages (65). It is not yet clear whether similar particles which appear transiently in the plasma of apparently healthy individuals following a fatty meal are associated with an increase in their predisposition to atherosclerosis. We need to gather more information on the extent to which the postprandial levels of such remnants vary from subject to subject.

3.4 Type IV hyperlipoproteinemia

According to Fredrickson's classification system, Type IV hyperlipoproteinemia is associated with accumulation in the plasma of VLDL of normal composition. The condition is therefore readily detectable on the basis of the particle's density, size and charge (Fig. 8); and, as might be expected, the plasma concentration of apolipoprotein CIII, a major C apoprotein in VLDL, is raised approximately three fold. Metabolic studies undertaken in several laboratories (66–68) have suggested that the disease represents an heterogenous collection of defects, some patients apparently oversynthesising the VLDL while others are unable to catabolise it normally. This is consistent with the apparent genetic heterogeneity which characterises the condition (Tab. 2). In some individuals, designated as suffering from familial hypertriglyceridemia the phenotype is transmitted faithfully in apparent autosomal dominant fashion (69). The more common genetic form of hyperlipidemia, familial combined hyperlipidemia, is also transmitted by autosomal dominant inheritance, but may be expressed as either hypertriglyceridemia or hypercholesterolemia in different members of the same family (69). Analysis of informative matings suggested that the different lipid phenotypes originated from variable expression of a single autosomal dominant gene, rather than from segregation of two separate genes such as one elevating the level of cholesterol and the other elevating the level of triglyceride. At present we have no way of distinguishing familial hypertriglyceridemia from familial combined hyperlipidemia except by laborious classical family studies although there is preliminary information to suggest that apolipoprotein measurements may help here (39). In a small pilot study it was found that the molar ratio of apolipoprotein B/CIII was substantially higher in subjects with familial combined hyperlipidemia. This report requires to be confirmed.

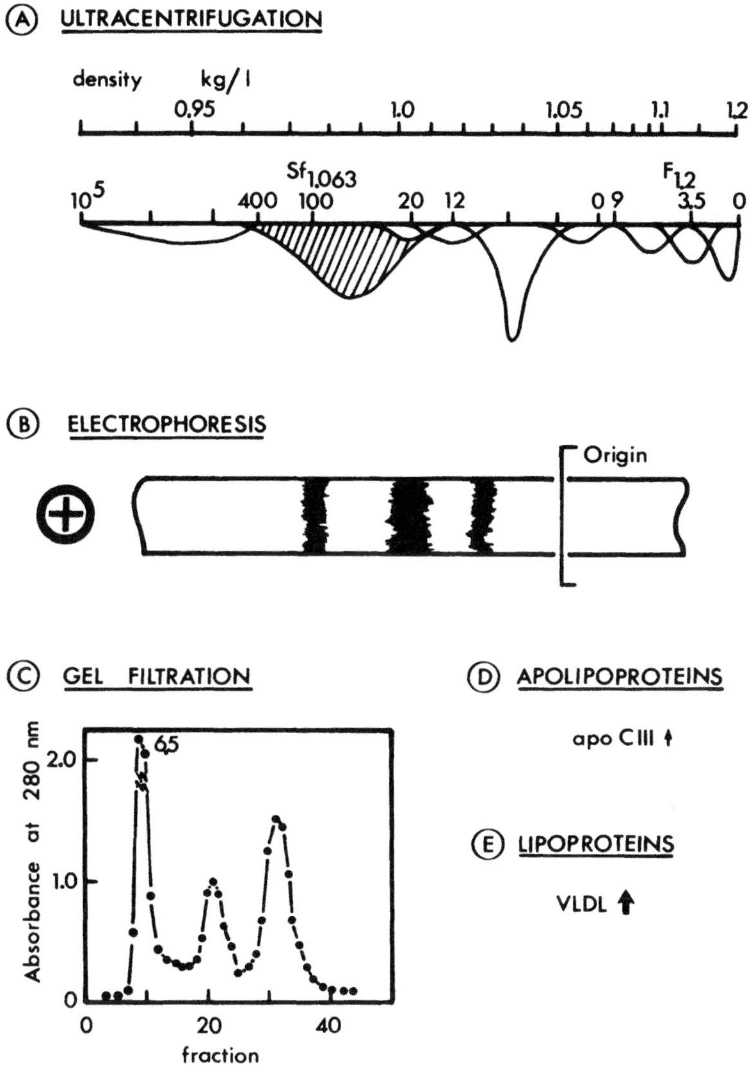

Fig. 8. Type IV hyperlipoproteinemia.
This rubric describes patients who because of oversynthesis or defective catabolism of VLDL accumulate this lipoprotein in their plasma. The pre-β band on their electrophoretogram is therefore accentuated and the accumulated VLDL increases the magnitude of the excluded peak (peak 1, figure 3) on the 6% agarose gel profile. Apolipoprotein C, found predominantly in VLDL, is increased in its plasma concentration. The lipoprotein profile of this patient was

	mmol/l
Total triglyceride	6.3
Total cholesterol	6.3
VLDL cholesterol	0.7
LDL cholesterol	4.5
HDL cholesterol	1.0

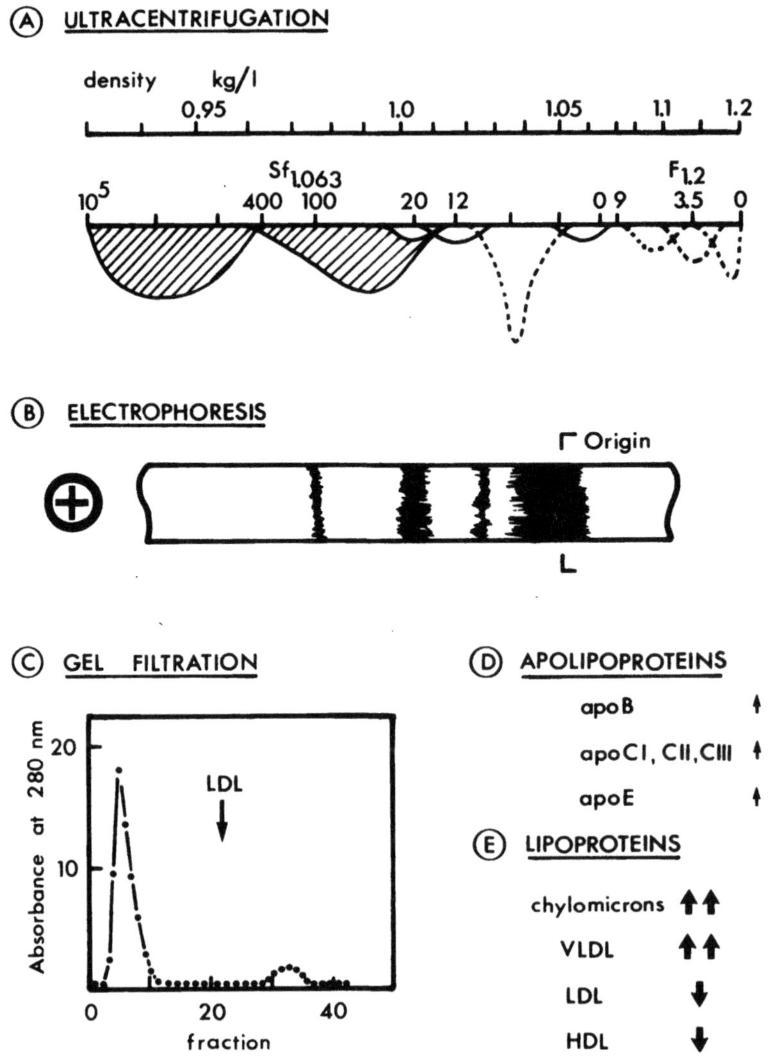

Fig. 9. Type V hyperlipoproteinemia.
Plasma accumulation of chylomicrons and VLDL is the hallmark of individuals with Type V hyperlipoproteinemia. The group is heterogenous in terms of etiology; most exhibit defective catabolism of triglyceride-rich particles and in some cases there is also evidence of oversynthesis. The accumulated particles are too large to penetrate the matrix of the electrophoretogram (B) and the gel filtration column (C) and their content of apolipoproteins B, C and E results in plasma enrichment with these components. LDL and HDL levels in the circulation are typically reduced. The lipoprotein profile in this subject was

	mmol/l
Total triglyceride	14.75
Total cholesterol	6.9
VLDL cholesterol	3.6
LDL cholesterol	1.4
HDL cholesterol	1.9

In Type IV hyperlipoproteinemia, where VLDL levels are very high, plasma LDL and HDL are commonly subnormal and indeed there seems to be an inverse correlation between the plasma concentrations of these higher density fractions and VLDL. This reflects the precursor-product relationship which exists between them (Chapter 3) since, during delipidation of VLDL its core lipid and Apo B transfers into the LDL density interval while its remaining surface coat components redistribute to HDL. It remains to be determined whether the reported increase in coronary risk in Type IV hyperlipoproteinemic subjects comes from their high circulating VLDL levels or from a reduction in the apparently protective HDL fraction.

3.5 Type V hyperlipoproteinemia

Although primary Type V hyperlipoproteinemia is rare, the phenotype is a common sequel (31) to overindulgence in alcohol (Chapter 13) and to diabetes mellitus (Chapter 9); and as noted earlier, it may result (41) from apolipoprotein CII deficiency (Tab. 2). It is therefore a disease more commonly seen in adult life and is characterised biochemically by the simultaneous accumulation of chylomicrons and VLDL in the plasma (Fig. 9). Clinically, the condition shares several features in common with Type I hyperlipoproteinemia, like hepatomegaly, predisposition to pancreatitis and eruptive xanthomata, suggesting that all of these phenomena derive from the massive hypertriglyceridemia that accompanies both diseases. As is the case of familial (Type I) hyperchylomicronemia, atherosclerosis appears not to be

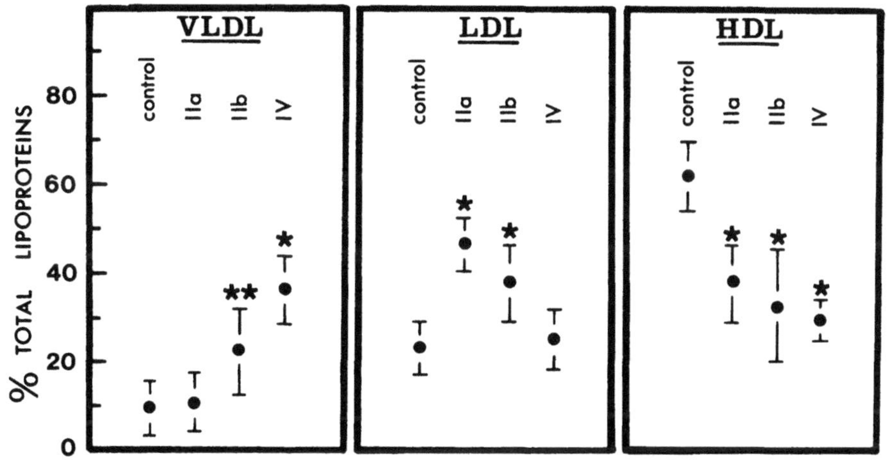

Fig. 10. Chromatographic distribution of lipoproteins in the common dyslipidemias.
Plasma lipoproteins from 12 controls, 8 Type IIa, 6 Type IIb and 6 Type IV subjects were fractionated by agarose gel filtration. The area of each peak was measured and its contribution to the overall lipoprotein profile calculated. The values represent means ± 1 SD. (* Significantly different from controls, $p < 0.001$; ** Significantly different from controls, $p < 0.005$.)

a feature of the condition, although this issue still remains contentious. The pattern of inheritance of Type V hyperlipoproteinemia is not completely clear (38), despite the fact that its familial nature is well documented. The data available in the literature suggest that it is transmitted in autosomal dominant fashion but why the trait may be variably expressed as Type IV or Type V hyperlipoproteinemia is not yet known.

Figure 10 summarises the results of a study of the merits of molecular sieve chromatography in the diagnosis and investigation of the common dyslipidemias (24). Type II hyperlipoproteinemia is distinguishable by its raised percentage of LDL (Peak 2 – Fig. 3) and reduced HDL. Moreover, it is possible, on the basis of the VLDL value, to subclassify the condition into IIa and IIb. Type IV hyperlipoproteinemia shows an increment in the mass of VLDL and a reduction in HDL. LDL is unchanged in this condition.

4 Phenotypes and Genotypes

Our need to resort to the above phenotypic classification of the hyperlipidemias is as much a reflection of our ignorance of the genetic factors associated with their development as it is an indication of our appreciation of the merits of the Fredrickson classification. Although that system is valuable in the clinical context, we use it merely until our understanding of the molecular defects responsible for the hyperlipidemias improves. Undoubtedly our progress in this direction will lead eventually to a new classification system based on genotype.

References

1. McFarlane, A. S. (1935). Biochem. J. *29*, 1175–1181.
2. Svedberg, T. & Pederson, K. O. (1940). The Ultracentifuge (R. H. Fowler and P. Kapitza, eds). International series of monographs on Physics, Clarendon Press, London.
3. Pederson, K. O. (1945). Ultracentrifugal studies on Serum and Serum Fractions. Almqvist and Wiksells, publishers. Uppsala, Sweden.
4. Gofman, J. W., Lindgren, F. T. & Elliott, H. (1949). J. Biol. Chem. *179*, 973–978.
5. De Lalla, O. & Gofman, J. W. (1954). Ultracentrifugal Analysis of Serum Lipoproteins. Methods of Biochemical Analysis *1*, 459–478.
6. Havel, R. J., Eder, H. A. & Bragdon, J. H. (1955). J. Clin. Invest. *34*, 1345–1356.
7. Lindgren, F. T. (1980). Ann. N. Y. Acad. Sci. *348*, 1–15.
8. Kunitake, S. T. & Kane, J. P. (1982). J. Lipid Res. *23*, 936–940.
9. Fainaru, M., Havel, R. J. & Imazumi, K. (1977). Biochem. Med. *17*, 347–355.
10. Edelstein, C., Pfaffinger, D. & Scanu, A. M. (1984). J. Lipid Res. *25*, 630–637.
11. Hatch, F. T. & Lees, R. S. (1968). Adv. Lipid Res. *6*, 1–68.
12. Fredrickson, D. S., Levy, R. I. & Lindgren, F. T. (1968). J. Clin. Invest. *47*, 2446–2457.
13. Noble, R. P. (1968). J. Lipid Res. *9*, 693–700.
14. Wong, R. A., Banchero, P. G., Jensen, L. C., Pan, S. S., Adamson, G. L. & Lindgren, F. T. (1977). J. Lab. Clin. Med. *89*, 1341–1348.

15. Bierman, E. L., Gordis, E. & Hamlin, J. T. (1962). J. Clin. Invest. *41*, 2214–2259.
16. Reissel, P. K., Hagopian, L. M. & Hatch, F. T. (1966). J. Lipid Res. *7*, 551–557.
17. Ressler, N., Springgate, R. & Kaufman, J. (1961). J. Chromatog. *6*, 409–417.
18. Naito, H. K., Wada, M., Erhart, L. A. & Lewis, L. A. (1973). Clin. Chem. *19*, 228–234.
19. Foo, Y., Tickner, T. R. & Cramp, D. G. (1979). Ann. Clin. Biochem. *16*, 44–46.
20. Levy, R. I. (1981). Clin. Chem. *27*, 653–662.
21. Margolis, S. (1967). J. Lipid Res. *8*, 501–507.
22. Rudel, L. L., Lee, J. A., Morris, M. D. & Felts, J. M. (1974). Biochem. J. *139*, 89–95.
23. Shepherd, J., Packard, C. J., Dryburgh, F. J. & Third, J. L. H. C. (1975) Clin. Chem. *21*, 1887–1891.
24. Shepherd, J. (1976). Clin. Chim. Acta *69*, 161–173.
25. Hara, I. & Okazaki, M. (1986). Methods in Enzymology *129*, 57–78.
26. Alaupovic, P. (1972). Protides Biol. Fluids Proc. Colloq. *19*, 9–19.
27. Alaupovic, P. (1982). La Ricerca Clin. Lab. *12*, 3–21.
28. Gustafson, A., Alaupovic, P. & Furman, R. H. (1966). Biochemistry *5*, 632–637.
29. Shore, V. and Shore, B. (1968). Biochemistry *7*, 3396–3401.
30. Shore, B. & Shore, V. (1969). Biochemistry *8*, 4510–4516.
31. Fredrickson, D. S., Levy, R. I. & Lees, R. S. (1967). N. Engl. J. Med. *276*, 34–44, 94–103, 148–156, 215–225, 273–281.
32. Lipid Research Clinics Program Manual of Laboratory Operations. Publication No NIH 75–628. US Dept of Health, Education and Welfare, Bethesda, MD.
33. Burstein, M. & Samaille, J. (1960). Clin. Chim. Acta *5*, 609–616.
34. Warnick, G. R. & Albers, J. J. (1978) J. Lipid Res. *19*, 65–76.
35. Friedewald, W. T., Levy, R. I. & Fredrickson, D. S. (1972). Clin. Chem. *18*, 499–502.
36. Beaumont, J. L., Carlson, L. A., Cooper, G. R., Fejfar, Z., Fredrickson, D. S. & Strasser, T. (1970). Bull. WHO *43*, 891–917.
37. Havel, R. J., Goldstein, J. L. & Brown, M. S. (1980). In: Metabolic Control and Disease (Bondy, P. K. and Rosenberg, L. E., eds.) pp. 383–494, W. B. Saunders, Philadelphia.
38. Nikkila, E. A. (1983). In: The Metabolic Basis of Inherited Disease (Stanbury, J. B., Wyngaarden, J. B., Fredrickson, D. S., Goldstein, J. L. and Brown, M. S., eds.) pp. 622–642, McGraw-Hill, New York.
39. Alaupovic, P., McConathy, W. J. & Fesmire, J. (1983). U. S. Department of Health and Human Services, NIH Publication No 83–1266, pp. 27–53.
40. Havel, R. J. & Gordon, R. S. (1960). J. Clin. Invest. *39*, 1777–1785.
41. Breckenridge, W. C., Little, J. A., Steiner, G., Chow, A. & Poapst, M. (1978). N. Engl. J. Med. *298*, 1265–1269.
42. Goldstein, J. L. & Brown, M. S. (1983). In: The Metabolic Basis of Inherited Disease (Stanbury, J. B., Wyngaarden, J. B., Fredrickson, D. S., Goldstein, J. L. and Brown, M. S., eds.) pp. 672–712, McGraw-Hill, New York.
43. Goldstein, J. L. & Brown, M. S. (1979). Ann. Rev. Genet. *13*, 259–289.
44. Langer, T., Strober, W. & Levy, R. I. (1972). J. Clin. Invest. *51*, 1528–1536.
45. Packard, C. J., Third, J. L. H. C., Shepherd, J., Lorimer, A. R., Morgan, H. G. & Lawrie, T. D. V. (1976). Metabolism *25*, 995–1006.
46. Goldstein, J. L. & Brown, M. S. (1977). Ann. Rev. Biochem. *46*, 897–930.
47. Packard, C. J. & Shepherd, J. (1983). Atheroscler. Revs. *11*, 29–63.
48. Shepherd, J., Packard, C. J., Bicker, S., Lawrie, T. D. V. & Morgan, H. G. (1980). New Engl. J. Med. *302*, 1219–1222.
49. Bilheimer, D. W., Grundy, S. M., Brown, M. S. & Goldstein, J. L. (1983). Proc. Natl. Acad. Sci. USA. *80*, 4124–4128.
50. Tolleshaug, H., Hobgood, K. K., Brown, M. S. & Goldstein, J. L. (1983). Cell *32*, 941–951.
51. Russell, D. W., Schneider, W. J., Yamamoto, T., Luskey, K. L., Brown, M. S & Goldstein, J. L. (1984). Cell. *37*, 577–585.
52. Davis, C. G., Lehrman, M. A., Russell, D. W., Anderson, R. G. W., Brown, M. S. & Goldstein, J. L. (1986) Cell *45*, 15–24.
53. Goldstein, J. L., Sobhani, M. K., Faust, J. R. & Brown, M. S. (1976). Cell *9*, 195–203.
54. Bilheimer, D. W., Ho, Y. K., Brown, M. S., Anderson, R. G. W. & Goldstein, J. L. (1978). J. Clin. Invest. *61*, 678–696.
55. Lees, A. M. & Lees, R. S. (1983). Proc. Natl. Acad. Sci. USA *80*, 5098–5102.

56. Cuthbert, J. A., East, C. A., Bilheimer, D. W. & Lipsky, P. E. (1986). New Engl. J. Med. *314*, 879–883.
57. Brown, M. S., Goldstein, J. L. & Fredrickson, D. S. (1983). In: The Metabolic Basis of Inherited Disease (Stanbury, J. B., Wyngaarden, J. B., Fredrickson, D. S., Goldstein, J. L. & Brown, M. S., eds.) pp. 655–671, McGraw-Hill, New York.
58. Weisgraber, K. H., Rall, S. C. & Mahley, R. W. (1981). J. Biol. Chem. *256*, 9077–9083.
59. Rall, S. C. Weisgraber, K. H. & Mahley, R. W. (1982). J. Biol. Chem. *257*, 4171–4178.
60. Rall, S. C., Weisgraber, K. H., Innerarity, T. L. & Mahley, R. W. (1982). Proc. Natl. Acad. Sci. USA *79*, 4696–4700.
61. Weisgraber, K. H., Innerarity, T. L. & Mahley, R. W. (1982). J. Biol. Chem. *257*, 2518–2521.
62. Utermann, G., Prium, N. & Steinmetz, A. (1979). Clin. Genet. *15*, 63–72.
63. Hazzard, W. R., Warnick, G. R., Utermann, G. & Albers, J. J. (1981). Metabolism *30*, 79–89.
64. Zilversmit, D. B. (1973). Circ. Res. *33*, 633–637.
65. Mahley, R. W., Innerarity, T. L., Brown, M. S., Ho, Y. K. & Goldstein, J. L. (1980). J. Lipid Res. *21*, 970–980.
66. Grundy, S. M. (1984). In: Treatment of Hyperlipoproteinaemia (Carlson, L. A. and Olsson, A. G., eds.) pp. 21–34, Raven Press, New York.
67. Reardon, M. W., Fidge, N. H. & Nestel, P. J. (1979). J. Clin. Invest. 61, 850–860.
68. Packard, C. J., Shepherd, J., Joerns, S., Gotto, A. M. & Taunton, O. D. (1980). Metabolism *29*, 2123–222.
69. Goldstein, J. L., Schrott, H. G., Hazzard, W. R., Bierman, E. L., Motulsky, A. G., Campbell, E. D. & Levinski, M. J. (1973). J. Clin. Invest. *52*, 1544–1568.

Chapter 2
Chemical and Physical Properties of Lipoproteins

G. M. Kostner and P. Laggner

1 Compositional Analysis

1.1 Introduction

As pointed out in Chapter 1, human plasma lipoproteins (Lp) consist of a great variety of lipid-protein complexes varying in their protein and lipid composition. Consequently, lipoproteins differ in their physico-chemical properties, notably in hydrated density, electric charge, carbohydrate content and immunochemical reactivity. These differences are the basis of methods for separation and purification. Despite the fact that column chromatography using ion exchangers, molecular sieves, lectins, immunoadsorbers and many others are frequently in use today for subfractionation of Lp-classes, the preparative ultracentrifuge remains the basic instrument for preparation of large quantities of lipoproteins. Because of existing concentration maxima and minima in the density range of $\rho < 1$ up to 1.25 kg/l which spans the lipoprotein spectrum, lipoproteins are divided into 4 – 8 main density classes. It must however be borne in mind that each of these main classes is in itself heterogenous. Subfractionation can be achieved by column chromatography or equilibrium banding in a density gradient using the ultracentrifuge. Since it is beyond the scope of this chapter to adequately address all these many subclasses, we will focus in this report on the main Lp-density classes, references being given for subfractions thereof wherever necessary.

In addition it must be stressed at this point, that lipoprotein composition is subject to changes postprandially, and as a result of hormonal influence or the presence of intercurrent disease. The composition given here refers primarily to lipoproteins isolated from fasting normolipemic healthy individuals. Only a few abnormal lipoproteins such as Lp-X or Lp-E will be mentioned as well. For more detailed information the reader is advised to consult some of the more specific reviews and monographs (1 – 4).

1.2 Plasma concentration of the main Lp-density classes

In the civilized western population, the major fasting plasma lipoprotein is low density lipoprotein (LDL) followed by the high density (HDL) and very low density fractions (VLDL). (Table 1). Plasma lipoprotein concentrations differ significantly between males and pre-menopausal females. In addition, plasma LDL and VLDL increase and HDL decreases with age.

Lp(a), a lipoprotein originally believed to represent a genetic variant of LDL exhibits the greatest variation ranging from < 10 mg/l up to extremes of 2.0 g/l and more.

Table 1. Concentration of lipoproteins in normal fasting human plasma.

Lipo-proteins:	Chylo-microns	VLDL	LDL	Lp (a)	HDL-E	HDL_2	HDL_3	VHDL
g/l (m)	< 0.1	0.5–2.0	2.0–3.5	0.01–0.5	< 0.05	0.5–1.0	1.0–2.0	0.1–0.2
g/l (f)	< 0.1	0.5–1.5	2.0–3.0	0.01–0.5	< 0.05	1.0–1.5	1.0–2.0	0.1–0.2

m: males; f: females.

1.3 Lipid distribution

For metabolic considerations it is advisable to divide lipoproteins according to their major lipid into triglyceride (TG-)rich, cholesterol- and cholesteryl ester (CE)-rich and phospholipid/protein rich lipoproteins (5).

1.3.1 **TG-rich lipoproteins**

The TG-rich lipoproteins are secreted in nascent form by the liver and intestine into the circulation and have a relatively short residence time there. They undergo continuous modification by enzyme action, lipid and apolipoprotein exchange and transfer and partial loss of constituents by selective organ uptake. Chylomicrons are usually absent in fasting plasma. They represent dietary fat packaged and secreted by the intestine into lymph. Upon entering the bloodstream, they gain additional proteins e. g. albumin and apolipoproteins.

VLDL if isolated from normal fasting plasma, exhibit flotation constants (S_f) in the range of 100–200 S^{-1}; the composition given here therefore refers to VLDL isolated either postprandially or from the plasma of fasting Type IV or Type V hyperlipoproteinemic individuals.

Table 2 shows the chemical compositions of chylomicron and VLDL. Chylomicrons consist of approximately 90% TG, 3% cholesterol + CE, 5% phospholipids and 2–3% proteins. Plasma containing noticeable amounts of chylomicrons is turbid but can be cleared by overnight storage at 4 °C or by centrifugation for 20 min at 30 000 × *g*. These chylomicrons exhibit S_f of 400–10^5.

The whole VLDL class is prepared by spinning the chylomicron-free plasma at its own density for 18 h at 120 000 × *g*. Under these conditions, VLDL with S_f of approximately 30–300 float to the top of the tube. For subfractionation into more homogeneous fractions, the use of swinging bucket rotors or zonal rotors using a density gradient is required (6).

Table 2. Average composition of plasma lipoproteins

Lipoproteins	Size (nm)	Density range (kg/l)	S_f (1.063) F (1.210)	Mass (MDa)	% of lipoprotein mass Protein	PL	FC**	CE	TG
Chylomicrons	$< 10^4$	< 1.00	$< 10^5$	> 150	2	5	1	2	90
VLDL	50	< 1.006	20–400	5–130	10	16	7	13	54
IDL	30	< 1.019	10–20	3.5	17	20	9	34	20
LDL	21	< 1.063	0–10	2.5	23	21	11	41	4
Lp (a)	25	< 1.090	20	5.5	34	18	9	36	3
HDL-E	12	< 1.125	10	0.5	35	34	10	20	1
HDL_2	10	< 1.125	4–9	0.36	42	35	5	13	5
HDL_3	8	< 1.210	0–4	0.20	56	23	3	15	3
VHDL	7	< 1.25	–	0.15	72	20	2	5	1

* Values for protein include the carbohydrates
** free cholesterol

1.3.2 Cholesterol/CE-rich lipoproteins: IDL, LDL, Lp(a)

Most of the CE rich lipoproteins derive from the metabolism of VLDL. In hypercholesterolemia, part of the IDL or LDL are probably synthesized directly. The same is true for Lp (a) which is secreted in its mature form by the liver (7).

Approximately 50% of the mass of LDL is cholesterol, making the particle the major sterol transporter in the circulation (Table 2). 80% of LDL-cholesterol is esterified, primarily with linoleic acid. The CE of LDL exchanges with the triglyceride in the core of VLDL in a process which is mediated by specific exchange/transfer proteins (8). Thus in certain forms of hypertriglyceridemia or in liver disease, LDL may become enriched with TG, reaching values of up to 20% of the total LDL mass.

IDL which are normally not found in significant amounts in fasting plasma are transient products of VLDL catabolism. The core lipids of IDL consist of CE and TG. In hyperlipoproteinemia and postprandially, IDL may be increased in plasma.

Together with chylomicron-remnants, IDL, LDL, and Lp (a) belong to the most atherogenic lipoproteins known today. Chemically Lp (a) seems to consist of an LDL particle with a specific antigen (a-protein) attached to it via disulfide bridges. Thus there are few if any differences between the lipid moieties of Lp (a) and LDL. The a-protein is very rich in carbohydrates, notably sialic acid, which increases the density of LDL towards that of the HDL_1/HDL_2 range (9).

1.3.3 Phospholipid/protein rich lipoproteins

These lipoproteins are believed to protect against atherosclerosis and vascular disease and are generally called high density lipoproteins (HDL). HDL comprise a great variety of lipoprotein subspecies which have been separated in the past by density gradient ultracentrifugation and column chromatography, notably using hydroxyapatite (10). The physiological significance of all these HDL subclasses is mostly obscure. There are three main HDL fractions which should be considered here: HDL_2, HDL_3 and HDL-E. HDL_2 is formed in a complex series of reactions involving HDL_3, TG-rich lipoproteins, LCAT, lipoprotein lipase, lipid transfer proteins and possibly others (11). During its passage through the liver, HDL loses part of its lipid complement and is shifted to the HDL_3 density region again. Thus not only the absolute amount and ratio of HDL_2/HDL_3 depends to a great extent upon the activity of all these enzymes and transfer proteins but also the chemical composition is subject to changes caused by dietary, hormonal and metabolic influences. Table 2 gives the average composition of HDL subclasses without taking account of the well known differences between males and females, vegetarians and meat eaters, women on oral contraceptives, alcoholics and many others. In addition it outlines the compositional analysis of HDL-E. HDL-E is a minor subfraction of HDL found predominantly in the HDL_1 density region in quantities between 20 – 50 mg/l (12). Nonetheless it seems to play a key role in reverse cholesterol transport from periphery to the liver. Part or all of it is synthesized in macrophages aimed at the export of stored CE. HDL-E has a great affinity for the (B/E-) and E-receptors on liver cells and is very effectively catabolized by this organ.

1.3.4 The phospholipid (PL) content of plasma lipoproteins

PL are freely exchangeable not only between different Lp-classes but also between lipoproteins and cell membranes and specific PL exchange/transfer proteins facilitate this process. Thus it is not surprising that the composition of individual PL do not differ to a great extent among various Lp-classes except for those with a very fast catabolism (e. g. chylomicrons). The most prominent PL in lipoproteins with a proportion of approximately 70% is lecithin followed by sphingomyelin (10 – 20%), lysolecithin (3 – 5%), phosphatidyl-ethanolamine (2 – 6%) phosphatidylserine and inositol (1 – 2% each) Lysolecithin is formed during the action of LCAT, and part of it is taken up by albumin and quickly removed by the liver. The PL are very important surface constituents which stabilize lipoprotein structure.

1.4 Lipoproteins occurring in disease states

As mentioned already, many Lp-classes occur in fasting normal plasma in very small quantities but may accumulate to a significant extent if their catabolism is impaired. Among those are chylomicrons, IDL, HDL-E and others. In hyperlipoproteinemia, lipoprotein composition may deviate quite strikingly from normal as lipid-exchange/transfer processes force them to another equilibrium; these lipoproteins are not considered as "pathological Lp" here and will not be treated at this point. In this paragraph are listed only those lipoproteins, which seem to be absent from normal plasma, but occur in certain diseases or inborn metabolic disorders.

Here again, numerous reports appear in the literature which deal with such lipoproteins, found for example, in patients with thyroid dysfunction, renal disease, diabetes mellitus (see chapter 9 – 13) (13, 14). We will focus here on the most important of these which either have proved to be of diagnostic value, or may be of relevance in atherogenesis.

1.4.1 Lipoproteins occurring in obstructive jaundice and in primary and secondary LCAT deficiency: Lp-X, Lp-Y, Lp-E

Most of the early work in this field was published by Seidel et al. (15). Individuals suffering from extrahepatic or intrahepatic cholestasis exhibit a gross abnormality of their plasma Lp-pattern. In the $\rho < 1.063$ fraction Lp-X and Lp-Y occur. Lp-X may reach concentrations of 10 g/l and more. Lp-X has a vesicular structure and appears as stacked discs on electron microscopy (16). This lipoprotein is characterized by a high content of lecithin and free cholesterol (FC). The characteristic feature of this "LDL" is the absence of apolipoprotein B (Apo B). Manzato et al. (17) have found that mixing bile derived from the liver (not the gallbladder) with serum gives rise to the formation of artificial Lp-X which is very similar to, if not identical with the natural one. On the basis of these findings these authors postulated that regurgitation of bile into the blood stream might represent the actual pathway of Lp-X synthesis. Since on the other hand Lp-X and Lp-E are also found in conditions where cholestasis is not apparent (LCAT-deficiency, infusion of Intralipid) it is tempting to extend the theory of Seidel and to postulate that the presence of these abnormal PL-FC rich lipoproteins reflect a situation where accumulation is caused either by an increased supply or due to an impaired removal e. g. lack of LCAT. The chemical composition of these lipoproteins is shown in Table 3. Another lipoprotein occurring in cholestasis and LCAT deficiency has been called Lp-Y (18). Chemically and physico-chemically it resembles β-VLDL or IDL; depending on the degree of LCAT deficiency, Lp-Y has a higher fraction of FC as compared to β-VLDL or IDL. The mechanism involved in Lp-Y synthesis has not been elucidated completely but it may be related to reduced activity of hepatic lipase in liver diseases.

Table 3. Composition of abnormal plasma lipoproteins*

Lipoprotein	Size (nm)	Density range (kg/l)	Mass (MDa)	% of lipoprotein mass Protein	PL	FC	CE	TG
Lp-X	35	1.019–1.063	5–10	6	61	25	5	3
Lp-Y	30	1.019–1.063	3–5	27	20	8	10	35
Lp-E	15	1.063–1.125	0.5–1	28	46	23	2	1
β-VLDL	30	< 1.006	40–80	12	24	6	25	35

* There are significant variations in the composition of these lipoproteins among individual donors.

Lp-E is a lipoprotein which is very rich in Apo E. It was first noticed in LCAT deficient plasma by Utermann et al. (19) and later also in patients with liver disease or secondary LCAT deficiency. The mechanism of Lp-E synthesis is not known. In electron microscopy, most of the Lp-E particles exhibit disc-like or rouleaux structure similar to Lp-X. A spherical counterpart of Lp-E is found in dogs and other species and may also appear in the plasma of humans fed a high cholesterol diet (12). This lipoprotein has a similar protein composition to that of Lp-E but is rich in CE and might be related more to HDL-E. The apolipoprotein composition of these abnormal lipoproteins is shown in Table 4. The major apolipoproteins in Lp-X are Apo C, in Lp-Y, Apo B and in Lp-E, Apo E. It must however be stressed at this point that these compositions may vary remarkably among individuals reflecting genetic influences, variations in nutritional status, drug therapy and onset of disease.

Table 4. Apolipoprotein content of abnormal plasma lipoproteins (Values are given in % of total apolipoprotein mass)

Lipoprotein	Apo AI	Apo AII	Apo B	Apo C*	Apo D	Apo E	Albumin	Others
Lp-X	1–2	< 1	0	50	2–4	0***	30	10–15
Lp-Y	< 1	< 1	85	10	< 1	2–4	trace	0
Lp-E	20	10**	0	3	< 0.1	60	0	5

* Sum of Apo CI + CII + CIII
** Most of the Apo AII in Lp-E is complexed via -S-S- bridges with Apo E
*** Some Lp-X preparations do contain up to 5% Apo E.

1.4.2 **β-VLDL**

In familial dys-β-lipoproteinemia (also known as Type-III hyperlipoproteinemia) the VLDL-LDL conversion process is impaired due to the absence of a functionally intact Apo E, and abnormal intermediate products called "β-VLDL" accumulate in the plasma. β-VLDL float at a density of 1.006 kg/l but unlike normal VLDL, migrate like β-globulins on electrophoresis. This Lp-fraction might be heterogeneous as it also contains variable amounts of chylomicron-remnants whose clearance is also impaired in Type-III hyperlipoproteinemia. β-VLDL are relatively rich in CE (Table 3) and are avidly taken up by a specific cell surface receptor found on macrophages and endothelial cells (20). They are thought to be very atherogenic.

1.5 Apolipoprotein content of plasma lipoproteins

The number of non-identical apolipoproteins found in human plasma not including genetic isoforms is approaching 20 (for special reviews see Refs. 1–5, 21, 22). Table 5 lists only those lipoproteins whose apolipoproteins are well established and fairly well characterized. Apolipoproteins have various functions: They stabilize lipid emulsions (Apo B, Apo A and Apo C), they serve as cofactors and modulators of enzymatic reactions (Apo C, Apo AI), they manage the export of lipids out of cells (Apo B, Apo E) and they direct lipids to target organs by specific receptor interaction (Apo B, Apo E and Apo A). In addition, apolipoproteins may have other specific functions e. g. in interfering with platelet aggregation (Apo H, Ref. 23). There are, however, still numerous apolipoproteins whose function either is completely unknown (Apo (a), Apo AIV, Apo F, Apo G) or whose actual physiological role remains to be established (Apo AII, Apo CIII, Apo B_{48}).

Table 5. Apolipoproteins of human plasma

Apo-lipoprotein	Major Lp-Class	Mass (kDa)	Concentration in plasma (mg/l)	Function
AI	HDL	28.5	1200–1400	LCAT activation, HDL-receptors
AII	HDL_3	17	350–500	Activation of liver lipase?
AIV	Chylomicron, HDL	46	< 50	unknown
B_{100}	VLDL, LDL	550	700–900	VLDL secretion, B/E receptors
B_{48}	Chylomicron, β-VLDL	265	< 50	Lipid absorption
CI	Chylomicron, VLDL	6.5	50–80	LCAT activation?
CII	Chylomicron, VLDL	8.8	30–70	LpL activation
$CIII_{0-3}$	Chylomicron, VLDL	8.9	100–120	Remnant uptake by the liver
D	HDL_3	29	80–100	LCAT/CE-exchange reaction
E	Chylomicron, VLDL, HDL-E	34	30–50	B/E and E-receptor, reverse cholesterol transport
F	HDL	30	< 50	unknown
G	VHDL	72	< 50	unknown
H	Chylomicron, bottom	55	150–300	regulates procoagulant activity
(a)	Lp (a)	650	variable	unknown

The M_r of apolipoproteins varies to a great extent ranging from 6500 (Apo CI) to approx. 650 000 (Apo (a)). In early days the existence of so called "lipoprotein families" was postulated. A Lp-family was characterized as a functional and metabolic entity and their protein moieties named in alphabetical order, Apo A, Apo B, ..., Apo H; the constitutive polypeptides of these families were designated with Roman numbers (e. g. AI, AII, ...). Although this concept greatly facilitated the understanding of the whole Lp-system and helped to avoid confusion in studying Lp-metabolism, it turned out in later years that the Lp-system is much more complex than originally believed. Almost any lipoprotein density class isolated from human plasma contains many if not all known apolipoproteins sometimes even if only in minute amounts. This might be due to adsorption caused by lipid–lipid or lipid–protein interactions, or by artifacts created during lipoprotein preparation. Lp-families in the classical sense containing only polypeptides of their own kind have been truly prepared by consecutive purification steps (24, 25) but these possibly also represent artifacts. Theoretical considerations certainly favor the view that any lipoprotein in the circulation might contain numerous polypeptides, because apolipoproteins have been shown to exchange freely from one lipoprotein particle to another, reaching a steady state once enzymatic reactions and lipid exchange/transfer processes have come to equilibrium. Thus what we are observing by analyzing Lp-fractions must be considered as a static interruption of a dynamic equilibrium which has been disturbed during the isolation procedure. It is therefore desirable to derive compositional data by isolating the lipoprotein without delay from fasting plasma after immediate blocking of enzymatic reactions. We have approached this by the use of zonal ultracentrifugation (26); these data are shown in Table 6. It should be stressed at this point again that the Lp-system must be considered as a continous spectrum of macromolecules with concentration maxima and minima. Moving from one side of a given maximum to the other results in identification of lipoproteins with somewhat different apolipoprotein contents.

Table 6. Apolipoprotein content of the main plasma lipoproteins (Values are given as % of the protein mass)

Lipo-protein	Apo AI	Apo AII	Apo B	Apo C	Apo D	Apo E	Apo (a)
Chylomicrons	5–20*	3–5	20–30**	50–60	n. d.	6–9	n. d.
VLDL	1–2	< 1	50–60	30–40	< 0.1	5–8	n. d.
IDL	< 1	< 1	80–90	10–15	< 0.1	1–3	n. d.
LDL	0.5	0.25	95	3	< 0.1	0.5	n. d.
Lp (a)	trace	trace	80–90	trace	< 0.1	trace	10–20
HDL-E	63	10	0	5	0	22	n. d.
HDL_2	85	8	***	7	0.5	***	***
HDL_3	72	22	0	5	1.5	trace	0

* Lymph chylomicrons probably contain large amounts of Apo AI which are lost upon entry into the blood stream.

** Most of the chylomicron Apo B consists of B_{48}.

*** HDL_2 is free of Apo E, Apo B and Apo (a) if these lipoproteins are specifically removed e. g. by heparin sepharose column chromatography.

The lipoproteins with $\rho < 1.063$ kg/l characteristically contain largely Apo B and Apo C. The higher the TG content the greater is the fraction of Apo C. Thus chylomicrons are the lipoproteins with the highest Apo C content. Chylomicrons, unlike the other TG-rich lipoproteins are characterized by the presence of Apo B_{48} whose molecular mass is only 48% of that of Apo B found in LDL (Apo B_{100}). It is currently believed that Apo B_{48} is formed by post-transcriptional modification as the Apo B genes in liver and intestine seem to be identical. One question of utmost importance which still remains to be answered is whether or not Apo B_{48} might be responsible for the short residence time of chylomicron remnants in the circulation.

The TG-rich lipoproteins in addition to Apo B and Apo C contain Apo E, Apo AIV, Apo H and others. It is of interest to note that by mixing artificial TG emulsions (e. g. Intralipid) with plasma, all the apolipoproteins except for Apo B associate with Intralipid; the newly formed particle has a strong resemblance to native chylomicrons. A similar process might be operative *in vivo* upon entry of native chylomicrons into the blood stream.

As the density of apolipoprotein B containing particles increases, the TG content and the non Apo B protein content decreases. IDL may still be relatively rich in

Apo E. The LDL fraction, however, contains no more than 5% of Apo E proteins. The main LDL peak, subfractionated by a variety of methods, varies not only in its lipid but also in its apolipoprotein composition (27–29).

Thus in general one may say that during the action of lipoprotein lipase on TG-rich lipoproteins, most non-Apo B proteins are lost, probably to the HDL density class. There are however also processes where LDL regain non-Apo B proteins. During the action of LCAT, surface lipids are lost and are replaced by Apo C and Apo A (11). Metabolic investigations and receptor studies are best performed with pure Lp-families e. g. Lp B. Unfortunately it is impossible to prepare Lp B solely by ultracentrifugation or column chromatography. We have recently shown that Lp B free of traces of Apo C, Apo E, Apo A and albumin can only be obtained by specific immunoabsorption (30).

Moving to densities greater than 1.063 kg/l one encounters several HDL particle types. The HDL_1 class is the most heterogeneous, containing Lp B (31), Lp (a), HDL-E and small amounts of AI/AII containing HDL. Lp B-HDL represents a small fraction of LDL which is shifted towards a density > 1.063 kg/l due to lipid loss. In hypertriglyceridemia, the mean density of LDL is shifted towards the HDL region due to the exchange of LDL-CE for TG followed by TG hydrolysis probably caused by the action of hepatic lipase. In this circumstance then Lp B may significantly contaminate the HDL fraction. Lp B-HDL must be differentiated from Lp (a). Lp (a) might be considered as an Lp B with an additional apolipoprotein (a) linked via disulfide bridges. There are, in addition, variable amounts of apparently lipid-free Apo (a) proteins which appear in the serum at $\rho > 1.22$ kg/l. Their total mass may amount to more than 10% of total Apo (a).

HDL-E represents a rapidly turning over fraction of plasma derived either from macrophages or from degradation of chylomicrons and VLDL.

The major subfractions of HDL are HDL_2 and HDL_3. HDL_2 have a significantly higher Apo AI/Apo AII ratio than HDL_3; the ratio decreases as the HDL density rises (26). HDL_2 has also a higher content of Apo C and Apo E; Apo D on the other hand concentrates in HDL_3 and VHDL.

1.6 Human apolipoprotein isoforms and mutants

Isoelectric focusing (IEF) is one of the most effective methods for separation of macromolecules with minor charge differences. We applied this technique to whole serum in 1969 and separated 8–10 intact lipoproteins differing in their apolipoprotein content (32). This method has been refined by Luley et al. (33) who obtained more than 10 lipoprotein bands by IEF on agarose gels. Today, analytical IEF is more commonly applied to apolipoproteins rather than intact lipoproteins. For this purpose, lipoproteins should be separated from other plasma proteins and delipidated prior to isoelectric focusing in the presence of 8 mol/l urea. Experience is required in order to obtain reproducible IEF patterns, and anyone who intends to undertake such analyses is advised to consult specialist literature (34).

IEF separates apolipoproteins according to charge differences. Two kinds of isoelectric subspecies are obtained: 1. chemical polymorphic isoforms and 2. genetic allotypes.

1.6.1 Chemical polymorphic isoforms of apolipoproteins

This polymorphism is caused by charge differences due to different degrees of sialylation, the presence of additional amino acids, or *in vitro* modification of proteins by carbamylation, oxidation and other reactions. The best known examples here are Apo CIII which exists in four forms (called CIII-0, CIII-1, CIII-2 and CIII-3) with an increasing sialic acid content of 0–3 moles per mole of protein. As Apo CIII has a small molecular mass, the differences of 1–3 negative charge units are sufficient to separate these isoforms by normal polyacrylamide gel electrophoresis. Similar isoforms are known from Apo E. Here unfortunately there is confusion in the literature over nomenclature. One mole of Apo E may contain from 0–2 or possibly even 3 moles of sialic acid with charge differences of 1–3 (35). Since these chemical isoforms overlap with genetic polymorphic forms differing by 1–3 positive charge units, normal IEF may yield erroneous results in screening for mutant Apo E isoforms. In such cases treatment with neuraminidase is advisable.

A completely different cause of polypeptide heterogeneity is found for Apo AI and Apo AII. Apo AI focuses in 6 isoforms with p*I*'s between 5.0–5.9 (36). AI_4 and AI_5 are the major forms found in plasma deriving from "nascent" AI_2 and AI_3. These latter isoforms have been identified as pro-Apo AI, which has a signal peptide attached to it. Upon entry into circulation pro-Apo AI are converted to mature Apo AI by the action of a signal peptidase. The identity of AI_1 and AI_6, two minor fractions, is not yet clear but it seems feasible that these isoforms may either reflect biosynthetic products from different organs or deamidation products.

Apo AII separates also into 6 isoforms (37). AII_0 with a p*I* of 4.90 is the major one found in plasma. There is a small amount of pro-AII (p*I* 6.79) and several minor fractions with p*I*'s of 4.20–5.17. Some of the isoforms in addition are sialylated to a variable degree. It is believed that some of the Apo AII isoforms represent deamidated isoproteins.

1.6.2 Apolipoprotein mutants

Genetic allotypes for Lp B were described as long ago as 1961. Genetic polymorphism of Apo B was studied extensively between 1960–1975 but publications from there on ceased. Thus little is known about differences of Apo B allotypes as far as chemistry is concerned. The new techniques of molecular biology will certainly shed some light on this topic in due time.

Apo B mutants. Lp (a) was thought for a long time to represent an additional genetic variant of Lp B but as mentioned above, this lipoprotein is completely independent of Lp B.

Genetic polymorphic forms of Apo B have now been located within the so called Ag-system (Ag stands for antigen). Ten Ag-factors have been detected so far for Apo B on the basis of the identification of xenoantibodies from multitransfused individuals (for a review see Ref. 9). These are expressed by five closely linked loci with codominant alleles. Other models of inheritance, however, cannot be excluded. The differences in Apo B from these Ag alleles is not clear but is believed to lie in variations of primary structure. Differences in carbohydrate content have not been detected.

Apo E mutants. The next apolipoprotein mutants in chronological order (Apo E) were been detected by Utermann (38). Since the binding of Apo E to specific liver receptors plays a key role in the conversion of VLDL to LDL, Apo E mutants with structural changes in the receptor binding region (amino acids 140–160) lead to familial dys-β-lipoproteinemia; only if additional genetic aberrations are expressed, will Type-III hyperlipoproteinemia develop. Since these early findings 5 major and several subtypes of genetic Apo E-isoforms have been described (reviewed by Utermann, Ref. 39, and Assmann, Ref. 40). The wild type with a gene frequency of approx. 0.8 is called E_3. The first characterized Apo E mutant was E_2 with a point mutation of Arg-158 to Cys. Today more than 10 Apo E mutants are described in the literature (Table 7). Some of the Apo E mutants exhibit increased, some reduced and others virtually zero binding to the B/E receptor. Epidemiological studies revealed a strong positive correlation between plasma LDL-cholesterol concentrations and the binding affinity of Apo E to the B-E/E-receptors. Thus in the absence of other genetic disorders or secondary hyperlipoproteinemias, individuals who are homozygous for Apo E_4 show the highest, E_3 intermediate and E_2 the lowest plasma LDL-cholesterol values (41). This has important implications for the development of atherosclerosis and myocardial infarction.

The isoforms E_2* and E_2** focus like E_2 but exhibit other amino acid replacements as compared to the classical E_2 mutant. Similarly there exist so far two E_3 isoforms (E_3* and E_3/Leiden) which behave normally on IEF but are characterized by variations in the primary structure of the binding region. E_3/Leiden is associated with the classical Type-III hyperlipoproteinemia; an abnormal Apo E structure in this case could only be demonstrated by micro sequencing of the Apo E peptide.

Table 7 Apolipoprotein-E mutants of human plasma (Most of the data are from Ref. (39))

Mutant	Substitution	Gene frequency[1]	Receptor binding	Association with HLp
E-1	127:Gly→Asp; 158 Arg→CYS	rare	↓↓↓	Hyper-TG
E-1 Bethesda	?	rare	?	Type-III
E-2	158: Arg→Cys	0.077	↓↓↓	Type-III
E-2*	145: Arg→Cys	rare	↓↓	Type-III
E-2**	146: Lys→Gln	rare	↓↓	Type-III
E-3	Wild type	0.773	normal	no
E-3*	112: Cys→Arg; 142: Arg→Cys	rare	↓↓	Type-III
E-3Leiden	No Cys	rare	↓↓	Type-III
E-3(1)	99: Ala→Thr; 152: Ala→Pro	rare	?	?
E-4	112: Cys→Arg	0.150	normal	? Type-V
E-5	Lower M_r	rare	?	?

[1] Gene frequency refers to the German population

Apo AI mutants. The first Apo AI mutant described was Apo AI-Milano (42). 5 affected heterozygous family members have been studied: Apo AI-Milano is the product of a point mutation (Arg-173 Cys). As a consequence dimeric Apo AI proteins are formed via -S-S- bridges yielding a molecular mass of 56 kDα. This

gives rise to the formation of 5 additional bands on isoelectric focusing gels. Affected individuals exhibit significant hypertriglyceridemia and hypo-alphalipoproteinemia. Their HDL particles are larger than normal but LCAT and post-heparin lipolytic activity are within the normal range. Despite the marked reduction of HDL-cholesterol to approximately 0.2 – 0.4 mmol/l, no signs of premature atherosclerosis were apparent in affected individuals.
A number of additional Apo AI mutants have been described in the literature (reviewed by Mahley et al., Ref. 43). In the german population some 6 – 8 different Apo AI mutants have been detected so far, appearing with a very low gene frequency (44). One of the several mutants (Table 8) seems to be of particular interest since it shows a deletion of amino acid 107, lysine. As a consequence, the formation of hydrophobic associates is impaired (45). The Apo AI mutant Giessen on the other hand with a point mutation of Pro-143 → Arg showed a reduced capacity to activate LCAT. This might reflect the fact that Pro-143 is situated in a putative β-turn between two amphipathic helices and therefore is of importance for structural and functional properties of Apo AI (46).

Table 8. Apolipoprotein AI mutants from human plasma (Most of the data are from Ref. (43))

Mutant	Charge relative to normal Apo AI	Substitution
Milano	−1	173: Arg→Cys
Münster 3(A)	+1	103: Asp→Asn
Münster 3(B)	+1	4: Pro→Arg
Münster 3(C)	+1	3: Pro→His
Münster 3(D)	+1	213: Asp→Gly
Giessen	+1	143: Pro→Arg
Marburg (Münster 2A)	−1	107: Lys→0
Münster 2(B)	−1	158: Ala→Glu
Münster 4	+2	198: Glu→Lys
Norway	+2	136: Glu→Lys

Other apolipoprotein mutants. It is likely that there are many mutants of all plasma apolipoproteins. Most of them will be silent, i. e. they are not associated with any disease or impairment of lipoprotein metabolism. Others, even if they cause such impairments, may be hardly detectable by current methods. Recently a very interesting method has been described by Jabs et al. (47) aimed at the detection of cryptic mutants which do not differ in their electrical charge. This method holds great promise for the detection of a variety of different apolipoprotein mutants in due time. Currently, additional genetic isoforms are known from Apo AII (48) Apo AIV (49) and Apo CII (50).

2 Physical Analysis

2.1 Lipoproteins as a microemulsion system

In simple physical terms, lipoproteins can be regarded as *microemulsion particles* which facilitate solubilisation of water-insoluble lipids in the aqueous medium of blood plasma. Such particles consist of two domains: a *core* of apolar, lipidic material which is surrounded by a *shell* of polar hydrophilic components, like charged lipid headgroups, amino acid side-chains, and sometimes sugars. The interface between the apolar core and polar shell is generally formed by *amphipathic* molecules, which consist of hydrophilic and hydrophobic domains; classical examples of such amphiphilic molecules are the phospholipids, but also some parts of the apoproteins have been found to have distinctly amphipathic character. This simple concept, i.e. that the lipoproteins are optimized by their nature to the purpose of solubilizing apolar material in aqueous medium has two important theoretical consequences with respect to their structure: Firstly, to fulfil the requirement of a minimal surface-to-volume-ratio they are expected to be spherical. Secondly, the size of the particles should increase with the relative amount of apolar lipids, since the surface-to-volume-ratio decreases when the particle size increases. It will be seen from the following that both of these theoretical principles are verified in nature. A quantitative theory based upon these thermodynamic principles indeed reproduces very well the observed particle size range of lipoproteins (51). An exception proving the rule is given by the abnormal lipoprotein X (Lp-X) which consists largely of phospholipids and cholesterol, and lacks neutral lipids, and consequently contains the lipid bilayer as its structural principle.

However general this microemulsion concept may be, it is not satisfactory because it fails entirely to explain the existence of major and minor lipoprotein classes and to relate individual lipoprotein species to their specific function in metabolism. Specificity is a matter of molecular properties – their structure, affinities and dynamics – and therefore a description in molecular terms is required for the discussion of function.

2.2 Properties of the constituents

2.2.1 Apoproteins

A large number of investigations have been performed to study the conformational properties of apoproteins in model complexes with defined lipid species (for reviews see e.g. Refs. 52–55). From these emerged the significant notion that *amphipathic* α-helices are widespread structural elements, especially in the A- and C-proteins. Figure 1 shows a model representation of such a domain in the CIII protein. An important exception is the B-protein, the dominant protein constituent of LDL, which contains very little, if any, α-helical conformation. As this protein has for a long time posed severe difficulties with regard to its isolation and characterization (56) little more is known about its conformation. The complete elucidation of the protein sequence of Apo B_{100} comprising one polypeptide chain with 4536 amino

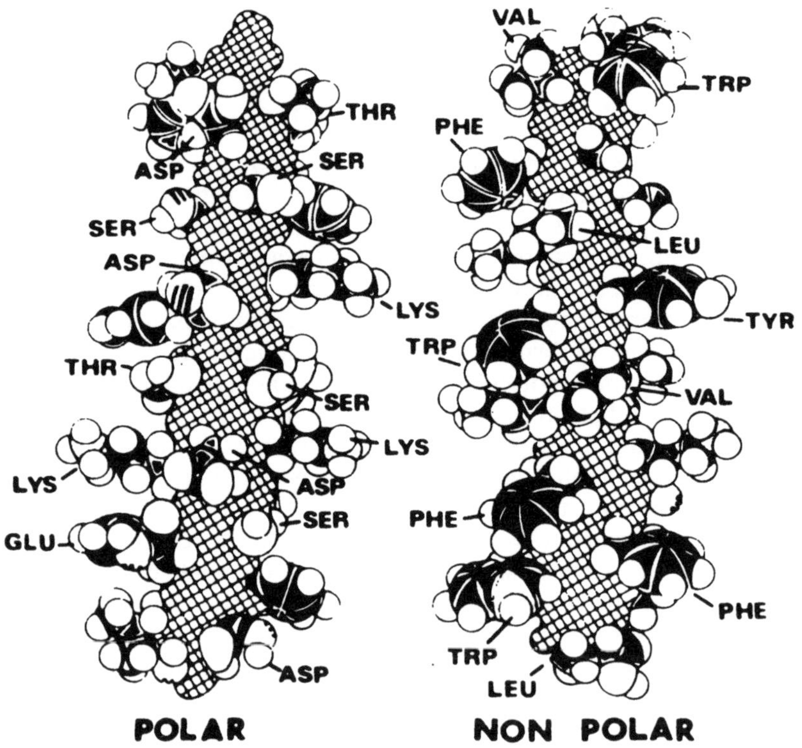

Fig. 1. The two faces of an amphipathic helical segment of apolipoprotein C-III as predicted by Segrest et al. (94). The model suggests an interaction between the paired acidic and basic residues and the zwitterionic polar head groups of phospholipids. The apolar face, with the tyr-, phe-, and trp-residues could interact with the fatty acyl chains of the lipids (from J. T. Sparrow and A. M. Gotto, Jr., Ann. N. Y. Acid. Sci. 348, 187–207, 1980, with permission).

acids and the identification of specific structural and functional domains has been reported recently by two independent groups of investigators (57).

From the known amino acid composition and sequence data it is also suggested that the protein constituents of lipoproteins fall into the category of peripheral proteins rather than into that of integral proteins, in terms of the concept used in membrane biophysics (58). From this it is to be expected that they are predominantly arranged at the outer polar/apolar interface of lipoprotein particles.

2.2.2 **Lipids**

Figure 2 shows representative chemical structures and approximate dimensions of the lipid constituents together with a schematic picture of their preferred aggregations in aqueous medium. *Phospholipids*, of which the phosphatidylcholines (lecithins) are the major class in lipoproteins, are classical amphipathic lipids which form bilayer arrangements in water, or monolayers on polar-apolar interfaces. They

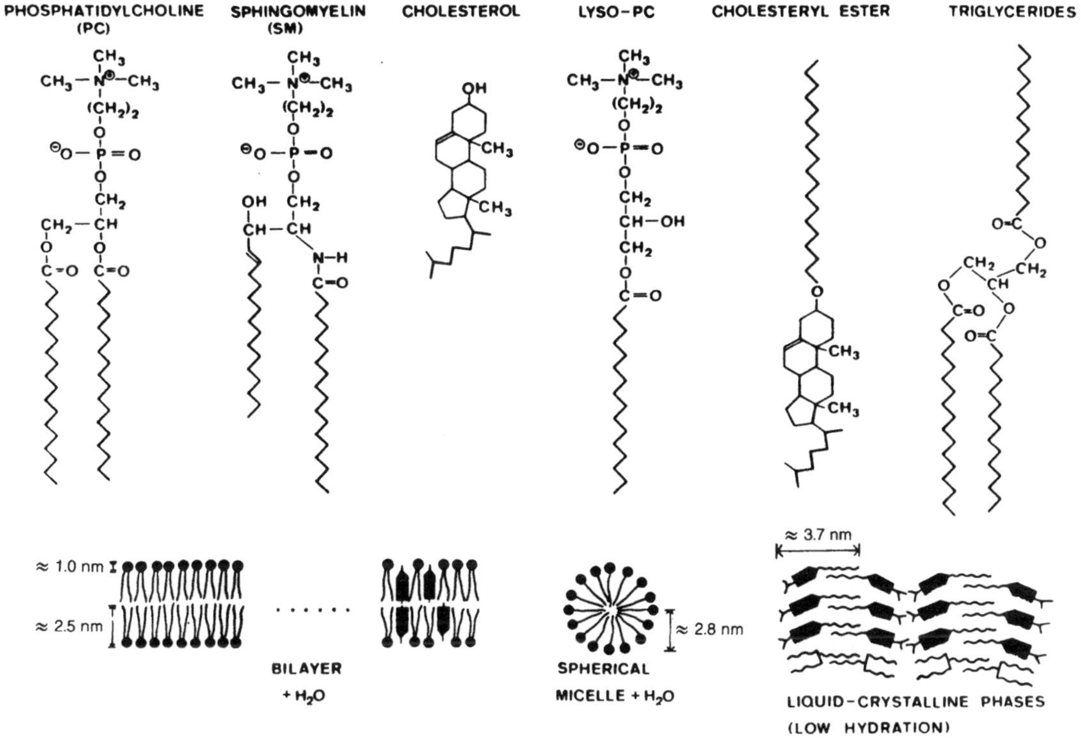

Fig. 2. The major lipid species of lipoproteins. The schematic drawings at the bottom show the preferred forms of aggregation and the approximate dimensions.

also have a high capacity for taking up unesterified ("free") *cholesterol* within monolayers or bilayers, such that the cholestane ring system is intercalated parallel to the phospholipid hydrocarbon chains, and the 3β-hydroxyl group is in a position close to the glycerol "backbone" region of the phospholipids. This interaction leads to considerable modifications in the molecular dynamics of the phospholipid hydrocarbon chains. From model studies it appears that there is a limit to the capacity of phospholipid bilayers for cholesterol at equimolarity. The compositions listed in Table 2 show that this limit is reached in VLDL, IDL and LDL, whereas the amount of free cholesterol in the HDL subclasses is far below saturation. While it has to be respected that this notion is obtained from pure lipid model systems which have no morphological similarity to lipoprotein particles, these facts are still interesting in relation to the documented role of HDL and their precursors as "scavengers" of cholesterol *in vivo* (59).

The apolar lipids, *cholesteryl esters* and *triglycerides* do not form hydrated phases on their own in aqueous environment, but separate as pure, unhydrated phases. In phospholipid bilayers they are only poorly soluble (60, 61). In lipoproteins, and in particular in LDL and larger classes, they are the dominating constituents. One central question of lipoprotein structure is, therefore, how they are incorporated in microdisperse particles. Of particular relevance to the structure of

LDL, where they are the dominating components, is the property of cholesteryl esters to form thermotropic liquid crystalline phases and to undergo various structural transitions before transforming into an isotropic oily state (62).

2.3 Methods of physico-chemical analysis

2.3.1 Possibilities of different approaches

It is beyond the scope of this review to adequately discuss the contributions made by all individual physical techniques to the present knowledge of lipoproteins. However, it shall briefly be summarised what kind of information could be gained by the major conceptually different groups of methods with their specific resolution ranges in time and space.

Of the hydrodynamic methods, *analytical ultracentrifugation* has played a prominent role in lipoprotein research. Apart from providing an operational basis for lipoprotein classification into ranges of buoyant densities or flotation rate coefficients (see Tables 2 and 3), the determination of molecular masses and frictional coefficients of the hydrodynamic particles provided important hints to size and shape, although in this latter respect with lower resolution and considerable uncertainties, since shape anisotropy and solvation effects are not easily disentangled.

The direct visualization of lipoprotein particles by *electron microscopy* has obviously had a profound and lasting impact (Fig. 3). Practically all lipoprotein subfractions have been investigated by this method. It confirmed and substantiated many of the properties inferred from hydrodynamic studies, in particular isotropic shapes and the average particle dimensions. Difficulties arise from the rather drastic environmental conditions that have to be used for most of the common contrast enhancing techniques, like negative staining, fixation and shadowing by metal vapor, which can at least in principle, create artifacts. In the case of negative staining with sodium phosphotungstate which is very frequently used, Gong et. al. (63) investigated the question whether it had any irreversible effects on HDL samples and found that they were practically negligible. They pointed out, however, that the dimensions obtained from this staining technique involve considerable error-limits, since phosphotungstate may also partially enter the hydrophilic particle shells and thus lead to an underestimation of sizes. In some cases the grain structure of negative stain seemed to indicate a subunit structure for the particles, which has even served as the basis for some earlier structural models, featuring discrete subunits. On the basis of the more recent information from X-ray small-angle scattering, however, they had to be discarded and referred to as artifacts from the electron microscopic preparation. The potentially less destructive methods of freeze fracturing and freeze etching have found application so far only with the larger LDL particles (64), since small sub-10 nm HDL particles are difficult to resolve by this technique. On the whole, with the present state of the art, it has to be accepted that structural details of sizes below about 3 nm are very difficult to resolve reproducibly. Furthermore, the methods of electron microscopy only reveal the particle contours, and the internal structure remains elusive.

The most powerful approaches to analyze both the external and internal particle structure are the *diffraction methods. X-ray crystal structure analysis*, as the only

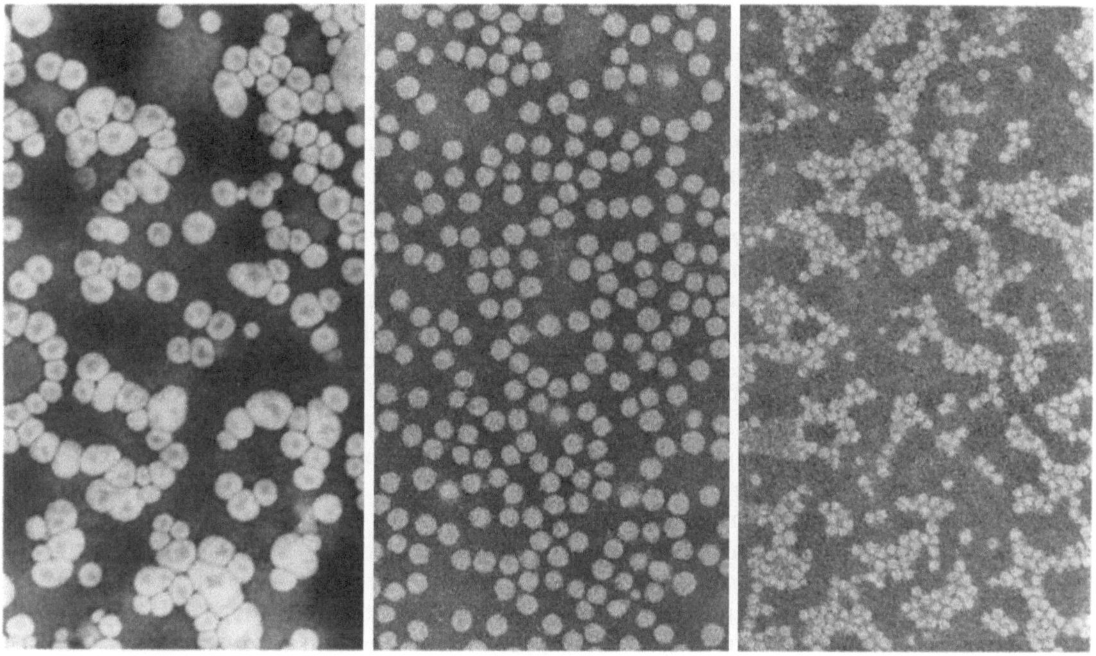

Fig. 3. Typical images of lipoprotein density classes as obtained by negative stain electron microscopy. Left: VLDL center: LDL; right: HDL_3. (Courtesy of Dr. Ingolic, ZFE, Graz)

method to provide three-dimensional information on the level of atomic resolution, has so far not been able to enter this field, simply because no single crystals suitable for diffraction analysis have been obtained from any plasma lipoprotein species. This certainly reflects the two fundamental differences between lipoproteins and pure proteins – their microheterogeneity in terms of lipid content and probably also size, and their internal "*fluidity*". However, these are not principal obstacles, and it is to be hoped that refined techniques of lipoprotein purification and cryocrystallization might eventually overcome the problems. This expectation is nurtured by the fact that lipoprotein complexes of the yolk-system in amphibian eggs are present in microcrystalline form and, therefore, lend themselves to crystal structure analysis (65).

A great deal of the present knowledge on the internal structure of lipoproteins comes from *X-ray and neutron small-angle scattering* on dilute, "monodisperse" samples (for reviews see refs. 66–68). These methods are based upon the same principles as crystal diffraction: while the latter involves the condition of perfect molecular order in a crystal lattice, which essentially serves as a signal amplifier for the individual particle structure, diffuse small angle scattering requires ideal disorder, i. e. completely random orientations of the individual particles in a dilute solution. Each particle gives rise to a scattering function, which is exclusively governed by its internal electron density distribution and hence its structure, and the overall effect of small-angle scattering is simply the sum of the individual particle scattering effects giving an average over all possible particle orientations. This implies a certain loss of three-dimensional and directional information. However, the experimental result provides access to a number of very important integral parameters, among others the molecular mass, particle size and volume; furthermore, the scattering curve contains information on the internal particle structure which is not accessible by any other technique, and therefore serves as an important basis for modelling the mutual arrangement of lipids and proteins within lipoproteins. In general, the resolution is about one order of magnitude lower than that of crystal structure analysis, but still better than that of e. g. electron microscopy. It can also be seen as a specific advantage of this technique that it works in a quasi native environment without requiring staining or other potentially destructive factors. The question of radiation damage has not yet been approached systematically in this field, but no signs of drifts in the scattering patterns have been observed so far even with prolonged irradiation, and analytical tests after X-ray exposure have not shown any signs of alterations.

Neutron small-angle scattering is conceptually similar to X-ray small-angle scattering and provides the same kind of information. The reason for its application, and its advantage, lies in the fact that the scattering power for hydrogen and deuterium atoms are extremely different, much more so than for instance between a heavy metal atom, and carbon or oxygen in X-ray diffraction. This opens two very attractive possibilities: Firstly, selective isomorphous replacement of ^{1}H by ^{2}H within certain specified chemical components of lipoproteins leads to a drastic contrast enhancement of these domains and therefore their location within the complex can be determined. Secondly, by changing the H_2O/D_2O composition of the solvent buffer, the entire range of individual scattering powers of lipids and proteins can be matched, so that by a series of experiments in selected H_2O/D_2O buffers, the domains of different scattering power can be studied individually. Both approaches have been fruitfully applied to answer specific structural questions in lipoprotein research which were not accessible by X-ray methods (69–71).

The conformations of the different molecular species, their dynamics and mutual interactions have been the subject of extensive studies by various *spectroscopic methods*. The *optical activity* of proteins has been used to study apoprotein conformation by optical rotatory dispersion and circular dichroism in the ultraviolet region.

The intrinsic *fluorescence* of tryptophan in the apoproteins has been used to study their accessibility to quenching agents added to the solvent buffer (72, 73), providing information on whether or not the tryptophan fluorophors are located at or near the particle surface. Another interesting use of this intrinsic fluorescence is the determination of the degree of energy transfer between tryptophan and artificial, fluorescent cholesterol analogs and their oleate esters, which can be introduced into lipoproteins (74). Since the energy transfer is strongly distance dependent (it decays with the sixth power of the distance) it can be used as a "molecular yardstick" to determine distances between the different fluorophores. Another possibility of fluorescence spectroscopy is given by the motion-dependent polarization of fluorescence (75). This fluorescence polarization anisotropy of lipophilic probes has found widespread application in biomembrane research to study the "microviscosity" of the lipid environment. Applications to lipoproteins (76–78) however, have yielded relatively little information, mainly because of the small size of their lipid domains per particle, which can hardly be treated in the sense of phases with uniform motional properties.

Nuclear magnetic resonance (NMR) and *electron spin resonance* (ESR) have been used successfully to define the polarity and mobility of the environment of either a specific kind of atomic nuclei (in the case of NMR) or a stable free radical introduced in the form of a spin label (in the case of ESR). For a review on these methods the reader is referred to (79).

In general, the specific answers obtained by spectroscopic techniques were found most valuable in the context of the overall structural information from e. g. X-ray methods, since by themselves they are not able to provide longer-range three-dimensional information. Therefore, the structural models which in some cases were extrapolated from spectroscopic results, before X-ray data had been available, were later found to be incorrect, although based upon correct experimental results.

Calorimetric techniques, such as *differential scanning calorimetry* (DSC) and *mixing microcalorimetry* have been important tools to determine thermal stability and thermodynamics of interaction between lipids and proteins in intact and reconstituted lipoproteins. DSC methods were especially fruitful in characterizing protein stability in lipoproteins, and in detecting phase-transitions of cholesteryl esters in LDL. Microcalorimetry was used in many studies to study the enthalpies involved in lipid-protein interaction in model systems.

2.3.2 **Limitations**

The physical description of a lipoprotein particle is an extremely complex task, much more so than for instance with "simple", globular proteins. This is due to the following general facts:

Heterogeneity: It has to be accepted that even the most refined subfractionation of lipoproteins does not result in an ideally homogeneous sample consisting of only one chemically and physically discrete particle species. This is due to the fact that lipoproteins are not thermodynamically stable or metastable products of a genetically determined and enzymatically controlled biosynthesis, but rather are intermediates in a dynamic, multiple-equilibrium catabolic process which, at best, reaches steady state *in vivo*, but not equilibrium. A system such as the lipoproteins, which is designed for mass transport and exchange as its primary function, is inherently a non-equilibrium system, which implies a broad distribution of compositional and structural states for the individual particles. For physico-chemical studies this has the consequence that the analysis of a given sample leads to an average result which necessarily falls short of atomic resolution – the standard for molecular structure. Moreover, in the process of isolation of a non-equilibrium particle from its environment, which is necessary to prepare suitable samples for physico-chemical analysis, the possibility of alterations to the particle properties has to be considered.

Internal molecular motions: The lipids of lipoproteins are generally in a "fluid" state at physiological temperatures (an important exception will be discussed in the context of LDL). In particular for those lipoprotein classes in which lipids are the dominating constituents this means that the entire particle structure is highly flexible and affected by internal thermal motion with relatively large amplitudes. From the methodological point of view this has the positive consequence that such spectroscopic techniques, which are sensitive to, and rely upon, motional averaging of molecules and residues (such as e. g. nuclear magnetic resonance) are applicable to probe the local environment of the mobile elements and to quantify the different mobilities of groups giving rise to a signal. On the other hand, it also implies the loss of discrete three-dimensional structure information.

As a result of these two factors, the strategy for physico-chemical analysis has to respect the notion, that a complete description of the lipoprotein system as the sum of its individual parts is impossible; and even if it were possible with the advent of new methods, it would be insufficient inasmuch as lipoproteins are parts of a non-equilibrium, constant-flow system.

Undoubtedly, nevertheless, the physico-chemical approach has given very important insights into the nature of lipoproteins which shall be summarized in the following.

2.4 Physical properties of the major lipoprotein classes

2.4.1 High density lipoproteins

This density class, within the relative mass range of about $2-5 \times 10^5$ Da consists of particles ranging in size between about 10 to 15 nm. It is now generally accepted, mainly on the basis of X-ray small-angle scattering evidence, that the particle shapes are highly isotropic, close to being spherical. Both the shapes and the internal structural organization do not fundamentally vary among the different subclasses of HDL studied so far. This is demonstrated by the X-ray scattering curves reduced to unit particle size and the electron-pair distance distribution $p(r)$ shown in Figure 4. Evidence for quasi-spherical shape comes from three features of these results:

1. The very well-defined secondary maxima and minima in the intensity curves which agree well with the notion of sphericity;
2. The same is reflected by the very steep decay of the $p(r)$ function towards the maximum particle dimension;
3. From the comparison between the smallest possible diameters (i. e. those of spheres) calculated from the excluded volume radii of gyration and the unambiguously determined maximum particle sizes from the $p(r)$ functions: the theoretical radii for spheres are 5.4 nm and 6.1 nm, respectively, which agrees very well with the maximum diameters of 10–11 nm for HDL_3, and 12–13 nm for HDL_2.

Numerous other subfractions obtained by hydroxylapatite column chromatography have been investigated in the same manner by the authors and the results were all in full agreement with this picture.

The principles of the internal particle structure are reflected by the radial electron density distribution obtained from X-ray scattering data (Fig. 4c). In all cases the core is of low electron density, similar to that expected for a mixture of hydrocarbon chains and cholesterol rings, and the surrounding, outer shell is of high electron

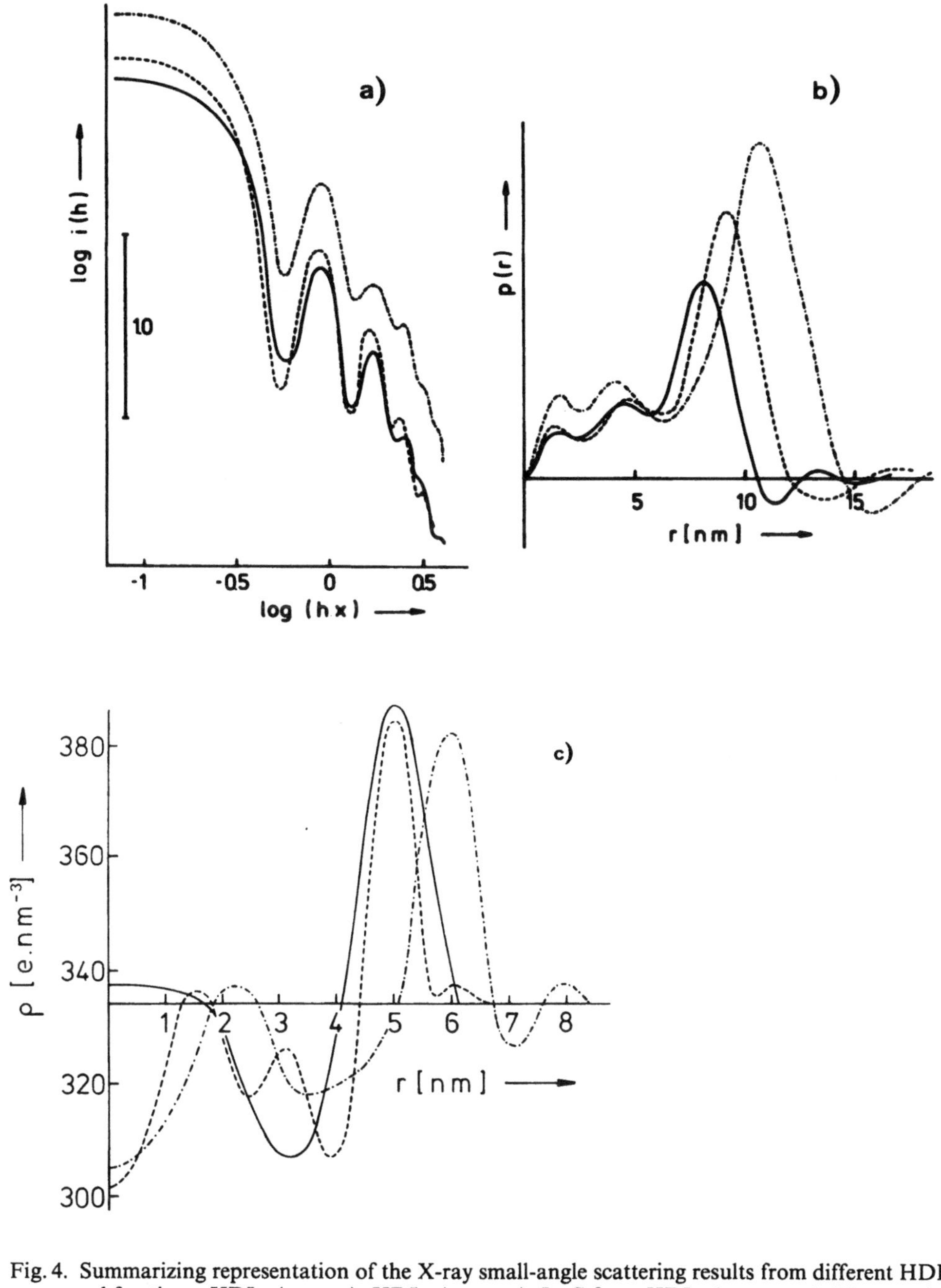

Fig. 4. Summarizing representation of the X-ray small-angle scattering results from different HDL subfractions: HDL_3 (———), HDL_2 (— — —), LpC from HDL_2 (— · — · —). The angular scale factors x are 1,0, 0.84 and 0.76 for HDL_3, HDL_2 and LpC, respectively, to account for their size differences. (a) scattering curves; (b) electron-pair distribution functions; (c) radial electron density distributions (from Laggner & Müller, ref. 66, with the publishers permission).

density, which, however, is considerably lower than the values expected for compact protein and glycerophosphorylcholine groups (450 e/nm^3). On this basis, and with the known molecular dimensions and relative concentrations of the molecular constituents, the structural model schematically shown in Figure 5 has been derived.

In this model the core is formed by a radial arrangement of the cholesteryl esters in extended conformation into which the hydrocarbon chains of the phospholipids and the unesterified cholesterol are interdigitated. The surface is formed by a mixed monolayer of phospholipid polar headgroups and protein. The thickness of this outer high-electron density shell is in all cases about 1.5 – 2.0 nm which agrees well with the diameter of the contour cylinder for an α-helix.

The model shown in Figure 5 has some interesting implications regarding the size spectrum of the HDL particles. The smallest observed core radius is 3.7 nm, in the case of HDL_3, which is very close to the length of an extended cholesteryl ester (e. g. oleate) molecule. In this case full interdigitation has to take place between the cholesteryl esters and the phospholipid hydrocarbon chains. A smaller particle than that cannot be envisaged. This agrees with the fact that the HDL_3 subclass

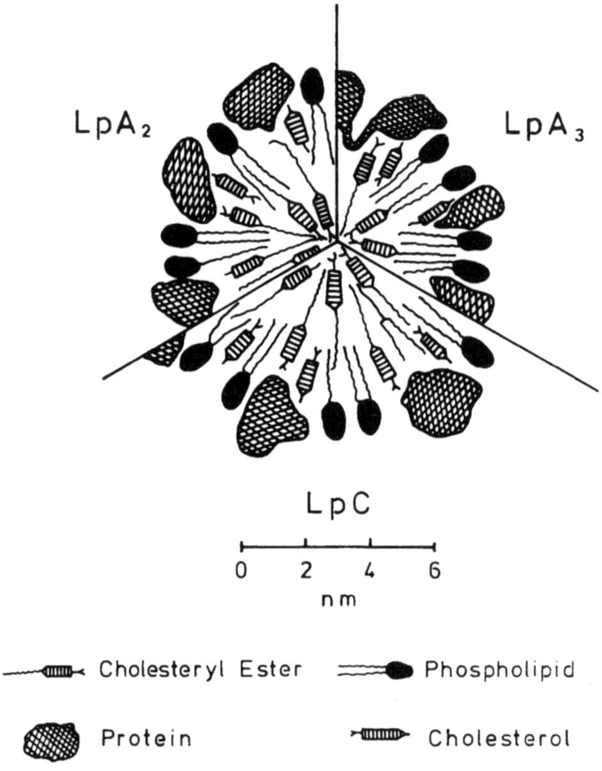

Fig. 5. Schematic model of the molecular arrangement of HDL components. The size differences are accommodated by different degrees of interdigitation between phospholipid hydrocarbon chains and cholesteryl esters arranged radially in the core. LpA_3 and LpA_2 are HDL_3 and HDL_2 subclasses containing only apolipoproteins A; LpC is an HDL_2 fraction rich in Apo C-protein (from Laggner & Müller, ref. 66, with the publishers permission).

represents the smallest lipoprotein particles of any natural abundance. The largest core radius, on the other hand, has been observed in the so-called LpC fraction of HDL_2 (80). Here, the value of 5.1 nm corresponds to the length of a cholesteryl ester plus an *all-trans* $(CH_2)_{11}$-chain, which leaves about the length of six methylene groups for interdigitation. This, again, appears to be a limiting case since the HDL spectrum normally does not contain larger particles. Thus, the natural size limits of the HDL density class confirms in a logical way the interdigitation model with radially arranged lipid molecules.

A brief reflection is warranted here on the frequently held opinion that the protein-lipid arrangement on the surface of HDL particles can be described in terms of globular protein particles embedded in a "sea of lipids". This view, adopted from the Singer-Nicholson model for biological membranes (58) has been postulated from NMR-spectroscopic data on HDL (81). On present knowledge, only the dynamic aspects of such a model are correct, i. e. most of the molecules and groups are mobile. However, from a structural point of view, the notion of a sea of lipids is difficult to sustain in that the numbers of phospholipid molecules per particle (about 40–60) are hardly able to constitute a uniformly liquid environment. Furthermore the existence of globular protein structure at the surface is not borne out by the X-ray results. They favor a two dimensional cage-like arrangement of helical rod elements into which the phospholipid headgroups are intercalated. This latter model is also confirmed by X-ray studies on recombined model particles of HDL apoproteins and phospholipids (82).

Protein conformation: It has early been recognized that the protein moieties of HDL possess a relatively high α-helical content. Optical rotatory dispersion and circular dichroism led to estimates of 65–70% α-helix in HDL_2 and HDL_3 samples (83–85). This conformation appears to require the intact HDL particle structure since the α-helicities were found to decrease upon various perturbing treatments such as heating to 80 °C or delipidation. Details on the conformations of distinct apolipoprotein species, such as AI, AII, and C-peptides, in intact lipoproteins are not known, since the individual optical effects cannot be singled out from the composite spectra. Numerous studies have been performed on the structure of isolated apolipoproteins and their recombination products with different lipids. For Apo AI it was found that the conformation is extremely variable within different environments. In its delipidated form, the α-helical content is about 55% (86) which agrees well with predictions based on the Chou-Fasman rules (87) from the primary structure. The α-helix content increases upon the binding of lipids: phospholipids, like egg yolk lecithin or lysolecithin raise the value about 70%, and the further addition of cholesteryl ester to 83% α-helix. These findings suggest that Apo AI may also change the details of its conformation according to the size of the lipoprotein particle: on the surface of a highly curved particle like HDL_3 it may be different from its structure on the much larger VLDL particle, where the surface is already essentially flat.

The relatively low structural stability of Apo AI is also reflected by its thermotropic behavior: differential scanning calorimetry has shown a broad, reversible transition between 43 °C and 71 °C, with mid-point at 54 °C, and a transition enthalpy of 10.0 J/g, which is much lower than the enthalpies normally found with globularly folded proteins (88). This correlates well with the anomalously high degree of solvent exposure of hydrophobic amino acid side chains shown by fluorescence spectroscopy (89), and the low concentration of urea or guanidinium hydrochloride required for its unfolding (90). Both of these characteristics are certainly favorable to the function of AI in binding and solubilizing a wide variety on nonpolar or amphipathic ligands.

Apo AII also shows conformational instability by the strong temperature dependence of its optical activity and by the low (1 mol/l) concentration of guanidine-HCl required for the total elimination of ordered secondary structure (90). In delipidated form it contains only about 35% of α-helix and the interaction with phospholipids and cholesteryl esters increase this value by about 15–25% (91).

The *C-apoproteins*, Apo CI, CII and CIII, have relatively poorly defined secondary structure in the delipidated state, the exception being Apo CI which shows about 55% α-helix (92, 93). Interaction with lipids, however, in all three cases was found to lead to a considerable increase in α-helicity.

Despite the abundant information from many recombination studies on apolipoproteins, the prime question on the specific features that make the apolipoproteins bind to the surface of globular lipid-micelles rather than become involved in binding to membranes, still remains largely unanswered in detail. One plausible proposal is that they are capable of forming lateral amphipathic helices, as first pointed out by Segrest et al. (94), which enables them to bind to zwitterionic phospholipid headgroups by featuring clearly defined polar and apolar faces and by generating at least one ion pair by the juxtaposition of acidic and basic residues at the polar face (see Fig. 1). However, other factors, such as the frequency of β-turn structures, general hydrophobicity and segmental length may assimilate about one fifth their weight in cholesterol without affecting the major structural characteristics. Similarly, phospholipids are readily exchangeable between HDL subclasses and with other lipoprotein classes and cell membranes. This important aspect will be dealt with in detail in Chapter 3 of this volume.

2.4.2 **Low density lipoproteins**

LDL is distinguished from all other lipoprotein classes by its high relative proportion of cholesteryl esters: almost half of its mass consists of this type of apolar lipid. As a second characteristic feature, the majority of its protein part is made up of one specific apoprotein, Apo B (see Table 5), whose molecular mass was until recently in dispute, but it is known now that 513 kDa is the appropriate figure, making this protein among the longest single polypeptide chains known in globular particles.

The molecular masses reported for LDL from human serum, in the buoyant density range of 1.063 and 1.090 kg/l, are between 2.2 and 3.0×10^6 Da. The particles are highly isotropic, quasi-spherical, and with diameters between 22 and 25 nm. Slightly lower or higher figures are sometimes reported, but it is uncertain to what extent they are influenced by the methods of preparation and/or analysis. The overall particle shape is close to spherical, as demonstrated by X-ray and neutron small-angle scattering, electron microscopy and hydrodynamic studies (97).

The outer shell is formed by a monolayer of protein and phospholipids, and probably most of the free cholesterol in a manner similar to one half of a fluid bilayer membrane. Studies on the details of protein-lipid arrangement in LDL have long been hindered by the extreme difficulties encountered in attempting to solubilize the protein component Apo B in a form that would allow controlled reconstitution experiments with defined lipid systems (56, 98). Consequently, the information available on protein conformation and protein-lipid interaction in LDL is still very fragmentary. The very fact of its resistance to common solubilization procedures indicates that Apo B binds relatively strongly to lipids. The most likely candidates for this interaction are the phospholipids. Both proton and ^{31}P-NMR studies have shown that about 20 – 30% of the polar phospholipid headgroups are immobilized presumably by phospholipid-protein interaction (99 – 101). The majority of the polar headgroups, however, are free from such constraints and accessible to spectroscopic quenching agents, such as chelated Mn^{2+} or Fe^{3+} ions. This accessibility of the phospholipids agrees also qualitatively with the results of enzymatic cleavage studies: digestion by phospholipase A_2 is practically quantitative and rapid (102), while the action of phospholipase C is biphasic (103). With respect to the stability of the LDL particles it is also noteworthy that both phospholipase digestion and proteolysis by trypsin do not result in the breakdown of the overall spherical shapes and sizes (104, 105). As a note of caution, however, it has to be emphasized that interpretation of such studies is limited by the possibility of microheterogeneity in LDL samples: for any result falling short of 100% reactivity (or susceptibility) the possibility must be carefully considered, that there exists a fraction of completely non-reacting particles within the sample.

Very little structural information is available so far on the protein moiety in LDL. A neutron scattering study on LDL with selectively deuterated phospholipid headgroups has shown that the average radius of gyration of the protein moiety is about 11 nm and thus about 0.8 nm higher than that of the phospholipid headgroups (70). This correlates well with the notion that some headgroups may be buried underneath the protein.

It is now generally accepted that all of the protein resides in the outer monolayer of LDL. Confirming evidence for this notion comes again mainly from accessibility studies. Fluorescence quenching experiments have shown that more than 80% of the intrinsic tryptophan residues are accessible to quenching by iodide although the fluorophores in LDL appear to be more immersed in the lipid environment (73) than is the case in HDL.

The protein conformation of Apo B, studied by its optical activity, differs significantly from that of most other apoproteins in that it contains only minor fractions of α-helix but considerable amounts of β-structure. Furthermore it appears to be highly sensitive to environmental changes, such as lipid depletion, addition of denaturants in small concentrations and temperature variation. On the whole it appears that Apo B has a high degree of conformational flexibility which may well have a functional significance in the adaptation to the varying demands of lipid transport.

Thermotropism of LDL and its cholesteryl ester core: A key observation to the characterization of LDL was the finding of a thermotropic transition around 30 °C which arises from a cooperative order-disorder transformation of the apolar core of LDL (108, 109). Most significantly this is reflected by the X-ray small-angle scattering patterns (Fig. 6) which show, in addition to the monotonously decaying

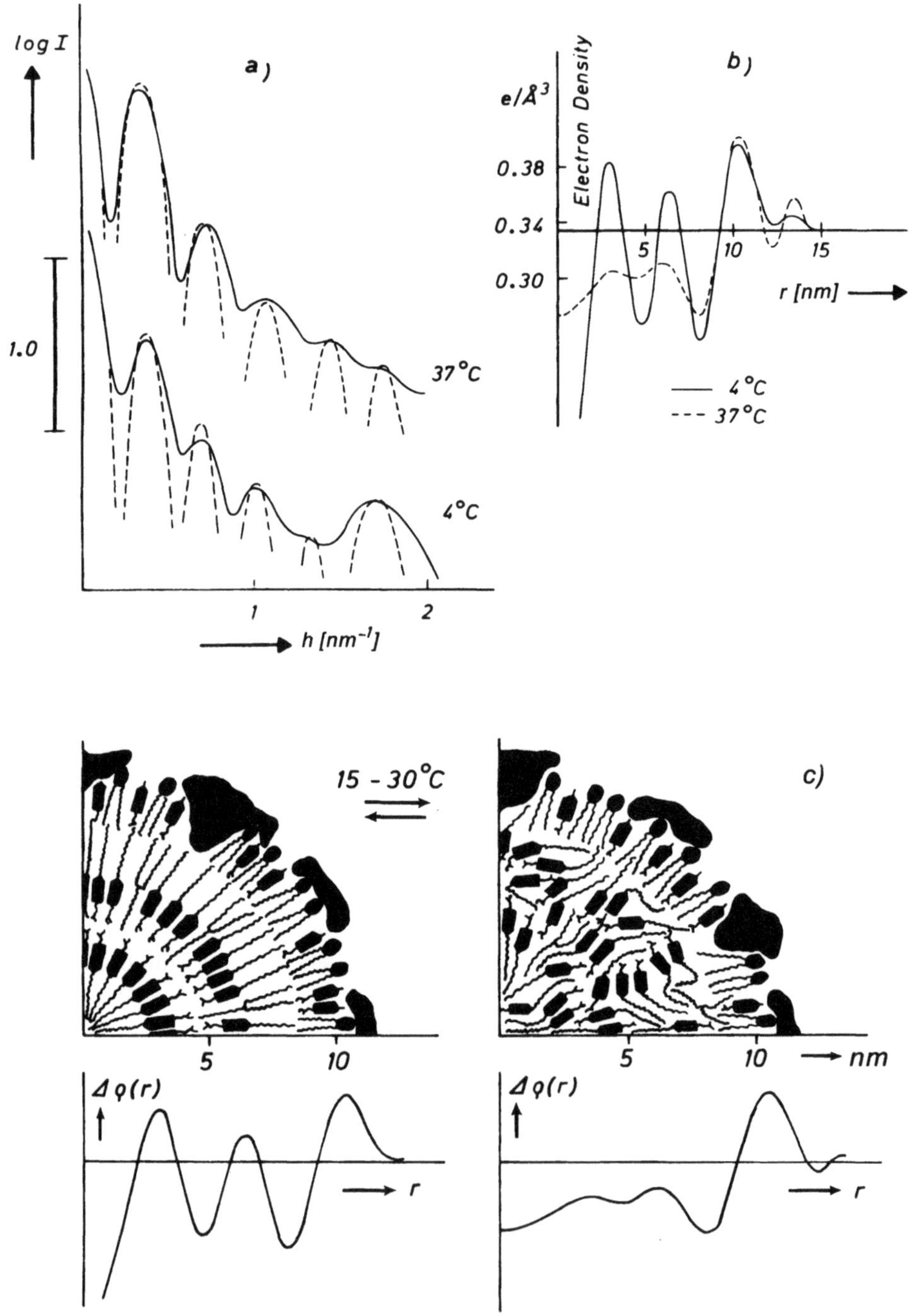

Fig. 6. Top: X-ray small-angle scattering curves (a) and radial electron density distributions (b) of LDL at 4° and 37°, respectively. The thermotropism of the concentric layered arrangement of cholesteryl esters in the particle core is schematically depicted in the model (bottom) (from Laggner & Müller, ref. 66, with the publishers permission).

secondary maxima related to the overall spherical shape, at low temperatures a very pronounced maximum at spacings of about 3.7 nm^{-1}, similar to that observed in pure cholesteryl ester phases. The corresponding features in the radial electron density distribution are two maxima around 3.5 and 6.5 nm which can be attributed to the cholestane ring structures (110, 111). Thus, at low temperatures, LDL contains two concentric core layers of radially arranged cholesteryl ester molecules. A more detailed study by neutron scattering using selectively deuterated cholesteryl esters has confirmed this view and indicated that the cholesteryl esters are arranged such that the sterol rings within each layer point towards the surface while the hydrocarbon chains point to the core (71). The small amounts of triglycerides present in LDL adapt to this structure and act only as weak impurities to the cooperativity of the cholesteryl esters. The transition width and midpoint temperatures are affected by the relative CE/TG ratio such that higher TG content leads to a broadening and decrease of the transition temperature (109, 112). However, other factors such as particle size and degree of unsaturation of the fatty acyl chains may also be effective in determining the transition characteristics.

It has been suggested that thermal transition may influence the rates of LDL catabolism and hence be of relevance to physiological cholesterol clearance. In experiments on rhesus monkeys, such correlations to atherogenesis have been positively confirmed (113). In normal humans, the mean transition of isolated LDL is around $30 \pm 2\,°C$; however, strong variations are observed in individuals ranging from about 24 °C up to 40 °C. Considering that the blood temperature in peripheral vessels may transiently drop to about 25 °C even under non extreme conditions, and the frequent occurence of xanthomatous tissues on exposed body sites such as knuckles or eyelids lends further weight to the suspicion that the cholesteryl ester transition may have patho-physiological significance. Clearly, however, this requires large-scale screening of controlled subjects to reach the state of proof.

Obviously in this context it is of interest to know whether the surface structure of LDL is affected by core transition. X-ray and neutron scattering have not shown any major changes in the surface layer, but these methods are certainly not suited to resolve fine molecular rearrangements. EPR studies using various kinds of spin labels introduced in the polar phospholipid monolayer, or attached to the protein, have shown significant spectral changes concomitantly to the core transition (114). These results have been qualitatively confirmed by EPR studies of the capacity of LDL to bind Mn^{2+} ions (115). Moreover, model studies on microemulsions of phospholipids and cholesteryl esters have shown that coupling between the thermal properties of the core and surface constituents takes place (116). Thus the requirement for a surface recognition of the core-lipid state in LDL appears to be verified.

2.4.3 **Very low density lipoproteins**

Owing to the increasing microheterogeneity with particle size, the class of VLDL is still very ill-defined from a detailed structural point of view. Electron microscopy and hydrodynamic data have shown that the principle of spherical structure with the typical core-surface partition of apolar (triglycerides and cholesteryl esters) and polar (proteins, phospholipids, free cholesterol) components is largely verified in this density class (117). Moreover, the core displays the dynamic features to be expected of particles which are largely composed of apolar and partly unsaturated lipids. In normal VLDL, the proportion of cholesteryl esters is too low to exert cooperative melting phenomena as in LDL. However, in studies on cholesteryl ester rich VLDL (118, 119) it has been demonstrated that a similar layered arrangement of radially extended CE and TG molecules with discrete thermotropic cooperativity as described above for LDL can exist. Furthermore it has been shown that the molecular mobilities in the surface monolayer are strongly influenced by the cholesteryl ester content of the core.

2.4.4 Lipoproteins lacking cholesteryl esters

According to the above scheme the molecular structure and phase behavior of cholesteryl esters is an important factor in determining size and internal structure of lipoproteins. This is further demonstrated by the structures of lipoprotein like particles lacking cholesteryl esters, amply demonstrated by numerous studies on reconstitution products of apoproteins and phospholipids. A natural system, where the same is the case, occurs in the abnormal lipoprotein Lp-X (Fig. 7) which is abundant in the serum of patients with obstructive jaundice or LCAT deficiency. Small-angle X-ray scattering has been used to characterize the structure of these

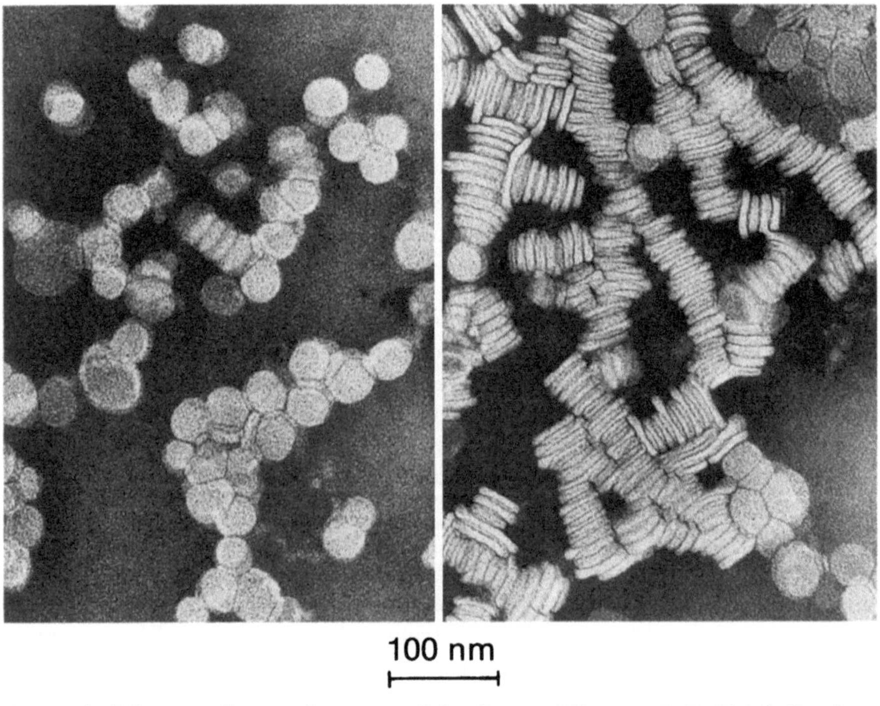

Fig. 7. Two typical electron microscopic aspects of the abnormal lipoprotein LpX. left: "rouleaux" structure of laterally stacked, "squashed" vesicles; right: single-standing spherical vesicles. (Negative stain preparation, phosphotungstate, courtesy of Dr. Ingolic, ZFE, Graz).

particles. It was found to resemble closely the closed unilamellar vesicle structure of artificial phospholipid/cholesterol bilayers (120). Thus, the lipid bilayer appears to be the determining structured element of lipoproteins in the absence of cholesteryl esters.

Acknowledgements

The original work of the authors cited here has been supported by grants no. 5891 (G. M. K.) and 5264 (P. L.) of the Österreichischer Fonds zur Förderung der Wissenschaftlichen Forschung and grant no. 2473 of the Jubiläumsfonds der Österreichischen Nationalbank.

References

1. Day, C. E. & Levy R. S. (1976) Low Density Lipoproteins, Plenum, New York, London.
2. Scanu, A. M. & Landsberger, F. R. (1989) Lipoprotein Structure, Ann. N. Y. Acad. Sci. *348*, 1–436.
3. Day, C. E. (1981) High Density Lipoproteins, Dekker, New York, Basel.
4. Assmann, G. (1982) Lipid Metabolism and Atherosclerosis, Schattauer, Stuttgart, New York.
5. Kostner, G. M. (1981) Medical Laboratory *9*, 1–10.
6. Patsch, W., Patsch J. R., Kostner, G. M., Sailer, S. & Braunsteiner H. (1978) J. Biol. Chem. *253*, 4911–4916.
7. Krempler, F., Kostner, G., Bolzano, K. & Sandhofer, F. (1979) Biochim. Biophys. Acta *575*, 63–70.
8. Bartner, P. J., Hopkins, G. J. & Calvert, G. D. (1982) Biochem. J. *208*, 1–7.
9. Kostner, G. M. (1976) in: Low Density Lipoproteins (Day, C. E. & Levy, R. I., eds.) pp. 229–269, Plenum, New York.
10. Kostner, G. M. & Laggner, P. (1979) NIH Report on Methodology, NIH Publ. Nr. 79–1661, 343–356.
11. Dieplinger, H., Zechner, R. & Kostner, G. M. (1985) J. Lipid Res. *25*, 283–287.
12. Mahley, R. W., Weisgraber, K. H., Bersot, T. B. & Innerarity, T. L. (1978) in: High Density Lipoproteins and Atherosclerosis (Gotto A. M. & Miller, N. E., eds.) Elsevier, Amsterdam pp. 149–154.
13. Seidel, D. (1982) XI Int. Congr. Clin. Chem. Kaiser, E. (Gabl, F. Müller, M. M. & Bayer M., eds.) pp. 279–297, de Gruyter, Berlin, New York.
14. Kostner, G. M. & Schernthaner, G. (1985) Monogr. Atheroscler. *13*, 12–24.
15. Seidel, D. (1977) Klin. Wochenschr. *55*, 611–623.
16. Hauser, H., Kostner, G. M. Müller, M. & Skrabal, P. (1977) Biochim. Biophys. Acta *489*, 247–261.
17. Manzato, E., Fellin, R., Baggio, G., Neubeck, W. & Seidel, D. (1976) J. Clin. Invest. *57*, 1248–1260.
18. Kostner, G. M. Laggner, P., Prexl, W. & Holasek, A. (1976) Biochem. J. *157*, 401–407.
19. Utermann, G., Menzel, H. J. & Langer, K. H. (1974) FEBS Lett. *45*, 29–32.
20. Brown, M. S. & Goldstein, J. L. (1983) Ann. Rev. Biochem. *52*, 223–261.
21. Schwandt, P. (1982) Klin. Wochenschr. *60*, 637–649.
22. Kostner, G. M. (1983) Adv. Lipid Res. *20*, 1–43.
23. Nimpf, J., Wurm, H. & Kostner, G. M. (1985) Thrombosis Res. *54*, 397–401.
24. Suenram, A., McConnathy W. J. & Alaupovic, P. (1979) Lipids *14*, 505–510.
25. Alaupovic, P. (1981) Canad. J. Biochem. *59*, 565–579.
26. Kostner, G., Patsch, J., Sailer, S., Braunsteiner, H. & Holasek, A. (1974) Europ. J. Biochem. *45*, 611–621.
27. Shen, M. S., Krauss, R. M., Lindgren, F. T. & Forte, T. M. (1981) J. Lipid Res. *22*, 236–244.
28. Lee, D. M. & Downs, D. (1982) J. Lipid Res. *23*, 14–27.
29. Kuchinskiene, Z. & Carlsson, L. A. (1982) J. Lipid Res. *23*, 762–769.

30. Zechner, R., Moser, R. & Kostner, G. M. (1986) J. Lipid Res. (in press).
31. Kostner, G. M. (1981) in: High Density Lipoproteins (Day, C. E., ed.) pp. 1–40, Dekker, New York.
32. Kostner, G. M., Albert, W. & Holasek, A. (1969) Hoppe Seyler's Z. Physiol. Chemie *350*, 1347–1352.
33. Luley, C. H., Puchois, P. & Kloer, H. U. (1985) Clin. Chim. Acta *146*, 185–193.
34. Bouthillier, D., Sing, C. F. & Davignon, J. (1983) J. Lipid Res. *24*, 1060–1069.
35. Mahley, R. W., Innrearity, T. L., Rall, S. C. & Weisgraber, K. H. (1984) J. Lipid Res. *25*, 1277–1294.
36. Ghiselli, G., Schaefer, E. J., Light, J. A. & Brewer, H. B. (1983) J. Lipid Res. *24*, 731–736.
37. Lackner, K. L., Edge, S. B., Gregg, R. G., Hoeg, J. M. & Brewer, H. B. (1985) J. Biol. Chem. *260*, 703–706.
38. Utermann, G., Hees, M. & Steinmetz, A. (1977) Nature *269*, 604–607.
39. Utermann, G. (1986) Proc. 4th Europ. Symp. Metabol. (Crepaldi, G., ed.) (in press).
40. Assmann, G. & Lenzen, H. J. (1985) Internist *26*, 692–700.
41. Assmann, G., Schmitz, G., Menzel, H. J. & Schulte, H. (1984) Clin. Chem. *30*, 641–643.
42. Franceschini, G. & Sirtori, C. R. (1980), J. Clin. Invest. *66*, 892–900.
43. Mahley, R. W., Innerarity, T. L., Rall, S. C. & Weisgraber, K. H. (1984), J. Lipid Res. *25*, 1277–1294.
44. Assmann, G., & Menzel, H. J. (1984) J. Clin. Chem. Clin. Biochem. *22*, 585–589.
45. Ponsin, G., Gotto, A. M., Utermann, G. & Pownall, H. J. (1986) Biochem. Biophys. Res. Commun. (in press).
46. Utermann, G., Haas, J., Steinmetz, A. Paetzold, R., Rall, R. C., Weisgraber, K. H. & Mahley, R. W. (1984) Eur. J. Biochem. *144*, 325–331.
47. Jabs, H. U., Assmann, G., Greifendorf, D. & Benninghoven, A. (1986) J. Lipid Res. (in press).
48. Schmitz, G., Ilsemann, K. & Assmann, G. (1983) J. Lipid Res. *24*, 1021–1029.
49. Menzel, H. J., Kladetzky, R. G. & Assmann, G. (1982) J. Lipid Res. *23*, 915–922.
50. Havel, R. J., Kotite, L. & Kane, J. P. (1979) Biochem. Med. *21*, 121–128.
51. Shen, B. W., Scanu, A. M. & Kézdy, F. (1977) Proc. Natl. Acad. Sci. USA *74*, 837–841.
52. Gennis, R. B. & Jonas, A. (1977) Ann. Rev. Biophys. Bioeng. *6*, 195–238.
53. Morrisett, J. D., Jackson, R. L. & Gotto, A. M. Jr. (1977) Biochim. Biophys. Acta *472*, 93–133.
54. Osborne, J. C. & Brewer, H. B. (1977) Advances Protein Chem. *31*, 253–337.
55. Laggner, P. (1981) in: High Density Lipoproteins (Day, C. E., ed.) pp. 43–72, Marcel Dekker, New York.
56. Zampighi, G., Reynolds, J. A. & Watt, R. M. (1980) J. Cell. Biol. *78*, 555–561.
57. (a) Yang, C. Y., Chen, S.-H., Giantureo, S., Bradley, W. A., Sparrow, J. T., Tanimura, M., Li, W.-H., Sparrow, D. A., Deloof, H., Rosseneu, M., Lee, F.-S., Gu, Z.-W., Gotto, A. M., Jr., & Chan, L. (1986) Nature *323*, 738–742.
(b) Knott, T. J., Pease, R. J., Powell, L. M., Wallis, S. C., Rall Jr., S. C., Innerarity, T. L., Blackhart, B., Taylor, W. H., Marcel, Y., Milene, R., Johnson, D., Fuller, M., Lusis, A. J., McCarthy, B. J., Mahley, R. W., Levy-Wilson, B. & Scott, J. (1986) Nature *323*, 734–738.
58. Singer, S. J. (1971) in: Structure and Function of Biological Membranes (Rothfield, L. I., ed.) pp. 145–222, Academic Press, New York.
59. Nestel, P. J. & Miller, N. E. (1981) in: High Density Lipoproteins (Day, C. E., ed.) pp. 281–306, Marcel Dekker, New York.
60. Loomis, C. R., Janiak, M. J., Small, D. M. & Shipley, G. G. (1974) J. Mol. Biol. *86*, 309–324.
61. Loomis, C. R., Janiak, M. J., Small, D. M. & Shipley, G. G. (1974) J. Mol. Biol. *86*, 325–339.
62. Small, D. M. (1970) in: Surface Chemistry of Biological Systems (Blank, M. ed.) pp. 55–82, Plenum, New York.
63. Gong, E. C., Forte, T. M. & Nichols, A. V. (1970) Physiol. Chem. Phys. *2*, 180–188.
64. Gulik-Krzywicki, T., Yates, M. & Aggerbeck, L. P. (1979) J. Mol. Biol. *131*, 475–484.
65. Ross, J., Wrenn, R. F., Ohlendorf, D. H. & Banaszak, L. J. (1980) Ann. N.Y. Acad. Sci. *348*, 408–418.
66. Laggner, P. & Müller, K. (1978) Quart. Rev. Biophys. *11*, 371–425.
67. Laggner, P. (1984) in: Surfactants in Solution (Mittal, K. L. & Lindman, B., eds.) Vol. 1, pp. 259–275, Plenum, New York.

68. Laggner, P. (1982) in: Small Angle X-Ray Scattering (Glatter, O. & Kratky, O., eds.) pp. 329–349, Academic Press, London.
69. Atkinson, D., Small, D. M. & Shipley, G. G. (1980) Ann. Acad. Sci. *348*, 284–298.
70. Laggner, P., Kostner, G., Rakusch, U. & Worcester, D. (1981) J. Biol. Chem. *256*, 11832–11839.
71. Laggner, P., Kostner, G., Degovics, G. & Worcester, D. (1984) Proc. Natl. Acad. Sci. USA. *81*, 4389–4393.
72. Jonas, A. (1973) Biochemistry *12*, 4503–4507.
73. Badley, R. A. (1975) Biochim. Biophys. Acta *379*, 517–528.
74. Smith, R. J. M. & Green, C. (1974) Biochem. J. *137*, 413–415.
75. Shinitzky, M. & Barenholz, Y. (1974) J. Biol. Chem. *249*, 2652–2657.
76. Jonas, A. & Jung, R. W. (1975) Biochem. Biophys. Res. Commun. *66*, 651–657.
77. Jonas, A. (1977) Biochim. Biophys. Acta *486*, 10–22.
78. Rosseneu, M., Soetewey, F. Vercaemst, R., Lievens, M.-J. & Peeters, H. (1978) Protides Biol. Fluids 25, 47–54.
79. Brasure, E. & Henderson, T. O. (1981) in: High Density Lipoproteins (Day, C. E., ed.), pp. 73–93, Marcel Dekker, New York.
80. Bjorklund, R. & Katz, S. (1956) J. Am. Chem. Soc. *78*, 2122–2126.
81. Assmann, G. & Brewer, H. B. Jr. (1974) Proc. Natl. Acad. Sci. USA. *71*, 1534–1538.
82. Laggner, P., Gott, A. M. & Morrisett, J. D. (1979) Biochemistry *18*, 164–171.
83. Scanu, A. M. (1965) Proc. Natl. Acad. Sci. USA. *54*, 1699–1705.
84. Scanu, A. M. & Hirz, R. (1968) Proc. Natl. Acad. Sci. USA. *59*, 890–894.
85. Scanu, A. (1969) Biochim. Biophys. Acta *181*, 268–274.
86. Lux, S. E., Hirz, R., Shrager, R. I. & Gotto, A. M. (1972) J. Biol. Chem. *247*, 2598–2606.
87. Chou, P. Y. & Fasman, G. D. (1974) Biochemistry *13*, 222–245.
88. Tall, A. R., Small, D. M., Shipley, G. G. & Lees, R. S. (1975) Proc. Natl. Acad. Sci. USA. *72*, 4940–4942.
89. Jonas, A. (1973) Biochemistry *12*, 4503–4507.
90. Reynolds, J. A. (1976) J. Biol. Chem. *251*, 6013–6015.
91. Lux, S. E., Hirz, R., Shrager, R. I., Gotto, A. M. (1972) J. Biol. Chem. *247*, 2598–2606.
92. Brown, W. V., Levy, R. I., & Fredrickson, D. S. (1970) J. Biol. Chem. *245*, 6588–6594.
93. Jackson, R. L., Morrisett, J. D., Sparrow, J. T., Segrest, J. P., Pownall, H. J., Smith, L. C., Hoff, H. F. & Gotto, A. M. Jr. (1974) J. Biol. Chem. *249*, 5314–5320.
94. Segrest, J. P. (1977) Chem. Phys. Lipids *18*, 7–22.
95. Brainard, J. R., Knapp, R. D., Patsch, J. R., Gotto, A. M. Jr. & Morrisett, J. D. (1980) Ann. N.Y. Acad. Sci. *348*, 299–317.
96. Lund-Katz, S. & Phillips, M. C. (1984) Biochemistry *23*, 1130–1138.
97. Laggner, P. (1976) in: Low Density Lipoproteins (Day, G. E. & Levy, R. S., eds.) pp. 49–69, Plenum, New York.
98. Dmawan, S. & Reynolds, J. A. (1983) Biochemistry *22*, 3660–3664.
99. Hamilton, J. A., Talkowsky, C., Childers, R. F., Williams, E., Allerhand, A. & Cordes, E. H. (1974) J. Biol. Chem. *249*, 4872–4875.
100. Yeagle, P. L., Langdon, R. G. & Martin, R. B. (1977) Biochemistry *16*, 3487–3491.
101. Henderson, T. O., Krusky, A. W., Davis, L. G., Glonek, T. & Scanu, A. M. (1975) Biochemistry *14*, 1915–1920.
102. Aggerbeck, L. P., Kezdy, F. J. & Scanu, A. M. (1976) J. Biol. Chem. *251*, 3823–3828.
103. Chapman, M. J. & Goldstein, S. (1976) Fed. Proc. 35, 379.
104. Chapman, M. J., Goldstein, S., Mills, G. L. (1978) Eur. J. Biochem. *87*, 375–488.
105. Triplet, R. B. & Fisher, R. W. (1978) J. Lipid. Res. *19*, 478–488.
106. Laggner, P., Goldstein, S. & Chapman, M. J. (1978) Biochim. Biophys. Res. Commun. *82*, 1332–1336.
107. Smith, R., Dawson, J. R. & Tanfod, C. (1972) J. Biol. Chem. *247*, 3376–3381.
108. Deckelbaum, R. J., Shipley, G. G., Small, D. M., Lees, R. S. & George, P. K. (1975) Science *190*, 392–396.
109. Deckelbaum, R. J., Shipley, G. G. & Small, D. M. (1977) J. Biol. Chem. *252*, 744–752.
110. Laggner, P., Degovics, G., Müller, K. W., Glatter, O., Kratky, O., Kostner, G. & Holasek, A. (1977) Hoppe Seyler's Z. Physiol. Chem. 358, 771–779.
111. Atkinson, D., Deckelbaum, R., Small, D. M. & Shipley, G. G. (1977) Proc. Natl. Acad. Sci. USA. *74*, 1042–1047.

112. Zechner, R., Kostner, G. M., Dieplinger, H., Degovics, G. & Laggner, P. (1984) Chem. Phys. Lipids *36*, 111–119.
113. Tall, A. R., Small, D. M., Atkinson, D. & Rudel, L. L. (1978) J. Clin. Invest. *62*, 1354–1363.
114. Laggner, P. & Kostner, G. (1978) Eur. J. Biochem. *84*, 227–232.
115. Herak, J. N., Pifat, G., Brnjas-Kraljevic, Knipping, G. & Jurgens, G. (1984) Bioscience Rep. *4*, 559–564.
116. Ginsburg, G. S., Small, D. M. & Atkinson, D. (1982) J. Biol. Chem. *257*, 8216–8227.
117. Johnson, J. D., Taskinen, M.-R., Matsuoka, N. & Jackson, R. J. (1980) J. Biol. Chem. *255*, 3461–3465.
118. Kroon, P. A. & Seidenberg, J. (1982) Biochemistry *21*, 6483–6488.
119. Morrisett, J. D., Gaubatz, J. W., Tarver, A. P., Allen, J. K., Pownall, M., Laggner, P. & Hamilton, J. A. (1984) Biochemistry *23*, 5343–5352.
120. Laggner, P., Glatter, O., Müller, K., Kratky, O., Kostner, G. & Holasek, A. (1977) Eur. J. Biochem. *77*, 165–171.

Chapter 3
Lipoprotein Metabolism

J. Shepherd and Ch. J. Packard

1 Introduction

Despite the difficulties inherent in conveying hydrophobic lipids through the bloodstream, the biological benefits of these substances as sources of energy and precursors of cell components has led to the evolution of a complex plasma lipoprotein transport system through which gram quantities of cholesterol, triglyceride and phospholipid pass each day. A wide variety of tissues make demands on this pool. The adrenal glands and gonads avidly assimilate lipoprotein cholesterol for the production of steroid hormones; and rapidly dividing intestinal villus cells take up the sterol for membrane synthesis. Metabolically active tissues like skeletal muscle use plasma triglyceride for energy production, while in times of surfeit this lipid is directed into adipocytes for storage. Two organs, the liver and intestine, play a particularly important role in corporeal lipid metabolism, and are together reponsible for the majority of lipoprotein synthesis and catabolism. In the plasma, lipid transport is regulated by specific apolipoproteins (Chapter 2), lipoprotein receptors (Chapter 5), lipolytic enzymes (Chapter 6) and transfer proteins (Chapter 7) which act in concert to maintain the balance of cholesterol and triglyceride homeostasis in tissues and plasma; their malfunction may cause or contribute to the development of dyslipidemia. The roles which they play in lipoprotein metabolism are depicted in Figure 1.

2 Metabolism of Dietary Lipid

Each day we ingest about 1 g of cholesterol (in the form of free and esterified sterol) and about 100 g of triglyceride. Digestive enzymes in the intestinal lumen hydrolyse the lipid esters to release free cholesterol, fatty acids and mono- and diglycerides. These amphipathic molecules form water-soluble micelles which carry the lipid to absorptive sites in the duodenum (1). Under normal circumstances, triglyceride absorption is virtually quantitative, while only about 50% of the cholesterol is taken up, the remainder being lost in the feces (2). Within the enterocyte, absorbed cholesterol is rapidly re-esterified by the action of the mem-

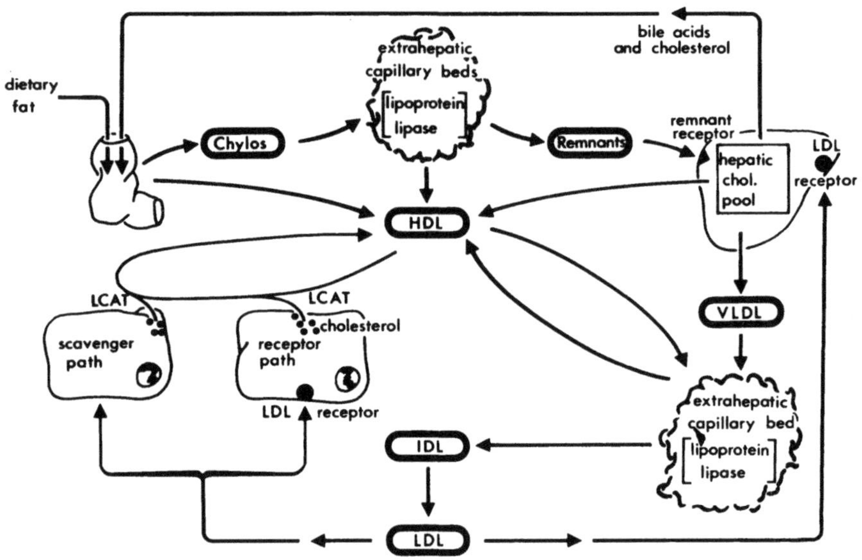

Fig. 1. Lipoprotein metabolism.
Following absorption, dietary fat is packaged into large triglyceride-rich chylomicron particles within the enterocyte and secreted into the circulation. There, lipolysis reduces the particle's triglyceride core and makes redundant part of its surface coat which is shed to HDL. The remnants produced in the process are rapidly assimilated by a receptor-mediated mechanism in the liver. In the fasting state, very low density lipoprotein replace chylomicrons as the major triglyceride transporter. They, too, are subject to tissue lipolysis which degrades them to low density lipoproteins via an intermediate species (IDL). The LDL is removed by receptors present on liver and peripheral tissues. When these are saturated, an alternative, scavenger pathway becomes dominant.

brane bound cytoplasmic enzyme acyl Coenzyme A:cholesterol acyltransferase (ACAT) (2) and packaged with reconstituted triglyceride into large, lipid rich chylomicron particles containing about 1% protein. These appear in abundance in the intestinal mucosal cells following a meal and are secreted into lacteals within the integument of the small bowel. The availability of a specific structural apolipoprotein known as apolipoprotein B (Apo B) is a prerequisite for the production of such particles. Subjects with the inherited disorder known as abetalipoproteinemia (3) are unable to synthesise this protein and normally fail to secrete chylomicrons from their intestinal cells following a fatty meal. The condition results in a failure to thrive in infancy and is associated with specific retinal and erythrocyte defects which are progressive and derive from defective absorption of fat soluble vitamins.

The B protein found in chylomicrons is now thought to be a modified, lower molecular weight form of the Apo B synthesised in liver and secreted in VLDL. Thus the former is termed B_{48} and the latter B_{100} in the nomenclature of Kane et al. (4). Other proteins are also present on the chylomicron surface. Some (apolipoproteins AI, AII, AIV and possibly C) are elaborated in the gut and secreted with the chylomicron (5) while additional peptides are acquired from interstitial fluid HDL. The latter include Apo C and Apo E, both of which participate in subsequent processing of the particle (5–7).

A number of rapid changes take place in the chylomicron particle when it enters the plasma. Within the capillary beds of skeletal muscle and adipose tissues it is exposed to the lipolytic action of endothelium-bound lipoprotein lipase which hydrolyses its triglyceride core, releasing free fatty acids and mono- and diglycerides for energy production or storage (8). Sequential delipidation steps result in the generation of a so-called "chylomicron remnant". This particle is structurally distinct from its parent (9), not only because it is the product of the degradation process outlined above but also because it acquires lipid, particularly cholesteryl ester, from high density lipoprotein (HDL) via the agency of cholesteryl ester transfer protein (CETP). This is exchanged for triglyceride which is transferred in the opposite direction (10). The extent of this process, which effectively enriches chylomicron remnants in sterol esters, is governed by the residence time of the chylomicron in the circulation. Rapid chylomicron clearance, resulting from highly efficient lipolysis, gives the particle little opportunity to acquire cholesteryl ester. In addition to these alterations in its hydrophobic core, the surface coat of the remnant shows substantial differences from that of its parent. All of the Apo B initially present on the particle is retained; but the relative amounts of the other proteins are altered. Apo AI and Apo AII are transferred to HDL (11, 12) while Apo AIV, upon release from the chylomicron, seems to exist free in the plasma since its affinity for other lipoproteins is low (12). As noted above the C protein content initially rises as a result of transfer from HDL (13). This is of particular significance as far as Apo CII is concerned since the latter is an obligatory cofactor for lipoprotein lipase (LpL), increasing the particle's affinity for the enzyme. The role of Apo CIII the major C peptide on the chylomicron surface is less clear although it may be responsible for regulating lipolysis of the particle (14) or delaying its hepatic clearance (15). As the chylomicron progresses down the lipolytic cascade the C proteins are transferred gradually back into the reservoir within HDL. Apolipoprotein E, which is acquired by the particle simultaneously with Apo C is not involved in lipolysis (16), but thereafter plays a role in triggering hepatic assimilation of the remnant (17).

The liver is responsible for the efficient and rapid clearance of cholesteryl ester rich chylomicron remnants from the circulation. Hepatic parenchymal cells contain, on their surfaces, receptors which recognise and bind to particular charged domains of the remnant's Apo E. This interaction results in delivery of the particle to lysosomal digestion within the liver. The key role played by Apo E in this process is evident from studies performed on individuals who express a mutant form of the E protein in which the conformation and charge of the receptor-binding domain is altered (see Chapter 2). They exhibit a delayed clearance of the remnants from the bloodstream (18). The metabolism of chylomicrons is depicted schematically in the general model of lipoprotein metabolism shown in Figure 1 which emphasises the bifunctionality of the process – the delivery of dietary triglyceride to skeletal muscle and adipose tissue and of cholesterol to the liver. The activity of lipoprotein lipase in a tissue is constantly changing and is under hormonal control. Thus the enzyme has some of the properties of a membrane receptor in determining the direction of triglyceride flow.

3 Hepatic Lipid Metabolism

The liver, although it has the ability to make the cholesterol and triglyceride that it needs prefers to utilise preformed components that are available either from the diet as outlined above or from adipose tissue by the fatty acid/albumin transport mechanism. Mobilisation of depot fat in adipocytes by the intracellular pathway depicted in Figure 2 is under the humoral control of agents like adrenaline (19). The fatty acids released from the cell into the bloodstream are extracted rapidly by a number of tissues including the liver which is able to use them as substrate for the synthesis of phospholipid and triglyceride. Under conditions in which their availability is limited, the liver can generate all the fatty acid it requires from small molecular weight precursors (20). Similarly a cholesterol deficit can be met by endogenous synthesis, following activation of the rate limiting enzyme 3-hydroxy-3-methylglutaryl Coenzyme A reductase (HMG CoA reductase) (21).

Lipids synthesised in the liver have a number of fates (2, 22). First, a significant proportion of the cholesterol and triglyceride is secreted in the form of very low density lipoprotein (VLDL), the major vehicle for plasma triglyceride transport in the post absorptive state. Secondly, lipid surplus to requirements may be stored (temporarily) as cytoplasmic oil droplets within the hepatocyte. And, thirdly, the

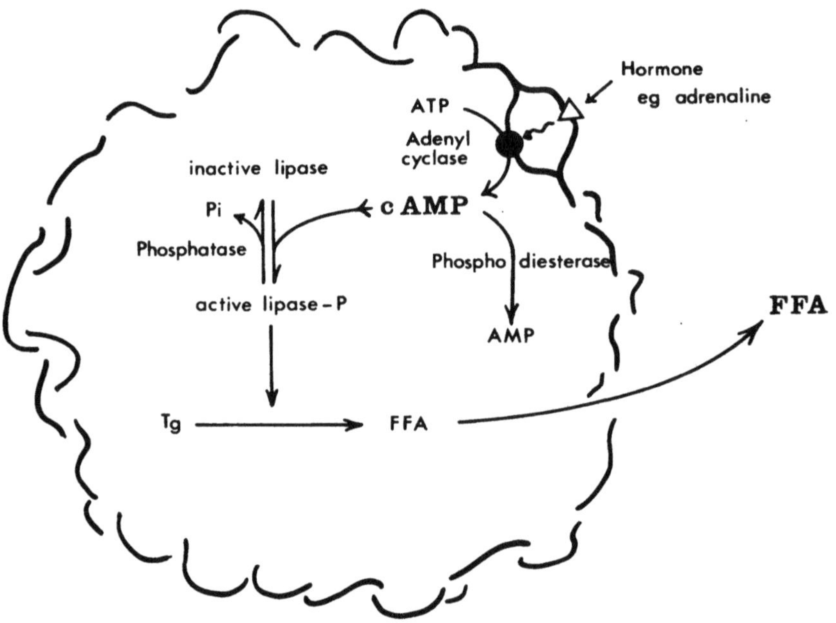

Fig. 2. Mobilisation of depot fat in adipocytes.
Hormonal stimulation of receptors on the adipocyte membrane activates adenyl cyclase and initiates cAMP production. This second messenger promotes phosphorylation and hence activation of intracellular triglyceride lipase; fatty acid generation and release into plasma ensues. Downregulation of the process may be achieved via the actions of two intracellular phosphohydrolases.

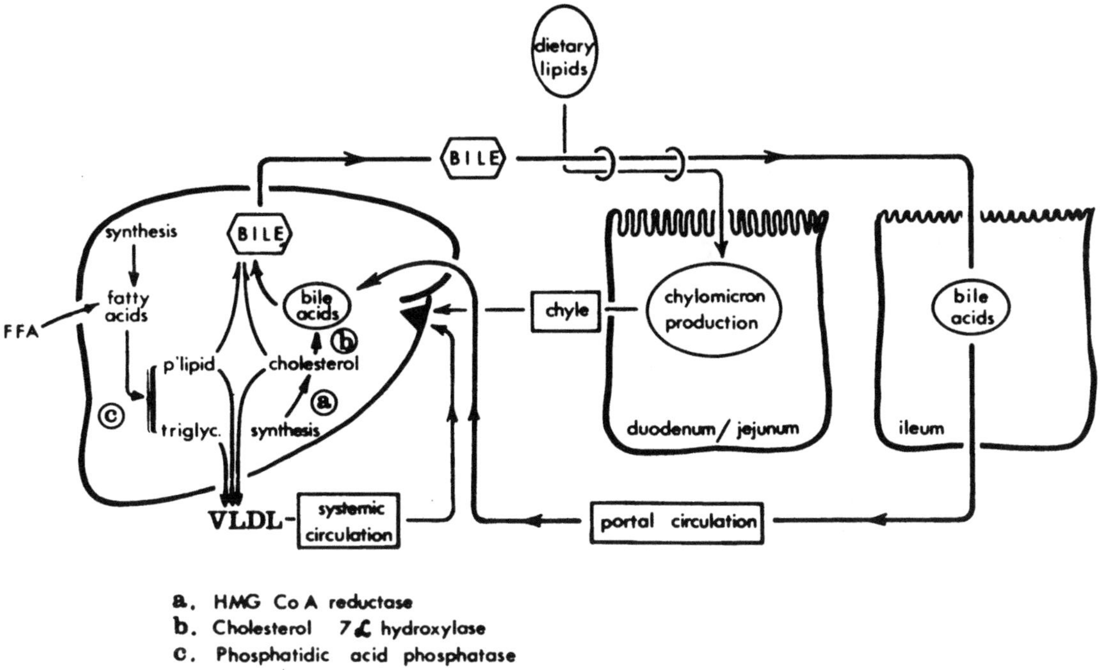

Fig. 3. Hepatic integration of corporeal lipid metabolism.
Lipid metabolism within the liver is integrated via three key regulatory enzymes.

(a) HMG CoA reductase, the pacemaker enzyme in cholesterol biosynthesis, upon demand promotes sterol production from small molecular weight precursors like acetate.
(b) Cholesterol 7α-hydroxylase initiates the oxidation of cholesterol to form the bile acids required for dietary fat absorption.
(c) Phosphatidic acid phosphatase stands at the branch point between production of triglyceride and phospholipid and determines the relative rates of synthesis of these two classes of lipid.

All three enzymes are probably coordinately regulated by the level of bile acids in the hepatocyte.

liver cell has the unique ability (23) to eliminate cholesterol into the bile, either unchanging or following oxidation to bile acids. Gram quantities of lipids flux through these pathways each day and so it is imperative for the hepatocellular economy that the mechanisms involved in their metabolism are precisely interregulated (Fig. 3). This is achieved by the coordinate control of certain key enzymes whose activity appears to be modulated by bile acids.

These compounds produced in the liver and secreted into the intestinal lumen to aid in digestion are reabsorbed in the terminal ileum and extracted by the liver from the portal blood supply. Their return to their site of origin in this 'enterohepatic' circulation produces a number of effects (24):

1. Cholesterol 7α hydroxylase, the rate limiting enzyme in bile acid synthesis, is downregulated.

2. HMG CoA reductase, the enzyme which governs cholesterol production, is also suppressed, although whether this is a direct or indirect effect of bile acids is a matter of controversy.
3. Phosphatidic acid phospatase, which stands at the branching point between the pathways of triglyceride and phospholipid synthesis (Fig. 4), is inhibited, favoring production of the latter.

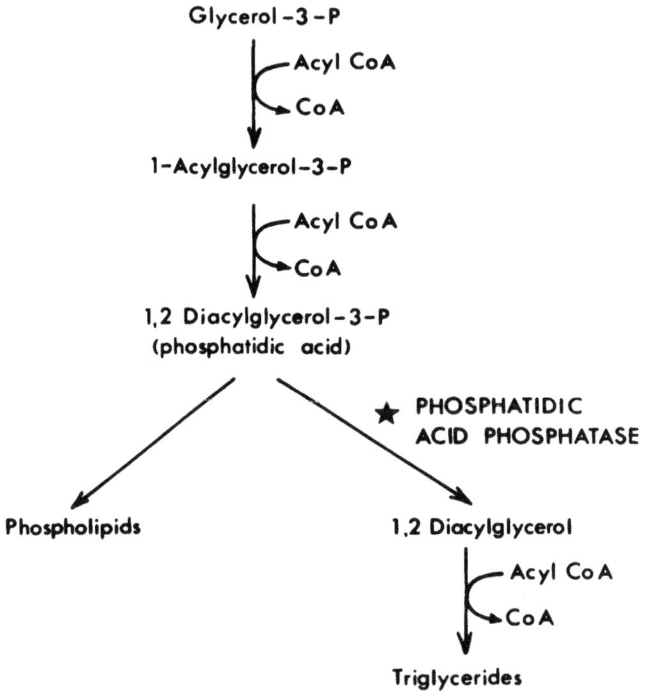

Fig. 4. Regulatory role of phosphatidic acid phosphatase.
Phospholipids and triglycerides are generated from a common phosphatidic acid precursor whose dephosphorylation via phosphatidic acid phosphatase commits it to triglyceride production. When the enzyme is inhibited, phospholipids are produced instead.

The net result of these regulatory events is to ensure that, as bile acids flow back to the liver, they suppresses their own *de novo* synthesis and at the same time stimulate adequate co-secretion of the emulsifier, lecithin, into the bile.

The enterohepatic flow of bile acids follows the cycle of dietary intake, and so strong diurnal variations are induced in all of these control mechanisms as evident from detailed studies of rats kept under conditions of controlled lighting and feeding (1, 23, 25). When food is administered during the dark period it triggers a release of bile acids from the hepatobiliary tree. This produces a zenith in the synthesis of hepatic triglyceride, cholesterol and bile acids, and secretion of very low density lipoprotein into the bloodstream also reaches a peak (26). The above phenomena are reversed in the middle of the light period when the liver is replete with recycled bile acids. However, it should be noted that although this scheme provides a plausible explanation for the basic laboratory observations, it is incomplete since even continuously fasted animals show periodicity in hepatic synthesis (25); and in man such diurnal rhythms are less well documented, although there is good evidence to indicate that they occur (27).

It is clear from the above argument that medical or surgical interruption of the enterohepatic circulation of bile acids will have profound effects on hepatic lipid metabolism. Such manoeuvers can therefore be used to advantage in the treatment of hypercholesterolemia (see Chapter 17).

4 The Metabolism of Apolipoprotein B Containing Lipoproteins

In the post absorptive state, VLDL replaces chylomicrons as the main vehicle of triglyceride transport in the plasma. Apolipoprotein B is integral to the particle and essential for its normal secretion. This protein is of higher molecular weight than of the gut (and as noted above is called B_{100}, using the centile nomenclature), and shares a number of antigenic sites in common with B_{48}. The B_{100} protein has now been sequenced using recombinant DNA technology (28–30). It is a large single polypeptide of 4336 amino acids which has numerous repeat sequences. Following synthesis in the rough endoplasmic reticulum, the protein associates with a triglyceride/phospholipid droplet in its passage to the Golgi apparatus where it becomes glycosylated before being secreted (31). Examination of VLDL particles isolated from purified Golgi membranes has shown that they are heterogenous, encompassing a size range from 40–70 nm, and therefore do not have a stoichiometrically definable structure (31). This information accords with complementary metabolic evidence that the liver has the capacity to elaborate a range of VLDL particles whose size and lipid composition may vary in response to changes in nutritional status. In addition to Apo B_{100}, nascent VLDL particles contain a few molecules of Apo C and Apo E which are augmented by transfer from HDL when the lipoprotein makes contact with the plasma.

Triglyceride-rich VLDL enter a metabolic cascade similar to that described for chylomicrons, and indeed they both compete for the same lipolytic sites on skeletal muscle and adipocytes. Studies designed to examine the metabolism of VLDL have focused largely on its B protein moiety. Early work with radioiodinated VLDL of density less than 1.006 kg/l (32) indicated that Apo B was transferred through an intermediate lipoprotein fraction (called intermediate density lipoprotein (IDL)) to low density lipoprotein (LDL). In the process, the particle's core is hydrolysed by lipoprotein lipase (8), a reaction which requires participation of Apo CII on the particle. Again, in parallel with chylomicron metabolism, cholesteryl esters (33) are acquired from other lipoproteins (principally HDL) by exchange, while surface coat Apo C's are transferred in the opposite direction (32). LDL therefore represents a "remnant" of VLDL catabolism in which the triglyceride core is virtually eliminated

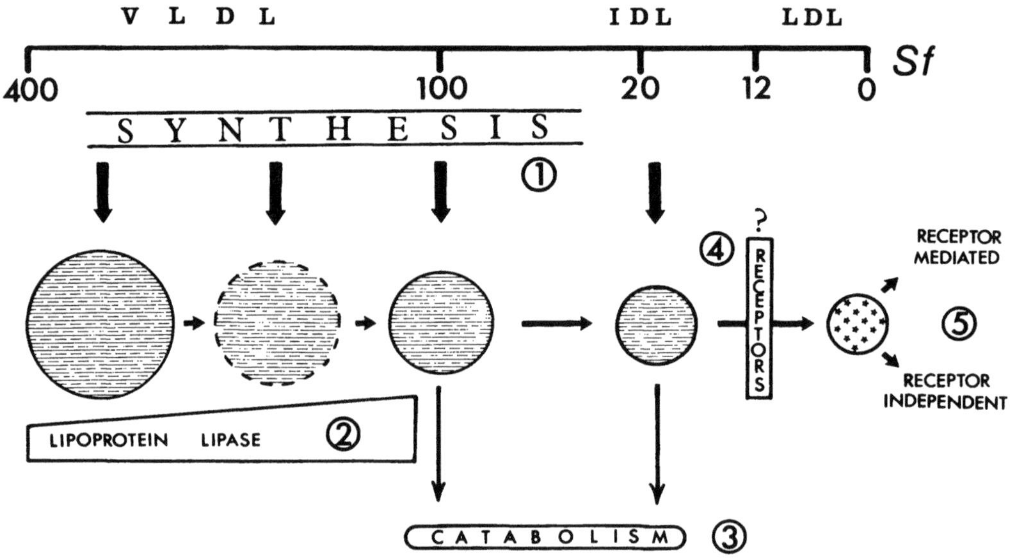

Fig. 5. Regulation of apolipoprotein B_{100} metabolism.
The flux of apolipoprotein B_{100} through the plasma is subject to a number of regulatory events:

(1) The liver is able to modulate both the rate of VLDL production and the size of particle produced in the process.
(2) Lipoprotein lipase preferentially acts on large triglyceride-rich particles, reducing them to lipid-depleted, cholesteryl ester enriched remnants.
(3) The latter may be removed directly from the circulation or
(4) may be converted to low density lipoprotein (LDL) by a process involving receptors.
(5) LDL itself is catabolised by at least two pathways, one of which requires the agency of the high affinity 'LDL' receptor.

and Apo B remains as the sole protein component. In most subjects, whether normolipemic or not, the rate of synthesis of B protein into VLDL (34) exceeds that into LDL. Not all VLDL particles are therefore destined to complete the cascade conversion to LDL. This phenomenon has been examined (35, 36) in detail using procedures which permit subfractionation of the VLDL spectrum in order to follow the fates of particles of narrower compositional range. The evidence which emerged indicated that lipolysis of large VLDL (S_f 60–400) generated remnants of S_f 12–60 (i.e. within the small VLDL/IDL flotation interval), the majority of which were removed directly from the plasma without appearing in LDL (Fig. 5). The latter seemed to come from rapid and quantitative transformation of small VLDL particles which had been secreted directly by the liver.

Both of the major plasma lipolytic enzymes, lipoprotein lipase and hepatic lipase (Chapter 6) have been implicated in the metabolism of Apo B_{100} containing particles (37). It is known that the affinity of lipoprotein lipase is higher for large triglyceride-rich particles than for smaller remnants (37), whereas hepatic lipase, which expresses both triglyceride lipase and phospholipase activities, seems to favor smaller VLDL and IDL particles. The complementary roles of these lipases are perhaps best exemplified in studies of patients who, for genetic reasons (38, 39), lack the enzymes or in animal experiments in which enzyme activity was inhibited using antibodies (40, 41). Under such conditions, absence of lipoprotein lipase results in accumulation of particles of S_f greater than 100 which contain both B_{100} and B_{48} (i.e. are of gut and liver origin respectively). Smaller, denser Apo B containing lipoproteins virtually disappear from the plasma. Conversely, antibody induced inhibition of hepatic lipase in cynomolgus monkeys (41) causes the accumulation of smaller VLDL and IDL (of S_f 12–20); and LDL becomes relatively enriched in triglyceride. Similarly, individuals with hepatic lipase deficiency express high plasma concentrations of VLDL (S_f 20–60) and IDL (S_f 12–20) which are accompanied by reduced circulating levels of LDL (38). This pattern is reminiscent of that seen in Type III hyperlipoproteinemia which of course arises by a completely different mechanism.

4.1 The role of receptors in apolipoprotein B metabolism

The LDL or Apo B/E receptor originally described by Goldstein and Brown (42) is now known to influence not only the metabolism of LDL but also that of a number of other B protein containing particles in the plasma, including VLDL and IDL. This 96,000 dalton membrane protein (discussed in detail in Chapter 5) appears on the cytoplasmic membranes of cells in response to their need for cholesterol. It recognises specific positively charged domains on apolipoproteins B and E and is responsible for the facilitated endocytosis of particles which contain these proteins. The affinity of E containing lipoproteins for the receptor is higher by an order of magnitude than that of those containing only Apo B (17), an observation which has important implications with regard to the rates at which particles with different proportions of B and E are cleared from the circulation. The receptor possesses multiple acidic binding sites at its amino terminus which generate electrostatic attraction for complementary positively charged domains on lipoprotein particles (43). Disruption of the interaction, which can be achieved by chemical modification of the lipoprotein itself, may be usefully employed to provide information on the role of receptors in its metabolism (44). For example, when the positive charges of LDL Apo B are neutralised, the particle is no longer recognised by the receptor *in vitro*; and *in vivo* its clearance from the plasma is delayed (45) in proportion to the extent of involvement of the receptor in its catabolism (Fig. 6). The receptor appears to be responsible for about one half of LDL catabolism in normal individuals, but for much less in receptor deficient familial hypercholesterolemic subjects.

The activity of the receptor pathway in cultured cells is closely regulated by intracellular sterol levels (42). When cells are replete with cholesterol, receptor activity on their membranes falls. In contrast, stimulation of cholesterol utilisation, as seen, for example, in ACTH treated adrenal cortical cells in culture (46), leads to a rapid increase in receptor expression. Therefore, the receptor would appear to occupy a pivotal role in sterol metabolism since it is not only an indicator of intracellular sterol needs but also by virtue of its quantitative importance in lipoprotein catabolism regulates the level of the extracellular sterol pool.

The susceptibility of the LDL receptor pathway to regulation has important clinical ramifications (47). When intracellular cholesterol levels in the liver are

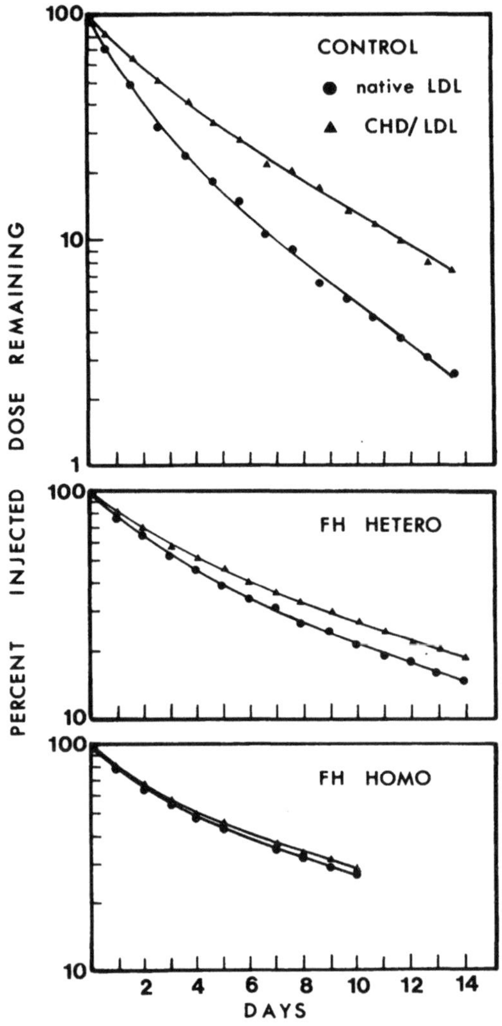

Fig. 6. Metabolism of native and cyclohexanedione-treated LDL in control and hypercholesterolemic subjects.
The involvement of receptors in LDL metabolism is evident from the observation that modification of the arginine residues on the apoprotein moiety of the particle delays its removal from the plasma by interfering with receptor-ligand interaction. When compared to controls, subjects with partial or complete receptor deficiency (i. e. familial hypercholesterolemic heterozygotes or homozygotes respectively) demonstrate reduced or negligible difference in their clearance of the native and receptor-blocked tracers.

reduced by interruption of the enterohepatic circulation or by inhibition of hepatic sterol synthesis, the LDL receptor number on hepatocytes (which contain more than half of the total body receptor complement) rises substantially, increasing the rate at which LDL is extracted from the circulation and so diminishing plasma cholesterol levels. Activation of the receptor pathway by this mechanism can in fact circumvent the genetic defect in heterozygous familial hypercholesterolemia by promoting the activity of the residual receptors which exist on the liver cell membranes of such individuals (48). The Lipid Research Clinics Coronary Primary Prevention Trial has shown that such a manoeuver confers important benefits upon patients who receive this treatment (49). The reductions in coronary risk which they experience may actually come from redirection of LDL catabolism away from less well controlled, receptor-independent and potentially proatherogenic pathways, possibly located in cells of the reticuloendothelial system (see below).

The LDL receptor therefore forms the focus for the physiological regulation of plasma LDL levels. But there is now growing evidence to suggest that its involvement in lipoprotein metabolism must be substantially broader than this. For example, in homozygous familial hypercholesterolemic subjects not only are plasma LDL levels raised, but there is also a several fold increase in VLDL and IDL (50); and the VLDL which accumulates (so called β VLDL) is cholesteryl ester rich and triglyceride poor, characteristic of remnants of triglyceride-rich particles. In kinetic studies of affected individuals, VLDL is cleared from the plasma more slowly than normal, and generates LDL less effectively (51) suggesting a role for the receptor in the conversion of VLDL to LDL. Studies designed to investigate this possibility further, using arginine-modified VLDL, have produced results consistent with this hypothesis (52). Lipolysis of large triglyceride-rich VLDL to smaller remnants – a step mediated by lipoprotein lipase – is not affected by chemical modification of the particle. However, subsequent metabolism of the remnant is retarded and the rate of appearance of LDL reduced. So, these observations indicate that the receptor should be inserted in mechanistic terms, into the delipidation cascade as indicated in Figure 5.

If this is true then VLDL, IDL and LDL particles possess on their surfaces two potential ligands (Apo B_{100} and Apo E) which can interact with either or both of the receptors, the E and B/E receptors on liver. The individual importance of these components is yet to be determined but some inferences can be drawn from the information already available on disease states like dysbetalipoproteinemia and familial hypercholesterolemia. We know that chylomicron remnant catabolism is normal in the Watanabe Heritable Hyperlipemic (WHHL) animal model of the latter (53), even although B/E receptors have been deleted. Dysbetalipoproteinemics, however, who exhibit a defect in their E protein ligand, fail to clear both chylomicrons and VLDL remnants from their circulation. So, while large triglyceride-rich VLDL and chylomicrons may share common metabolic pathways which involve hepatic catabolism as the end stage, their different behavior in these two disease states suggests that they interact with distinct receptors – the chylomicron remnant is removed via the E receptor while VLDL remnants bind preferentially to the B/E receptor. However, it is still possible that a proportion of VLDL remnants are bound and cleared by the E receptor. Certainly, in homozygous familial hypercholesterolemia, B and E containing particles like VLDL remnants and IDL leave the circulation substantially faster than LDL (50, 51) whose catabolism is restricted to the LDL receptor. When both LDL receptor deficiency and defective E protein binding coincide in a single individual the levels of VLDL remnants and IDL in the plasma are even higher (54), suggesting again that both the E and B/E receptors contribute to the clearance of these particles.

5 High Density Lipoprotein Metabolism

In addition to the B protein containing particles whose metabolism is outlined above, the plasma lipoprotein spectrum also encompasses particles whose main protein component is apolipoprotein A. They lie in the density interval 1.063 – 1.210 kg/l and are the smallest of the lipoproteins, having a molecular weight not much different from plasma components like α_2-macroglobulin. A typical high density lipoprotein (HDL) particle is about 7 nm in diameter and has a volume of 130 nm^3. In contrast, the volume of the average VLDL is approximately 390,000 nm^3, which would enable almost 3000 HDL particles to be incorporated inside it. HDL is therefore, in terms of particle numbers, the most abundant lipoprotein in the circulation. This fact often escapes our attention when we consider the metabolic relationships which exist between lipoproteins. Moreover, it has profound implications relative to the structure of HDL in that each lipid or protein component found within the spectrum cannot be represented within every particle. In other words, HDL is an heterogeneous mixture of particles endowed with a diversity of metabolic properties. To some extent this feature may be a reflection of the sundry sources of origin of the lipoprotein (Fig. 7). The recent expansion of our understanding of the functions of HDL have come largely from appreciation of its heterogeneity and from the development of techniques (e.g. immunochemical precipitation, molecular sieve electrophoresis) designed to exploit it.

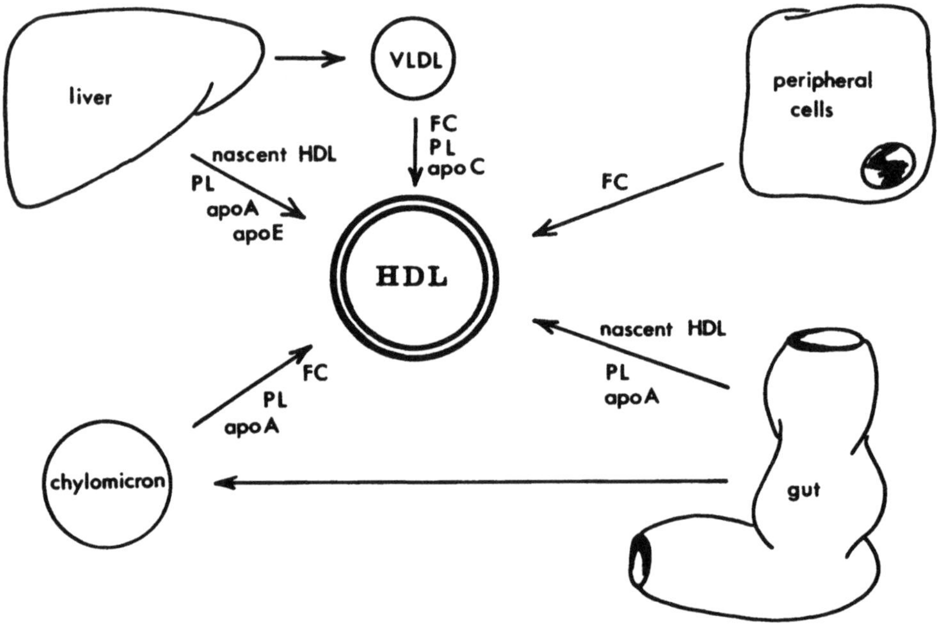

Fig. 7. Origins of HDL.
The high density lipoprotein fraction not only is synthesised *de novo* by intestine and liver but also acquires components from lipoproteins in the circulation and from peripheral tissues.

HDL represents an amalgam of diverse components which come together following:

(1) direct secretion by the liver and intestine;
(2) transfer from other lipoproteins;
(3) transfer from peripheral tissues.

The major HDL proteins, apolipoproteins AI and AII (Chapter 2) are elaborated in precursor form in the liver and intestine. Estimates of the contribution of each of these organs, at least in rats, suggest that they share roughly equally in the process of synthesis (55). Certainly, intestinal A protein production is prodigious as evidenced from its copious appearance in the urine of chyluric patients (56). Some of the protein is secreted in association with chylomicrons and appears in the HDL density range following lipolysis of these particles (5). Consequently, after a fatty meal, the levels of Apo AI and Apo AII in the plasma rise. In addition to this, the intestine is capable of elaborating and secreting small Apo A containing particles directly into the HDL density interval (57); a process which is continuous even in the fasting state. Studies using perfused intestine preparations (57) have shown that under conditions where lecithin:cholesterol acyltransferase (LCAT) activity is blocked, a discoidal particle is produced rich in free cholesterol and phospholipid and with Apo A distributed around its circumference. Particles with a similar structure are also isolable from lymph (58), a body fluid which is naturally low in LCAT activity. Exposure of such particles to the enzymatic and transfer activities of plasma transforms them into mature spherical HDL (59). It is of interest to note that the A proteins lost from the surface of chylomicrons during lipolysis are thought also to appear initially in bilayer discs which undergo rapid conversion in the plasma to mature HDL spheres (60). Electron micrographic studies have shown that lipolysis, by reducing the core volume of the chylomicron renders a proportion of its surface coat redundant. The monolayer responds by folding upon itself to form a frond-like bilayer which, shed into the plasma, comes under the influence of LCAT and is converted to mature HDL. The central role of LCAT in the maturation of HDL is evident from studies of those rare patients (61) who present with an inherited deficiency of the enzyme. Their HDL circulates as bilayer discs which readily undergo transformation *in vitro* to normal HDL by the addition of LCAT.

As noted above, the liver and intestine share equally in the synthesis of HDL. Perfusion studies in rats have shown that the primary hepatic secretion product is again discoidal (62) and rich in both apolipoproteins AI and E. Exposure to LCAT induces its rapid conversion to HDL. In the process, Apo AI is retained while E is lost to lipoproteins of lower density (61).

The action of LCAT upon discoidal HDL triggers a change in the disposition of cholesterol within the particle. The enzyme catalyses the transfer of a fatty acid from lecithin to the hydroxyl residue on cholesterol, leading to the generation of lysolecithin and cholesteryl ester (61). The polarities of these products are fundamentally different from those of their precursors. Lysolecithin is more hydrophilic and dissociates readily from the lipoprotein into the aqueous environment. On the other hand, esterification of the sterol increases its hydrophobicity and causes it to partition into the non polar interior of the particle. The surface site vacated by the sterol is then available to accept additional cholesterol molecules either from other lipoproteins or from cell surfaces (60). Thus, LCAT has a dual action. It facilitates the sequestration of cholesterol within the hydrophobic core of HDL, generating in the process a chemical potential gradient which leads to continuing uptake of the sterol. Even mature HDL particles have a free cholesterol/phospholipid ratio which is less than that of VLDL and LDL (Table 2, Chapter 2). Therefore it is not surprising that HDL is a good acceptor of tissue cholesterol whereas the latter are not. Acquisition of cholesterol by HDL proceeds until the nascent disc is fully transformed into a pseudomicellar spherical structure. Obviously, this process would be limited by the capacity of the HDL particles were it not possible for the sterol ester to undergo facilitated transfer to less dense lipoproteins via the agency of a neutral lipid transfer protein. This protein exchanges the cholesteryl ester in the core of HDL for triglyceride acquired from VLDL or chylomicrons. The triglyceride enriched HDL then becomes more susceptible to intravascular lipolysis which renders it smaller and denser (63). Thus the size of circulating HDL depends on the balance between the opposing processes of cholesterol esterification (which increases particle size) and lipolytic digestion (which reduces it – see Chapter 6 and 7).

The dominant components of mature plasma HDL are its proteins which constitute about 50% of its mass and cover approximately 90% of its surface. They are responsible for its interaction with enzymes (Apo AI is a cofactor for LCAT) and may govern its recognition by receptors on cell membranes. Added to the structural heterogeneity within HDL is the diversity which characterises its proteins. Of particular interest are the isoforms of apolipoproteins AI and E (which comprise 6 and 3 major variants, respectively). The Apo AI polymorphism arises largely from posttranslational proteolytic cleavage of the protein (64). The nascent protein first appears within the cell in a prepro-form (65) with a 24 amino acid extension to its amino terminus (Fig. 8). 18 amino acids are cleaved within the cell to release a proprotein which is secreted into the circulation. There it undergoes slower proteolytic transformation to the mature protein, with release of the six amino acid propeptide. On average, this final step takes about six hours. Pro Apo AI is more basic than its mature product and is distinguishable by isoelectric focussing, even

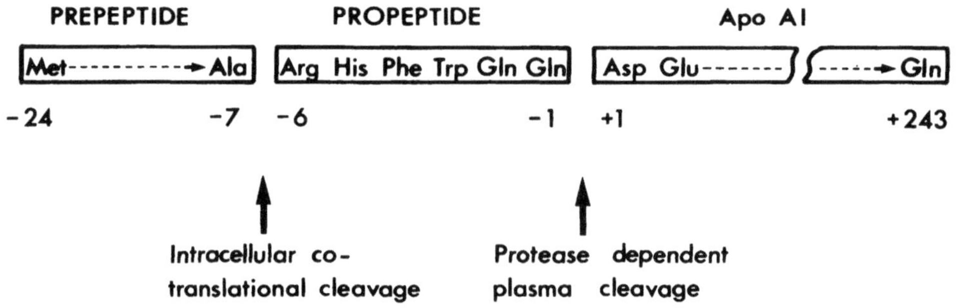

Fig. 8. Biosynthesis of apolipoprotein AI.
Apolipoprotein AI is synthesised with a 24 amino acid N-terminal extension. Eighteen of these are cleaved cotranslationally releasing a propeptide which is secreted into the plasma. There, a protease cleaves the remaining hexapeptide to yield the mature apoprotein.

although it normally only represents 1–2% of the total Apo AI mass. In patients who express the HDL defect known as Tangier Disease, pro Apo AI constitutes a much larger proportion of total Apo AI, a finding which led initially to the proposal (66) that the disease results from defective pro Apo AI maturation. However, kinetic studies have shown (67) that the rates of production and clearance of the proprotein are normal. Hypercatabolism of the mature form appears to be responsible for the relative increment in the mass of pro AI in the circulation. Apo AII, the other main HDL protein, is also synthesised in prepro-form. However, both pre- and pro-peptides are cleaved intracellularly, and only the mature form is found in measurable amounts in the plasma (56). We do not as yet know whether there is any link between A protein maturation and the formation of nascent or mature HDL.

Despite the fact that Apo E represents only a minor component of HDL, its presence on a limited number of these particles may have a profound influence on their metabolic behavior. Particles which contain the protein can be separated (69) from the bulk of HDL by affinity chromatography (ultracentrifugal separation procedures tend to result in their denaturation). They are larger than average and bind with high affinity both to the LDL (or Apo B/E) receptor and to the Apo E receptor on hepatocytes. So, the presence of the E protein on the particles opens up the possibility that they may selectively feed into catabolic pathway controlled by the activities of these two receptors (17). It has already been noted that Apo E exists in a variety of isoforms (the major representatives being E_2, E_3 and E_4 – Chapter 2) which seem to influence the amount of Apo E associated with HDL, in that individuals with the E_2 phenotype transport more of the protein in that density interval than do those with E_4. One of two processes may contribute to this phenomenon. Either the E_2 containing particles may bind less efficiently to the receptor and therefore accumulate in the plasma (70). Or the E_2 itself may form covalent disulfide bonds with Apo AII, trapping the protein in HDL in a form which is not a good ligand for the receptor (71). Indeed, the E-AII complex has been identified by some (71) as a physiological mechanism for delaying the catabolism of E containing particles.

Early analytical ultracentrifugation studies made it clear that HDL was not monodisperse (72) but rather existed as two or more populations which were originally designated HDL_1, HDL_2 and HDL_3. It transpired that "HDL_1" actually represented a mixture, the main component of which was a variant of LDL (Lp(a) – see Chapter 2). Of the other two, HDL_3 ($\rho = 1.125-1.210$ kg/l) was major and HDL_2 ($\rho = 1.063-1.125$ kg/l) minor in mass terms, but not in clinical significance. Further work in the Donner Laboratory in Berkeley has further resolved HDL into a number of species (HDL_{2a}, HDL_{2b}, HDL_{3a}, HDL_{3b} and HDL_{3c}) using gradient gel electrophoresis (73). The importance of each fraction is at present the subject of intensive investigation. The polydispersity which exists in HDL is good evidence for the existence within the size constraints of this fraction of a limited number of thermodynamically stable, distinct entities; and we are now beginning to relate changes in metabolism to alterations in HDL size. For example, in hypertriglyceridemic individuals, HDL is a small, dense particle which is relatively depleted in cholesteryl ester (63). As noted above, it seems to be generated by transfer of a proportion of its core cholesteryl ester into the large mass of circulating VLDL in exchange for triglyceride. The action of hepatic lipase on this particle then causes it to shrink as triglyceride and phospholipid are hydrolysed from its core and coat respectively. At the opposite end of the spectrum, in patients suffering from abetalipoproteinemia, plasma triglyceride levels are extremely low, and the circulating HDL is large and cholesteryl ester and Apo E enriched (79).

Normal individuals do not remotely approach either of the above extremes, although throughout the day they do exhibit cyclical variations in plasma triglyceride which impact on the distribution of HDL and, indirectly, on coronary heart disease risk. Patsch and his colleagues (75) have shown that individuals who possess an efficient chylomicron clearing mechanism show only minor increments in the flotation rates and negligible changes in the masses of their plasma HDL subfractions following a fatty meal. In such patients, HDL_2 and HDL_3 become enriched in phospholipid which is lost gradually, probably through the action of hepatic lipase, as the lipemia resolves. HDL probably acts in this situation as a single acceptor of chylomicron coat phospholipid, shed during hydrolysis of the triglyceride-rich particle. At the other end of the normal spectrum exist individuals whose efficiency for clearing chylomicrons is low. They show marked and prolonged lipemia following a fatty meal. The increased residence time of triglyceride in their plasma facilitates its transfer into HDL in exchange for cholesteryl ester. Phospholipid is also transferred from the chylomicron to HDL during the lipolytic process (76) so that these denser particles acquire both coat and core components. This results in an increase in their susceptibility to hepatic lipase mediated hydrolysis which leads to a shrinkage in the size of the particles and an increase in their density. The magnitude of this change is so great that particles which initially appeared in the HDL_2 density interval now isolate as HDL_3 (75). Such individuals therefore present, even in the fasting state, with a lower HDL_2/HDL_3 ratio than is found in those whose lipolytic processes are efficient. Reduced HDL levels, and in particular low HDL_2/HDL_3 ratios have been equated with increased coronary heart disease risk (77). It is possible that this finding is not of primary importance but is merely a reflection of the status of triglyceride-rich particle metabolism. Certainly, in our laboratory (78) we have found that individuals with angiographically proven coronary heart disease not only have lower HDL_2/HDL_3 ratios in their plasma but also suffer from a defect in their ability to clear chylomicrons from their bloodstream.

Apolipoproteins AI and AII are not distributed equally across the HDL density interval – the lighter and larger HDL_2 are enriched in Apo AI relative to AII. Pharmacologic manipulation of the HDL subfraction distribution with drugs like nicotinic acid results in changes in plasma AI/AII of the order of 25% (79). In light of the size of HDL it is impossible to envisage how such a change could take place if all particles have the same Apo AI and Apo AII content. Indeed, immunological studies have clearly shown that some lipoproteins ((AI)HDL) within the HDL density interval contain AI but no AII while others possess both proteins ((AI+AII)HDL) (80, 81). The (AI)HDL particles appear at both extremes of the HDL spectrum and represent the bulk of the HDL_2 class. (AI+AII)HDL, on the other hand, occur primarily in HDL_3. Both particle types appear to be thermodynamically stable in that they do not equilibrate *in vitro*, although *in vivo* they may participate in a slow exchange of protein. Their physiological relevance is highlighted in their differential responses to a number of lipid lowering agents. Nicotinic acid has been shown to selectively increase (AI)HDL and reduce (AI+AII)HDL. Probucol, on the other hand, produces a specific decrement in (AI)HDL (81). The precise roles that these two HDL particles play in lipid metabolism and their cardioprotective influence is not yet established, although preliminary evidence would suggest that there is a negative association between predisposition to myocardial infarction and the level of (AI)HDL in the plasma.

6 The Role of the Macrophage in Lipoprotein Metabolism

Recent electron micrographic studies performed by Ross and his colleagues (82) have shown that the monocyte/macrophage plays an important role in the earliest stages of the developing atherosclerotic lesion. Primates fed a high cholesterol diet demonstrated, within a few days, an increase in the adhesiveness of their blood borne monocytes to the vascular endothelium, particularly at sites of high shear. With the passage of time the cells migrated between neighboring endothelial cells and took up residence in the intimal space where they began to accumulate cytoplasmic lipid inclusions, forming the focus of development of the early fatty streak. As the lesion progressed, the integrity of the overlying endothelium became compromised so that surface cells were lost, exposing the subendothelial matrix. The accretion of platelets at these sites, the authors suggested, might then accelerate development of the injury by the production and release of platelet derived growth factor which has both chemotactic and growth promoting properties. Such changes probably wax and wane, even in normal arteries, but appear more likely to progress when plasma cholesterol levels are high, presumably as a result of an imbalance in the rates of cellular influx and efflux of sterol, mediated by Apo B and Apo A containing lipoproteins respectively.

The macrophage is a multifunctional cell which expresses a bewildering array of binding sites/receptor activities on its cytoplasmic membrane. Some of these are potentially of great importance in the pathogenesis of atherosclerosis; and four distinct lipoprotein receptor activities are now acknowledged to exist on such cells (83).

1. The LDL (Apo B/E) receptor

The classical Goldstein and Brown receptor first identified in fibroblasts, is also found in significant numbers on blood borne monocytes. As these cells mature to tissue macrophages, their receptor activity declines so much that it may become undetectable (84). This is certainly true for peritoneal macrophages which, when incubated with high concentrations of LDL singularly fail to develop cytoplasmic cholesteryl ester inclusions (85). Consequently, many now believe that the LDL receptor is unlikely to be involved in the process of cholesteryl ester acquisition in the arterial wall, even although there is no *a priori* evidence to suggest that the peritoneal macrophage is truly representative of its arterial wall counterpart, given the diversity of functionality of the monocyte/macrophage system. Notwithstanding, for the moment, attention has been diverted from the LDL receptor to other binding activities which appear to have the ability in the presence of appropriate ligands to assimilate sterol from the environment.

2. The modified LDL receptor

The fact that macrophages from homozygous familial hypercholesterolemic subjects are replete with cholesterol despite their lack of LDL receptors (86) indicates that there must be a separate pathway of cholesteryl ester assimilation by these cells. This premise led to the postulate that modification of the surface of the lipoprotein may be a prerequisite of its uptake. Acetylation, among a variety of modification procedures, was found to result in an LDL particle that was avidly bound to the surface of the macrophage in a process which appeared to be receptor-mediated (85). Since the pathway was not regulated, the availability of abundant acetyl LDL resulted in excessive cholesteryl ester storage in the macrophage and the generation of foam cells. Recent studies have shown that the receptor is a protein with a molecular mass of approximately 240,000 daltons (87). Obviously, acetyl LDL is not its physiological ligand, but several alternative candidates have been proposed. When LDL is incubated with malondialdehyde, a byproduct of prostaglandin synthesis, it acquires the macrophage binding characteristics of the acetylated lipoprotein (88). Although it is inconceivable that this compound would even reach a plasma level sufficiently to modify LDL, it is possible that under certain circumstances (e.g. during platelet aggregation) its local concentration might rise sufficiently at potential sites of atherogenesis. Direct assimilation of the modified lipoprotein by macrophages in the locality would of course preclude its identification in circulating blood. Another potential source of modified LDL has been identified by Henriksen et al. (89). They noted that endothelial cells in culture possessed the ability to generate an oxidised form of LDL that was assimilated by the macrophage's acetyl LDL receptor. Again, excessive cholesteryl ester storage occurred and the cells adopted a typical foamy morphology. Local production of such a lipoprotein in the subendothelial arterial space may well occur during transendothelial migration of the lipoprotein. Indeed, Hoff and his colleagues (90) have reported the isolation of such a modified LDL particle from atherosclerotic arterial lesions.

3. The β-VLDL receptor

In several animal models of atherosclerosis and in Type III hyperlipoproteinemia, the predominant cholesterol transporting lipoprotein in the plasma is not LDL, but a modified form of VLDL (called β-VLDL). This particle has a characteristically reduced electrophoretic mobility as its name would indicate. When it is added to macrophages in culture it has the ability to cause cholesteryl ester deposition (91) and foam cell formation. Indeed, it is the only particle isolable from human plasma which possesses this property. In reality, β-VLDL is a mixture of two lipoprotein particle types — chylomicron remnants from the intestine (containing Apo B_{48}) and hepatogenous VLDL remnants (containing Apo B_{100}). The former is a much more potent foam cell generator (92). Normally chylomicron remnants are rapidly extracted from the circulation and do not appear in fasting plasma. But, as has already been noted, subjects with angiographically positive coronary heart disease catabolise chylomicrons and their remnants ineffectively; and in fact retinyl palmitate, a marker of remnant removal, can actually be used to differentiate between diseased and healthy individuals (78). These data suggest that chylomicron remnant removal may be of major pathophysiological significance in atherosclerosis and it is salutory to recognise that studies which use fasting subjects ignore this potentially important pathway.

4. The HDL receptor

Epidemiological evidence has long supported the view that the cardioprotective function of HDL may derive from its ability to mediate reverse cholesterol transport from peripheral sites to the liver (93). The postulated scheme of events outline in Figure 9 starts with the known ability of the lipoprotein to extract sterol from cultured cells. Macrophages, charged with cholesteryl ester by prior incubation with β-VLDL or acetyl LDL readily release their sterol to HDL in the medium (94). The assimilation process requires that cholesteryl ester within the cells be first hydrolysed to free sterol which diffuses to the outer leaf of the cell membrane and

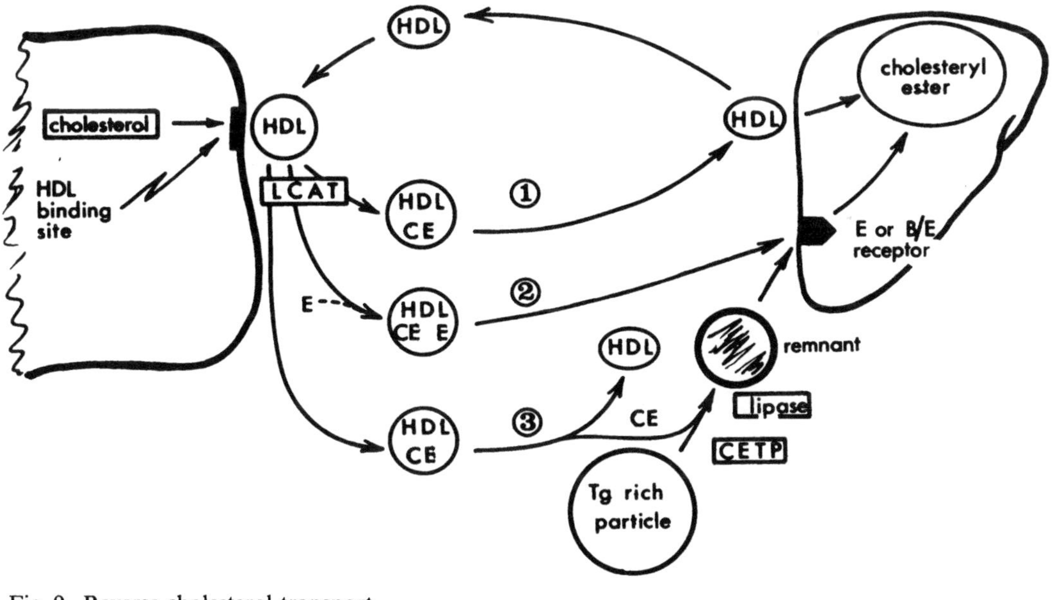

Fig. 9. Reverse cholesterol transport.
HDL plays a key role in cholesterol transport from the periphery to the liver. It avidly assimilates free sterol from cell membranes, possibly following interaction with an HDL receptor site. The cholesterol is trapped within the particle as a result of LCAT mediated esterification. Delivery of the ester to the liver may occur by several routes including

(1) Binding of HDL to a liver receptor with transfer of cholesteryl ester across the hepatocyte membrane.
(2) acquisition of Apo E which then promotes the particle's hepatic uptake by the Apo E or Apo B/E (LDL) receptor.
(3) transfer of the ester to triglyceride rich particles en route to the liver.

from there is transferred to HDL by desorption/resorption or following direct binding of the lipoprotein. At present there is continuing debate regarding the form and even the existence of the HDL receptor. Saturable binding of HDL to cells like fibroblast, smooth muscle cells and macrophages certainly occurs and can be upregulated by loading the cells with sterol (95). But, as yet, the nature of the binding site has not been elucidated.

After assimilating the free sterol (Fig. 9) the HDL, under the influence of lecithin:cholesterol acyltransferase undergoes a structural rearrangement which increases its size. Concomitantly, the particle acquires Apo E either from other plasma lipoproteins or from cells like macrophages (96, 97) in the environment (which have been shown to synthesise the protein). The net result is a cholesteryl ester laden particle containing a ligand which will bind with high affinity to receptors on hepatocyte membranes making the sterol available for fecal excretion (97). In animals like the rat which use HDL as their major cholesterol transporter in the plasma this mechanism may be the prime means for delivery of sterol to the liver. Such species possess HDL which is enriched in cholesteryl ester and Apo E. But the significance of this process as the sole means of reverse cholesterol transport in man is doubtful since

a) The amount of human HDL taken up and degraded by the liver each day is inadequate.

b) Individuals with mutant Apo E that will not bind to receptors do not usually develop premature atherosclerosis.

Consequently, the human species may have developed a modified transport mechanism, for the sterol. This is envisaged to involve the transfer of cholesterol esters acquired by HDL to Apo B/Apo E containing particles via a plasma cholesterol ester transfer protein. The latter would then represent the final mechanism for sterol delivery to the liver.

References

1. Danielson, H. & Sjovall, J. (1975). Bile acid metabolism. Ann. Rev. Biochem. *44*, 233–253.
2. Norum, K. R., Berg, T., Helgerud, P. & Drevon, C. A. (1983). Transport of cholesterol. Physiol. Rev. *63*, 1343–1397.
3. Herbert, P. N., Assmann, G., Gotto, A. M. & Fredrickson, D. S. (1983). Familial lipoprotein deficiency. In: Metabolic Basis of Inherited Disease (Stanbury, J. B., Wyngaarden, J. B., Fredrickson, D. S., Goldstein, J. L. & Brown, M. S., eds.) pp. 589–621, McGraw-Hill, New York.
4. Kane, J. P., Hardman, D. A. & Paulus, H. E. (1980). Heterogeneity of apolipoprotein B: isolation of a new species from human chylomicrons. Proc. Natl. Acad. Sci. *77*, 2465–2469.
5. Green, P. H. R. & Glickman, P. M. (1981). Intestinal lipoprotein metabolism. J. Lipid Res. *22*, 1153–1173.
6. Imaizumi, K., Fainaru, M. & Havel, R. J. (1978). Composition of mesenteric lymph chylomicrons in the rat and alterations produced upon exposure of chylomicrons to blood serum and serum proteins. J. Lipid Res. *19*, 712–722.
7. Green, P. H. R., Glickman, R. M., Sandel, C. D., Blum, C. B. & Tall, A. R. (1979). Human intestinal lipoproteins: studies in chyluric subjects. J. Clin. Invest. *64*, 233–242.
8. Nilsson-Ehle, P., Garfunkel, A. S. & Schotz, M. C. (1980). Lipolytic enzymes and plasma lipoprotein metabolism. Ann. Rev. Biochem. *49*, 667–693.

9. Mjos, O. D., Faergeman, O., Hamilton, R. L. & Havel, R. J. (1975). Characterisation of remnants produced during the metabolism of triglyceride-rich lipoproteins of blood plasma and intestinal lymph in the rat. J. Clin. Invest. *56*, 603–615.
10. Zilversmit, D. B. (1984). Lipid transfer proteins. J. Lipid Res. *25*, 1563–1569.
11. Schaeffer, E. J., Jenkins, L. L. & Brewer, H. B. (1978). Human chylomicron apolipoprotein catabolism. Biochem. Biophys. Res. Commun. *80*, 405–412.
12. Tall, A. R., Green, P. H. R., Glickman, R. M. & Riley, J. W. (1979). Metabolic fate of chylomicron phospholipids and apoproteins in the rat. J. Clin. Invest. *64*, 977–989.
13. Havel, R. J., Kane, J. P. & Kashyap, M. L. (1973). Interchange of apolipoproteins between chylomicrons and high density lipoproteins during alimentary lipemia in man. J. Clin. Invest. *52*, 32–38.
14. Ginsberg, H. N., Ngoc-Anh Le, Goldberg, I. J. R. & Brown, W. V. (1986). Apolipoprotein B metabolism in subjects with deficiency of apolipoproteins CIII and AI. J. Clin. Invest. *78*, 1287–1295.
15. Windler, E., Chao, Y. & Havel, R. J. (1980). Regeneration of the hepatic uptake of triglyceride-rich lipoproteins in the rat. J. Biol. Chem. *255*, 8303–8307.
16. Blum, C. B. (1982). Dynamics of apolipoprotein E metabolism in humans. J. Lipid Res. *23*, 1308–1316.
17. Mahley, R. W. & Innerarity, T. L. (1983). Lipoprotein receptors and cholesterol homeostasis. Biochim. Biophys. Acta. *737*, 197–222.
18. Hazzard, W. R. & Bierman, E. L. (1975). Broad β disease versus endogenous hypertriglyceridemia: levels and lipid composition of chylomicrons and very low density lipoproteins during fat-free feeding and alimentary lipemia. Metabolism *84*, 817–828.
19. Krebs, E. G. & Beavo, J. A. (1979). Phosphorylation-dephosphorylation of enzymes. Ann. Rev. Biochem. *48*, 923–959.
20. Barter, P. J. & Kevin, F. C. (1972). Precursor of plasma triglyceride fatty acid in humans. Effect of glucose consumption, clofibrate administration and alcoholic fatty liver. Metabolism *21*, 117–124.
21. Goldstein, J. L. & Brown, M. S. (1984). Progress in understanding the LDL receptor and HMG CoA reductase, two membrane proteins that regulate the plasma cholesterol. J. Lipid Res. *25*, 1450–1462.
22. Packard, C. J. & Shepherd, J. (1986). Cholesterol 7α hydroxylase: involvement in hepatobiliary axis and regulation of plasma lipoprotein levels. In: Cholesterol 7α hydroxylase (Fears, R. and Sabine, J. R., eds.), pp. 147–165, CRC Press, Boca Raton, Fla.
23. Dietschy, J. M. & Wilson, J. D. (1970). Regulation of cholesterol metabolism. N. Engl. J. Med. *282*, 1128–1138, 1179–1183, 1241–1249.
24. Angelin, B., Bjorkheim, I. & Einarsson, K. (1986). Cholesterol 7α hydroxylase and bile acid synthesis in relation to triglyceride and lipoprotein metabolism. In: Cholesterol 7α hydroxylase (Fears, R. and Sabine, J. R. eds.) pp. 167–177, CRC Press, Boca Raton, Fla.
25. Myant, N. B. & Mitropoulos, K. A. (1977). Cholesterol 7α hydroxylase. J. Lipid Res. *18*, 135–153.
26. Goh, E. H. & Heimberg, M. (1979). Relationship between the activity of hepatic 3-hydroxy-3-methylglutaryl Coenzyme A reductase and secretion of very low density lipoprotein cholesterol in the isolated perfused liver and in the intact rat. Biochem. J. *184*, 1–6.
27. Parker, T. S., McNamara, D. J., Brown, C., Garrigan, P., Kolb, R., Batwin, H. & Ahrens, E. H. (1982). Mevalonic acid in human plasma: relationship of concentration and circadean rhythm to cholesterol synthesis rates in man. Proc. Natl. Acad. Sci. USA, *79*, 3037–3041.
28. Knott, T. J., Rall, S. C., Innerarity, T. L. et al. (1985). Human apolipoprotein B: structure of carboxy-terminal domains, sites of gene expression, and chromosomal localisation. Science *230*, 37–43.
29. Carlsson, P., Olofsson, S. O., Bondjers, G., Darnfoss, C., Wilklund, O. & Bjursell, G. (1985). Molecular cloning of human apolipoprotein B cDNA. Nucleic Acids Res. *13*, 8813–8826.
30. Chan, L., van Tuinen, P., Ledbetter, D. H., Daijer, S. P., Gotto, A. M. & Chen, S. H. (1985). The human apolipoprotein B-100 gene: a highly polymorphic gene that maps to the short arm of chromosome 2. Biochim. Biophys. Res. Commun. *133*, 248–255.
31. Hamilton, R. L. (1983). Hepatic secretion of nascent plasma lipoproteins. In "Plasma protein secretion by the liver" (Glauman,

H., Peters, T. and Redman, C. eds) pp. 357–374, Academic Press, London.

32. Berman, M., Hall, M., Levy, R. I., Eisenberg, S., Bilheimer, D. W., Phair, R. D. & Goebel, R. H. (1978). Metabolism of Apo B and Apo C lipoproteins in man: kinetic studies in normal and hyperlipoproteinemic subjects. J. Lipid Res. *19*, 38–56.
33. Eisenberg, S. (1985). Preferential enrichment of large sized very low density lipoprotein populations with transferred cholesteryl esters. J. Lipid Res. *26*, 487–493.
34. Janus, E. D., Nicoll, A. M., Turner, P. R., Magill, P. & Lewis, B. (1980). Kinetic bases of the primary hyperlipidaemias: studies of apolipoprotein B turnover in genetically defined subjects. Eur. J. Clin. Invest. *10*, 161–172.
35. Packard, C. J., Munro, A., Lorimer, A. R., Gotto, A. M. & Shepherd, J. (1984). Metabolism of apolipoprotein B in large triglyceride-rich very low density lipoproteins of normal and hypertriglyceridemic subjects. J. Clin. Invest. *74*, 2178–2192.
36. Shepherd, J., Packard, C. J., Stewart, J. M., Atmeh, R. F., Clark, R. S., Boag, D. E., Carr, K., Lorimer, A. R., Ballantyne, D., Morgan, H. G. & Lawrie, T. D. V. (1984). Apolipoprotein A and B (Sf 100–400) metabolism during bezafibrate therapy in hypertriglyceridemic subjects. J. Clin. Invest. *74*, 2164–2177.
37. Nicoll, A. & Lewis, B. (1980). Evaluation of the roles of lipoprotein lipase and hepatic lipase in lipoprotein metabolism: in vivo and in vitro studies in man. Eur. J. Clin. Invest. *10*, 487–495.
38. Carlsson, L. A., Holmqvist, L. & Nilsson-Ehle, P. (1986). Deficiency of hepatic lipase activity in post heparin plasma in familial hyperalphatriglyceridemia. Acta Med. Scand. *219*, 435–447.
39. Nikkila, E. A. (1983). Familial lipoprotein lipase deficiency. In: Metabolic basis of inherited disease (Stanbury, J. B., Wyngaarden, J. B., Frederickson, D. S., Goldstein, J. L. and Brown, M. S., eds.) pp. 622–642, McGraw-Hill, New York.
40. Baker, S. R., Patsch, J. R., Forte, T. & Bensadoun, A. (1981). Plasma lipoprotein changes resulting from immunologically blocked lipolysis. J. Lipid Res. *22*, 443–451.
41. Goldberg, I. J., Ngoc-Anh Le, Paterniti, J. R., Ginsberg, H. N., Lindgren, F. T. & Brown, W. V. (1982). Lipoprotein metabolism during acute inhibition of hepatic triglyceride lipase in the Cynomolgus monkey. J. Clin. Invest. *70*, 1184–1192.
42. Goldstein, J. L. & Brown, M. S. (1977). The low density lipoprotein pathway and its relation to atherosclerosis. Ann. Rev. Biochem. *46*, 897–930.
43. Yamamoto, T., Davis, C. G., Brown, M. S., Schneider, W. J., Casey, M. L. & Goldstein, J. L. (1984). The human LDL receptor: a cysteine-rich protein with multiple Alu sequences in its mRNA. Cell *39*, 27–38.
44. Mahley, R. W., Weisgraber, K. H., Melchior, J. W., Innerarity, T. L. & Holcombe, K. S. (1980). Inhibition of receptor mediated clearance of lysine and arginine modified lipoproteins from the plasma of rats and monkeys. Proc. Natl. Acad. Sci. USA *77*, 225–229.
45. Shepherd, J., Bicker, S., Lorimer, A. R. & Packard, C. J. (1979). Receptor mediated low density lipoprotein catabolism in man. J. Lipid Res. *20*, 999–1006.
46. Brown, M. S. & Goldstein, J. L. (1978). General scheme for regulation of cholesterol metabolism in mammalian cells. In: Disturbances in lipid and lipoprotein metabolism (Dietschy, J. M., Gotto, A. M. & Ontko, J. A., eds.) pp. 173–180, American Physiological Society, Bethesda, MD.
47. Packard, C. J. & Shepherd, J. (1982). The hepatobiliary axis and lipoprotein metabolism. J. Lipid Res. *23*, 1081–1098.
48. Shepherd, J., Packard, C. J., Bicker, S., Lawrie, T. D. V. & Morgan, H. G. (1980). Cholestyramine promotes receptor mediated low density lipoprotein catabolism. New Engl. J. Med. *302*, 1219–1222.
49. The Lipid Research Clinics Program, The Lipid Research Clinics Coronary Primary Prevention Trial results. I. Reduction in the incidence of coronary heart disease. JAMA *251*, 351–346.
50. Soutar, A. K., Myant, N. B. & Thompson, G. R. (1982). The metabolism of very low density and intermediate density lipoproteins in patients with familial hypercholesterolaemia. Atherosclerosis *43*, 217–231.
51. Demant, T., Shepherd, J. & Packard, C. J. Unpublished observations.
52. Packard, C. J., Boag, D. E., Clegg, R., Bedford, D. K. & Shepherd, J. (1985). Effects of 1,2-cyclohexanedione modification on the metabolism of very low density lipoprotein apoprotein B: potential role of receptors in intermediate density lipoprotein catabolism. J. Lipid Res. *26*, 1058–1067.

53. Kita, T., Goldstein, J. L., Brown, M. S., Watanabe, Y., Hormicer, C. A. & Havel, R. J. (1982). Hepatic uptake of chylomicron remnants in WHHL rabbits. Proc. Natl. Acad. Sci. USA *79*, 652–663.
54. Hazzard, W. R., Miller, N. E., Albers, J. J., Warnick, G. R., Baron, P. & Lewis, B. (1981). Association of isoapolipoprotein E_3 deficiency with heterozygous FH: implications for lipoprotein physiology. Lancet i, 298–301.
55. Wu, A-L. & Windmueller, H. G. (1979). Relative contributions by liver and intestine to individual plasma apolipoproteins in the rat. J. Biol. Chem. *254*, 7316–7322.
56. Green, P. H. R., Glickman, R. M., Sandek, C. D., Blum, C. B. & Tall, A. R. (1979). Human intestinal lipoproteins: studies in chyluric subjects. J. Clin. Invest. *64*, 233–242.
57. Green, P. H. R., Tall, A. R. & Glickman, R. M. (1978). Rat intestine secretes discoidal high density lipoprotein. J. Clin. Invest. *61*, 528–534.
58. Dory, L., Boquet, L. M., Hamilton, R. L., Sloop, C. H. & Roheim, P. S. (1985). Heterogeneity of dog intestinal fluid (peripheral lymph high density lipoproteins). J. Lipid Res. *26*, 519–527.
59. Dory, L., Sloop, C. H., Boquet, L. M., Hamilton, R. L. & Roheim, P. S. (1983). Lecithin cholesterol acyltransferase mediated modification of discoidal peripheral lymph high density lipoproteins. Proc. Natl. Acad. Sci. USA *80*, 3489–3493.
60. Tall, A. R. & Small, D. S. (1978). Plasma high density lipoproteins. New Engl. J. Med. *299*, 1232–1236.
61. Norum, K. R. (1984). Familial lecithin: cholesterol acyl transferase deficiency. In: Clinical and metabolic aspects of high density lipoproteins (Miller, N. E. and Miller, G. J., eds.) pp. 297–318, Elsevier, Amsterdam.
62. Hamilton, R. L., Williams, M. C. Fielding, C. J. & Havel, R. J. (1976). Discoidal bilayer structure of nascent high density lipoproteins from perfused rat liver. J. Clin. Invest. *58*, 667–680.
63. Deckelbaum, R. J., Olivecrona, T. & Eisenberg, S. (1984). Plasma lipoproteins in hyperlipidemia: roles of neutral lipid exchange and lipase. In: Treatment of hyperlipoproteinemia (Carlson, L. A. and Olsson, A. G., eds.) pp. 85–93, Raven Press, New York.
64. Brewer, H. B., Fairwell, T., Kay, L., Meng, M., Ronan, R., Law, S. & Ligne, J. D. (1983). Human plasma pro Apo AI: Isolation and amino terminal sequence. Biochim. Biophys. Res. Commun. *113*, 626–632.
65. Law, S. W., Gray, G. & Brewer, H. B. (1983). cDNA cloning of human Apo AI: amino acid sequence of preproapo AI. Biochim. Biophys. Res. Commun. *112*, 257–264.
66. Zannis, V. I., Lees, A. M., Lees, R. S. & Breslow, J. L. (1982). Abnormal Apo AI isoprotein composition in patients with Tangier disease. J. Biol. Chem. *257*, 4978–4986.
67. Bojanovski, D., Gregg, R. E. & Brewer, H. B.(1984). In vitro conversion of pro Apo $AI_{Tangier}$ to mature Apo $AI_{Tangier}$. J. Biol. Chem. *259*, 6049–6051.
68. Gordon, J. I., Budelier, K. A., Suus, H. F., Edelstein, C., Scanu, A. M. & Strauss, A. W. (1983). Biosynthesis of human prepro apolipoprotein AII. J. Biol. Chem. *25*, 14054–14059.
69. Weisgraber, K. H. & Mahley, R. W. (1980). Subfractionation of human high density lipoproteins by heparin-sepharose affinity chromatography. J. Lipid Res. *21*, 316–325.
70. Rall, S. C., Weisgraber, K. H., Innerarity, T. L. & Mahley, R. W. (1983). Identical structural and receptor binding defects in apolipoprotein E_2 in hypo-, normo- and hypercholesterolemia dysbetalipoproteinemia. J. Clin. Invest. *71*, 1023–1031.
71. Innerarity, T. L., Mahley, R. W., Weisgraber, K. H. & Bersot, T. P. (1978). Apolipoprotein (E-AII) complex of human plasma lipoproteins. J. Biol. Chem. *253*, 6289–6295.
72. De Lalla, O. F., Elliott, H. A. & Gofman, J. W. (1954). Ultracentrifugal studies of high density serum lipoproteins in clinically healthy adults. Am. J. Physiol. *179*, 333–337.
73. Blanche, P. J., Gong, E. L., Forte, T. M. & Nichols, A. V. (1981). Characterization of human low density lipoproteins by gradient gel electrophoresis. Biochim. Biophys. Acta *665*, 408–419.
74. Deckelbaum, R. J., Eisenberg, S., Oschry, Y., Cooper, M. & Blum, C. B. (1982). Abnormal high density lipoproteins of abetalipoproteinemia. J. Lipid Res. *23*, 1274–1282.
75. Patsch, J. R., Prasad, S., S., Gotto, A. M. & Bengttson-Olivecrona, G. (1984). Post-

prandial lipemia. J. Clin. Invest. *74*, 2017–2023.

76. Schaefer, E. J., Wetzel, M. G., Bengtsson, G., Slow, R. O., Brewer, H. B. & Olivecrona, T. (1982). Transfer of human lymph chylomicron constituents to other lipoprotein density gradients during in vitro lipolysis. J. Lipid Res. *23*, 1259–1273.
77. Gofman, J. W., Young, W. & Tandy, R. (1966). Ischaemic heart disease, atherosclerosis and longevity. Circulation *34*, 679–697.
78. Simpson, H. W., Shepherd, J. & Packard, C. J., unpublished observations.
79. Shepherd, J., Packard, C. J., Patsch, J. R., Gotto, A. M. & Taunton, O. D. (1979). Effects of nicotinic acid therapy on plasma high density lipoprotein subfraction distribution and composition on apolipoprotein A metabolism. J. Clin. Invest. *63*, 858–867.
80. Cheung, M. C. & Albers, J. J. (1982). Distribution of high density lipoprotein particles with different apoprotein composition. particles with AI and AII and particles with AI but no AII. J. Lipid Res. *23*, 747–753.
81. Atmeh, R. F., Shepherd, J. & Packard, C. J. (1983). Subpopulations of apolipoprotein AI in human high density lipoproteins. Biochim. Biophys. Acta *751*, 175–188.
82. Fagiotto, A., Ross, R. & Harker, L. (1984). Studies of hypercholesterolemia in the nonhuman primate. Arteriosclerosis *4*, 323–340.
83. Brown, M. S. & Goldstein, J. L. (1983). Lipoprotein metabolism in the macrophage. Implications for cholesterol deposition in atherosclerosis. Ann. Rev. Biochem. *52*, 223–261.
84. Schechter, I., Fogelman, A. M., Haberland, M. E., Seager, J., Hokom, M. & Edwards, P. A. (1981). The metabolism of native and malondialdehyde altered low density lipoproteins by human monocyte macrophages. J. Lipid Res. *22*, 63–70.
85. Goldstein, J. L., Ho, Y. K., Basu, S. K. & Brown, M. S. (1979). Binding site on macrophages that mediate uptake and degradation of acetylated low density lipoprotein producing massive cholesterol deposition. Proc. Natl. Acad. Sci. USA *76*, 333–337.
86. Goldstein, J. L. & Brown, M. S. (1978). Familial hypercholesterolemia; pathogenesis of a receptor disease. Johns Hopkins Med. J. *143*, 8–16.
87. Dresel, H. A., Fredrick, E., D. P., Schettler, G. & Sinn, H. (1985). Characterization of binding sites for acetylated low density lipoprotein in the rat liver in vivo and in vitro. The EMBO Journal 4, 1157–1162.
88. Fogelman, A. M., Schechter, I., Seager, S., Hokom, M., Child, J. S. & Edwards, P. A. (1980). Malondialdehyde alteration of low density lipoproteins leads to cholesteryl ester accumulation in human monocyte-ophages. Proc. Natl. Acad. Sci. USA *77*, 2214–2218.
89. Henriksen, T., Mahoney, E. M. & Steinberg, D. (1981). Enhanced macrophage degradation of low density lipoprotein previously incubated with cultured endothelial cells: Recognition by receptors for acetylated low density lipoproteins. Proc. Natl. Acad. Sci. USA. *78*, 6499–6503.
90. Morton, R. E., West, G. A. & Hoff, H. F. (1984). Properties of an LDL-like lipoprotein isolated from atherosclerotic lesions. Arteriosclerosis *4*, 534a.
91. Mahley, R. W., Innerarity, T. L., Brown, M. S., Ho, Y. K. & Goldstein, J. L. (1980). cholesteryl ester synthesis in macrophages: stimulation by β very low density lipoproteins from cholesterol-fed animals of several species. J. Lipid Res. *21*, 970–980.
92. Fainaru, M., Mahley, R. W., Hamilton, R. L. & Innerarity, T. L. (1982). Structural and metabolic heterogeneity of very low density lipoproteins from cholesterol fed dogs and from humans with Type III hyperlipoproteinemia. J. Lipid Res. *23*, 702–714.
93. Gordon, T., Castelli, W. P., Hjortland, M. C., Kannel, W. B. & Dawber, T. R. (1977). High density lipoprotein as a protective factor against coronary heart disease. Framingham study. Amer. J. Med. *62*, 707–714.
94. Ho, Y. K., Brown, M. S. & Goldstein, J. L. (1980). Hydrolysis and excretion of cytoplasmic cholesteryl esters by macrophages: stimulation by high density lipoprotein and other agents. J. Lipid Res. *21*, 391–398.
95. Oram, J. F., Brinton, E. A. & Bierman, E. A. (1983). Regulation of high density lipoprotein receptor activity in cultured human skin fibroblasts and human arterial smooth muscle cells. J. Clin. Invest. *72*, 1611–1621.
96. Basu, S. K., Goldstein, J. L. & Brown, M. S. (1983). Independent pathways for secretion of cholesterol and apolipoprotein E by macrophages. Science *219*, 871–873.
97. Koo, C., Innerarity, T. L. & Mahley, R. W. (1984). Obligatory role of cholesterol and apolipoprotein E in the formation of large cholesterol enriched and receptor active high density lipoproteins. J. Biol. Chem. *260*, 11934–11943.

Chapter 4
Laboratory Measurement of Plasma Lipids and Lipoproteins

J.-C. Fruchart, S. Marcovina and P. Puchois

1 Introduction

The increased morbidity and mortality associated with hyperlipoproteinemia makes early diagnosis and precise identification of the condition a matter of some urgency.

Classically, lipids are evaluated in serum or plasma collected into EDTA after a period of fasting of not less than 12 hours, by the following three-step procedure: 1. examination of the appearance of the specimen, 2. analysis of serum or plasma lipids and 3. lipoprotein quantitation.

The specimen may either be clear or turbid, the latter reflecting either deficient catabolism of triglyceride-rich lipoproteins (chylomicrons and VLDL) or increased synthesis of VLDL by the liver. If the serum is allowed to stand at +4 °C in the refrigerator any chylomicrons present will float to the surface of the sample and be readily identified there within 24 hours.

2 Plasma Lipid Analysis

Serum lipids form an heterogeneous group of hydrophobic compounds incorporated into hydrophilic lipoprotein complexes which are responsible for their transport and distribution. The circulating lipoprotein complexes can be denatured by adding alcohol to the serum which facilitates subsequent extraction of the lipids using an appropriate solvent. Lipids are highly diversified chemical compounds. Most of them contain fatty acids esterified either with glycerol or sterols; and some contain charged units of choline, ethanolamine or serine which render them highly polar.

The lipid content of blood serum is approximately 7 g/l and comprises:

1. *1.5 to 2.6 g/l of phospholipids*. The major constituents of phospholipids are:

lecithin	69%
sphingomyelin	19%
lysolecithin	7%
phosphatidyl-ethanolamine	5%

The percentages of these constituents were determined on the basis of phosphoglyceride phosphorus measurement. These polar lipids, more particularly lecithins, contribute actively to maintaining the solubility of more hydrophobic lipoprotein lipids.
2. *1.5 to 2.8 g/l (3.9 to 7.2 mmol/l) of total cholesterol.* This amount can fluctuate with age and sex. Approximately 70% of this sterol in the plasma is esterified (principally with oleate and linoleate).
3. *0.7 to 1.6 g/l (0.8 to 1.8 mmol/l) of glycerides.* 85% of these are triglycerides while the remainder represents intermediates (diglyceride and monoglyceride) in the metabolism of triglycerides.
4. *0.11 to 0.23 g/l (0.387 to 0.810 mmol/l) of free fatty acids (FFA).* In spite of their low concentration (5% of total fatty acids), FFA represent, because of their high turnover rates, an appreciable source of energy. They diffuse easily into cells from serum albumin, and derive in the fasting state from the hydrolysis of adipose tissue triglycerides.

2.1 Isolation and separation of serum lipid constituents

Lipids are released from lipoproteins by denaturation of the particles in weak acid, alkali, or alcohol (ethanol or isopropanol) (1).

The lipids are then extracted with an organic solvent. These methods have grown obsolete in routine clinical biochemistry for measuring total lipids, but remain, however, useful as procedures for extracting lipids from serum (and also from tissues or feces) prior to their qualitative analysis.

The lipid constituents are fractionated by adsorption chromatography on silica columns or, more easily, by thin layer chromatography on silica gel. The lipid extract in chloroform or isopropanol is concentrated and applied to the top of the column or the bottom of the chromatoplate. From the column, elution is obtained using solvents of increasing polarity. With a chromatoplate, the solvent systems are used according to the substances to be separated:

- neutral lipids: either petroleum/ether/acetic acid (80 : 20 : 2; by vol.)
- phospholipids: chloroform/methanol/water (65 : 35 : 4; by vol.)

Detection is carried out using gaseous iodine (which produces a brown colour), copper acetate in hot phosphoric acid (black colour) or by specific staining for cholesterol, phosphorus etc.

Thin layer chromatography enables the overall lipid fractions to be visualized, and approximate quantitative determination can be carried out densitometrically.

2.2 Determination of cholesterol

Cholesterol can be measured by a large number of chromatographic, colorimetric or enzymatic procedures (1,2).

2.2.1 Chromatographic methods

Gas liquid chromatography has been successfully applied by many authors to cholesterol measurement. When carried out on capillary columns, this procedure enables cholesterol trimethylsilyl (TMS) ether to be separated from closely related compounds such as for example cholestanol which is not separable from cholesterol on a conventional column (3). The "Société Française de Biologie Clinique" has recently proposed a reference method for measuring blood cholesterol using capillary gas liquid chromatography coupled with mass fragmentography. This method enables quantitative analysis to be carried out on a few microlitres of serum.

2.2.2 Colorimetric methods

Chromogenic reactions of cholesterol. The colorimetric methods are generally based on the use of two chromogenic reactions (1, 2):

- Liebermann-Burchard's reaction: development of a green colour in the presence of acetic anhydride and sulfuric acid.
- Zak's reaction: development of a red colour in the presence of acetic acid, ferric chloride and sulfuric acid.

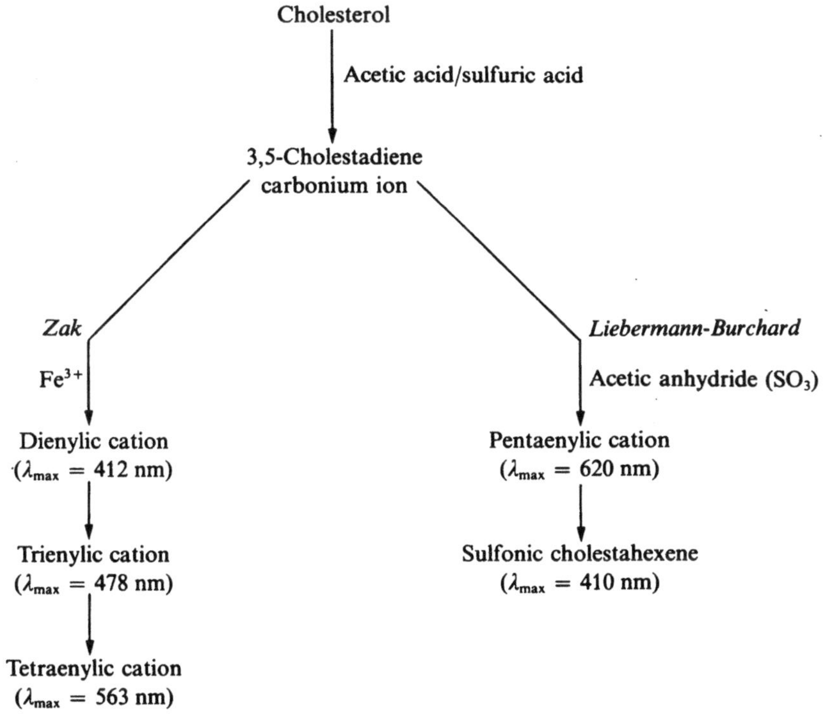

Fig. 1. Mechanism of Liebermann-Burchard's and Zak's reactions (ref. 187).

The reaction mechanism is common to these methods (Fig. 1). After dehydration of cholesterol, 3,5-cholestadiene is oxidised with Fe(III) ions or sulfuric acid to form a homologous series of polyenic ions. Both reactions show progressive color development with time: the final colored derivative measured by Zak's reaction is usually the red cholestate-trienylic cation which has a maximum absorbance at 563 nm. The colored derivative determined by Liebermann-Burchard's reaction on the other hand is the green pentaenylic cation whose maximum absorbance is at 620 nm. Many modifications of these reactions have been described. Zak's reaction can be sensitized by adding *o*-phthalaldehyde or iron(II) sulfate. The colored reactions used for measuring cholesterol are not specific. These group reactions can be given by structurally analogous sterols or by other substances. Many factors such as the development time of the color, the test temperature and primarily the presence of bilirubin in the reaction medium, have an influence on Liebermann-Burchard's reaction. If the reaction is carried out in chloroform, cholesterol esters give a deeper color than free cholesterol. Zak's reaction is more sensitive but still less specific than Liebermann-Burchard's reaction. The interference of many drugs adds to that of bilirubin and other sterols. At a wavelength of 560 nm, the absorbance for cholesterol and cholesterol esters is identical.

Methods of determination. The colorimetric assay of cholesterol in biological fluids is undertaken in three steps: extraction, saponification, and color development. The method of Abell et al. (4) and its variants are usually considered to be reference procedures but are difficult to apply and cannot be used routinely.

In *two-steps procedures* which omit the saponification step, lipids are extracted by solvents that simultaneously break the lipoprotein bonds and extract cholesterol. Bilirubin interference can be prevented by adding adsorbents. When the Liebermann-Burchard reaction is carried out on a chloroform extract, a positive error is introduced because of the higher color-producing potential of minor contaminating steroids.

These two-steps methods have been partly automated and give results which correlate well with those obtained using the reference methods.

The *direct methods* in current use are technically simple, easy to automate, but open to many sources of errors.

The Liebermann-Burchard reaction has been modified to permit its direct application to serum. Specifically, a stabilizing reagent such as arylsulfonic acids dissolved in acetic acid or sodium sulfate is added to permit resolution of the dissociated proteins in the reagent.

These methods lack specificity and need a stabilized temperature for color development. In addition bilirubin interference has to be suppressed by using a blank or by reading out at two different wavelengths.

The direct iron(III) chloride methods are generally difficult to control. A number of analysts, however, have obtained results comparable with those achieved with the method of Abell by using a single reagent (containing sulfuric acid, ethyl acetate and ferric chloride) and by standardizing the different assay times or pre-precipitating the interfering proteins and chromogens with a solution of ferric acetate and uranyl acetate before running the reaction on the supernatant.

Generally speaking, the direct methods supplied as reagent kits lack specificity and reproducibility and give results which are in most cases 10% higher than those achieved with the chromatographic reference method.

2.2.3 **Enzymatic methods**

Several microorganism enzymes are capable of degrading cholesterol and can be usefully employed in its measurement (1, 2). Flegg (5) pioneered the use of cholesterol dehydrogenase (this enzyme has a 3β-hydroxysteroid-oxido-reductase-like and an isomerase-like activity) of Nocardia erythropolis. After enzymatic saponification of the serum extracts, Flegg measured the Δ^4-cholestenone produced, by its absorbance at 240 nm. Richmond (6) isolated a "cholesterol:oxygen-oxidoreductase" from another species of Nocardia and measured cholesterol in the serum by the production of H_2O_2. Hernandez and Chaikoff (7) and Hyun et al. (8) have isolated a cholesterol esterase from swine pancreas or pancreatic juice. When these enzymes were amalgamated, it was possible to devise an assay method for total serum cholesterol based on the following two reactions:

1. Cholesterol ester $\xrightarrow[+ H_2O]{\text{cholesterol esterase}}$ cholesterol + fatty acids
2. Cholesterol $\xrightarrow[+ O_2]{\text{cholesterol oxidase}}$ Δ^4-cholestenone + H_2O_2

The final procedure for measuring hydrogen peroxide can vary. One uses the Trinder reaction (9). In the presence of 4-aminopyrine, phenol and a peroxidase, a quinoneimine forms which gives a maximum absorbance at 500 nm.

Other authors use a sequence of reactions in which hydrogen peroxide oxidizes methanol to formaldehyde which then reacts with ammonium ions and acetyl acetone to form a lutidine assayable either colorimetrically or fluorimetrically (2) (Hantzch's reaction).

There are also proposals in which hydrogen peroxide is allowed to react with:

- homovanillic acid in the presence of a peroxidase. A fluorescent product can be measured fluorimetrically.
- iodide which is transformed into iodine and assayed photometrically.

Physicochemical methods have also been developed: for example an automatic potentiometric technique is described for measuring hydrogen peroxide (10, 11). Noma and Nakayama (10) have outlined a rapid process based on polarographic determination of the oxygen consumed during the transformation of cholesterol into Δ^4-cholestenone.

The enzymatic methods for measuring cholesterol are generally highly reproducible (C. V. $< 2\%$) and the results are well correlated with those achieved with reference methods (12–14). The specificity of the enzymatic methods is better than that of chemical procedures. There is no interference by bilirubin or hemoglobin, and the other serum sterols have a negligible influence on the results. Drug interferences are minimal.

2.2.4 **Selection of standards**

Primary standards of cholesterol must be pure ($\geqslant 99\%$) and are better stored at $-20\,°C$ in the absence of light, in the presence of a drying substance or under argon. For the enzymatic methods, it is appropriate to use a cholesterol ester or aqueous solutions of cholesterol stabilized with a detergent. It is, however, preferable to adopt as a secondary standard a lipoprotein-enriched serum sample, the cholesterol content of which has been previously determined by a reference method (2, 15).

2.2.5 **Reference values and physiological variations**

Recent studies have shown that it is becoming increasingly difficult to define a reference value for cholesterol. The results vary according to the procedure used and depend on race, sex, age, eating habits and the life-style of the population. Studies of age-related physiological variations have been generally consistent. Cholesterol increases progressively with age, reaching a maximum at approximately 60 years. Moreover, serum cholesterol concentrations in women of childbearing age are significantly lower than those of age matched men. Following the menopause, cholesterol levels rise to values which exceed those found in males. It should be noted that fixing the upper limit of serum cholesterol and triglyceride concentrations at two standard deviations above the value of the overall population mean for each age group is arbitrary and carries no implications with regard to physiological needs or health. This reported "normal value" in Western countries, may be abnormal from the viewpoint of the incidence of coronary heart disease. Indeed, the consensus view of clinical specialists who convened recently at the National Institute of Health was that an action limit of 5.2 mmol/l for cholesterol be adopted in adults.

2.2.6 **Determination of free and esterified cholesterol**

Cholesterol is present in serum in both free (approximately 30%) and esterified form (essentially cholesterol oleate and linoleate). These can be separated by thin layer chromatography or high performance liquid chromatography. It is easier, however, if an assessment of free cholesterol is required to carry out the enzymatic assay of the sterol in the absence of cholesterol esterase. The esterified cholesterol can then be obtained by difference from the total cholesterol (2, 16, 17).

2.3 Determination of glycerides

2.3.1 **Colorimetric or fluorimetric methods**

Glyceride determination is made in three steps (1):

- extraction of glycerides,
- alkaline hydrolysis,
- measurement of glycerol released.

One of two approaches may be used for the extraction.

The older is based on conventional lipid extraction methods (chloroform/methanol, isopropanol). They require subsequent elimination of interfering concomitantly extracted substances by adsorption onto:

- zeolite, silicic acid, albumin, florisil or diatomaceous earths which capture phospholipids,
- lime and copper sulfate in alkaline medium which eliminate glucose and organic acids,
- bentonite which absorbs bilirubin.

In the more recently developed techniques, glycerides are extracted selectively by phase partition between two solvents (isopropanol-nonane, isopropanol-heptane). Glycerides are generally hydrolysed by saponification with alcoholic potassium hydroxide or transesterified using sodium methylate. The glycerol released is evaluated either directly or in most cases after periodic acid oxidation to formaldehyde. Many colorimetric or fluorimetric methods have been described to measure the formaldehyde formed. The most currently used is the Hantzch's reaction in which formaldehyde is condensed with acetyl-acetone and ammonium ions to form a yellow-colored, fluorescent 3,5-diacetyl-1,4-dihydrolutidine. Lutidine can be assayed colorimetrically at 410 nm or fluorimetrically (Table 1).

Table 1. Colorimetric or fluorimetric methods for measuring glycerides

1. Extraction of Glycerides

- Extraction of lipids (chloroform/methanol, isopropanol) then subseqeunt elimination of interfering substances.
- Selective extraction of glycerides by phase partition
  (isopropanol/nonane)
  (isopropanol/heptane)

2. Alkaline Hydrolysis (saponification, transesterification)

3. Measurement of Glycerol Released

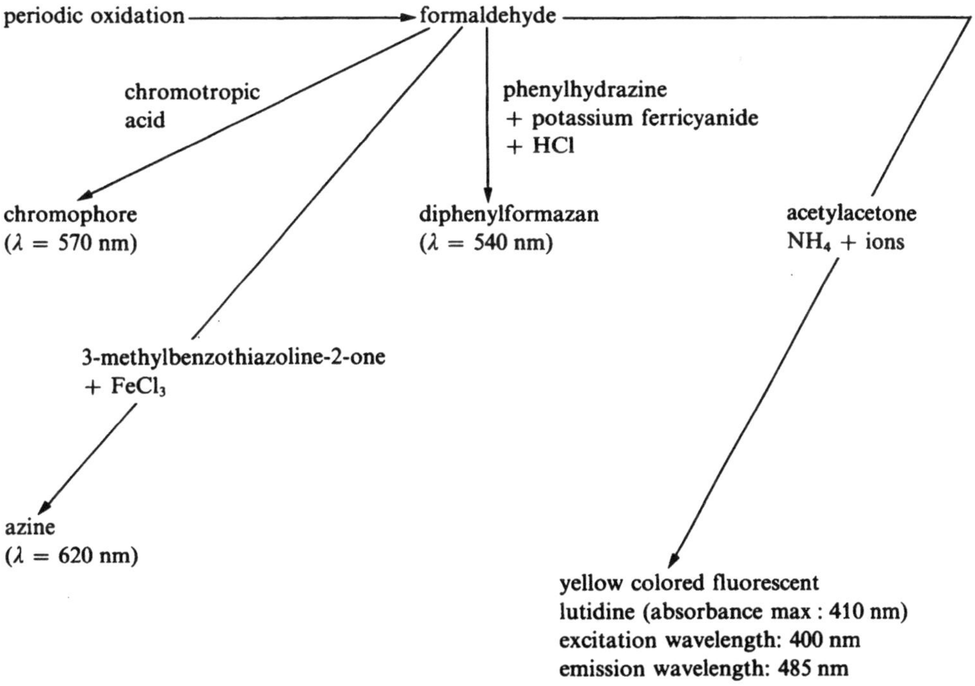

2.3.2 Enzymatic methods

The enzymatic methods are based on the measurement of glycerol released by alkaline hydrolysis of lipoprotein glycerides (18–21) following lipid extraction. The recent introduction of lipase as hydrolytic agent in association with a protease has largely simplified the methodology. Three reactions can be used for measuring glycerol enzymatically (Table 2). The most commonly applied method involves phosphorylation of the glycerol with ATP in the presence of glycerol kinase followed by measurement of the ADP thus formed using a series of enzymatic reactions with pyruvate kinase and lactate dehydrogenase. The three reactions are run in buffered medium (pH 7.2); the glycerol content of the sample is proportional to the quantity of NAPH + H^+ consumed.

Table 2. Enzymatic methods for measuring glycerides

1. Alkaline or Enzymatic Hydrolysis of Glycerides
2. Enzymatic Measurement of Glycerol Released

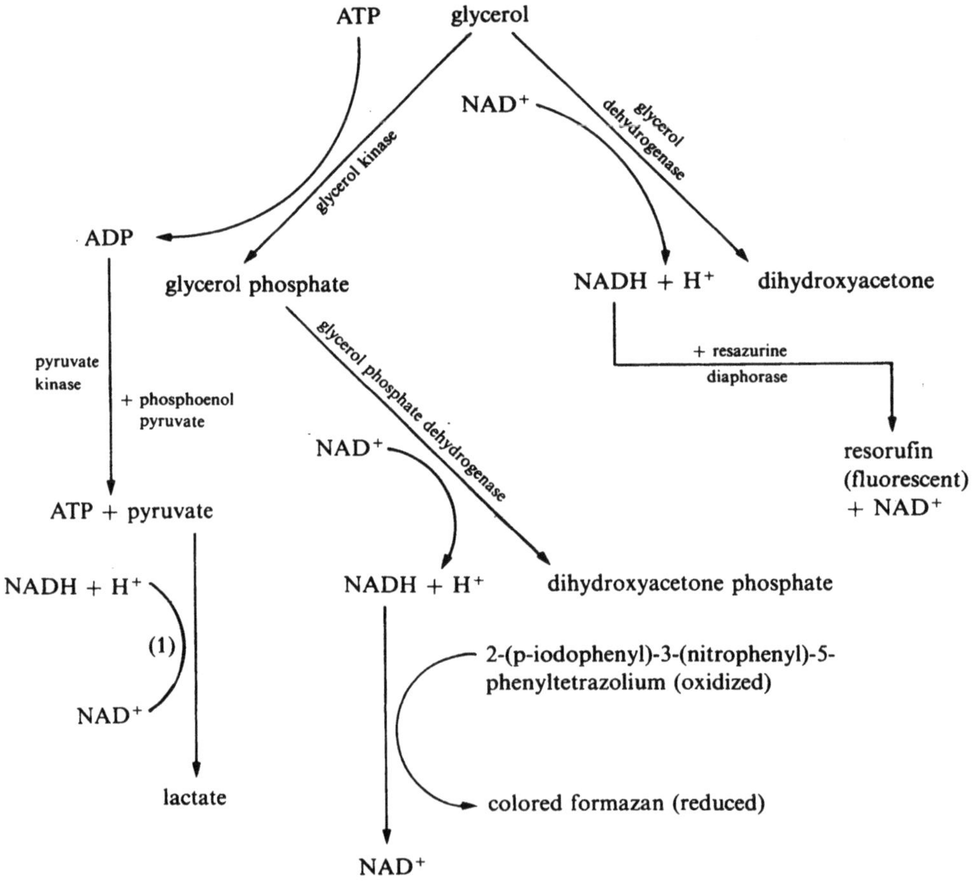

(1) NADH consumption measured by absorptiometry (340 nm) or fluorimetry (λ_{ex} = 355 nm, λ_{fl} = 460 nm)

2.3.3 Selection of standards

In order to eliminate the interference of preexisting free glycerol, it is necessary to subtract a blank value determined by running the reaction without previous hydrolysis.

Pure triglycerides (purity ≥ 99%) are recommended as the primary standard in the assay. When constructing the calibration curve allowance needs to be made for the molecular mass of the triglyceride selected. With entirely enzymatic methods, triolein can be used, provided that it is emulsified ultrasonically in the presence of gum arabic. Many authors prefer, however, to use glycerol as calibrating standard. Because glycerol is highly hygroscopic, these solutions need to be titrated iodometrically. They do not constitute real primary standards since the hydrolysis step is not tested. Consequently, it is preferable to use lipoprotein-enriched control serum samples, the titer of which has been controlled by reference methods.

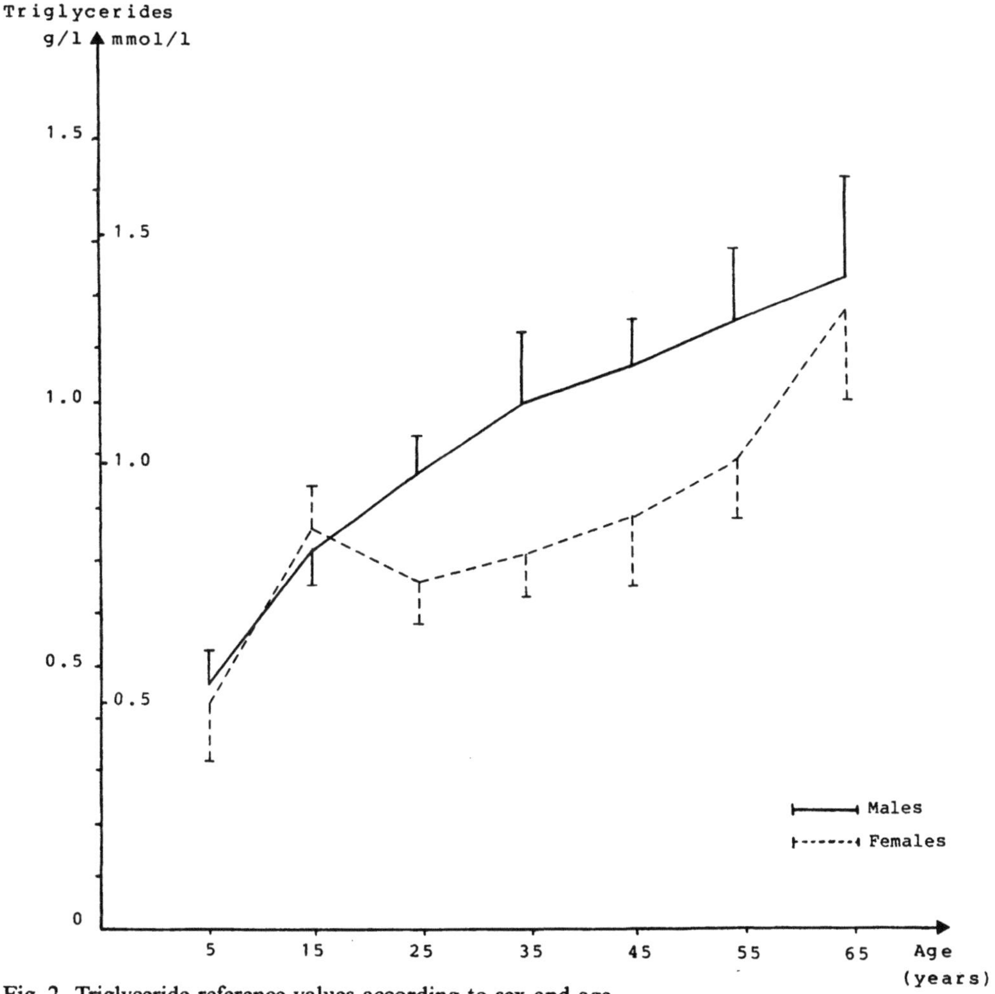

Fig. 2. Triglyceride reference values according to sex and age.

2.3.4 Reference values and physiological variations

To define serum triglyceride concentration in a reference population allowance needs to be made for race, sex, age, eating habits and life-style of the individual. Most studies have shown that the mean serum triglyceride concentration lies within the range 1.0 ± 0.4 mmol/l. It increases with age and is higher in men than in women (Fig. 2). It also increases during pregnancy or following oral contraceptive treatment.

2.4 Determination of phospholipids

2.4.1 Determination of total phospholipids

Serum phospholipids represented by lecithins, phosphatidylethanolamines, sphingomyelins and lysolecithins, are generally measured on the basis of their phosphorus content (1, 22). After isolation by lipid extraction or lipoprotein precipitation, serum phospholipids are carbonised and the released phosphorus assayed by formation of a phosphomolybdic or phosphomolybdovanadic complex. A number of methods have been developed to permit this measurement to be made directly on the lipid extract.

2.4.2 Determination of phosphatidylcholine

A method has recently been proposed to determine serum cholinic phospholipids enzymatically. This reaction is schematically illustrated in Table 3 (23–25). To improve its specificity, other authors have proposed that lecithins be hydrolysed with phospholipase C, and sphingomyelins with sphingomyelinase (Table 4). Phosphorylcholine released is hydrolyzed by alkaline phosphatase, with release of choline and phosphate ions which can be easily measured.

Table 3. Enzymatic measurement of choline-containing phospholipids

1. Hydrolysis of phospholipids

$$\text{lecithin} + H_2O \xrightarrow{\text{phospholipase D}} \text{phosphatidic acid} + \text{choline}$$

$$\text{lysolecithin} + H_2O \xrightarrow{\text{phospholipase D}} \text{lysophosphatidic acid} + \text{choline}$$

$$\text{sphingomyelin} + H_2O \xrightarrow{\text{phospholipase D}} N\text{-acylsphingosylphosphate} + \text{choline}$$

2. Oxidation of choline

$$\text{choline} + 2\,O_2 \xrightarrow{\text{choline oxidase D}} \text{betain} + 2\,H_2O_2$$

3. Indicator reaction

$$2\,H_2O_2 + \text{4-aminoantipyrine} + \text{phenol} \xrightarrow{\text{peroxidase}} \text{quinone diimine dye} + H_2O \ (\lambda_{max} = 500\ \text{nm})$$

Table 4. Differential enzymatic measurement of lecithins and sphingomyelins in human serum

1. Specific part of the procedure

$$\text{lecithins} + H_2O \xrightarrow{\text{phospholipase}} \text{diglyceride} + \text{phosphorylcholine}$$

$$\text{sphingomyelins} + H_2O \xrightarrow{\text{sphingomyelinase D}} \text{N-acylsphingosine} + \text{phosphorylcholine}$$

2. $\text{phosphorylcholine} + H_2O \xrightarrow{\text{alkaline phosphatase}} \text{choline} + \text{phosphate}$

3. $\text{choline} + 2\,H_2O \xrightarrow{\text{choline oxidase}} \text{betain} + 2\,H_2O_2$

4. $2\,H_2O_2 + \text{4-aminoantipyrine} + \text{phenol} \xrightarrow{\text{peroxidase}} \text{quinone diimine dye} + H_2O\ (\lambda_{max} = 500\ \text{nm})$

2.5 Determination of non esterified fatty acids

These acids are usually transported in the plasma bound to albumin or lipoproteins. They are often termed free fatty acids. They principally consist of oleic: (38%), palmitic (28%), stearic (15%) and linoleic (9%) acids.

Many methods have been proposed for measuring non esterified fatty acids in serum. The oldest procedures require previous extraction and make use of their acid property or of their capacity to form chloroform-soluble copper or cobalt soaps. These titrimetric or colorimetric methods although automated remain relatively impracticable (1).

The most recent are enzymatic methods based on a series of reactions as illustrated schematically in Table 5 (26, 27). These methods are sensitive, accurate, specific and require only small amounts of serum (5 μl): they are particularly suitable for dynamic studies.

Table 5. Enzymatic measurement of unesterified fatty acids (ref. 188)

1. Auxiliary reactions

$$\text{RCOOH} + \text{ATP} + \text{CoA} \xrightarrow{\text{acyl-CoA synthetase}} \text{RCO-CoA} + \text{AMP} + PP_i$$

$$\text{RCO-CoA} + O_2 \xrightarrow{\text{acyl-CoA oxidase}} \text{enoyl-CoA} + H_2O_2$$

2. Indicator reaction

$$2\,H_2O_2 + \text{MEHA} + \text{4-aminoantipyrine} \xrightarrow{\text{peroxidase}} \text{quinonic compound}\ (\lambda_{max} = 550\,\text{nm}) + 4\,H_2O$$

(MEHA = 3-methyl-*N*-ethyl-*N*-(β-hydroxyethyl)-aniline)

3 Lipoprotein Quantitation

Lipoprotein quantitation is essential for the diagnosis and treatment of dyslipoproteinemias (1). The concept of the lipoprotein was originally proposed by Macheboeuf (28). However, it was only in 1941 that Blix et al. succeeded in separating by electrophoresis two fractions termed alpha 1 and beta lipoproteins (29). In the 1950's, Gofman et al. made use of the established low hydrated density of plasma lipoproteins to define a new basis for their separation into several density classes (30) (chylomicrons, VLDL, LDL, and HDL). Burstein (31) proposed selective precipitation methods for isolation and quantitation of the various plasma lipoprotein fractions.

Most metabolic, clinical and epidemiological studies carried out over the past twenty years have relied on physicochemical separation of the plasma lipoproteins. Recent work has, however, established that fractions isolated by these techniques actually represent a mixture of particles that vary (Chapter 2) with regard to their apolipoprotein composition (32–34). Beyond the methods used for the physicochemical analysis of lipoproteins, special emphasis will be laid on data recently obtained from immunological characterization of these macromolecules.

3.1 Analysis of lipoproteins: methods based on physico-chemical criteria

Lipoproteins can be separated according to:

- their electrical charge,
- their hydrated density,
- their size,
- their interaction with polyanions or lectins.

3.1.1 Methods based on the mobility of lipoproteins in an electric field

The qualitative analysis of lipoprotein distribution can be carried out easily and rapidly using zone electrophoresis with a resolving power which varies according to the quality of the support used (1).

Paper electrophoresis in albumin buffer which had enabled hyperlipoproteinemias to be typed, is now only of historical interest. With cellulose acetate or agarose gel, lipoproteins are separated according to their charge, enabling the major lipoprotein classes to be distinguished on the basis of their increasing electrophoretic mobility: β, pre-β and α lipoproteins corresponding to LDL, VLDL and HDL, respectively. In addition, other subfractions can be separated on the following supports: slow-moving pre-β lipoproteins on agarose gel (35), and double α lipoproteins bands on cellulose acetate (36); the significance of these fractions remains controversial.

On the other hand, lipoprotein Lp(a), which has many of the properties of LDL, migrates with pre-β lipoproteins and cannot be distinguished from them by electrophoresis. Likewise, if albumin is added to the electrophoresis buffer it causes chylomicrons in serum to migrate (possibly with trailing) close behind the pre-β lipoprotein fractions. Normally they would be trapped in the interstices of the support medium or remain at the origin of the electrophoretogram.

Agar gel electrophoresis (37) is ideally suited to the examination of LpX, an abnormal lipoprotein present in the serum of subjects with obstructive jaundice.

LpX, which normally does not migrate, is induced by electroendosmosis to shift towards the cathode, separating it from the other lipoproteins. Its subsequent identification can then be carried out by precipitation with either polyanions or an anti-LpX immune serum.

A polyacrylamide gel matrix (38) permits particle separation on the combined principles of electrophoresis and molecular sieving. As a result, the migration of VLDL and LDL is reversed when compared to that obtained on the other supports. Separation is in most cases carried out on the following discontinuous gradient polyacrylamide gel:

- an upper 2% acrylamide gel to retain possible chylomicrons,
- a concentrating gel containing 3.24% acrylamide,
- a separation gel containing 3% acrylamide.

These gels are prepared in tubes or on plastic films (acrylamide agarose sheets). Lipoproteins are pre-stained with either acetylated or succinylated Sudan black or with tetrazolium nitroblue diformazan. In the course of the migration, chylomicrons separate clearly from VLDL and remain in the 2% acrylamide gel. VLDL passes through the 2% acrylamide gel to be localized at the upper limit of the concentration gel. LDL forms a band distinctly separated from VLDL; and the diffuse area contiguous to the anode corresponds to HDL. Some hyperlipidemic serum samples may present an additional lipoprotein Lp(a) band above that of LDL, while in other VLDL-rich sera, a distinct extra band is found below VLDL on the surface of the separation gel. This band is present in most cases on electrophoretic patterns of serum samples with abnormally elevated triglyceride concentrations and is closely related to intermediate density lipoproteins (IDL) which are formed during the conversion of VLDL into LDL.

Discontinuous polyacrylamide gels permit excellent separation of the different lipoproteins and can be used for typing hyperlipoproteinemias. They also may be used in a screening test for type III hyperlipoproteinemia. Whereas on agarose gel the abnormal lipoprotein of type III hyperlipoproteinemia seems to adopt β migration, on polyacrylamide gel it behaves like VLDL or IDL. In this case, the LDL band is almost absent in spite of hypercholesterolemia.

It should be borne in mind that electrophoretic separation of lipoproteins permits only qualitative analysis of particle distribution. It is possible to quantify the bands obtained by densitometry but dye uptake by each lipoprotein is variable and correlates poorly with concentration. Therefore the percentage values obtained for each lipoprotein do not correspond absolutely to its concentration and may mislead physicians. For that reason, a lipoproteinogram should be interpreted only with full knowledge of the results of the lipoprotein profile.

Zone electrophoresis is a useful method for separating lipoproteins. It may under some circumstances enable quantitative analysis of the lipoproteins. Indeed some authors have described methods to recover the fractions separated on polyacrylamide gel used in discontinuous or continuous gradient (39) for the quantitation of cholesterol in each fraction by gas liquid chromatography on capillary columns. This method, although selective and accurate, is time consuming and has no place in the routine laboratory. An alternative method has been described by Muniz which consists in separating pre stained lipoproteins on disc polyacrylamide gel and thereafter quantifying them by densitometric integration (40). This approach should be used with caution since no direct relationship exists between the lipoprotein concentration and the quantity of dye fixed.

Other authors have proposed separating the lipoproteins by agarose gel or cellulose acetate electrophoresis, and then evaluating enzymatically their cholesterol content *in situ* (41–43). Such methods require only small volumes of samples, but are, however, poorly reproducible and provide results which are sometimes in disagreement with those obtained with a reference method. Seidel et al. (44) have described a method which consists first in separating lipoproteins on agarose gel, then quantifying them specifically by selective precipitation and densitometry. β and α lipoproteins are easily and accurately measured by this approach (45).

Isotachophoresis, which separates macromolecules according to their specific mobility and not to their size has been applied to the separation of lipoproteins (46). The high resolving power of the method enables lipoproteins (and in particular HDL) to be better separated than with conventional electrophoresis.

More recently, isoelectric focusing of plasma lipoproteins on agarose gel has permitted differentiation within HDL of several molecular entities (47, 48) with defined protein and lipid composition. Up to now, however these methods have only been of qualitative value.

3.1.2 Methods based on the hydrated density of lipoproteins

Being rich in lipids, lipoproteins are lighter than proteins; their hydrated density varies according to the lipid/protein ratio in the molecule and in inverse order to their diameter. This characteristic permits their fractionation by ultracentrifugation (1).

In the fixed angle ultracentrifugal separation procedure, the object is to increase in a step-wise fashion the solvent density of the plasma by progressive addition of an appropriate salt like potassium bromide. This results in the separation of apparently discrete fractions of increasing density. Thereafter, the cholesterol content of the separated lipoprotein can be determined directly. This procedure has several advantages. It is simple, robust, relatively rapid and can handle a large number of samples. However the fixed angle ultracentrifugal separation procedure is not without disadvantages. One weakness lies in its reliance on discrete lipoprotein separation at fixed densities which fails to take account of possible interindividual variations in particle composition or of changes which might occur in response to diseases. Moreover, consideration must be given to the fact that gradient formation occurs in the background solute during the centrifugation run (49). In fact, it is now generally accepted that this fixed density technique is not strictly accurate, since different separation conditions in various rotors may, unless care is taken, result in the selection of an inappropriate density interval. This, coupled with the problem of wall effects and the gradient destabilization which occurs during rotor acceleration and deceleration has led to the development of other procedures which allow monitoring of the separation of the lipoproteins, prior to their fractionation.

Technological improvements in rotor specification and performance and the introduction of titanium in their construction have permitted the development of techniques for the rate zonal ultracentrifugal separation of lipoproteins. Basically, in this procedure a salt gradient is established in the rotor prior to introduction of the plasma sample whose lipoproteins are then separated by flotation. The fractionation can be achieved either by attainment of equilibrium or on the basis of differential flotation. The latter is a function of the relative densities of particle and solute, and particle size. Since lipoproteins are pseudomicellar spheres, variations in shape can largely be ignored (1).

Two technologies based on different equipment have become popular for lipoprotein zonal separation. Large capacity, single chamber zonal rotors enable dynamic separations to be made without subjecting the gradient to reorientation. A discontinuous gradient of sodium bromide is formed in the slowly spinning rotor prior to introducing the sample to its periphery (1,50). Following high-speed centrifugation the rotor is decelerated to loading speed and its contents expelled through a continuous monitoring system and collected into fractions as required. This procedure is ideal for preparative separations using large sample volumes and has the advantage that the spectral distribution obtained is continuous. Wall effects are eliminated and in a relatively short time, the lipoprotein subfractions can be prepared in amounts suitable for subsequent chemical or metabolic studies. However, only one sample can be analysed every centrifugation run and the procedure is not suitable for the examination of small volumes. Moreover, considerable sample dilution occurs and the lipoproteins become exposed, if only for a short time, to substantially higher ionic strengths than employed in fixed angle centrifugation. This may lead to artificial changes in particle structure or composition (51, 54).

The second approach to rate zonal separation of lipoproteins depends on the use of rotors with hinged sample buckets which rotate radially during acceleration and deceleration to eliminate any sample remixing which may occur at these times as a result of gradient reorientation. The rotor geometry permits simultaneous separations of up to six samples and the principle is essentially as described above (1, 52, 53). The sample is usually placed at the bottom of the separating tube and above it is constructed a discontinuous salt gradient. Following centrifugation the separated lipoproteins must be collected very carefully.

3.1.3 **Procedures based on lipoprotein size**

High performance liquid chromatography (HPLC) separates the major lipoprotein classes according to their size (55). This method, however, is expensive and at present largely restricted to specialized research laboratories. But alternative procedures separating them on the basis of size may have applications in the clinical setting. For example, Gambert et al. (56) and Blanche et al. (57) have been able to quantify lipoproteins according to their size using quantitative polyacrylamide gel electrophoresis.

3.1.4 **Procedures based on selective lipoprotein precipitation**

These methods were mainly developed in the laboratory of Burstein with the object either of isolating lipoproteins from large serum sample volumes (58) or of eliminating LDL and VLDL from small sample volumes so as to determine HDL cholesterol easily. In essence low and very low density lipoproteins are precipitated from the sample and measurement of cholesterol in the supernatant gives an index of the plasma HDL concentration. The precipitating reagent may be either organic (e. g. heparin) or non organic (sodium phosphotungstate) polyanions in the presence of divalent cations, or non ionic macromolecules such as polyvinylpyrrolidone, polyethylene-glycol ($M_r = 6000$) or concanavalin A (a polypeptide extracted from jack beans) (31).

In these procedures, it is assumed that all lipoproteins containing Apo B are precipitated while those devoid of Apo B and of density above 1.063 kg/l remain in solution (31). Many authors have studied the specificity of this procedure using normolipidemic or dyslipoproteinemic plasma or serum samples. The results which they obtained are almost identical with those derived from ultracentrifugation (1, 31, 59–62).

It is noteworthy that most comparative investigations were based on the measurement of cholesterol. Some precipitating agents were shown, however, to interfere variably with the cholesterol assay methods, thus giving a bias to the results (63, 64). Moreover, these procedures seem to result in the separation of different mixtures of lipoprotein particles. Even if the cholesterol content of the fractions isolated by precipitation and by ultracentrifugation are similar, they do not necessarily, represent the same lipoprotein particles (59, 65–67). The refined molecular characterization based on the apoprotein content (67, 68) of precipitated and non precipitated lipoproteins suggest that there is not an ideal precipitating method. Although HDL's isolated by ultracentrifugation and selective precipitation have in common a number of physicochemical and immunological properties, they are not strictly identical from the molecular viewpoint. Likewise, the specificity (33, 68) of precipitation methods used to isolate HDL_2 and HDL_3 (69) or LDL (70, 71) may be questioned on the basis of recent studies (72–74). However, because they are easy to carry out, these methods have enabled us, with some reservations to speculate on the *in vivo* metabolism of lipoproteins.

3.2 Methods based on the evaluation of lipoprotein content

Because of their chemical and immunological specificity, apolipoproteins are ideal markers for characterizing lipoproteins.

In the seventies, emphasizing the improved specificity of apoproteins, Alaupovic et al. proposed the use of an "apoprotein profile" for typing dyslipoproteinemias (32–34, 75, 76). Many authors have subsequently shown that apoprotein determination is one of the best ways so far available of revealing changes in the distribution of lipoprotein particles.

Ideally, these particles should be measured directly. The recently developed immunometric methods based on a specific reaction with conventional or monoclonal antibodies (77–79) enable particles to be analysed at the molecular level and should permit considerable progress in the analysis of serum lipoproteins.

3.2.1 Measurement of apolipoproteins

General aspects

Two groups of methods (non immunological and immunological) have been proposed in the past few years for the measurement of apolipoproteins (1).

The *non immunological methods* involve separating, according to their physicochemical properties (size, charge, isoelectric point) the proteins of partially purified and delipidated lipoproteins. After the separation, the various fractions are stained and quantified densitometrically. These methods give only an approximate evaluation of the apolipoprotein concentration but are particularly useful for the identification of polymorphism.

A variety of *immunological methods* can be conveniently applied to the determination of apolipoproteins provided that specific antisera and appropriate standards are available. These procedures, however, give rise to a number of problems:

- in contrast with other plasma proteins, apolipoproteins are integral to the lipid mass of lipoproteins and their antigenic determinants may be more or less concealed inside the molecule by lipids or other substances;
- when a polyclonal immune serum is prepared against an apolipoprotein purified after delipidation, it contains a population of antibodies that is heterogenous with regard to its nature and specificity. This antiserum not infrequently interacts weakly with the native lipoprotein. For that reason, a number of authors recommend that plasma samples be delipidated or treated with detergents before assay. Pretreatment of samples may be avoided if specific antibodies of the determinants present on the surface of native lipoproteins, are used. It should be borne in mind that dilution of the sample can by itself lead to some physicochemical effects.
- the accessibility of the antigenic determinants on lipoproteins generally depends on their size and composition, both of which are constantly changing. Monoclonal antibodies have been recently described which are directed against antigenic determinants constantly accessible on the surface of lipoproteins (pan-antibodies) (80).
- during sampling or storage, an apoprotein may be partially degradated or split, thereby altering its immunological properties;
- purified apoproteins are able to aggregate in aqueous solution. Their preparation and storage require the addition of dissociating agents such as urea or guanidine hydrochloride. For the assay, it is preferable to use as a standard a pool of plasma to which has been added preservatives. The apoprotein content of this pool would have already been determined.
- a number of immunological assay methods such as radial immunodiffusion, electroimmunodiffusion, liquid immunoprecipitation or radioimmunoassay require standards and samples comparable in size. Apoproteins, however, are incorporated in molecules, whose size and heterogeneity vary according to the physiological or pathological state of the patient.
- one of the major obstacles to the development of apoprotein immunological assays is the variability of immune sera and its consequences on standardization. Two recent developments should improve the reliability of assay procedures: the application of recently obtained monoclonal antibodies that can be used individually or in a selected mixture and the establishment of standardization committees (Like the Expert Panel on Apolipoproteins of the International Federation of Clinical Chemistry or the Lipid Apoprotein Subcommittee-Standardization Committee of the International Union of Immunochemical Societies IUIS) (81).

Measurement of the main human serum apolipoproteins

The main apolipoproteins determined in human serum (or plasma) are apoproteins AI, AII, B, CII, CIII, E, and Lp(a).

Apolipoprotein AI (Apo AI)

Apolipoprotein AI is found essentially in the lipoprotein particles present in HDL. Some contain Apo AI and no Apo AII while others contain both Apo AI and Apo AII (82).

Determination of total Apo AI. Apo AI is the major constituent of serum apoproteins in normal subjects. Total Apo AI is usually measured using conventional immunological procedures. Its determination requires preparation of specific immune sera and appropriate standards (83).

Purification of Apo AI and preparation of a specific antiserum: Apo AI is purified from high density lipoproteins separated by flotation ultracentrifugation. Following delipidation of the HDL, its proteins are isolated by the conventional methods of gel filtration or ion exchange chromatography or by preparative isoelectrofocusing. To prevent the formation of aggregates, urea or guanidine needs to be added to the buffers. The spatial structure of the Apo AI thus obtained differs from that of Apo AI incorporated in lipoproteins. More particularly, the percentage of helix is much lower in the absence of lipids. The immunogenicity of delipidated Apo AI is thus different from that of the native apoprotein.

Antisera prepared from rabbits, goats or sheep react variably with Apo AI in lipoproteins. The quality of the immunological reaction depends not only on the number and the diversity of the antigenic sites accessible on the lipoprotein particles, but also on the quantity and the affinity of each specific antibody present in the immune serum. This raises the problem of standardization of the immune serum used in the assays (83).

A number of authors have recently suggested that it might be possible to isolate by affinity chromatography "surface" antibodies specific for Apo AI directly accessible in the native lipoproteins. This selection of antibodies is interesting per se but is defeated by the lack of standardization of the affinity chromatographic procedure.

The development and the accurate characterization and selection of anti Apo AI monoclonal antibodies directed against accessible epitopes in the lipoproteins could resolve most problems related to the measurement of these apoproteins.

Since the immunological reactivity of Apo AI differs according to whether it is free or associated with lipids, it is preferable to use a standard similar in form to the sample. The standard serum added with preservatives or possibly lyophilized should be calibrated against a preparation of HDL_3, the Apo AI content of which has been predetermined by combining the measurement of total proteins with polyacrylamide gel electrophoresis. Alaupovic et al. (84) have recently shown that if a serum standard is used the problems which arise as a result of variability of antisera can be largely minimized.

Measurement of Apo AI by gel immunodiffusion: Radial immunodiffusion and electroimmunodiffusion, because of their simplicity, have been used widely for measuring the level of Apo AI in lipoproteins. Both methods can be applied irrespective of the lipid levels, although prior delipidation or denaturation of the lipoproteins has been recommended by a number of authors. However, it would be preferable to use precipitating antibodies (for example a mixture of monoclonal antibodies) that are specific for surface epitopes only and not to pretreat samples. In this regard, it has recently been proposed that two combined monoclonal antibodies may be effective in the determination of Apo AI giving values that correlate well with the results obtained with the use of polyclonal antisera. In this instance, pretreatment of the samples is not required (85). In order that the standards and test samples may be comparable in size, it is preferable to use secondary standards (sera with preservatives added and stored at −20 °C or lyophilized standards).

Apo AI immunoprecipitation assay: Liquid immunoprecipitation has been widely used for measuring Apo AI. The resulting immune complexes are detected either nephelometrically or turbidimetrically using kinetic or end point assays. Addition of detergents or guanidine hydrochloride was employed to make antigenic sites accessible. This pretreatment of samples may, however, generate problems, particularly when a secondary serum sample is used, and the presence of Apo AI in triglyceride-rich lipoproteins may, when hypertriglyceridemic sera are being analysed, result in the formation of large-sized immune complexes which introduce a positive bias to the assay results.

Apo AI determination by radioimmunology or immunoenzymology: The radioimmunological assay of Apo AI is generally based on competition between the Apo AI of the sample and a ^{125}I-Apo AI standard for binding sites on a specific antiserum. One of the problems of this determination is that the antigenic structure of the labelled ligand may differ from that of the apoprotein in the sample. In addition, the labelled ligand may bind non specifically to the assay tubes. To circumvent these difficulties, Mao et al. have added to their reaction medium a non ionic detergent, but no consensus has been reached on whether the use of such additives or previous delipidation of the sample is advisable. The separation of the bound and unbound forms of the marker is generally obtained by adding a second antibody or protein A (83).

Immunoenzymology is based on enzymatic markers. Methods proposed for measuring Apo AI use a sandwich binding technique in which anti Apo AI antibodies purified by affinity chromatography are linked to a solid substrate. Following addition of the sample to be assayed, the antigen retained is vizualized by adding purified anti Apo AI antibodies labelled with an enzyme. Although a number of authors have suggested that it is preferable to partially delipidate the samples in order to detect all the concealed or apparent epitopes of the molecule, it is possible to assay the antigenic sites accessible in the native molecules without pretreating the samples and the serum standard (84). The sensitivity and specificity of immunoenzymology are close to those of radioimmunology. It can be entirely automated and is applicable to all sera, hyperlipidemic or not. The antigen requires to be neither adsorbed nor labelled, avoiding any possibility of alteration of epitopes. Like radioimmunology, immunoenzymology is convenient to use with monoclonal antibodies (86−88).

Analysis of isoforms of Apo AI. Apo AI is present in plasma in different forms separable by isoelectrofocusing or two-dimensional electrophoresis. Several polymorphic forms can be differentiated in the plasma of apparently healthy subjects. The most positively charged form corresponds to a precursor of the mature Apo AI, and is designated pro Apo AI. It normally represents less than 3% of the total apoprotein and is characterized by the presence of an additional hexapeptide at the N-terminus of the protein. The other forms differ in their charge but not in their molecular mass (89 – 93).

Analysis of abnormal Apo AI. A few genetic variants of Apo AI have been recently detected as a result of variation in their isoelectric points (see Chapter 2). Designated initially by the name of the town in which they were discovered, they are now classified according to the molecular abnormality, if known (89, 93, 94). Examples:

Apo AI Marburg	Apo AI (Lys 107→0)
Apo AI Giessen	Apo AI (Pro 143→Arg)

Isoelectrofocusing or two-dimensional electrophoresis generally secure the diagnosis, although in the future, the development of molecular probes will enable variants to be detected even if they have a normal pI value.

Reference values. Apo AI serum or plasma concentrations increase during the first years of life to reach a threshold value (110 – 160 mg/dl) which remains relatively steady irrespective of age. For a given age range, mean values are appreciably higher in women than in men.

Apolipoprotein AII (Apo AII)

Apolipoprotein AII is mainly found in lipoprotein particles present in the HDL density interval.

Determination of total Apo AII. Apo AII is usually measured by immunological methods. The appreciable variations which exist in the values reported in the literature suggest that this assay is not without problems and needs to be standardized (95 – 100).

Purification of Apo AII and preparation of a specific immune serum: Apo AII consist of two identical peptide chains of molecular mass 8350, bound together by a disulfide bridge. It can be purified, from HDL_3, following dissociation of its Apo AI by incubation in the presence of guanidine hydrochloride. After dialysis and delipidation, Apo AII is isolated in urea-containing medium by ion exchange chromatography possibly coupled with polyacrylamide gel preparative electrophoresis or gel isoelectricfocusing. Although most Apo AII in the plasma is dimeric, small amounts of monomer are also found there, raising the problem of potential differences in antigenicity between both forms of the apoprotein. Immune sera prepared against Apo AII may differ in quality depending on the animal immunized. The immune sera prepared from rabbits are usually specific for the antigenic sites of Apo AII exposed on plasma HDL while those obtained from sheep react better with the delipidated apoprotein. Recently, Bhatnagar et al. (101) reported the existence of three principal epitopes on Apo AII localized between residues 4 – 23, 31 – 46, and 60 – 77. The sites are assumed to be exposed on HDL since delipidation of the samples does not affect the values obtained in the assays.

The availability of anti Apo AII monoclonal antibodies should solve the problems which arise from the variability of conventional antisera.

In order to remove possible differences of immunological reactivity between the native and delipidated forms of Apo AII, it is preferable to use a serum standard prepared by the method described under the determination of Apo AI.

Immunoprecipitation assay: Liquid immunoprecipitation has been used to measure Apo AII, detection of the immune complexes being made nephelometrically or turbidimetrically (95 – 97). Most authors recommend pretreatment of the samples so as to enable the antigenic sites to be better exposed. For AII, however, this may cause a number of problems. Total delipidation can result in the appearance of aggregates which alter the immunological properties of Apo AII. Rosseneu et al. (96) have recently suggested that sodium cholate be added, enabling the antigenic sites to be maximally accessible without denaturation.

Radioimmunological or immunoenzymological assays (100 – 104): Radioimmunological measurements of Apo AII are generally based on competition of the protein in the sample with a ^{125}I Apo AII standard, against a specific antiserum. Special attention should be given to the selection of reagents. With most antisera, very few differences are observed between the values obtained before and after delipidation, indicating that most epitopes are accessible. Conflicting studies have, however, been reported by a number of authors using sheep antisera: Musliner et al. (102) found a significantly increased immunoreactivity following delipidation while Mordasini et al. (100) demonstrated a loss of reactivity for delipidated HDL compared to native HDL. The quality of the tracer antigen is also critical. According to Goldberg et al. (103), in order to achieve identical curves for the standard and the samples tested, ^{125}I Apo AII requires, before use, to be purified by passage down a gel filtration column. Reduced or carboxymethylated Apo AII was shown to have a greater antibody affinity than the native dimeric protein. This difference of immunoreactivity could be a source of error since the monomeric form is found both in plasma samples and in the standards.

Immunoenzymology is suitable for measuring Apo AII. The measurement can be either competitive (104) or non competitive (105); and monoclonal antibodies permit better standardization (106).

Analysis of Apo AII isoforms. Human plasma contains several isoforms of Apo AII with similar molecular mass and immunological properties but with different pI values. These forms of Apo AII can be separated by two-dimensional electrophoresis. The structural differences generating such mutants are not yet clearly understood (89, 90, 91, 94, 107, 108).

Reference values. Human Apo AII concentration after an overnight fast ranges from 25 to 45 mg/dl. No appreciable change with age and sex is noted.

Apolipoprotein B (Apo B).

Apolipoprotein B is a structural protein found in low density and very low density lipoproteins and in Lp(a). Polyacrylamide gel electrophoresis in sodium dodecylsulfate medium enables two forms to be separated in human plasma:

- a preponderant form or Apo B_{100}, found in lipoprotein particles present in VLDL, IDL, LDL, Lp(a). Normal plasma concentration of Apo B_{100} in adult subjects ranges from 70 to 130 mg/dl;

- a minor form or Apo B_{48}, the molecular mass of which is 48% of that of Apo B_{100}. Apo B_{48} is found in chylomicrons and intestinal VLDL and is normally present only in trace amounts in human plasma samples taken after a 18 hour period of fasting.

Determination of total Apo B.

Chemical procedure by insolubilization

When detached from lipids, apolipoprotein B autooxidises and becomes water-insoluble. This specific property has been used to develop a procedure for measuring Apo B in isolated lipoproteins (1).

Immunological procedure

Most antibodies against Apo B are prepared using an LDL subfraction containing only this protein as the antigen. The determination is then carried out by a variety of methods (1, 109).

Gel diffusion: Radial immunodiffusion has been used to measure Apo B even, although the latter appears in a spectrum of particles of varying size which will inevitably show different rates of diffusion through the support gel. As a consequence, the results obtained with this method are falsely low in hyperlipidemic subjects. Sniderman et al. (110) made use of the differences in size between LDL and VLDL to propose a specific procedure for the measurement of plasma LDL Apo B by radial immunodiffusion in a limited time in a gel of known porosity. Using a selected mixture of monoclonal antibodies instead of polyclonal immune serum, Marcovina et al. (111) have recently published a radial-immune diffusion method with a reduced assay time which gives results that correlate with those obtained either with polyclonal antibodies or by measurement of cholesterol. This approach presents distinct advantages for the standardization of the assays.

Electroimmunodiffusion: Electroimmunodiffusion, which is generally a rapid and an accurate method gives problems for the measurement of Apo B in hypertriglyceridemic samples. These may be circumvented by adding lipase to the sample prior to analysis of the protein. By hydrolysing triglycerides, the enzyme reduces the size of the VLDL to an extent which permits its diffusion. In the presence of albumin (to capture the fatty acids released), the results obtained after treatment are close to those achieved with a reference chemical method. The use of a mixture of monoclonal antibodies and an appropriate standard improves the quality of the assays (112). The development of a system permitting dehydration of the gel containing the antibodies and storage of the plates ready for use as dry plastic films, also increases the practicability of the method (113).

Immunoprecipitation: anti Apo B antibodies added in excess to plasma induces the formation of insoluble immune complexes the turbidity of which can be measured nephelometrically or turbidimetrically. This method is very rapid and can be automated. It is subject to strict requirements concerning the concentration ratios between antigens and antibodies. Many variants of the procedure have been described on the basis of differences in light source, detection angle, measurement mode (kinetic, end-point), addition of precipitation-promoting agents (eg polyethyleneglycol). In a large-scale study of the techniques presently available we have found that the results are affected by the presence of triglyceride-rich lipoproteins as soon as the serum triglyceride concentration exceeds 2.3 mmol/l. Treatment with detergents or lipolytic enzymes minimizes the influence of the initial turbidity of the serum but also unmasks concealed antigenic sites and alters the structure of lipoproteins (114–116).

Radioimmunology: Many methods have been described for measuring Apo B radioimmunologically (1). All of these procedures are based on the principle of competition of the Apo B to be determined with a labelled lipoprotein (generally LDL) for polyclonal or monoclonal antibodies. They vary according to labelling method and in the requirement for detergents be added to the incubation medium. Separation between free and bound forms of the labelled lipoprotein is generally obtained by immunoprecipitation using a second antibody. Curry et al. (117) comparing the results obtained by different methods with those achieved with a gravimetric reference procedure, found that the radioimmunological assays underestimate plasma Apo B. The discrepancy might result from differences between the antigen-antibody affinities when Apo B is bound to LDL (LDL is used as standard) or from the presence of different populations of triglyceride-rich lipoproteins in the sample. Moreover, the degree of dilution of the sample might modify the availability of the antigenic sites, and create problems in this regard. The use of monoclonal antibodies which react equally with all Apo B-containing lipoproteins should eliminate these disadvantages (80).

Marcovina et al. (111) have developed a precise and reproducible radioimmunoassay in which single or combined monoclonal antibodies are used. A pool formed by mixing four different clonal products can provide a controlled and standardized reagent that can be used in different assays for the immunochemical determination of Apo B in plasma. The use of monoclonal antibodies and a common secondary standard are essential for the standardization of the assay.

Immunoenzymology: Immunoenzymology can have a sensitivity close to that of radioimmunology and avoids the problems which come from using radio-labelled products. Early methods had the same technical problems as radioimmunology, but a recently developed descending phase method (antigen in excess) avoids dilution of the samples (118). A non competitive sandwich method has also been devised for the determination of Apo B on the surface of different lipoprotein particles provided that the epitopes corresponding to the antibodies are accessible.

Although requiring high dilution of the samples to be determined, this method offers other advantages (119). It is sensitive, specific, can be automated, is applicable to normal and hyperlipidemic sera, uses monoclonal antibodies (80) and employs safe and stable reagents.

Analysis of Apo B variants. Immunological separation of Apo B_{48} from Apo B_{100} has proved impossible until now. Among the monoclonal antibodies obtained by different authors, some react only with Apo B_{100} but all those that are specific for Apo B_{48} react also with Apo B_{100}.

In order to distinguish between the two proteins, it is necessary to use a physicochemical method which discriminates on the basis of their molecular mass (like polyacrylamide gel electrophoresis). No easy, accurate, specific method has been described so far for measuring Apo B_{48}.

Apo B isoforms. The existence of abnormal Apo B is still hypothesized. No molecular variant has been identified. Retrospective epidemiological studies carried out recently in subjects having undergone coronarography showed that some epitopes identified by particular monoclonal antibodies were present in larger amounts in subjects with atherosclerotic coronary arteries (120). It remains to be proven however, whether this difference is related to the existence of abnormal epitopes or to variation in the accessibility of epitopes normally present in the Apo B of lipoproteins.

Reference values. The total plasma Apo B concentration in adult subjects after an overnight fast ranges from 70 to 130 mg/dl. The Apo B_{100} form represents the majority, Apo B_{48} being present only in trace amounts.

Apolipoprotein CII (Apo CII)

Apolipoprotein CII is found essentially in lipoprotein particles present within the VLDL and HDL density intervals.

Determination of total Apo CII. The determination of total human serum or plasma Apo CII is generally carried out by immunological methods. It requires preparation of specific antisera and appropriate standards (1).

Purification of Apo CII and preparation of a specific immune serum: Apo CII is usually purified from delipidated VLDL, ion exchange chromatography, chromatofocusing and isoelectricfocusing enabling a highly enriched fraction to be obtained which is, however, not infrequently contaminated by Apo CIII. To eliminate this, it is essential to expose the preparation to affinity chromatography on gel columns containing anti Apo CIII antibodies. Gesquière et al. (121) have recently proposed the use of a synthetic antigen that reproduces the structure of part of Apo CII. Immune sera are normally prepared from rabbits, goats or sheep. Apo CII is a poor antigen, but its potential can be improved by coupling it to a non homologous protein. The development of anti Apo CII monoclonal or oligoclonal antibodies directed against epitopes always accessible in lipoproteins should permit the development of more reproducible assays (121). The primary standard is a purified Apo CII solution whose protein content has been determined by chemical methods or aminoacids analysis. As for the other apoproteins, it is recommended to use in the routine analysis a secondary serum standard (serum with preservative added and stored at -20°C).

Apo CII determination by gel immunodiffusion : Electroimmunodiffusion is a rapid, easy and inexpensive method which has been adapted for the determination of Apo CII (122). It has, however, the disadvantage of requiring much antibody and of lacking sensitivity. Hypertriglyceridemic samples need to be delipidated before analysis. Radial immunodiffusion is less sensitive and more influenced by variation in the size of lipoproteins. For that reason, it is not in common use for this assay.

Apo CII determination by radioimmunology or immunenzymology (123–126): Because of their sensitivity, these are the methods of choice for measuring Apo CII. The radioimmunological determination is based on the competition of Apo CII in the samples with ^{125}I Apo CII for the specific antiserum. The methods differ in the nature of the agent used for precipitating the antigen-antibody complex. Most authors do not pretreat the samples. Some, however, prefer using a detergent to ensure that all the antigenic sites are exposed, particularly if the measurement is carried out on lipoproteins with high triglyceride content (126). Immunenzymology is also very well adapted for the measurement of Apo CII in serum (127) or in lipoprotein fractions (124). Carlson and Holmquist (127) have devised a method in which alkaline phosphatase is used for labelling VLDL; the Apo CII in the samples is then allowed to compete with Apo CII of the conjugate for an antibody fixed on a solid support. A non competitive method identical with that described for the other apolipoproteins has also been proposed (128).

Analysis of Apo CII variants and of abnormal Apo CII. Apo CII occurs in plasma in different forms (89, 90, 129). The major isoform, can be separated from other minor components, by isoelectricfocusing or two-dimensional electrophoresis. In Apo CII deficiency, the level of the protein is greatly reduced but a spot can be localized by two-dimensional electrophoresis in an abnormal position.

Reference interval. Most authors consider 1.5 – 4.3 mg/dl as normal physiological interval for Apo CII in fasting adults. For children, mean values are slightly lower. In normolipidemic subjects, 70% of the protein is present in the high density lipoprotein fraction and 30% in low or very low density lipoprotein particles (125).

Apolipoprotein CIII (Apo CIII)

Like apolipoprotein CII, apolipoprotein CIII is distributed among very low, low and high density lipoprotein particles.

Determination of total Apo CIII. In most cases, human serum or plasma Apo CIII is measured by immunological procedures using antibodies and appropriate standards (1).

Purification of Apo CIII and preparation of a specific immune serum: the various sialylated forms of Apo CIII that exist in the plasma are easily obtained in the pure state, from delipidated VLDL, by ion exchange chromatography (130, 131) or isoelectric focusing (130, 132) and present the same antigenicity (133). Synthetic peptides reproducing part of Apo CIII and coupled to a carrying protein can also be used as antigen.

Antisera are prepared from rabbits, goats or sheep. Most antibodies obtained from rabbits are directed against the C terminus of the apolipoprotein and their immunoreactivity against pure or lipoprotein bound protein seems to be the same (133, 134).

The Apo CIII primary standard is generally a purified solution of the protein whose concentration has been established by chemical means or aminoacid analysis (97). As for Apo CII, it is preferable to use in routine assays a secondary serum standard whose Apo CIII content is periodically controlled.

Apo CIII determination by gel immunodiffusion: Electroimmunodiffusion is perfectly adapted for the determination of Apo CIII in non delipidated serum (135). In case of hypertriglyceridemic samples, prior delipidation is required.

Apo CIII determination by immunoprecipitation: A method based on end-point nephelometric measurement of the insoluble immune complexes formed when plasma or lipoprotein fractions are added to an anti Apo CIII immune serum has been proposed by Burry et al. (136). Delipidation of the samples is not required but the authors recommend the use of a non ionic detergent for hypertriglyceridemic sera in order to reduce the initial turbidity of the sample and avoid high blanks.

Apo CIII determination by radioimmunology or immunoenzymology: Because of their sensitivity, these methods are highly suitable for measuring Apo CIII in plasma or in lipoprotein fractions. The radioimmunological determination is based on competition of the Apo CIII in the sample with ^{125}I Apo CIII for specific antibodies (137, 138). Pretreatment of the samples does not affect the results (125, 137).

Immunoenzymology is the method of choice for determining Apo CIII. The methods proposed are based on the insolubilization on polystyrene microplates of anti Apo CIII antibodies purified by affinity chromatography. The measurement may be competitive (124, 127) or non competitive (131, 133) and requires no pretreatment of the samples.

Analysis of Apo CIII variants. Apo CIII is present in plasma in three isoforms: Apo $CIII_0$, Apo $CIII_1$ and Apo $CIII_2$ with pI values of 4.9, 4.65 and 4.5, respectively. The two major forms, Apo $CIII_1$ and Apo $CIII_2$, contain 1 and 2 sialic acid residues, respectively, while the desialylated form is minor. The three isoforms are easily separated by isoelectrofocusing (130, 139) or by two-dimensional electrophoresis (89). A fourth isoform, Apo $CIII_3$, has been recently isolated by Jabs and Assmann (140).

Reference interval. The normal physiological values of serum or plasma Apo CIII in fasting adult subjects range from 6.2 to 14.8 mg/dl. In children, the mean values are lower and increase progressively with age (109). In normolipidemic subjects, approximately 80% of Apo CIII are present in high density, 7% in low density and 13% in very low density lipoprotein particles. The distribution of the isoforms in the very low density fraction averages 15.6%, 48.4% and 36% for Apo $CIII_0$, Apo $CIII_1$ and Apo $CIII_2$, respectively (130).

Apolipoprotein E (Apo E)

Apolipoprotein E is found largely in HDL although small amounts are detectable in VLDL and LDL.

Measurement of total Apo E. Total human serum or plasma Apo E are normally measured immunologically (1).

Purification of Apo E and preparation of a specific immune serum: Apo E is a glycoprotein which exists in several isoforms differing from one another by one aminoacid (141). It is usually purified from delipidated VLDL in a series of steps involving gel filtration or ion exchange chromatography, chromatofocusing or affinity chromatography on sepharose-heparin (142, 143). The immunological reactivity of the isoforms is identical (144). Polyclonal antibodies are usually prepared from rabbits, goats or sheep. They do not appear to discriminate between the free or lipoprotein-associated protein (144, 145). Oligoclonal antibodies have been obtained using as antigen a synthetic peptide reproducing part of Apo E, coupled with tetanus toxoid (146). Milne et al. (147) obtained and characterized in mice five specific monoclonal antibodies of Apo E which are specific for all the isoforms. The primary standard used in the Apo E assay is usually a purified solution of the protein whose concentration has been determined by Lowry's method or by aminoacid analysis; the secondary standard is a serum or plasma stored in the frozen or lyophilized state and standardized as a function of the primary standard.

Determination of Apo E by gel immunodiffusion: Radial immunodiffusion and principally electroimmunodiffusion are commonly used to measure total Apo E in serum and in lipoprotein fractions (148). Under their experimental conditions and with the antisera used, Curry et al. (148) have reported that pretreatment of samples is not required.

Determination of Apo E by immunoprecipitation: Weisweiler and Schwandt (149) have described an immunoephelometric end-point procedure for the quantitative determination of total Apo E in human serum. The authors recommend addition of a detergent in order to reduce the background noise of hypertriglyceridemic sera.

Determination of Apo E by radioimmunology or immunoenzymology: Both methods are appropriate for the measurement of Apo E in plasma or in previously fractionated lipoproteins.

Radioimmunological determination (150–152) is usually based on competition of Apo E in the sample with ^{125}I Apo E for polyclonal (150, 151) or monoclonal (152) antibodies. Sample delipidation does not seem to be required (151).

The immunoenzymatic assay is carried out after linking Apo E antibodies to microplates or polystyrene tubes. The assay may be competitive (127) or non competitive (144) and seems to be practicable without pretreating the samples.

Analysis of Apo E isoforms. Three common alleles exist for the structural gene of Apo E which codes for 3 major isoforms (E_2, E_3, E_4) in plasma. This polymorphism determines 6 phenotypes that can be detected by one- (153) or two-dimensional (154) isoelectrofocusing of VLDL or whole plasma (155).

Analysis of abnormal Apo E. Many other genetic variants of Apo E have been described (154) recently. Some of them such as Apo E_1 or Apo E_5 can be located by isoelectricfocusing or by two-dimensional electrophoresis (156, 157). It should, however, be borne in mind that structurally different Apo E may have the same charge. To identify them, more sophisticated methods need to be applied such as examination of the aminoacid sequence, mass spectrometric analysis of the peptides obtained by proteolysis (140), or recombinant DNA methodology.

Reference interval. Reference values of serum or plasma Apo E in fasting adult subjects range from 1.6 to 9.2 mg/dl (144). The lack of standardization of the various methods explains the different reference values quoted in the literature. According to Blum et al. (158), 26% of Apo E are present in very low density lipoprotein particles, 27% in low density and 47% in the high density lipoprotein particles.

Apolipoprotein (a)

Apolipoprotein (a) or Apo(a) is present in widely varying concentrations in the plasma of healthy individuals. It is found within the density intervals 1.05 to 1.08 kg/l being linked to LDL Apo B via a disulfide bridge and non covalent bonds (159). It can also be found in the free state in plasma (160).

Measurement of Apo(a). An immunological assay is used for Apo(a) which requires preparation of a specific immune serum and of a standard.

Preparation of an anti Apo(a) immune serum: A variety of protocols have been published for the purification of Apo(a). Most of them are based on the isolation of lipoproteins of density ranging from 1.05 to 1.08 kg/l, by sequential ultracentrifugation followed by gel filtration chromatography (161) or preparative electrophoresis (162).

In an alternative approach, Fless et al. (159) have recently isolated Apo(a) directly by density gradient zonal ultracentrifugation after having dissociated it from lipoprotein particles with dithiothreitol.

If Lp(a) is used as antigen, the antiserum obtained from rabbits or sheep needs to be absorbed with LP-B in order to eliminate anti Apo B antibodies (161). Recently obtained monoclonal antibodies specific to Apo(a) have opened the possibility of improving the standardization of the method (163).

Measurement of Lp(a) by gel immunodiffusion: Albers and Hazzard (161) were first to describe, in 1975, a radial immunological assay method but because of the high molecular mass of Apo(a), many authors prefer to accelerate the process by using electroimmunodiffusion (164). These procedures, however, lack sensitivity, their detection limit being of the order of 5 mg/dl.

Determination of Lp(a) by turbidimetry or nephelometry: In the presence of excess anti Apo(a) antibodies, addition of plasma Lp(a) induces formation of insoluble immune complexes. The turbidity obtained can be measured nephelometrically using kinetic or end-point determination (165). With lipemic serum, pretreatment with lipase is required.

Ngoc Vu Dac et al. (166) have recently proposed a latex-immunoassay based on direct agglutination by Lp(a) of latex particles covered with a specific antibody. This easy and reproducible method is sensitive, its working range being from 5 – 115 μg/dl. Moreover, it requires only very small amounts of antibody (in contrast to the nephelometric method).

Lp(a) measurement by radioimmunological assay: Albers et al. (167) have described a radioimmunological method based on competition between Apo(a) in the test sample and labeled Apo(a) for an anti Apo(a) antiserum. Separation of the free and bound forms of the labelled lipoprotein is obtained by immunoprecipitation. No pretreatment of the sample is required. However, because the concentration of this antigen in human plasma is so variable, the authors recommend that a preliminary radial immunodiffusion assay be performed in order to determine the appropriate dilution required for the test sample. Although this has the disadvantage of being time-consuming, the method overall is highly sensitive.

Reference values. Large interindividual variations are observed. Values from 0.5 – 200 mg/dl have been reported. Apo(a) serum concentration is not correlated with age or sex.

Measurement of lipoprotein particles defined according to their apoprotein composition

The existence of distinct lipoprotein particles within well defined density classes has been demonstrated by many methods and is the subject of several recent reviews (168 – 170). In order to elucidate their origin, metabolism, function and clinical significance, it is of considerable interest to develop methods for the quantitative determination of these lipoprotein particles.

Several methods have been proposed for the measurement of lipoprotein particles according to their apolipoprotein composition (170 – 172). An absorption crossed electroimmunoassay has been used by Atmeh et al. (172) to evaluate lipoprotein particles which contain both apolipoprotein AI and AII and lipoprotein particles which contain apolipoprotein AI but not apolipoprotein AII.

This method employs electrophoresis through two agarose gel layers, each of which contains different antibodies. The first layer, to which has been added antibodies to Apo AII, retains only lipoprotein which contain Apo AII while those which contain Apo AI but not Apo AII are measured in the upper layer which is impregnated with antibody to Apo AI (172, 173).

Immunoprecipitation in free solution, using antibodies to Apo AII, has also been used to evaluate these lipoprotein particles (171, 172). A sequential immunoprecipitation in free solution, using several successive antibodies was described by Alaupovic to measure lipoprotein particles of very low and low densities (174, 175).

Immunoaffinity chromatography may be employed to study lipoprotein particles defined by their apoprotein composition (172, 176, 177).

Recently a two site differential immunoenzymometric assay has been proposed which allows the direct measurement in plasma of different apolipoprotein associations (178 – 180). This method is convenient and reliable and the values obtained by it are in good agreement with the results of sequential immunoprecipitation studies.

The development of techniques to produce highly specific antibodies (monoclonal antibodies, antibodies directed against chemically synthesized short fragments of an apolipoprotein (181)) may offer the possibility of separating lipoprotein particles with an epitope expressed on the surface from particles in which the epitope is masked or altered in conformation (182 – 184). If functionally important epitopes are identified, a search for their expression in whole plasma or in lipoprotein subclasses under physiological and pathological states may become of great clinical importance (185, 186).

References

1. Mills, G. L., Lane, P. A. & Weech, P. K. (1984) A Guidebook to Lipoprotein Technique in Laboratory Techniques in Biochemistry and Molecular Biology, Vol. 14, (Burdon, R. H. and Van Knippenberg, P. H., eds.), Elsevier, Amsterdam.
2. Sodhi, H. S., Kudchodkar, B. J. & Mason, O. T. (1979) Clinical Methods in Study of Cholesterol Metabolism in Monographs on Atherosclerosis, Vol. 9, (Kritchevsky, D. and Pollak, O. J., eds.), Karger, Basel.
3. Gambert, P., Lallemant, C., Archambault, A., Maume, B. F. & Padieu, P. (1979) J. Chromatogr. *162*, 1 – 6.
4. Abell, L. L., Levy, B. B., Brodie, B. B. & Kendall, F. E. (1952) J. Biol. Chem. *195*, 357 – 366.
5. Flegg, H. M. (1973) Ann. Clin. Biochem. *10*, 79 – 84.

6. Richmond, W. (1973) Clin. Chem. *19*, 1350–1356.
7. Naito, H. K. (1984) In: Theory, Analysis and correlation (Kaplan, L. A., Pesce, A. J., eds.) pp. 1194–1207, Mosby, St Louis, Toronto.
8. Zak, B. (1977) Clin. Chem. *23*, 1201–1274.
9. Allain, C. C., Poon, L. C., Chan, C. G., Richmond, W. & Fir, P. C. (1974) Clin. Chem. *20*, 470–475.
10. Noma, A. & Nakayama, K. (1976) Clin. Chem. *22*, 336–340.
11. Kumer, A. & Christian, G. D. (1977) Clin. Chim. Acta *74*, 101–108.
12. Hinsch, W., Antonijevic, A. & Sindaram, P. V. (1981) J. Clin. Chem. Clin. Biochem. *19*, 307–308.
13. Nobbs, B. T., Smith, J. M. & Walker, A. W. (1977) Clin. Chim. Acta *79*, 391–397.
14. Van Gent, C. M., Van Der Voort, H. A., De Bruyn, A. M. & Klein, F. (1977) Clin. Chim. Acta *75*, 193–204.
15. De Bruyn, A. M., Klein, F., Van Gent, C. M. & Van der Voort, H. A. (1978) Clin. Chim. Acta *85*, 215–218.
16. Carlson, S. E. & Goldfarb, S. (1977) Clin. Chim. Acta *79*, 575–582.
17. Majkic, N. & Berker, I. (1977) Clin. Chim. Acta *80*, 121–131.
18. Sugiura, M., Oikawa, T., Hirano, K., Maeda, H., Yoskimura, H., Sugiyama, M. & Kuratsu, T. (1977) Clin. Chim. Acta *81*, 125–130.
19. Koehler, D. F., Steele, B. W., Azar, M. M., Kuba, K. & Dempsey, M. E. (1978) Clin. Chem. *24*, 326–329.
20. Jones Ouven, V. M. & Tai, O. H. (1977) Clin. Biochem. *10*, 210–212.
21. Kohlmeier, M. (1986) Clin. Chem. *32*, 63–66.
22. Daunizeau, A., Vanhoutte, G., Puchois, P., Parra, H. & Fruchart, J. C. (1981) Gaz. Méd. France *88/36*, 5121–5127.
23. Takayama, M., Itoh, S., Nagasaki, T. & Taminuzu, I. (1977) Clin. Chim. Acta *79*, 93–98.
24. Sagiura, M., Oikarwa, T. & Hirano, K. (1978) Clin. Chim. Acta *89*, 447–453.
25. Okazaki, M., Hagiwara, N. & Hara, I. (1981) J. Chromatography *231*, 13–24.
26. Shimizu, S., Inowe, K., Tani, Y. & Yamada, H. (1979) Analyt. Biochem. *98*, 341–345.
27. Okabe, H., Uji, Y., Nagashima, K. & Noma, A. (1980) Clin. Chem. *26*, 1540–1543.
28. Macheboeuf, M. & Rebeyrotte, P. (1949) Discuss. Chem. Soc. *6*, 62–70.
29. Blix, G., Tiselius, A. & Svensson, H. (1941) J. Biol. Chem. *137*, 485–494.
30. Gofman, J. W., Delalla, O., Glazier, F. & Nichols, A. V. (1954) Plasma *2*, 413–484.
31. Burstein, M. & Legmann, P. (1982) Lipoprotein precipitation. In: Monographs on Atherosclerosis (Clarkson, T. B., Kritcheusky, D., Pollak, O. J., eds.).
32. Alaupovic, P. (1980) in: Handbook of Electrophoresis. Lipoproteins: basic principles and concepts (Lewis, L. A., Opplt, J. D., eds.) pp. 27–46, CRC Press, Boca Raton, Florida.
33. Alaupovic, P. (1982) Ric. Clin. Lab. *12*, 3–21.
34. Alaupovic, P. (1972) Protides Biol. Fluids Proc. Colloq. *19*, 9–19.
35. De Gennes, J. L., Dairou, F., Hamon, P. & Truffert, J. (1981) Nouv. Presse méd. *10*, 691–695.
36. Fruchart, J. C., Sezille, G., Jaillard, J., Decoopman, E. & Desreumaux, C. (1974) Lille Méd. *19*, 84–86.
37. Seidel, D., Alaupovic, P. & Furman, R. (1969) J. Clin. Invest. *48*, 1211–1223.
38. Moulin, S., Dewailly, P., Sezille, G. & Fruchart, J. C. (1979) Clin. Chim. Acta *91*, 159–163.
39. Gambert, P., Lallement, C. & Padieu, P. (1980) Clin. Chim. Acta *100*, 99–105.
40. Muniz, N. (1977) Clin. Chem. *23*, 1826–1833.
41. Baillie, E. E., Welborn, P., Hill, J. & Sexton, B. (1980) Clin. Chem. *26*, 679–680.
42. Conlon, D. R., Blankstein, L. A. & Pasakarnis, P. A. (1979) Clin. Chem. *25*, 1965–1969.
43. Cobb, S. A. & Sanderd, J. L. (1978) J. Clin. Invest. *24*, 1116–1120.
44. Seidel, D., Wieland, H. & Ruppert, C. (1973) Clin. Chem. *19*, 731–741.
45. Parra, H., Cachera, C., Clavey, V., Daunizeau, A. & Fruchart, J. C. (1981) Innov. Tech. Biol. Méd. *2*, 189–203.
46. Bittolo, Bon, G., Cazzolato, G. & Avogaro, P. (1981) J. Lipid Res. *22*, 998–1002.
47. Marcel, Y. L., Weech, P. K., Nguyen, T. D., Milne, R. W. & McConathy, W. J. (1984) Eur. J. Biochem. *840*, 1–10.
48. Luley, C. H., Puchois, P. & Kloer, H. U. (1985) Clin. Chim. Acta *146*, 185–193.
49. Lindgren, F. T. & Jensen, L. C. (1972) in: Blood Lipids and Lipoproteins: Quantitation, Composition and Metabolism,

(Nelson, G. J., ed.) pp. 181–274, Wiley, New York.
50. Patsch, J. R., Sailer, S., Kostner, G., Sandhofer, F., Holasek, A. & Braunsteiner, H. (1974) J. Lipid Res. *15*, 356–366.
51. Kunitake, S. T. & Kane, J. P. (1982) J. Lipid Res. *23*, 936–940.
52. Shepherd, J., Caine, E. A., Bedford, D. K. & Packard, C. J. (1984) Analyst *109*, 347–351.
53. Demacker, P. N., Van Sommeren-Zondag, D. F., Stalenhoef, A. F., Stuyt, P. M. & Van't Laar, A. (1983) Clin. Chem. *29*, 656–663.
54. Chapman, M. J., Goldstein, S., Lagrange, D. & Lapland, P. M. (1981) J. Lipid Res. *22*, 339–358.
55. Okazaki, M., Hagiwara, N. & Hara, I. (1982) Biochem. *92*, 517–524.
56. Gambert, P., Lallemant, C., Andre, J. L., Athias, A., Bourquard, R., Pierson, M. & Padieu, P. (1981) Clin. Chim. Acta *110*, 295–300.
57. Blanche, P. J., Gong, E. L., Forte, T. M. & Nichols, A. V. (1981) Biochim. Biophys. Acta. *665*, 408–419.
58. Burstein, M., Scholnick, H. R. & Morfin, R. (1970) J. Lipid Res. *11*, 583–595.
59. Durrington, P. N. (1983) in: CRC Critical Reviews in Clinical Laboratory Sciences, *18/1*, 31–78.
60. Warnick, G. R., Nguyen, T. & Albers, A. A. (1985) Clin. Chem. *31*, 217–222.
61. Warnick, G. R., Cheung, M. C. & Albers, J. J. (1979) Clin. Chem. *25*, 596–604.
62. Demacker, P. N. M., Vos-Janssen, H. E., Hijmans, A. G. M., Van't Laar, A. & Jansen, P. (1980) Clin. Chem. *26*, 1780–1786.
63. McGowan, M. W., Artiss, J. D. & Zak, B. (1982) Microchem. J. *27*, 574–582.
64. Niedmann, P. D., Luthe, H., Wieland, H., Schaper, G. & Seidel, D. (1983) Klin. Wochenschr. *61*, 133–138.
65. Luley, C. H., Puchois, P. & Kloer, H. U. (1985) Clin. Chim. Acta *146*, 185–193.
66. King, M. E. & Rifai, N. (1984) Clin. Chem. *30*, 859.
67. Puchois, P., Luley, C., Kloer, H. U. & Alaupovic, P. (1983) Arteriosclerosis *3*, 5121.
68. Alaupovic, P. (1980) in: CRC Handbook of Electrophoresis (Lewis, L. A., Opplt, J. J., eds.) pp. 27–46, Vol. I, CRC Press, Boca Raton, Florida.
69. Gidez, L. I., Miller, G. J., Burstein, M., Slagle, S. & Eder, H. A. (1982) J. Lipid Res. *23*, 1206–1223.
70. Wieland, H. & Siedel, D. (1983) J. Lipid Res. *24*, 904–909.
71. Demacker, P. N., Hijmans, A. G., Brenninkmeijer, B. J., Jansen, A. P. & Van't Laar, A. (1984) Clin. Chem. *30*, 1797–1800.
72. Simpson, H. S. (1983) Med. Lab. Sci. *40*, 402–407.
73. Daerr, W. H., Windler, E. E. T., Rohwer, H. D. & Greten, H. (1983) Atherosclerosis *49*, 211–213.
74. Simpson, H. S., Ballantyne, F. C., Packard, C. J., Morgan, G. & Shepherd, J. (1982) Clin. Chem. *28*, 2040–2043.
75. Puchois, P., Alaupovic, P. & Fruchart, J. C. (1985) Ann. Biol. Clin. *43*, 831–840.
76. Alaupovic, P. (1983) Determination of plasma apolipoprotein profiles in patients with dyslipoproteinemias. In: 5ème Colloque Biologie Prospective (Masson, ed.), pp. 683–689.
77. Fruchart, J. C., Koffigan, M., Fievet, C., Cachera, C., Vu Dac, N., Gesquière, J. C. & Puchois, P. Molecular analysis of lipoproteins. Clinical applications. In: Proc. Atherosclerosis, Plenum Press, in press.
78. Puchois, P., Bertrand, M., Lablanche, J. M. & Fruchart, J. C. (1985) Circulation *72*, 199.
79. Fruchart, J. C., Kandoussi, A., Parsy, D., Koren, G. & Puchois, P. (1985) J. Clin. Chem. Biochem. *23*, 619.
80. Koren, E., Solter, D., Lee, D. M., Reiner, Z., McConathy, W. J., Dashti, N. & Alaupovic, P. (1986) Biochim. Biophys. Acta *876*, 91–100.
81. Cooper, G. R., Smith, S. J., Wiebe, D. A., Kuchmak, M. & Hannon, W. H. (1985) Clin. Chem. *30*, 223–228.
82. Alaupovic, P. (1983) Serum apolipoprotein profiles of dyslipoproteinemias. In: Dietary fats and healthy (Perkins, E. G., Visek, W. J., eds.) pp. 574–597, American Oil Chemist's Society, Champaign, IL.
83. Steinberg, K. K., Cooper, G. R., Graiser, S. R. & Rosseneu, M. (1983) Clin. Chem. *29*, 415–426.
84. Fesmire, J. D., McConathy, W. J. & Alaupovic, P. (1984) Clin. Chem. *30*, 712–716.
85. Marcovina, S. & Catapano, N. In: Atheroscleros. Plenum Press, in press.
86. Koren, E., Puchois, P., McConathy, W. J., Fesmire, J. D. & Alaupovic, P. (1985) Clin. Chim. *147*, 85–95.
87. Bury, S. & Rosseneu, M. (1985) Clin. Chem. *31*, 247–251.

88. Fruchart, J. C., Fievet, C. & Puchois, P. (1985) U. E. Bergmeyer, Vol. VIII, pp. 126–138.
89. Sprecher, D. L., Taam, L. & Brewer, H. B. Jr. (1984) Clin. Chem. *30*, 2084–2092.
90. Scanu, A. M., Byrne, R. E. & Edelstein, C. (1984) J. Lipid Res. *25*, 1593–1602.
91. Stoffel, W. (1984) J. Lipid Res. *25*, 1586–1592.
92. Ghisellini, M., Pecorari, M. & Calandra, S. (1986) Atherosclerosis *59*, 247–256.
93. Schaefer, E. J. (1984) Atherosclerosis *4*, 303.
94. Zannis, V. I. & Breslow, J. L. (1985) Adv. Hum. Genet. *14*, 125–215.
95. Schriewer, H., Emke, F. & Assman, G. (1985) J. Clin. Biochem. *23*, 355–379.
96. Rosseneu, M., Vinaimont, N., Musliner, T. A., Bernier, D., Herbert, P. N. & Belpaire, F. (1984) Clin. Chem. *30*, 234–237.
97. Heuck, J. B., Erbe, I., Frech, K., Lohse, P. & Munscher, J (1983) Clin. Chem. *23*, 1385–1388.
98. Cheung, M. C. & Albers, J. J. (1977) Clin. Invest. *60*, 43–50.
99. Curry, M. D., Alaupovic, P. & Suenram, C. A. (1976) Clin. Chem. *22*, 315–322.
100. Mordasini, R. C. & Riesen, W. F. (1980) J. Clin. Chem. Clin. Biochem. *18*, 917–920.
101. Bhatnagar, P. K., Mao, S. J. T., Gotto, A. M. & Sparrow, J. R. (1983) Peptides *4*, 343–349.
102. Musliner, T. A., Garner, P. A., Henderson, L. O. & Herbert, P. N. (1982) Atherosclerosis *2*, 160–169.
103. Goldberg, R. B., Karlin, J. B. & Juhn, D. L. (1980) J. Lipid Res. *21*, 902–912.
104. Dufaux, B., Ilsemann, K. & Asslmann, G. (1983) Clin. Chem. Clin. Biochim. *21*, 39–43.
105. Puchois, P., Kandoussi, A., Duriez, P. & Fruchart, J. C. (1986) J. Immunoassay *7*, 285–307.
106. Stein, E. A., Pesce, A. J. (1983) In: Proc. Workshop apolipoprotein quantification (K. Lippel, ed.) pp. 319–331, NIH Publication No 83–1266.
107. Knott, T. J., Priestley, L. M., Urdea, M. & Scott, J. (1984) Biochem. Biophys. Res. Commun. *120*, 734–740.
108. Gordon, J. I., Budelier, K. A., Sims, H. F., Edelstein, C., Scanu, A. M. & Strauss, A. W. (1983) J. Biol. Chem. *258*, 14054–14059.
109. Rosseneu, M., Vercaemst, R., Steinberg, K. K. & Cooper, G. R. (1983) Clin. Chem. *29*, 427–433.
110. Sniderman, A., Teng, B. & Jerry, M. (1975) J. Lipid Res. *16*, 465–467.
111. Marcovina et al. (1985) Clin. Chem. *31*, 1654–1658.
112. Pruvot, I., Fievet, C. & Fruchart, J. C. (1987) Clin. Chem., submitted.
113. Fruchart, J. C., Duthilleul, P. & Kora, I. (1981) Clin. Chim. Acta *114*, 123–126.
114. Weisweiler, P. & Schwandt, P. (1984) Clin. Chem. Clin. Biochem. *22*, 113–118.
115. Fievet, C., Dedonder, E. & Fruchart, J. C. (1979) Clin. Chim. Acta *25*, 405–408.
116. Heuck, C. C. & Schliert, G. (1979) Clin. Chem. *25*, 221–226.
117. Curry, M. D., Gustafon, A., Alaupovic, P. & McConathy, W. J. (1978) Clin. Chem. *24*, 280–288.
118. Fievet, C., Dedieu, A., Kandoussi, A., Luyéyé, I. & Fruchart, J. C. (1986) Clin. Chim. Acta *159*, 269–278.
119. Fievet, C., Koffigan, M., Ouvry, D., Marcovina, S., Moschetto, Y. & Fruchart, J. C. (1984) Clin. Chem. *30*, 98–100.
120. Fruchart, J. C., Parra, H., Kandoussi, A., Fievet, C., Lablanche, J. M. & Bertrand, M. (1985) Circulation *72*, Supp. III, 92.
121. Gesquière, J. C., Delpierre, C. C., Tartar, A. L., Cachera, C., Equagoo, K. V. & Fruchart, J. C. (1985) Clin. Chem. *31*, 784–785.
122. Curry, M. D., McConathy, W. J., Fesmire, J. C. & Alaupovic, P. (1980) Biochim. Biophys. Acta *617*, 503.
123. Kashyap, M. L., Srivastava, L. S., Chen, C. Y., Perisutti, G., Gamphell, M., Lutmer, R. F. & Glueck, C. J. (1977) Clin. Invest. *60*, 171.
124. Holmquist, L. (1980) J. Immunolog. Methods *34*, 243–251.
125. Gustafson, S., Ostlund-Lindqvist, A. M. & Vessby, B. (1984) Scand. J. Clin. Lab. Invest. *44*, 291–297.
126. Richard, B., Mao, S. J. T., Kawamoto, T. & LaRusso, N. F. (1983) J. Lipid Res. *24*, 391–401.
127. Carlson, L. A. & Holmqvist, L. (1982) Clin. Chim. Acta *124*, 163–178.
128. Fievet, C., Bouslama, A., Delpierre, C., Equagoo, V., Gesquière, J. C., Tartar, A., Cachera, C. & Fruchart, J. C. (1986) In: Recent Aspects on Diagnosis and Treatment of Lipoprotein Disorders Impact on Prevention of Atherosclerotic Diseases, International Symposium, Vienne, 21–23 august.
129. Fojo, S. S., Law, S. W. & Brewer, H. B. (1984) Proc. Natl. Acad. Sci. USA *81*, 6354–6357.

130. Bugugnani, M. J., Koffigan, M., Kora, I., Ouvry, D., Clavey, V. & Fruchart, J. C. (1984) Clin. Chem. *30*, 349–351.
131. Bury, J. & Rosseneu, M. (1985) J. Clin. Chem. Clin. Biochem. *23*, 63–68.
132. Marcel, Y. L., Bergseth, M. & Nestruck, A. C. (1979) Biochim. Biophys. Acta *573*, 175–183.
133. Parsy, D., Clavey, V., Fievet, C., Kora, I., Duriez, P. & Fruchart, J. C. (1985) Clin. Chem. *31*, 1632–1635.
134. Mao, S. J. T., Bhatnagar, P. K., Gotto, A. M. & Sparrow, J. T. (1980) Biochemistry *19*, 315–320.
135. Curry, M. D., McConathy, W. J., Fesmire, J. D. & Alaupovic, P. (1980) Biochim. Biophys. Acta *617*, 503–513.
136. Bury, J., De Keersgieter, W. & Rosseneu, M. (1985) Clin. Chim. Acta *145*, 249–258.
137. Schonfeld, G., George, P. K., Miller, J., Reilly, P. & Witztum, J. (1979) Metabolism, *28*, 1001–1010.
138. Kashyap, M. L., Srivastava, L. S., Hynd, B. A., Gartside, P. S. & Perisotti, G. (1981) J. Lipid Res. *22*, 800–810.
139. Daunizeau, A., Barkia, A., Sefraoui, M. & Fruchart J. C. 6ème Colloque de Biologie Prospective de Pont à Mousson, 30 septembre–4 octobre 1985.
140. Jabs, H. U. & Assmann, G. (1985) Circulation *72*, 199.
141. Weisgraber, K. H., Rall, S. C. & Mahley, R. W. (1981) J. Biol. Chem. *256*, 9077–9083.
142. Weisweiler, P. & Schwandt, P. (1982) Clin. Chim. Acta *124*, 45–50.
143. Shelburne, F. A. & Quarfordt, S. J. (1977) J. Clin. Invest. *60*, 944–950.
144. Koffigan, M., Kora, I., Clavey, V., Bard, J. M., Chapman, J. & Fruchart J. C. (1987) Clin. Chim. Acta, in press.
145. Curry, M. D., McConathy, W. J., Alaupovic, P., Ledford, J. H. & Popovic, M. (1976) Biochim. Biophys. Acta *439*, 413–425.
146. Gesquière, J. C., Delpierre, C., Tartar, A., Cachera, C., Equagoo, K. & Fruchart, J. C. (1985) Clin. Chem. *31*, 784.
147. Milne, R. W., Douste-Blazy, P., Retegui, L. & Marcel, Y. L. (1981) J. Clin. Invest. *66*, 111–117.
148. Curry, M. D., McConathy, W. J., Alaupovic, P., Ledford, J. H. & Popovic, M. A. (1976) Biochim. Biophys. Acta, *439*, 413–425.
149. Weisweiler, P. & Schwandt, P. (1983) J. Clin. Chem Clin. Biochem. *21*, 227–230.
150. Havel, R. J., Kotite, L., Vigne, J. L. et al. (1980) J. Clin. Invest. *66*, 1351–1362.
151. Blum, C. B., Aron, L. & Sciacca, R. (1980) J. Clin. Invest. *66*, 1240–1250.
152. Milne, R. W., Douste-Blazy, P., Retegui, L. & Marcel, Y. L. (1981) J. Clin. Invest. *68*, 111–117.
153. Bouthillier, D., Sing., C. F. & Davignon, J. (1983) J. Lipid Res. *24*, 1060–1069.
154. Zannis, V. & Breslow, J. L. (1981) Biochemistry *20*, 1033–1041.
155. Sprecher, D. L., Taam, L. & Brewer H. B. Jr. (1984) Clin. Chem. *30*, 2084–2092.
156. Weisgraber, K. H., Rall, S. C. Jr., Innerarity, T. L. et al. (1984) J. Clin. Invest. *73*, 1024–1033.
157. Yamamura, T., Yamamoto, A., Hiramori, K. & Nambu, S. (1984) Atherosclerosis *50*, 159–172.
158. Blum, C. B., Davis, P. A. & Forte, T. M. (1985) J. Lipid Res. *26*, 755–760.
159. Fless, G. M., Zun Mallen, M. E. & Scanu, A. (1985) J. Lipid Res. *26*, 1224–1229.
160. Reiner, Z. & McConathy, W. J. (1985) Circulation *72*, 364 (abstract 1453).
161. Albers, J. J. & Hazzard, W. R. (1974) Lipids *9*, 15–26.
162. Desreumaux, C., Fruchart, J. C., Dewailly, P., Jaillard, J. & Sezille, G. (1977) J. Chromatogr. *130*, 336–341.
163. Duvic, C. R., Smith, G., Sledge, W. E., Lee, L. T., Murray, M. D., Roheim, P. S., Gallaher, W. R. & Thompson, J. J. (1985) J. Lipid Res. *26*, 540–548.
164. Kostner, G. M., Avogaro, P., Cazzolato, G., Marth, E., Bittolo-Bon, G. & Qunici, G. B. (1981) Atherosclerosis, *38*, 51–61.
165. Cazzolato, G., Prakasch, G., Green, S. & Kostner, G. M. (1983) Clin. Chim. Acta *135*, 203–208.
166. Vu Dac, N., Chekkor, A., Parra, H., Duthilleul, P. & Fruchart J. C. (1985) J. Lipid Res. *26*, 267–269.
167. Albers, J. J., Adlophson, J. L. & Hazzard, W. R. (1977) J. Lipid Res. *18*, 331–338.
168. Alaupovic, P. (1980) in: CRC Handbook of Electrophoresis. Lipoproteins: Basis Principles and Concepts (L. A. Lewis, J. J. Opplt, eds.) pp. 27–46 CRC Press, Boca Raton, Florida.
169. Puchois, P., Alaupovic, P. & Fruchart, J. C. (1985) Ann. Biol. Clin. *43*, 831–840.
170. Alaupovic, P. (1984) in: Clinical and metabolic aspects of high density lipoproteins

(N. E. Miller, G. J. Miller eds.), pp. 1–43. Elsevier, Amsterdam.

171. Cheung, M. C. & Albers, J. J. (1982) J. Lipid Res. *23*, 747–753.
172. Atmeh, R. F., Shepherd, J. & Packard, C. J. (1983) Biochim. Biophys. Acta, *751*, 175–188.
173. McConathy, W. J., Alaupovic, P. & Fesmire, J. C., (1984) Various applications of electroimmunoassay for plasma apolipoprotein. In: Proc. Workshop on Apolipoprotein Quantification, NIH Monograph 831266, pp. 276–281.
174. Alaupovic, P. (1984) Mapping of lipoprotein particles of very low and low densities characterized by apolipoproteins B, C-I, C-II, C-III and E as their protein moieties. In: Journées du Gerli, Le Touquet.
175. Alaupovic, P. (1986) Identification of lipoprotein particles defined by their apolipoprotein composition in selected dyslipoproteinemic states and their significance in atherosclerosis. In: 3ème Colloque de la Fondation de Recherche sur l'Athérosclérose, p. 28, Bruxelles.
176. Cheung, M. C. & Albers, J. J. (1984) J. Biol. Chem. *259*, 12201–12221.
177. Fielding, C. J., Frohlich, J., Moser, K. & Fielding, P. D. (1982) Metabolism *31*, 1023–1028.
178. Puchois, P., Bertrand, M., Lablanche, J. M. & Fruchart J. C. (1985) Circulation, *72*, 795.
179. Fruchart, J. C., Kandoussi, A., Parsy, D., Koren, G. & Puchois, P. (1985) J. Clin. Chem. Clin. Biochem. *23*, 619.
180. Koren, E., Puchois, P., Alaupovic, P., Fesmire, J., Kandoussi, A., & Fruchart, J. C., (1987) Clin. Chem. *33*, 38–43.
181. Gesquière, J. C., Delpierre, C., Tartar, A., Cachera, C. & Equagoo, V. (1985) Clin. Chem. *31*, 784–785.
182. Tikkanen, M. J., Cole, T. G., Hahm, K. S. & Krul, E. S. (1984) Arteriosclerosis, *4*, 138–146.
183. Koren, E., Solter, D., Knight, C. & Alaupovic, P. (1983) Circulation *48*, 217.
184. Fruchart J. C. (1985) Oligoclonal and monoclonal antibody mapping of lipoprotein particles. In: 7th Internat. Symp. Atherosclerosis, Melbourne, october 6–10.
185. Watt, R. M., Carhart, R. L., Gentile, T. C. & Watt, T. S. (1985) Pharmac. Ther. *28*, 29–50.
186. Curtiss, L. K. (1986) application of monoclonal antibodies to issues of plasma lipoprotein. Structure and Function *17*, 291–308.
187. Burke, R. W., Diamondstone, B. I., Velapoldi, R. A. & Menis, O. (1974) Clin. Chem. *20*, 794–801.
188. Mulder, C., Schouten, J. A. & Popp-Snijders, C. (1983) J. Clin. Chem. Clin. Biochem. *21*, 823–827.

Chapter 5
Tissue-Lipoprotein Interactions: Lipoprotein Receptors

W. J. Schneider and U. Beisiegel

1 Introduction

The concept of receptor-mediated metabolism of lipoproteins emerged from studies initiated in 1973 on human skin fibroblasts grown in culture. These experiments were designed to elucidate the normal function of low density lipoprotein (LDL), about which little was known at that time. Biochemical studies showed that a specific cell surface receptor, the LDL receptor, mediates the binding, uptake and degradation of LDL, thus supplying almost all cells in the body with cholesterol. Detailed insight into the molecular mechanisms underlying this complex process was obtained from studies with fibroblasts derived from patients with the phenotype of homozygous familial hypercholesterolemia (FH). As in many other biological systems, the expression of a disease state in a defined cellular system was essential to the discovery of the causal factor: FH is now one of the best characterized genetic diseases at the molecular level. As will be outlined below, in FH several groups of mutations occur naturally in the structural gene for the LDL receptor which disrupt its normal function and lead to severe hypercholesterolemia, producing myocardial infarctions and premature atherosclerosis. Thus, the important role of lipoprotein receptors in normal physiology is underscored by the dramatic consequences of their functional absence.

The demonstration of receptors for LDL in extrahepatic as well as hepatic tissues has triggered intense research efforts into defining different specific receptor systems for other lipoprotein classes than LDL. Lipoproteins for which receptor-mediated transport mechanisms are thought to exist include normal metabolic products, such as high density lipoproteins (HDL) and chylomicron remnants (CMR), lipoproteins produced under extreme physiological conditions or in disease, such as β-migrating very low density lipoproteins (β-VLDL), and chemically altered lipoproteins, such as charge-modified LDL (e. g. acetyl-LDL). The number of investigations designed to clarify the physiologic role, if any, of various molecular entities capable of interaction with one or a group of these lipoproteins has recently increased dramatically, but much work remains to be done.

In this chapter, we will first consider physiological and molecular aspects of the normal LDL receptor and of mutant receptors in FH, and then attempt to summarize the current status of research in the field of other, less well defined receptor systems for lipoproteins.

2 The Receptor for Low Density Lipoprotein

2.1 Experimental systems

The expression of LDL receptors has been investigated in a variety of *in vitro* and *in vivo* settings. Initially, normal cultured human fibroblasts were shown to possess cell surface receptors that facilitate the uptake of LDL (1,2). This system now known as the LDL receptor pathway (3, 4, 5) has proved to be a powerful approach to delineating the sequence of reactions involved in the general process of endocytosis. Subsequently, the *in vivo* operation of this pathway was demonstrated in freshly isolated lymphocytes (6, 7), followed by investigations of the LDL receptor pathway in adrenal cortex (8). The presence of LDL receptors in hepatic tissues has been a longstanding issue, but it is now clear that the liver is the major site of receptor-mediated metabolism of LDL in animals (9–11) as well as man (12, 13). In the following sections, we shall outline how each of these systems has contributed to our knowledge about the LDL receptor. Then, we shall describe some findings in detail.

2.1.1 **Cultured human fibroblasts**

Uptake of LDL into fibroblasts occurs by a mechanism know as *receptor-mediated endocytosis*. In fact, the uptake process for LDL in fibroblasts was the first such system studied in detail which defined the distinct steps involved. To date, more than 25 specific receptors have been recognized to participate in receptor-mediated endocytosis for a variety of proteins. These receptors include transport systems for nutrient delivery to cells, such as LDL itself, the iron transport protein transferrin, and the vitamin B_{12} transport protein transcobalamin II. Non transport plasma proteins, including α_2-macroglobulin, asialoglycoproteins, immune complexes, and even lysosomal enzymes are taken up via receptor-mediated endocytosis as well. Moreover, certain toxins and viruses use this mechanism to invade cells, by binding opportunistically to receptors that normally function in the uptake of other substances. Receptors that mediate endocytosis of certain molecules, such as epidermal growth factor (EGF) and platelet-derived growth factor (PDGF), as well as classic polypeptide hormones (insulin, luteinizing hormone), also mediate their ligands' physiologic actions. It appears, however, that in these cases, cellular entry of the ligand frequently does not seem to be required for the action of the hormone or growth factor.

What are the steps in receptor-mediated endocytosis? We shall consider the sequence of events which leads to the uptake of LDL by human fibroblasts. Initially, LDL particles bind with high affinity to LDL receptors on the cell surface. These receptors spontaneously (i.e., even in the absence of LDL) and continuously move into specialized regions of the plasma membrane called coated pits (14, 15). Coated pits derive their name from their characteristic features: they constitute indented regions on the surface of the cell; and on the cytoplasmic side of the plasma membrane these pits are coated with an electron-dense fuzzy material. Pearse demonstrated that a single protein, clathrin, forms this cytoplasmic coat (16). Even though coated pits make up only 2% of the surface of a fibroblast cell, the kinetics of movement of LDL receptors into these pits are such that in the steady state, 50–80% of all receptors are localized there. The LDL, when bound to a receptor in a coated pit, then undergoes rapid internalization as follows: the coated pit invaginates deeper and deeper into the cytoplasm, finally pinching off into the cell's interior as a clathrin-coated vesicle. After shedding its coat the vesicle fuses with others to form so-called endosomes. The endosomal compartment gets acidified by ATP-driven proton pumps (17–19); under these conditions, the LDL parts from the receptor. Subsenquently, the LDL is delivered to lysosomes, where it is degraded, while the receptor cycles back to the cell surface, ready to bind and internalize new ligand molecules (20). One aspect of this process has been of great experimental value: whereas incubation of cells at 37 °C leads to lysosomal degradation of LDL only a few minutes after it has bound to the receptor, at 4 °C the process is retarded to such an extent that no invagination of coated pits and thus no internalization of LDL is observed. This allows accurate measurements of the number of cell surface receptors by simple assessment of the amount of LDL bound to cells after incubation at 4 °C (21). On the other hand, at 37 °C the cells' capacity to internalize and degrade surface-bound LDL can be determined by measuring cell-associated LDL, and degradation products in the medium. The exact experimental procedures for the measurement of binding, uptake and degradation of ^{125}I-labeled LDL by human fibroblasts are described in (22).

The LDL receptor pathway. This pathway, depicted in Figure 1, describes how cells, through receptor-mediated endocytosis of LDL, control their internal cholesterol concentration. The cholesterol derived from lysosomal hydrolysis of LDL cholesteryl esters mediates a complex array of feedback control mechanisms that

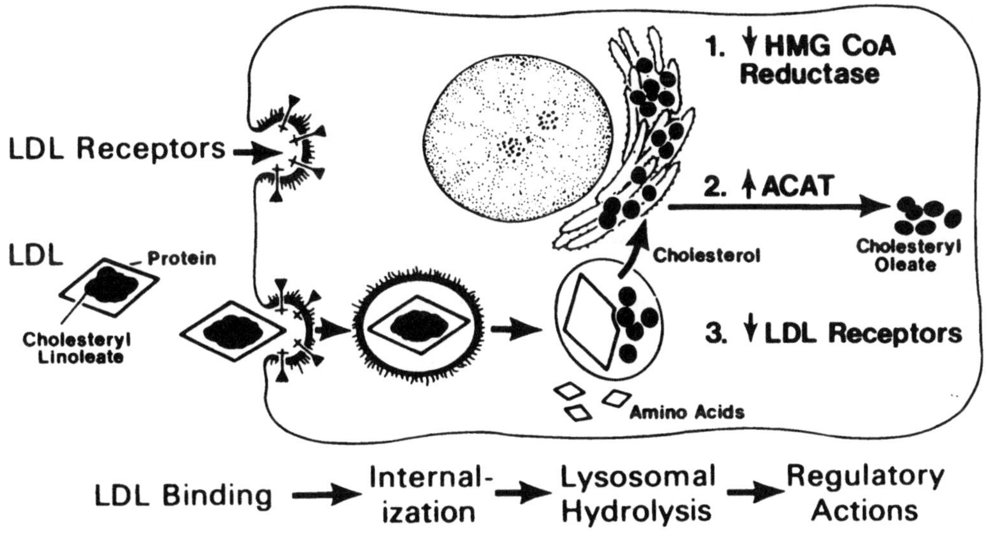

Fig. 1. Sequential steps in the LDL pathway in cultured mammalian cells. LDL denotes low density lipoprotein; HMG CoA reductase denotes 3-hydroxy-3-methylglutaryl CoA reductase; and ACAT denotes acyl-CoA:cholesterol acyltransferase

protects the cell from overaccumulation of cholesterol. First, the LDL-derived cholesterol suppresses the activity of the enzyme 3-hydroxy-3-methylglutaryl CoA reductase (HMG-CoA reductase), the rate-controlling enzyme in cholesterol biosynthesis, thereby turning off cholesterol synthesis in the cell (24). Second, the cholesterol activates a cholesterol-esterifying enzyme called acyl-CoA:cholesterol acyltransferase (ACAT) so that excess cholesterol can be stored as droplets of cholesteryl esters (25). Third, the cholesterol suppresses the synthesis of new LDL receptors, preventing further cellular entry of LDL which would overload the cell with cholesterol (26).

The overall effect of this regulatory system is to coordinate the intracellular and extracellular sources of cholesterol in order to maintain a constant level of cholesterol within the cell while the external supply in the form of lipoproteins undergoes fluctuation. Human fibroblasts and other mammalian cells are able to exist in the absence of lipoproteins because they can synthesize cholesterol from acetyl CoA. In contrast, when LDL is accessible, the cells primarily use the LDL receptor to acquire sterol and keep their own cholesterol synthesis suppressed (3–5).

Familial Hypercholesterolemia: Clinical consequences of the genetic disruption of the LDL receptor pathway. The delineation of the LDL receptor pathway in fibroblasts was greatly facilitated by studies of fibroblasts from a large number of patients with the typical phenotype of homozygous FH. In 1974, it was recognized that mutations affecting the function of the LDL receptor are responsible for FH (1, 2). Now, 13 years later, at least 11 different alleles at the LDL receptor locus have been identified in 104 fibroblast lines from patients with the clinical features of FH homozygotes. The results of these studies will be outlined in Section 2.4.

Here, we concentrate our discussion on the clinical findings in familial hypercholesterolemia. FH is characterized by three cardinal features: 1. a selective elevation in the plasma level of LDL; 2. deposition of cholesterol in abnormal sites, in particular in tendons (formation of xanthomas) and in arteries (forming atheromas); and 3. inheritance as an autosomal dominant trait with a gene dosage effect. This means that individuals who inherit two mutant alleles (FH homozygotes) are more severely affected than those with one mutant allele (FH heterozygotes) (23, 27). Heterozygotes occur at a frequency of about 1 in 500 persons, while homozygotes occur at a frequency of about 1 in one million persons among European and American populations. Through the delineation of the molecular defects in certain patients (see below), it is now clear that the severely affected offspring of a marriage between two heterozygotes can be either true homozygous or actually a heteroallelic genetic compound (23, 28–31). Thus, the term *homozygote* in most cases is not strictly correct in a genetic sense, yet is a convenient one and is generally applied to patients with two mutant alleles at the LDL receptor locus.

Familial hypercholesterolemia is the outstanding example of a single-gene mutation that regularly produces atherosclerosis. The deposition of LDL-cholesterol within the intima of major arteries in homozygotes becomes rapidly progressive, and leads to remarkably uniform clinical findings. Myocardial infarction, angina pectoris, and sudden death usually occur in homozygotes before the age of 15 (32).

The clinical features of heterozygous FH are more variable and less severe than those of the homozygous form. Heterozygous men have a 75% probability of experiencing a myocardial infarction before 60 years of age; in normal men, this risk is 15% (33, 34). Despite the expression of the same genetic abnormality and similarly elevated plasma LDL levels, heterozygous women suffer from coronary heart disease less often and at a later age than do heterozygous men. The risk for female heterozygotes to develop coronary artery disease by age 60 is about 45% as compared with 10% in unaffected females (33, 34).

Biochemical Characterization of FH in fibroblasts. The development of biochemical assays for each step in the LDL receptor pathway – i.e., LDL binding, uptake, and degradation as well as for the measurement of the activity of HMG-CoA reductase allowed for comparison of these parameters in normal and homozygous FH fibroblasts (1, 2, 4, 25, 35, 36).

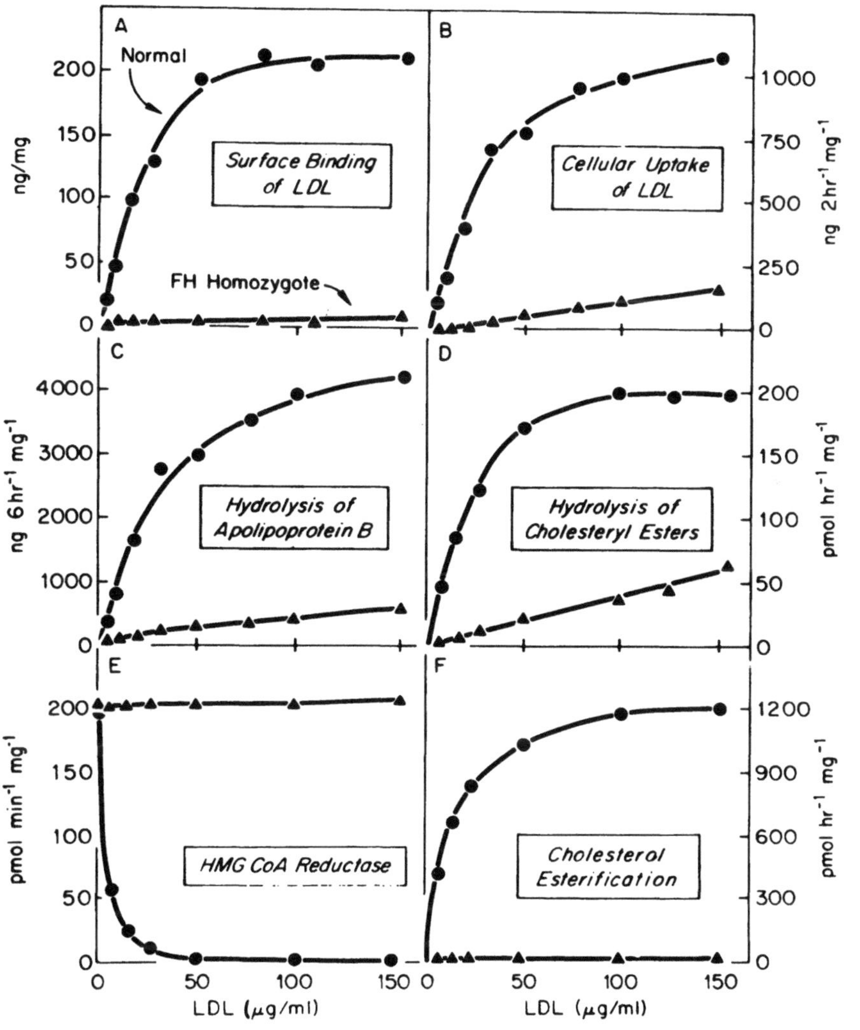

Fig. 2. Manifestations of low-density lipoprotein receptor actions in normal (●) and receptor-negative FH homozygote (▲) fibroblasts incubated with varying concentrations of LDL at 37 °C for 2 hours (A, B) or 6 hours (C–F). (From Ref. 3; © 1976 by the AAAS)

Figure 2 shows a typical set of data establishing the LDL receptor as the site of the defect in FH homozygote cells. In each panel, the effects of LDL are plotted as a function of its concentration in the culture medium. The first panel (A) shows the surface binding of ^{125}I-labeled LDL (^{125}I-LDL). In normal cells (●) the amount of radioactive LDL bound to the LDL receptor increases until all the receptor sites become occupied. Panel B shows that the rate of cellular uptake of ^{125}I-LDL parallels the amount of surface binding. The middle two panels, C and D, show the rates of hydrolysis for the degradation of apolipoprotein B, the protein component of LDL, and for the liberation of unesterified cholesterol from LDL cholesteryl esters, respectively. In normal cells, these rates are proportional to the surface binding and cellular uptake. Panels E and F show the regulatory actions of LDL that are dependent on receptor-mediated uptake of the lipoprotein. In normal cells, the LDL derived cholesterol leads to suppression of HMG CoA reductase activity and stimulation of cholesterol esterification.

The FH homozygote cells behave in striking contrast to normal cells. No surface binding of radioactive LDL can be detected (Panel A, ▲). As a result, high affinity cellular uptake of LDL is absent, and lysosomal hydrolysis of the lipoprotein cannot occur at normal rates (Panel C, D). Because they are unable to take up LDL through the receptor mechanism, the FH homozygote cells cannot utilize the cholesterol of the lipoprotein. Consequently, the activity of HMG-CoA reductase remains high, and LDL is also unable to stimulate cholesterol esterification (Panel E, F).

Hence, normal fibroblasts satisfy their need for cholesterol by receptor-mediated utilization of LDL, but FH homozygote cells must synthesize their own cholesterol in order to grow.

2.1.2 *In vivo* demonstration of LDL receptors — the lymphocyte model

In order to correlate the clinical finding of elevated LDL levels in FH patients with the biochemical finding that cultured FH homozygote cells cannot remove LDL from the medium efficiently, it was essential to demonstrate operational LDL receptors *in vivo*. This was first achieved by direct analysis of LDL receptor activity in circulating blood mononuclear cells. The expression of LDL receptors in normal cells, but not in lymphocytes from FH homozygotes was demonstrated by the use of ^{125}I-LDL (6, 7, 37, 38). Furthermore, the LDL receptor defect in blood cells from FH homozygotes was also demonstrated by a visual approach. To this end, cholesteryl esters are removed from the core of LDL and replaced with fluorescent cholesteryl esters (39). Cells are then incubated with the fluorescent LDL, and cells that take up fluorescent LDL are visualized by fluorescence microscopy (40). With this technique, it was demonstrated that more than 90% of freshly isolated normal lymphocytes were able to take up LDL, whereas none of the fluorescent LDL was internalized by lymphocytes from an FH homozygote (40).

Thus, the characteristics of the LDL receptor pathway and its genetic disruption are identical in cultured fibroblasts and in freshly isolated body cells. This provides for a convenient way to rapidly diagnose FH at the functional level, since under controlled conditions mononuclear cells from FH heterozygotes display half-normal receptor activity (7).

2.1.3 LDL receptor activity in the membrane fraction of isolated tissues

Physiologic LDL receptors can also be demonstrated by the use of membrane binding assays. Human or animal organs are homogenized, membranes isolated, and their ability to bind ^{125}I-LDL in a high-affinity, saturable fashion is determined. This assay was developed for membranes prepared from human fibroblasts (41), has been used extensively in a number of laboratories, and is described in its most updated form in (42). The membrane binding assay faithfully reflects the properties of the physiologic LDL receptor of intact cells in that it shows specificity, high affinity, reduced activity after previous exposure of cells to cholesterol, and absence in membranes prepared from FH homozygote cells (41).

Using this assay, Kovanen et al. (43) demonstrated that of all tissues, the specific LDL receptor activity is highest in the adrenal cortex and the ovarian corpus luteum. In other studies, these authors have shown that cultured adrenocortical cells utilize the LDL receptor pathway to extract cholesterol from the medium for conversion to steroid hormones (44). In contrast, much lower ^{125}I-LDL binding activity was found in membranes from adrenal medulla and the ovarian interstitium, tissues which do not produce large amounts of steroids. A similar distribution of LDL receptors was found with the help of the membrane binding assay in tissues of a human fetus (8). Thus, there is a tight correlation between a tissue's output of steroids *in vivo* and the number of LDL receptors measured *in vitro*.

2.1.4 Hepatic receptors

The membrane binding assay has proven useful in measuring LDL receptor activity in tissues with considerable levels of receptor expression. However, these studies could not provide a clear answer as to whether the liver possesses functional LDL receptors. Such direct evidence for hepatic LDL receptors initially came from a series of studies conducted in rats treated with pharmacologic doses of 17 α-ethinyl estradiol (45–47). In rats, removal of LDL from the plasma is increased dramatically following estrogen treatment due to a marked rise in the number of lipoprotein binding sites on liver cell membranes. In addition, VLDL and chylomicron remnants bind to receptors on liver membranes from estrogen-treated rats. Thus, at first sight, it appears possible that the LDL receptor is responsible for hepatic clearance of all plasma lipoproteins containing apoprotein B. The situation, however, is not so simple: in FH homozygotes, who lack LDL receptors, there is no evidence of delayed clearance of chylomicron remnants or VLDL remnants from the plasma (23, 48). These observations and many others, details of which would go far beyond the scope of the present review, have led to the conclusion that there are at least two kinds of hepatic lipoprotein receptors (Fig. 3): the LDL receptor, which is the primary transport mechanism for endogenous cholesterol, and the remnant receptor, its counterpart for exogenous cholesterol transport (49, 50). Two thirds of the LDL particles are metabolized by LDL receptors in liver and extrahepatic cells, with the liver accounting for 75% of the total expression of LDL receptors *in vivo* (49, 51–53). LDL receptor levels are effectively regulated. Hepatic LDL receptors are suppressed whenever the liver content of cholesterol increases or its demand for cholesterol is reduced (54–56). Conversely, the number of LDL receptors increases

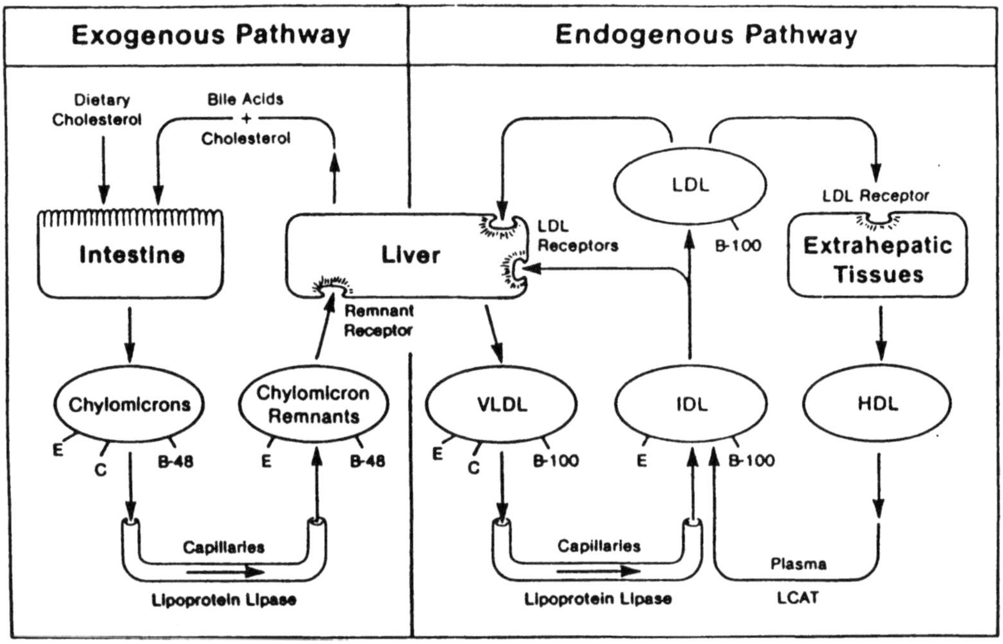

Fig. 3. Separate pathways for receptor-mediated metabolism of lipoproteins carrying endogenous and exogenous cholesterol. HDL denotes high-density lipoprotein, LCAT lecithin:cholesterol acyltransferase, LDL low density lipoprotein, IDL intermediate density lipoprotein, and VLDL very low density lipoprotein. The distinction between exogenous and endogenous cholesterol applies to the immediate source of the cholesterol in plasma lipoproteins. After the exogenous cholesterol has been delivered to the liver and has been secreted in VLDL, it is considered endogenous cholesterol. HDL is shown as the lipoprotein that removes cholesterol from extrahepatic cells. The cholesterol is delivered to IDL through the action of plasma LCAT and a cholesteryl ester-transfer protein. (From Ref. 50; reprinted with permission of The New England Journal of Medicine 309, 288 (1983).)

when hepatic cholesterol synthesis is blocked by the drugs compactin or mevinolin (57, 58), and when bile acid binding resins are given (59, 60). All of the changes in receptor activity alter the rate of uptake of LDL by the liver and cause reciprocal changes in the plasma LDL level.

In contrast, remnant receptors do not appear to be regulated. Metabolic manipulations that affect hepatic LDL receptors do not influence the hepatic uptake of chylomicron remnants, nor do they cause a significant change in the plasma level of these lipoproteins (54, 61). As mentioned above, in humans with the homozygous form of FH, LDL accumulates in plasma, but chylomicron remnants do not. Remnants can bind to LDL receptors *in vitro*, but functional and genetic evidence, as outlined very briefly here, indicates that *in vivo* these particles enter the liver by binding to the chylomicron remnant receptor.

From recent studies that will be summarized in Section 3, it appears that this receptor recognizes remnant particles by way of their constituent, apolipoprotein E (Apo E). Inasmuch as certain but not all Apo E-containing lipoproteins can also bind to the LDL receptor (53, 62, 63), the physiologic importance of hepatic binding proteins which are distinct from the LDL receptor but which have an affinity for Apo E has now become an issue that is receiving great attention. Some of the most recent results of this research will be described below (Section 3).

2.1.5 **The WHHL rabbit**

Much insight into the role of LDL receptors in the liver has been gained through the availability of an animal model for human FH. In 1973, Y. Watanabe identified a strain of rabbits with heritable hypercholesterolemia and hypertriglyceridemia, partially expressed in the heterozygous state and fully expressed in homozygotes (64). Aortic atherosclerosis develops in homozygous Watanabe Heritable Hyperlipidemic (WHHL) rabbits within the first few months of life. The primary defect in WHHL rabbits was identified as a failure to produce normal LDL receptors (65), resulting in elevated LDL levels very similar to the situation in human FH. Since rabbit LDL contains relatively high amounts of triglyceride, hypertriglyceridemia as well as hypercholesterolemia ensues. A deficiency of LDL receptors in WHHL rabbits has been shown in cultured fibroblasts, membranes of liver and adrenal glands (66), and cultured hepatocytes (67). From *in vivo* turnover studies with ^{125}I-labeled lipoproteins (68, 69) further evidence for the presence of two hepatic lipoprotein receptors (the LDL receptor and the chylomicron remnant receptor) was obtained. Chylomicrons and their remnants are cleared rapidly from the circulation, while VLDL, IDL, and LDL are removed at only about one third of the normal rate (68). Uptake of chylomicron remnants must therefore be mediated by a receptor distinct from the LDL receptor; the genetic defect in the WHHL rabbit affects only the LDL receptor. The mutant LDL receptors in WHHL rabbits are able to bind about 5% of the normal amount of LDL; their molecular defect has now been identified (70). Mutations giving rise to apparently the same receptor phenotype have been observed in families with FH and will be described in Section 2.4.

2.2 Structure of the normal human LDL receptor

The complete amino acid sequence of the human LDL receptor was derived from the nucleotide sequence of the full-length receptor cDNA. The strategy for the isolation of this cDNA is outlined in Ref. 71. The amino acid sequence, in combination with biochemical experiments (72), revealed that the mature receptor can be divided into five domains. A schematic model of the LDL receptor structure is shown in Figure 4. In the following, the distinguishing features of these domains will be highlighted.

First domain: ligand binding. The mature receptor (after cleavage of a 21-residue signal sequence) consists of 839 residues. The ligand binding domain (292 residues) is located at the NH_2-terminus of the receptor. These 292 amino acids, 42 of which are cysteines, exist in a tightly folded conformation held together by multiple disulfide bonds. Analysis of the amino acid sequence and exon/intron mapping

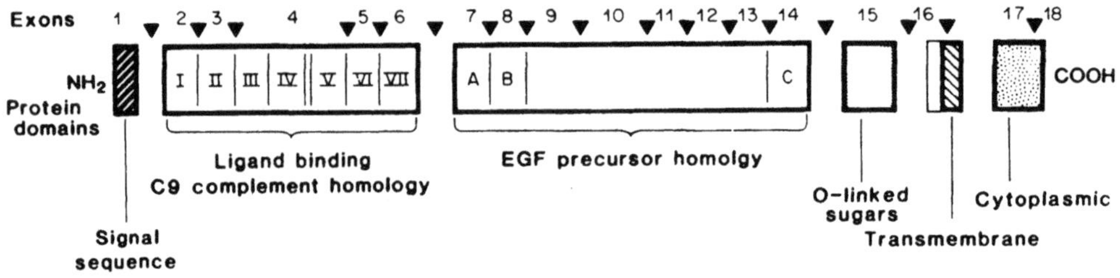

Fig. 4. Exon organization and protein domains in the human LDL receptor. The domains of the protein are delimited by thick black lines and are labeled in the lower portion of the figure. The 7 cysteine-rich, 40-amino acid repeats in the LDL binding domain are assigned Roman numerals I – VII. Repeats IV and V are separated by 8 amino acids. The 3 cysteine-rich repeats in the EGF precursor homology domain are lettered A – C. The positions at which introns interrupt the coding region are indicated by arrow heads. Exon numbers are shown between the arrow heads. (Reprinted from Ref. 73; © 1985 by the AAAS.)

Apo-E (Residues 140-150)	-His-Leu-Arg-Lys-Leu-Arg-Lys-Arg-Leu-Leu-Arg-
Apo-B (Residues 276-286)	-Thr-Thr-Arg-Leu-Thr-Arg-Lys-Arg-Gly-Leu-Lys-
Apo-B,E(LDL) Receptor (Consensus Sequence)	-Cys-<u>Asp</u>-X-X-X-<u>Asp</u>-Cys-X-Asp-Gly-Ser-<u>Asp</u>-<u>Glu</u>-

Fig. 5. Structural similarities between the receptor binding domain of Apo E and a region of Apo B. The homologous positively charged residues lysine and arginine are boxed in the sequence of Apo E and Apo B. The consensus sequence of the postulated ligand binding site of the Apo B,E (LDL) receptor, which is enriched in the negatively charged residues aspartic and glutamic acids (underlined), is shown for comparison. (From Ref. 75; © 1985 by the AAAS.)

data (73) suggest that the first domain is made up of seven repeats, each consisting of ≈ 40 amino acids. Within these 40 residues, 6 cysteines are spaced 4 – 7 amino acids apart and are located exactly in register for all repeats. At the COOH-terminus of each repeat unit, there is a cluster of negatively-charged amino acids. These sequences (Fig. 5) are complementary to positively charged sequences in the receptor's ligands, Apo E (74) and Apo B (75). It has been speculated (75, 76) that the negatively charged clusters of amino acids within the cysteine-rich repeat sequences constitute binding sites for Apo E and Apo B. Studies by Mahley and associates (53, 77, 78) and others (79, 80) strongly support this notion. In addition, the disulfide bridges may be important in stabilizing the binding site structure, in particular when the receptor encounters the acidic environment of the endosome during recycling (72).

Second domain: EGF precursor homology. This region, consisting of about 400 residues, is homologous to a portion of the extracellular domain of the precursor to epidermal growth factor (EGF) (71, 72). The growth factor itself is a peptide of 53 amino acids, which is derived by proteolytic cleavage from a large, membrane-bound precursor of 1217 residues (81, 82). The portion of the precursor facing the cytoplasm contains multiple homologous but not identical repeats of the EGF sequence, as well as unrelated spacer sequences. The EGF precursor homology region of the LDL receptor is flanked by 3 cysteine-rich repeats (A, B, and C in Fig. 4), which are highly homologous to four repeat sequences in the EGF precursor (73, 81, 83). Repeats A, B, and C of the LDL receptor are also homologous to certain proteins of the blood clotting system, such as Factor IX, Factor X, and protein C (73, 83).

All observations regarding the significance of the existence of these homologies seem to suggest that the homologous regions arose by duplication of an ancestral gene. These genes might have differentiated further by, e. g., acquisition of exons from other genes providing specific functions, such as nutrient delivery (LDL) or signalling of cell proliferation through the secretion of a peptide (EGF).

Third domain: O-linked sugars. There is a sequence of 58 amino acid residues that contains 18 serines or threonines immediately external to the plasma membrane (71). This region contains a large number of carbohydrate chains attached to the hydroxyl groups of Ser and Thr (72, 84). The elongation of these sugar chains during posttranslational processing of the receptor brings about the dramatic increase in apparent molecular weight that is observed in biosynthetic studies of radiolabeled receptor (see Section 2.3). Carbohydrate chains in O-linkage are also present on other parts of the receptor; these chains do not appear to be clustered as they are in the O-linked sugar domain (85). The role of O-linked carbohydrate in the function of the LDL receptor, if any, is unknown.

Fourth domain: membrane-spanning region. Twenty-two uncharged and hydrophobic amino acids constitute the only membrane-spanning portion of the receptor (72). There is a single intramembranous cysteine residue in the human receptor, which is replaced by an alanine in the bovine receptor (7). Since the two receptors function similarly, it appears likely that this cysteine residue is not involved in a disulfide bridge, but rather exists in a reduced state.

Fifth domain: cytoplasmic region. At the COOH-terminus of the LDL receptor, 50 amino acid residues project into the cytoplasm. The intracellular location of this tail region was demonstrated in proteolysis experiments by the use of an anti-peptide antibody directed against the COOH-terminus sequence (72). As outlined below (Section 2.4), the cytoplasmic region is crucial for clustering of LDL receptors in coated pits, a prerequisite for subsequent internalization of receptor/LDL complexes. Localization to coated pits may occur through interaction of the cytoplasmic tail with clathrin itself or with some clathrin-associated protein on the intracellular side of the plasma membrane. In search of a mediator site on the LDL receptor, the intracellular regions of other coated pit receptors have been investigated for common features, but no obvious structural element has been identified (76). Thus, while the function of the cytoplasmic tail of the LDL receptor is established, the mechanism of its action has yet to be elucidated.

2.3 The biosynthetic pathway of LDL receptors in normal human fibroblasts

The biosynthesis of LDL receptors can be conveniently studied by metabolic labeling of cultured cells with ^{35}S-methionine, followed by immunoprecipitation of receptor protein with a monoclonal anti-receptor antibody (86). Experimental procedures are detailed in Refs. 30 and 42. The normal human LDL receptor is initially synthesized as a precursor with an apparent M_r of 120000, as estimated by SDS polyacrylamide gel electrophoresis (SDS-PAGE). It contains asparagine-linked (N-linked) high mannose oligosaccharide chains and *N*-acetylgalactosamine (GalNAc) residues attached to serines and threonines by O-linkage (84). Between 30 and 60 min after synthesis, the LDL receptor precursor undergoes a sudden shift in apparent M_r from 120000 to 160000 in SDS-PAGE (30). This change in apparent M_r is due to the elongation of the O-linked chains (84) which temporally coincides with the maturation of the N-linked carbohydrate moiety (84). The calculated molecular mass of the protein portion of the receptor is 93102, and including the mature carbohydrate its M_r is $\approx$ 115000, yet those O-linked carbohydrate chains forming a cluster in a Ser/Thr rich region of the receptor (72) cause it to migrate to the position of a 160 kDa-protein in SDS-PAGE (84).

In other studies, it was shown that the 120 kDa precursor is localized intracellularly (31), and that the addition of N-linked carbohydrate chains is required neither for transport to the surface nor for the receptor's ability to bind LDL and to recycle. The role of O-linked carbohydrate in the receptor's cellular itinerary is now under investigation. A deletion of the clustered O-linked carbohydrates does not appear to impair the function of the LDL receptor (85). However, there is a small number of isolated chains of O-linked carbohydrates in addition to the clustered chains, and their significance for receptor function needs to be delineated (85).

2.4 Molecular defects in LDL receptors of patients with familial hypercholesterolemia

At least 11 different mutant alleles at the LDL receptor locus have been described (31, 32, 70, 87). Fibroblasts from 104 FH homozygotes were analyzed in terms of LDL receptor structure and function; these studies revealed that the mutations could be divided into four classes (for review, see Ref. 76). The mutant alleles are summarized in Table 1 and described in the following.

Class 1 mutations: No detectable precursor. These alleles, designated *R-O*, fail to express receptor proteins as measured by functional assays (binding of ^{125}I-LDL) or attempts to immunologically identify proteins with a variety of monoclonal and polyclonal antibodies directed against the LDL receptor. *R-O* alleles are the most frequent of the mutant alleles and probably account for about 40% of all mutations at the LDL receptor locus. This class most likely includes nonsense mutations that introduce termination codons early in the protein coding region, but point mutations in the promoter that block transcription, mutations in intron/exon junctions that lead to abnormal splicing of the mRNA, and large deletions may also be included in this class of mutant alleles. Examples are described in Refs. 30 and 31.

Table 1. Mutations at the LDL receptor locus that produce familial hypercholeresterolemia (FH)

Class of mutation	Allele designation[1]	Apparent receptor mass on SDS gels (kDa)		Receptor location				LDL binding to intact cells	Frequency in FH patients
				Intra-cellular	Plasma membrane		Extra-cellular		
		Pre-cursor	Mature		Coated pits	Non-coated regions			
Class 1: No detectable precursor	R-0	None	None					None	Common
Class 2: Precursor not processed	R-100	100	100	+				None	Rare; found in Lebanese
	R-120	120	120	+				None	Common
	R-135	135	135	+				None	Rare
Class 2 variant: Precursor processed slowly, mature receptor binds LDL poorly	R-120 $\xrightarrow{\text{slow}}$ 160 b^-	120	160	+	$(+)^2$			Reduced	Rare; found in Afrikaners and in WHHL rabbits
Class 3: Precursor processed normally, mature receptor binds LDL poorly	R-140 b^-	100	140		+			Reduced	Rare
	R-160 b^-	120	160		+			Reduced	Common
	R-210 b^-	170	210		+			Reduced	Rare
Class 4: Precursor processed normally, mature receptor binds LDL normally, but does not enter coated pits	R-150 i^-, sec	110	150			(+)	+	Normal binding; defective internalization	Rare
	R-160 i^-	120	160			+		Normal binding; defective internalization	Rare
	R-155 i^-	115	155			+			

[1] Allele designations are based on the apparent molecular mass (in kilodaltons) on SDS polyacrylamide gel electrophoresis of the mature form of the receptor, i.e. the predominant form observed after a 2h pulse followed by a 2h chase. b^- denotes defective LDL binding; i^- denotes defective LDL binding; sec denotes secretion from the cell.

[2] Symbols in parentheses refer to minor populations.

Class 2 mutations: precursor not processed. These alleles specify receptor precursors that are produced in normal or reduced amounts, have normal (120 kDa) or abnormal (100 kDa, 135 kDa) molecular mass, but do not undergo any apparent increase in molecular weight after synthesis. Receptors encoded by these alleles never reach the cell surface (31).

Most of the mutant receptor precursors in this class have an apparent M_r of 120 000 (allele designation, R-120) but precursors with apparent molecular weights of 100 000 (R-100 allele) and 135 000 (R-135 allele) have also been identified as members of this class (31). These M_r abnormalities are likely due to alterations in the length of the protein chain and do not result from changes in the carbohydrate moieties (31).

Variants of the class 2 mutations were observed in a black American family, in Afrikaners, and in WHHL rabbits (70). In these variants the receptor is produced as a precursor of apparently normal M_r that is processed to the mature form, but much slower than the normal precursor. Eventually, about 5 – 10% of the receptors reach the cell surface as 160 kDa mature proteins. Once on the cell surface, these receptors display a reduced ability to bind LDL; that means the mutation has disrupted both transport and binding capacity of the receptor (70). Very recently, a mutation in the WHHL rabbit receptor gene has been identified that causes the deletion of four amino acids in the LDL binding domain (T. Yamamoto, M. S. Brown, J. L. Goldstein et al., unpublished observations). While this might explain the inability of the mutant receptors to bind LDL normally, the reason for their delayed processing is less obvious. In any event, expression of the mutant gene in transformed cells will provide the answer as to whether this mutation is sufficient to bring about the WHHL phenotype.

The molecular basis of other defects in the class 2 mutations and their variants are not known. It seems likely that the failure of transport arises from some subtle changes in structure, similar to the one occurring in the WHHL rabbit.

Class 3 mutations: precursor processed, abnormal LDL binding. Receptors in this class of mutants reach the cell surface at normal rates, but are able to bind less than 15% of the normal amount of LDL (31, 32, 86). Most of class 3 receptors have normal M_r in SDS-PAGE (allele designation, R-160 b^-). Receptors with apparent M_r of 140 000 (R-140 b^-) (31) and 210 000 (R-210 b^-) (30) have been described as well. All mature receptor proteins in this class originate as precursors with apparent molecular weights that are 40 000 less, i.e., 120 000, 100 000, and 170 000, respectively. It appears that the abnormal molecular weights are due to alterations in the amino acid sequence and not to changes in carbohydrate content (84).

The structure of the LDL binding domain – 7 repeats of a 40 amino acid sequence – suggests an explanation for the abnormally sized, binding deficient receptors in class 3. The DNA encoding the repeat structure might be susceptible to deletion or duplication following mispairing and recombination of homologous regions during meiosis. Such deletions or duplications would change size and binding capacity of the receptors.

Class 4 mutations: Precursor processed, receptor binds LDL but does not cluster in coated pits. These mutations have been termed internalization-defective (88). From the description of the process of receptor-mediated endocytosis (see Section

2.1.1) it is apparent that clustering of receptors in coated pits is a prerequisite for subsequent internalization of ligand; clustering of LDL receptors is disrupted in class 4 mutations. To date, the molecular defects in five patients with this phenotype have been elucidated. The original example was patient J. D. (88). While he inherited a R-0 (class 1) allele from his mother, he inherited from his father a gene that produces a cell surface receptor of normal size and normal LDL binding capacity, but which is unable to carry the bound LDL into the cell (R-160 i$^-$ allele) (89). Thus, J. D. is a compound heterozygote with regard to his genotype at the LDL receptor locus.

It was suspected early that the R-160 i$^-$ allele is producing a receptor altered in its intracellular portion of 50 amino acid residues, because this region of the receptor is believed to mediate the clustering of receptors into coated pits through interaction with cytoplasmic coated-pit specific proteins (89). The molecular defect, indeed, was found to reside in the intracellular domain (90). A point mutation has converted the codon for tyrosine 807 into one for cysteine. Subsequently, four other patients with internalization defects have been identified and their genes analyzed. In one of them, the internalization-defective allele produces a receptor that lacks the intracellular domain and the membrane-spanning region, and the truncated protein is secreted from these cells (91). The secreted receptor has an apparent M_r of 150 000 (allele designation, R-150 i$^-$, sec). Another is a patient from Japan (92), the offspring of consanguineous parents, apparently homozygous for an R-150 i$^-$, sec allele. Two more internalization-defective alleles were identified and the mutant genes analyzed (87). In one (FH patients 682 and 683), the cytoplasmic domains consists of only two residues due to a point mutation that changes a tryptophan codon into a termination codon (87). The other mutant gene (found in patient FH 763) contains a four-base duplication, producing a shift in the reading frame. The cytoplasmic tail of this mutant receptor is made up of six of the normal plus eight additional amino acids instead of the normal 50 (87).

Taken together, the analysis of class 4 mutations at the LDL receptor locus strongly suggests that the signal for movement of the receptor into coated pits resides in its cytoplasmic domain.

In Section 2, we have summarized current knowledge of the LDL receptor at the structural and functional level. We have also discussed the mutations in the structural gene for the receptor and their functional consequences in FH. Readers interested in details of the structure of the mRNA and the gene for the LDL receptor are referred to recent papers by Brown, Goldstein and associates (73, 76 and 93).

We will now turn to other, as yet less well characterized lipoprotein receptor and binding site systems.

3 The Chylomicron Remnant Receptor

As mentioned above (Section 2.1.4), liver membranes possess at least two distinct lipoprotein receptors: the LDL receptor which binds Apo B and Apo E containing lipoproteins, and the remnant receptor which binds lipoprotein containing Apo E.

but not the Apo B containing LDL. Recently it has been shown that chylomicron remnants devoid of Apo E are removed by the rat liver inefficiently (94). Furthermore, receptors on dog liver membranes do not bind Apo E-deficient chylomicron remnants (95). Thus, the hepatic chylomicron remnant receptor appears to be specific for the E apolipoprotein. An Apo E-specific binding site has been identified recently in liver membranes of dogs, humans and swine (54, 96). This specific site, termed the Apo E receptor by Mahley and associates (54), may represent the functional receptor for chylomicron remnants. Since the emphasis in this review is on the physiologic aspect of lipoprotein receptors, this receptor will be referred to as the chylomicron remnant receptor.

Contrary to the LDL receptor, the chylomicron remnant receptor is resistant to metabolic regulation, as described in Section 2.1.4. This makes it very difficult to correlate biochemical findings, such as the characterization of an Apo E binding protein, with the *in vivo* role of such protein. Nevertheless, efforts are underway to isolate distinct Apo E-binding proteins from hepatic tissues and to demonstrate their receptor properties.

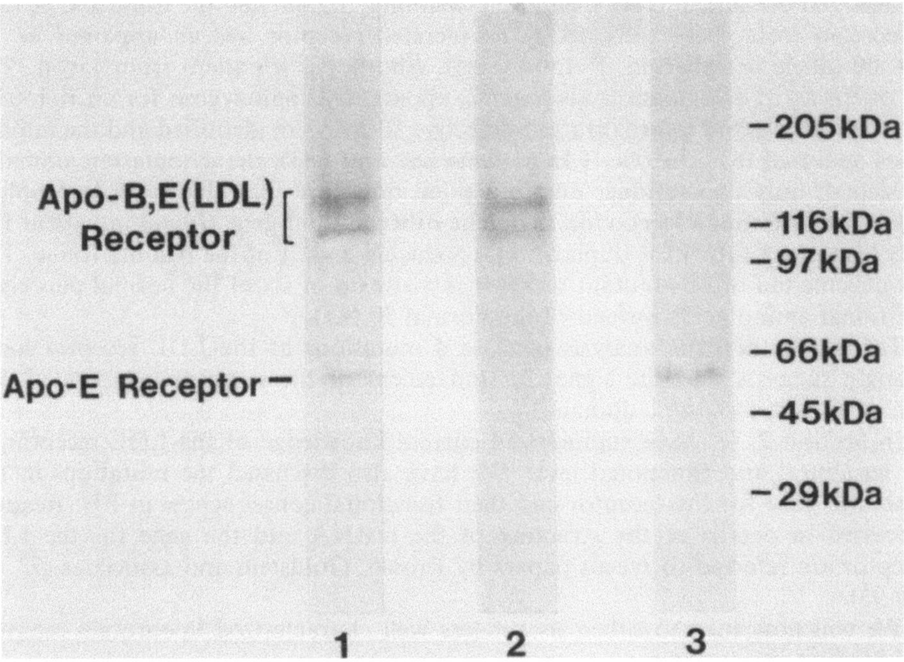

Fig. 6. Immunoblots of Apo B,E (LDL) and Apo E receptors of canine liver membranes. 300 µg of protein from DEAE-purified fractions (lane 1) or LDL-Sepharose-bound (lane 2) and -unbound fractions (lane 3) were applied to 7.5% SDS-polyacrylamide gels. The proteins were resuspended in buffer containing 62.5 mmol/l Tris-HCl, pH 6.8, 3% SDS, and 10% glycerol (without reducing agent) prior to application to the gel. The proteins were transferred to nitrocellulose paper and were incubated with 50 µg/ml of anti-Apo B,E (LDL) receptor IgG. Immunoreactive proteins were identified using an anti-rabbit ^{125}I-IgG. The position of migration of the LDL receptor and of the Apo E receptor are indicated. (From Ref. 97, with permission.)

In one study (97), a protein that binds Apo E, but not Apo B, has been purified from canine and human liver (Fig. 6). Liver membranes were subjected to extraction with the zwitterionic detergent 3-[(3-cholamidopropyl)dimethylammonio]propane sulfate, and soluble receptors partially purified by DEAE-cellulose chromatography. The partially purified preparation contained both LDL receptors and Apo E receptors which could be separated by removing the LDL receptors on an LDL-affinity column. The unbound material contained the Apo E receptor, which was purified by affinity chromatography on a column containing Apo E-HDL$_c$, a lipoprotein in which Apo E is the only protein component and which can be isolated from the plasma of hypercholesterolemic dogs (98). The Apo E receptor was found to be distinct from the LDL receptor: it had an apparent molecular weight of 56000, its migration of SDS polyacrylamide gel was identical under reducing and nonreducing conditions, and it did not bind LDL. Interestingly, antibodies prepared against the LDL receptor cross-reacted with the Apo E receptor, but not vice versa. The authors speculate that this is due to similarities in the apoprotein binding regions of the two proteins.

The isolation of an hepatic protein that specifically binds Apo E containing lipoproteins and the availability of specific antibodies should lead to the identification of its physiologic role. Future studies will demonstrate whether the 56 kDa protein is indeed the chylomicron remnant receptor. The same holds for a 36 kDa protein identified in rat liver membranes by ligand blotting with ^{125}I-Apo E-HDL$_c$ (99). Furthermore, the relation of this protein to the receptor described by Hui et al. (97), who found the 56 kDa protein in rat liver as well, remains to be elucidated.

4 Binding Sites for Lipoproteins that Cause Cholesterol Accumulation in Macrophages *In Vitro*

The macrophage has become an intense area of research since the discovery in 1980 (100) that mouse peritoneal macrophages possess a surface binding site that mediates the uptake and lysosomal degradation of rabbit β-migrating VLDL (β-VLDL). The first lipoprotein demonstrated to enter macrophages by receptor-mediated endocytosis had been *in vitro* acetylated human LDL (acetyl-LDL) one year earlier (101). The experimental systems for cholesterol-loading of macrophages based on these findings have provided a biochemical foundation on which to develop models for the *in vivo* formation of foam cells. Yet, after seven years we must still find the physiologic role, if any, for these receptor systems. In the following, only the most recent advances will be summarized. For an historical overview, the reader is referred to Ref. 102.

Receptor responsible for βVLDL binding to macrophages. It appears now that at least in mouse peritoneal macrophages, the β-VLDL receptor displays all the properties of the classical LDL receptor except that it has lost its high affinity for lipoproteins containing exclusively Apo B (LDL) and is resistant to regulation by intracellular cholesterol. Whatever the nature of this receptor, it most likely binds β-VLDL predominantly due to its Apo E content (103). This is supported by the finding that Apo E apparently is the ligand mediating the recognition of normal VLDL by receptors on human monocyte-macrophages (104). Currently, studies with macrophages from WHHL rabbits and humans with homozygous FH (105, 106) are aimed at establishing the genetic basis for the LDL/β-VLDL receptor diversity and biochemical characterization of the β-VLDL receptor protein.

Binding sites for acetylated LDL. In contrast to β-VLDL, acetyl-LDL has not been detected *in vivo*. However, high affinity binding sites for human acetyl-LDL have been implicated as "scavenger receptors" for LDL and physiologically modified LDL, and as a mechanism to defend against the accumulation of such altered lipoproteins in the plasma. This binding site has been demonstrated in rat hepatic membranes (107) and mouse peritoneal macrophages (101). In these tissues, this binding site appears to be represented by a protein of ≈ 250 kDa which displays all of the characteristics for binding demonstrated on mouse macrophages (107). In the latter report, the hepatic uptake of radiolabeled acetyl-LDL was shown to be sensitive to the same inhibitors as the *in vitro* binding. However, a direct demonstration that the 250 kDa-protein is involved in the scavenging of chemically modified LDL has not been achieved.

To summarize, while there is an intensive ongoing search for a role of receptors in the development of the atherosclerotic plaque *per se*, convincing evidence for such involvement is still lacking.

In the following section, we will consider the metabolism of HDL with emphasis on the possible involvement of specific receptors in HDL-cell surface interactions.

5 Receptors for High Density Lipoproteins

A great deal of interest currently centers on the metabolism of HDL, because a high negative correlation of serum HDL levels and the risk of coronary heart disease (CHD) has been demonstrated (108). Receptor-mediated cellular uptake of HDL has been proposed and very intensively investigated in the last few years. The interaction of HDL with cells is not only the first step in the catabolism of these dense lipoproteins, but HDL might also be responsible for reverse cholesterol transport from peripheral tissues to the liver (109). Both functions are important in the human organism and therefore the interactions of HDL with different tissues should be discussed separately.

The high density lipoprotein fraction is not homogenous, but can be divided into the subclasses HDL_1, HDL_2 and HDL_3 (Chapters 1 and 2 and Ref. 110). Depending on the apolipoprotein composition of the isolated HDL fraction, the metabolism of the particles might be different. For example, if Apo E is present, as in HDL_1 or HDL_c, the particles will be recognized by the Apo B, E-receptor (77) and the E-receptor (97) in the liver and subsequently catabolized. If, however, the HDL preparation is devoid of Apo E, the mainly Apo AI and AII-containing particles appear to have another specific recognition site on the cell surface. The

Table 2. Interaction of HDL with Four Groups of Tissues

Tissue	Species[1]	Cells	System	Ligand	Proposed mechanism	Ref.
Steroidogenic tissues	m	adrenal	in vivo	h-HDL (E-free)		(111)
	r	adrenal + gonads	in vivo	r-HDL	CE uptake only	(112)
	r	testes	membranes	h-HDL	Apoprot. + lipid	(142)
	r	adrenal cells	TC	h-HDL_3(E-free)	AI mediated CE uptake	(113–115)
Tissues involved in CHD	m	Macrophages	TC	c-HDL_c/β-VLDL[2]	–	(128)
	m	Macrophages	TC	HDL/acLDL	–	(126)
	m	Macrophages	TC	h-HDL	AI mediated	(129, 130)
	r	SMC	TC	r-HDL/h-HDL	–	(117)
	rab	SMC	TC	h-HDL_3	Lipid interact.	(124)
	p	EC	TC	h-HDL_3	Lipid interact.	(124)
	h	SMC	TC	h-HDL, VHDL	AI mediated	(125)
	h	EC	TC	h-HDL (E-free)	AI mediated	(118–120)
Liver	r	Hepatocytes	TC	r-HDL	AI mediated	(134)
	r	Hepatocytes	Membranes	h-HDL_3	AI mediated	(123)
	rab	Hepatocytes	TC	rab.-HDL_2	–	(135)
	p	Hepatocytes	Membranes	p-HDL (E-free)	–	(135)
	p	Hepatocytes	Membranes	p-HDL (E-free)	–	(136)
	h	Hepatocytes	Membranes	h-HDL (E-free)	AI and AII med.	(122)
	h	Hepatoma	TC	h-HDL	D and CI med.	(137)
Other cell lines	r	Intest. Mucosa	TC	h-HDL_3	AI mediated	(140)
	r	Kidney	in vivo	r-HDL	–	(112)
	r	Kidney	Membranes	r-HDL	Apo C-med.	(138)
	h	Fibroblasts	TC	h-HDL_3 (E-free)	AI mediated	(113)
	h	Fibroblasts	TC	h-HDL (E-free)	AI and AII med.	(118)
	h	Fibroblasts	TC	h-HDL (E-free)	AI and AII med.	(122)
	h	Fibroblasts	TC	h-HDL_3	Lipid interact.	(124)

TC = tissue culture; SMC = smooth muscle cells; EC = endothelial cells

[1] h = human; r = rat; p = porcine; rab = rabbit; c = canine; m = mouse

[2] from cholesterol-fed rabbits

functions of this binding site or receptor will be considered here. Of interest are interactions with:

1. Steroidogenic tissues
2. Peripheral tissues
3. Liver cells
4. Other cell types

In the following, the role and fate of HDL in these different settings are discussed.

5.1 HDL interaction with steroidogenic tissues

In mammals most of the body cells, including the cells in steroidogenic tissues (i.e. adrenal gland and gonads) possess LDL receptor activity for supplying them with cholesterol. In addition, these cells have the capacity for cholesterol biosynthesis. Studies employing both *in vivo* and *in vitro* models with mouse adrenal tissues showed that adrenal cells can specifically recognize HDL and use its cholesterol for their metabolism (111). In comparison with other tissues, adrenal glands and gonads have the highest HDL uptake on a unit per weight basis (112). The mechanism of this uptake is not fully understood.

In a series of papers (113–115) Fidge et al. described a surface receptor for HDL_3 (Apo E-free) on rat adrenal cortical cells which mediates HDL_3 uptake via receptosomes. This endosomal compartment does not fuse with lysosomes. The HDL_3 thus are not degraded, but supply cholesterol to the cells. This work provided evidence that Apo AI (and Apo AII?) are the major ligands involved in the recognition of HDL by cell surface components.

Glass et al. (112) have shown the preferential uptake by rat adrenal and ovary cells of cholesteryl esters from HDL_3 bound via Apo AI. In *in vivo* experiments rat adrenal and ovary cells accumulate cholesteryl esters independently of Apo AI uptake. Uptake of cholesteryl ester was 7-fold greater in the ovary cells.

To summarize the data, it is apparent that adrenal cells do have a binding site for HDL. This site most probably interacts with Apo AI. Most of the bound lipoprotein is not degraded, yet supplies cholesterol for cellular metabolism. The molecular mechanism by which HDL donates the cholesterol to the cells remains to be studied.

5.2 HDL interaction with peripheral cells

Peripheral body cells, i.e. fibroblasts, macrophages, endothelial cells, muscle cells, and others have been shown to interact with HDL and to be involved in the process termed "reverse cholesterol transport". Several different cell culture systems, including cell lines, have been studied in regard to this process. Chemical treatments that decrease HDL binding have been described. In 1975 Miller and Miller (109) suggested that HDL facilitates the uptake and transport of cholesterol from peripheral tissues to the liver for catabolism and excretion. More recently, Miller et al. demonstrated results from *in vivo* studies in rabbits which again suggested that HDL is involved in the reverse transport of cholesterol (116). Reverse cholesterol transport is an important concept in the study of coronary heart disease. The peripheral cells of interest to initiate these studies are smooth muscle cells (SMC), endothelial cells (EC) and macrophages, which are all involved in the formation of atherosclerotic plaques.

To date, studies on reverse cholesterol transport have been performed in tissue culture with several cell systems. The release of cholesterol from cells was measured in the presence and absence of various protein and phospholipid acceptors (117).

The most extensive studies with human SMC and other peripheral cells have been reported by Oram and collaborators (118 – 120). They showed Apo AI dependent binding of HDL without subsequent internalization and degradation. This specific, high affinity binding (Fig. 7) can be up-regulated in response to cholesterol loading of the cell. Thus, following treatment with acetylated LDL, an increase of intracellular cholesterol is seen which results in an increase of HDL binding to the cell. Kinetic analysis of these data propose an increase in the number of HDL receptors (119, 120).

The treatment of HDL with tetranitromethane (TNM) significantly reduces the binding of HDL to SMC, fibroblasts and liver cells (121 – 123). TNM alters the tyrosine residues of proteins by nitration. The effects of this apoprotein modification support the results of competition studies, which favor Apo AI as the possible ligand. The suggestion (124) that cell surface lipids are responsible for the binding could not be explained by these results.

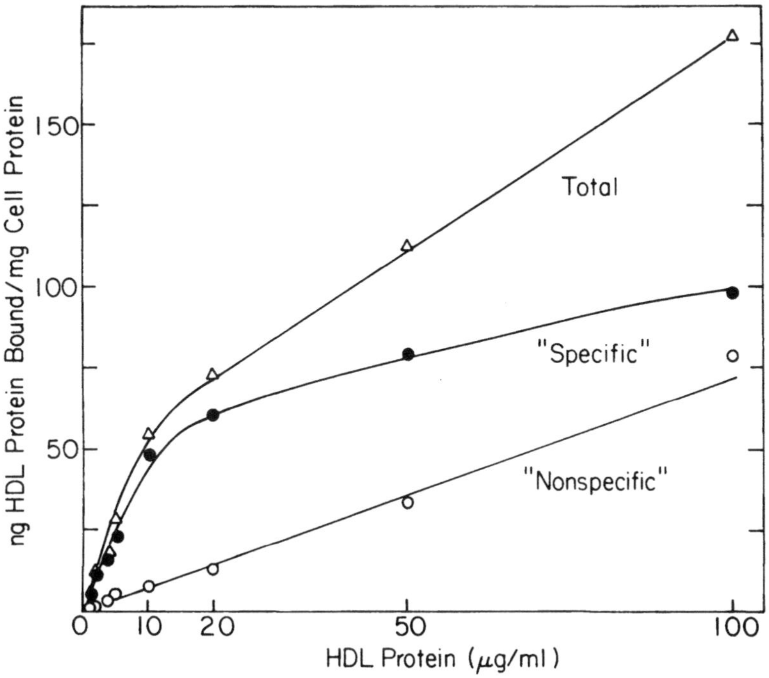

Fig. 7. Apo E-free ^{125}I-HDL binding to human arterial smooth muscle cells. Nonspecific binding was determined from incubation at each concentration of Apo E-free ^{125}I-HDL, in the presence of 1 mg/ml unlabeled Apo E-free HDL. Specific binding was calculated by subtraction of nonspecific from total binding. (From Ref. 118, by copyright permission of the American Society for Clinical Investigation.)

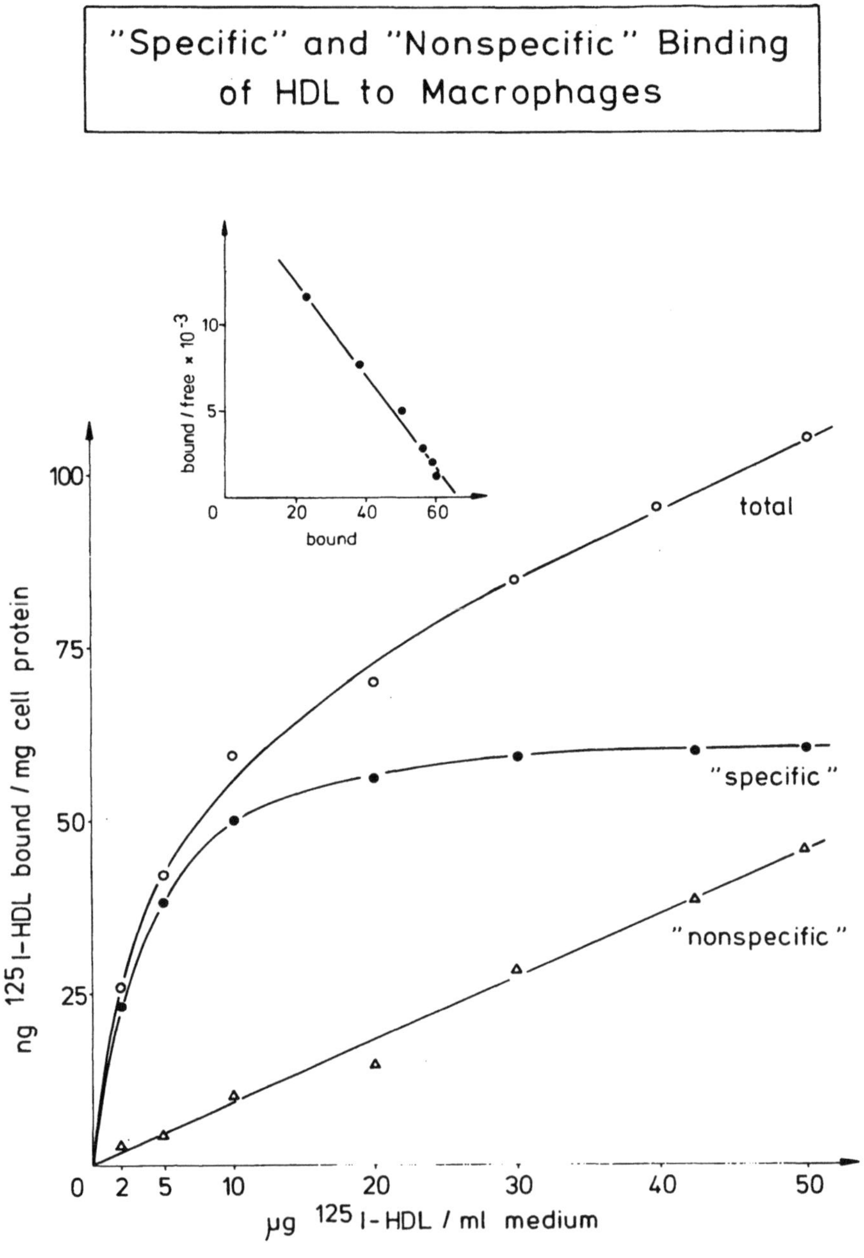

Fig. 8. The effects of excess unlabeled HDL_3 on the binding of $[^{125}I]HDL_3$ to cultured peritoneal macrophages. Macrophages were incubated at 4°C with increasing concentrations of $[^{125}]HDL_3$ plus (▲) or minus (○) at least 20-fold excess unlabeled HDL_3. Specific binding (●) was calculated by subtraction of the values for binding in the presence of excess unlabeled HDL_3 (non-specific binding) from the values for total binding. (From Ref. 130 with permission.)

Human endothelial cells bind Apo E-free HDL with high affinity in a similar way to SMC (125).

Macrophages have been considered to play an important role as cholesterol donors in reverse cholesterol transport. In 1980, Ho et al. (126) reported, following a series of studies by Werb and Cohn (127), that cholesterol excretion by macrophages does require HDL as effective acceptors in the culture medium. The same effect was described by Innerarity et al. (128). An induced massive accumulation of cholesteryl esters in mouse peritoneal macrophages resulted in the demonstration that HDL can remove cholesterol from these cells.

The most intensive studies with macrophages were carried out in the last two years by Schmitz et al. (129, 130). They analyzed the mechanism of reverse cholesterol transport out of macrophages. Unlike the results of the SMC experiments (118), Schmitz describes binding (Fig. 8), internalization, and intracellular transport of the HDL in endosomes. The endosomes do not fuse with lysosomes, but interact with intracellular lipid droplets. The internalized HDL are then resecreted. This implies that cholesterol uptake from macrophages is mediated by HDL via a receptor-dependent, so-called retroendocytotic pathway.

Once again, these studies point to Apo AI as the ligand responsible for specific cell surface binding of HDL.

During the regulation of cellular cholesterol metabolism it was shown (130) that the inhibition of ACAT enhances the number of HDL receptors and thereby promotes cholesterol efflux from macrophages. Inhibition of ACAT leads to an intracellular increase of free cholesterol which in turn becomes incorporated into HDL.

Very recently, Schmitz et al. (131) postulated that Tangier Disease may be due to a defect in intracellular membrane traffic, affecting the retroendocytosis of HDL. In this disease, the HDL seem to be diverted into the lysosomal compartment, where they are degraded instead of being loaded with cholesterol and resecreted. If these data can be confirmed in other Tangier patients, it would provide a very interesting model to study HDL receptors in macrophages.

Human skin fibroblasts are frequently used as a model system for peripheral cells. Several authors have also chosen them to study reverse cholesterol transport (113, 118, 122, 124) and could demonstrate cholesterol efflux out of fibroblasts when HDL is present as an acceptor in the medium.

In addition to the reverse cholesterol transport mediated by HDL discussed here, the efflux of cholesterol has been described by mediation of other acceptors, such as phospholipid vesicles and a component of lipoprotein deficient serum (132, 133).

5.3 HDL interaction with hepatic cells

HDL have been shown to bind to liver cells in rats, rabbits, pigs and humans. The HDL binding site on liver cells most probably serves the purpose of introducing HDL into the catabolic pathways that function in this organ.

Experiments to study HDL catabolism were performed using liver cells in suspension (134 – 136) and human hepatoma cells (137). In addition, binding studies were performed on liver cell membranes (122, 123). Rifici et al. (134) demonstrated a specific receptor for HDL on rat hepatocytes, with Apo AI as the binding protein. The HDL was bound, internalized and degraded. Two other investigators tried to compare the binding of HDL and LDL on hepatocytes: O'Malley (135) has shown the preferential uptake of HDL cholesterol over that of apoprotein, in analogy to steroidogenic cells. The cross competition between different lipoproteins in binding to pig hepatocytes discussed by Bachorik (136) is not well demonstrated and needs further substantiation.

In human hepatoma cells (Hep G2) Dashti et al. (137) have shown that HDL binding, uptake and degradation is mediated by a high affinity process. For the mechanism of binding, apolipoproteins D or CI were proposed as ligands rather than Apo AI. These results require further evaluation.

Human liver cell membranes were shown to bind Apo E-free HDL in a process unaffected by EDTA and following saturation kinetics (122). The binding appears to be mediated by Apo AI and was inhibited by anti-AI antibodies and tyrosine modification with TNM. The same has been demonstrated with rat liver membranes (123).

5.4 HDL interaction with other cell types

In addition to the above interactions, HDL appears to bind to kidney cells and cells from intestinal mucosa, as well as human cancer cell lines. Specific saturable binding of rat HDL to rat kidney membranes has been shown by van Tol et al. (138). The kidney is known to play an important role in HDL catabolism in the rat (139). A strong competition between the binding of HDL and VLDL to kidney membranes has been observed and the authors conclude that a common apolipoprotein might be responsible for the binding of both HDL and VLDL. The possibility of lipid-lipid-interaction was eliminated experimentally. For a variety of reasons, it was proposed that one of the C-apoproteins might mediate the binding of HDL and VLDL to rat kidney membranes. Glass (12) demonstrated in plasma turnover experiments in rats that in contrast to the preferential uptake of HDL-cholesterol ester in adrenal cells and gonads, the kidney took up almost no cholesterol, but 18% of the Apo AI. The studies which were presented do not establish the mechanism of this selective uptake of protein and lipid constituents of the HDL.

Rat intestinal cells possess HDL_3-specific, saturable binding sites on their basolateral plasma membranes. These receptors (140) differ from the LDL receptor. The authors suggest that the binding of HDL to intestinal cells may allow the transfer of protein or lipid constituents to newly secreted lipoproteins, especially to facilitate the formation of chylomicrons.

High affinity binding sites for HDL could be demonstrated on human cancer cells (141) (HEC-B-296). Subsequent to the binding, the protein component of HDL was degraded. This degradation did not occur in the lysosomal compartment, but on the external surface of the cell. A proteolytic enzyme was described that acted specifically on HDL. The biological significance of this degradation process remains to be defined.

5.5 Conclusions

The receptor(s) for Apo E-free HDL do(es) not appear to be as unique as the LDL receptor, but can be found with different functional consequences in various cell types. Two main functions of HDL emerge from the different studies. They are: 1. to supply cells with cholesterol, and 2. to facilitate the secretion of cellular cholesterol. In these functionally opposite processes the same apolipoprotein, Apo AI, appears to be involved. Glass (112) has asked the question "whether there can be two faces of a single binding event ... or are we dealing with two quite different receptors, both of which bind Apo AI?" In the first case, a common binding reaction could be postulated which might introduce different postreceptor events.

To summarize the studies on the different tissues and cell lines, four groups can be distinguished:

- Steroidogenic tissue; HDL supplies cholesterol for cellular metabolism
- Peripheral tissues; HDL mediates reverse cholesterol transport
- Liver cells; Location of HDL catabolism
- Other tissues and cells; Processes need to be further investigated and specified.

It is obvious that much more work has to be done to define the character and probe for the existence of a specific HDL receptor or several HDL binding sites. Apparently, functionally different processes can be brought about by the initial binding of HDL to cells.

References

1. Brown, M. S. & Goldstein, J. L. (1974) Prod. Natl. Acad. Sci. USA *71*, 788–792.
2. Goldstein, J. L. & Brown, M. S. (1974) J. Biol. Chem. *249*, 5153–5162.
3. Brown, M. S. & Goldstein, J. L. (1976) Science *191*, 150–154.
4. Goldstein, J. L. & Brown, M. S. (1976) Curr. Topics Cell. Reg. *11*, 147–181.
5. Goldstein, J. L. & Brown, M. S. (1977) Ann. Rev. Biochem. *46*, 897–930.
6. Ho, Y. K. Brown, M. S. & Bilheimer, D. W. (1976) J. Clin. Invest. *58*, 1465–1474.
7. Bilheimer, D. W., Ho, Y. K., Brown, M. S., Anderson, R. G. W. & Goldstein, J. L. (1978) J. Clin. Invest. *61*, 678–696.
8. Brown, M. S., Kovanen, P. T. & Goldstein, J. L. (1979) Recent Prog. Hormone Res. *35*, 215–257.
9. Kita, T., Goldstein, J. L., Brown, M. S., Watanabe, Y., Hornick, C. A. & Havel, R. J. (1982) Proc. Natl. Acad. Sci. USA *79*, 3623–3627.
10. Carew, T. E., Pittman, R. C. & Steinberg, D. (1982) J. Biol. Chem. *257*, 8001–8007.
11. Spady, D. K., Bilheimer, D. W. & Dietschy, J. M. (1983) Proc. Natl. Acad. Sci. USA *80*, 3499–3503.
12. Brown, M. S. & Goldstein, J. L. (1981) N. Engl. J. Med. *305*, 515–517.
13. Bilheimer, D. W., Goldstein, J. L., Grundy, S. M., Starzl, T. E. & Brown, M. S. (1984) N. Engl. J. Med. *311*, 1658–1664.
14. Basu, S. K., Goldstein, J. L., Anderson, R. G. W. & Brown, M. S. (1981) Cell *24*, 493–502.
15. Anderson, R. G. W., Brown, M. S., Beisiegel, U. & Goldstein, J. L. (1982) J. Cell. Biol. *93*, 523–531.
16. Pearse, B. M. F. (1975) J. Mol. Biol. *97*, 93–98.
17. Tycko, B. & Maxfield, F. R. (1982) Cell *28*, 643–651.
18. Helenius, A., Mellman, I., Wall, D. & Hubbard, A. (1983) Trends Biochem. Sci. *8*, 245–250.
19. Pastan, I. H. & Willingham, M. C. (1981) Science *214*, 504–509.
20. Brown, M. S., Anderson, R. G. W. & Goldstein, J. L. (1983) Cell *32*, 663–667.
21. Goldstein, J. L., Basu, S. K., Brunsched, G. Y. & Brown, M. S. (1976) Cell *7*, 85–92.

22. Goldstein, J. L., Basu, S. K. & Brown, M. S. (1983) Methods Enzymol. *98*, 241–260.
23. Fredrickson, D. S., Goldstein, J. L. & Brown, M. S. (1978) in: The Metabolic Basis of Inherited Disease (Wyngaarden, J. B. & Fredrickson, D. S., eds.) 4th ed., p. 604, McGraw-Hill, New York.
24. Brown, M. S., Dana, S. E. & Goldstein, J. L. (1974) J. Biol. Chem. *249*, 789–796.
25. Goldstein, J. L., Dana, S. E. & Brown, M. S. (1974) Proc. Natl. Acad. Sci. USA *71*, 4288–4292.
26. Brown, M. S. & Goldstein, J. L. (1975) Cell *6*, 307–316.
27. Goldstein, J. L. & Brown, M. S. (1978) Johns Hopkins Med. J. *143*, 8–16.
28. Khachadurian, A. K. (1964) Am. J. Med. *37*, 402–407.
29. Goldstein, J. L., Dana, S. E., Brunschede, G. Y. & Brown, M. S. (1975) Proc. Natl. Acad. Sci. USA *72*, 1092–1096.
30. Tolleshaug, H., Goldstein, J. L., Schneider, W. J., Brown, M. S. (1982) Cell *30*, 715–724.
31. Tolleshaug, H., Hobgood, K. K., Brown, M. S. & Goldstein, J. L. (1983) Cell *32*, 941–951.
32. Goldstein, J. L. & Brown, M. S. (1983) in: The Metabolic Basis of Inherited Disease (Stanbury, J. B., Wyngaardon, J. B., Fredrickson, D. S. et al., eds.) 5th ed. pp. 672–712, McGraw-Hill, New York.
33. Slack, J. & Nevin, N. C. (1968) J. Med. Genet. *5*, 4–8.
34. Stone, N. J., Levy, R. I., Fredrickson, D. S. & Verber. J. (1974) Circulation *49*, 476–488.
35. Brown, M. S., Dana, S. E. & Goldstein, J. L. (1973) Proc. Natl. Acad. Sci. USA *70*, 2162–2166.
36. Brown, M. S., Dana, S. E. & Goldstein, J. L. (1975) Proc. Natl. Acad. Sci. USA *72*, 2925–2929.
37. Brown, M. S., Kovanen, P. T. & Goldstein, J. L. (1980) Ann. N.Y. Acad. Sci. *348*, 48–68.
38. Ho, Y. K., Faust, J. R., Billheimer, D. W., Brown, M. S. & Goldstein, J. L. (1977) J. Exp. Med. *145*, 1531–1538.
39. Krieger, M. Brown, M. S., Faust, J. R. & Goldstein, J. L. (1978) J. Biol. Chem. *253*, 4093–4101.
40. Krieger, M., Smith, L. C., Anderson, R. G. W., Goldstein, J. L., Kao, Y. J., Pownall, H. J., Gotto, A. M., Jr. & Brown, M. S. (1979) J. Supramol. Structure *10*, 467–478.
41. Basu, S. K., Goldstein, J. L. & Brown, M. S. (1978) J. Biol. Chem. *253*, 3852–3856.
42. Schneider, W. J., Goldstein, J. L. & Brown, M. S. (1985) Methods Enzymol. *109*, 405–417.
43. Kovanen, P. T., Basu, S. K., Goldstein, J. L. & Brown, M. S. (1979) Endocrinology *104*, 610–616.
44. Kovanen, P. T., Faust, J. R., Brown, M. S. & Goldstein, J. L. (1979) Endocrinology *104*, 599–609.
45. Chao, Y.-S., Windler, e. E., Chen, G. C. & Havel, R. J. (1979) J. Biol. Chem. *254*, 11360–11366.
46. Kovanen, P. T., Brown, M. S. & Goldstein, J. L. (1979) J. Biol. Chem. *254*, 11367–11373.
47. Windler, E. E. T., Kovanen, P. T., Chao, Y.-S., Brown, M. S., Havel, R. J. & Goldstein, J. L. (1980) J. Biol. Chem. *255*, 10464–10471.
48. Brown, M. S., Kovanen, P. T. & Goldstein, J. L. (1981) Science *212*, 628–635.
49. Brown, M. S. & Goldstein, J. L. (1983) J. Clin. Invest. *72*, 743–747.
50. Goldstein, J. L., Kita, T. & Brown, M. S. (1983) N. Engl. J. Med. *309*, 288–296.
51. Pittman, R. C., Carew, T. E., Attie, A. D., Witztum, J. L., Watanabe, Y. & Steinberg, D. (1982) J. Biol. Chem. *257*, 7994–8000.
52. Spady, D. K., Bilheimer, D. W. & Dietschy, J. M. (1983) Proc. Natl. Acad. Sci. USA *80*, 3499–3503.
53. Mahley, R. W. & Innerarity, T. L. (1983) Biochim. Biophys. Acta *737*, 197–222.
54. Hui, D. Y., Innerarity, T. L. & Mahley, R. W. (1981) J. Biol. Chem. *256*, 5646–5655.
55. Kovanen, P. T., Brown, M. S., Basu, S. K., Bilheimer, D. W. & Goldstein, J. L. (1981) Proc. Natl. Acad. Sci. USA *78*, 1396–1400.
56. Angelin, B. C., Raviola, C. A., Innerarity, T. L. & Mahley, R. W. (1983) J. Clin. Invest. *71*, 816–831.
57. Goldstein, J. L. & Brown, M. S. (1982) Clin. Res. *30*, 417–426.
58. Bilheimer, D. W., Grundy, S. M., Brown, M. S. & Goldstein, J. L. (1983) Proc. Natl. Acad. Sci. USA *80*, 4124–4128.
59. Shepherd, J., Packard, C. J., Bicker, S., Lawrie, T. D. V. & Morgan, H. G. (1980) N. Engl. J. Med. *302*, 1219–1222.
60. Kovanen, P. T., Bilheimer, D. W., Goldstein, J. L., Jaramillo, J. J. & Brown, M.

S. (1981) Proc. Natl. Acad. Sci. USA *78*, 1194–1198.
61. Sherill, B. C. & Dietschy, J. M. (1978) J. Biol. Chem. *253*, 1859–1867.
62. Innerarity, T. L., Pitas, R. E. & Mahley, R. W. (1980) Biochemistry *19*, 4359–4365.
63. Kita, T., Goldstein, J. L., Brown, M. S., Watanabe, Y., Hornick, C. A. & Havel, R. J. (1982) Proc. Natl. Acad. Sci. USA *79*, 3623–3627.
64. Watanabe, Y. (1980) Atherosclerosis *36*, 261–268.
65. Tanzawa, K., Shimada, Y., Kuroda, M., Tsujita, Y., Arai, M. & Watanabe, Y. (1980) FEBS Lett. *118*, 81–84.
66. Kita, T., Brown, M. S., Watanabe, Y. & Goldstein, J. L. (1981) Proc. Natl. Acad. Sci. USA *78*, 2268–2272.
67. Attie, A. D., Pittman, R. C., Watanabe, Y. & Steinberg, D. (1981) J. Biol. Chem. *256*, 9789–9792.
68. Kita, T., Brown, M. S., Bilheimer, D. W. & Goldstein, J. L. (1982) Proc. Natl. Acad. Sci. USA *79*, 5693–5697.
69. Bilheimer, D. W., Watanabe, Y. & Kita, T. (1982) Proc. Natl. Acad. Sci. USA *79*, 3305–3309.
70. Schneider, W. J., Brown, M. S. & Goldstein, J. L. (1983) Mol. Biol. Med. *1*, 355–367.
71. Yamamoto, T., Davis, C. G., Brown, M. S., Schneider, W. J., Casey, M. L., Goldstein, J. L. & Russell, D. W. (1984) Cell *39*, 27–38.
72. Russell, D. W., Schneider, W. J., Yamamoto, T., Luskey, K. L., Brown, M. S. & Goldstein, J. L. (1984) Cell *37*, 577–585.
73. Südhof, T. C., Goldstein, J. L., Brown, M. S. & Russell, D. W. (1985) Science *228*, 815–822.
74. Innerarity, T. L., Weisgraber, K. H., Arnold, K. S., Rall, S. C., Jr. & Mahley, R. W. (1984) J. Biol. Chem. *259*, 7261–7267.
75. Knott, T. J., Rall, S. C., Jr., Innerarity, T. L., Jacobson, S. F., Urdea, M. S., Levy-Wilson, B., Powell, L. M., Pease, R. J., Eddy, R., Nakai, H., Byers, M., Priestley, L. M., Robertson, E., Rall, L. B., Betsholtz, C., Shows, T. B., Mahley, R. W. & Scott, J. (1985) Science *230*, 37–43.
76. Goldstein, J. L., Brown, M. S., Anderson, R. G. W., Russell, D. W. & Schneider, W. J. (1985) Ann. Rev. Cell. Biol. *1*, 1–39.
77. Innerarity, T. L. & Mahley, R. W. (1978) Biochemistry *17*, 1440–1447.
78. Pitas, R. E., Innerarity, T. L. & Mahley, R. W. (1980) J. Biol. Chem. *255*, 5454–5460.
79. Basu, S. K., Goldstein, J. L., Anderson, R. G. W. & Brown, M. S. (1976) Proc. Natl. Acad. Sci. USA *73*, 3178–3182.
80. Schneider, W. J., Beisiegel, U., Goldstein, J. L. & Brown, M. S. (1982) J. Biol. Chem. *257*, 2664–2673.
81. Scott, J., Urdea, M., Quiroga, M., Sanchez-Pascador, R., Fong, N., Selby, M., Rutter, W. J. & Bell, G. I. (1983) Science *221*, 236–240.
82. Gray, A., Dull, T. J. & Ullrich, A. (1983) Nature *303*, 722–725.
83. Doolittle, R. F., Feng, D.-F. & Johnson, M. S. (1984) Nature *307*, 558–566.
84. Cummings, R. D., Kornfeld, S., Schneider, W. J., Hobgood, K. K., Tolleshaug, H., Brown, M. S. & Goldstein, J. L. (1983) J. Biol. Chem. *258*, 15261–15273.
85. Davis, C. G., Elhammer, A., Russell, D. W., Schneider, W. J., Kornfeld, S., Brown, M. S. & Goldstein, J. L. (1986) J. Biol. Chem. *261*, 2828–2838.
86. Beisiegel, U., Schneider, W. J., Goldstein, J. L., Anderson, R. G. W. & Brown, M. S. (1981) J. Biol. Chem. *256*, 11923–11931.
87. Lehrman, M. A., Goldstein, J. L., Brown, M. S., Russell, D. W. & Schneider, W. J. (1985) Cell *41* , 735–743.
88. Brown, M. S. & Goldstein, J. L. (1976) Cell *9*, 663–674.
89. Goldstein, J. L., Brown, M. S. & Stone, N. J. (1977) Cell *12*, 629–641.
90. Davis, C. G., Lehrman, M. A., Russell, D. W., Anderson, R. G. W., Brown, M. S. & Goldstein, J. L. (1986) Cell *45*, 15–24.
91. Lehrman, M. A., Schneider, W. J., Südhof, T., Brown, M. S., Goldstein, J. L. & Russell, D. W. (1985) Science *227*, 140–146.
92. Miyake, Y., Tajima, S., Yamamura, T., & Yamamoto, A. (1981) Proc. Natl. Acad. Sci. USA *78*, 5151–5155.
93. Südhof, T. C., Russell, D. W., Goldstein, J. L., Brown, M. S., Sanchez-Pescador, R. & Bell, G. I. Science *228*, 893–895.
94. Arbeeny, C. M. & Rifici, V. A. (1984) J. Biol. Chem. *259*, 9662–9666.
95. Hui, D. Y., Innerarity, T. L., Milne, R. W., Manel, Y. L. & Mahley, R. W. (1984) J. Biol. Chem. *259*, 15060–15068.
96. Mahley, R. W., Hui, D. Y., Innerarity, T. L. & Weisgraber, K. H. (1981) J. Clin. Invest. *68*, 1197–1206.

97. Hui, D. Y., Brecht, W. J., Hall, E. A., Friedman, G., Innerarity, T. L. & Mahley, R. W. (1986) J. Biol. Chem. *261*, 4256–4267.
98. Mahley, R. W. & Weisgraber, K. H. (1974) Biochemistry *13*, 1964–1969.
99. Kinoshita, M., Teramoto, T., Kato, H., Hashimoto, Y., Naito, C., Toda, G. & Oka, H. (1985) J. Biochem. *97*, 1803–1806.
100. Goldstein, J. L., Ho, Y. K., Brown, M. S., Innerarity, T. L. & Mahley, R. W. (1980) J. Biol. Chem. *255*, 1839–1848.
101. Goldstein, J. L., Ho, Y. K., Basu, S. K. & Brown, M. S. (1979) Proc. Natl. Acad. Sci. USA *76*, 333–337.
102. Brown, M. S. & Goldstein, J. L. (1983) Ann. Rev. Biochem. *52*, 223–261.
103. Innerarity, T. L., Arnold, K. S., Weisgraber, K. H. & Mahley, R. W. (1986) Arteriosclerosis *6*, 114–122.
104. Wang-Iverson, P., Ginsberg, H. N., Peteanu, L. A., Le, N.-A. & Brown, W. V. (1985) Biochem. Biophys. Res. Commun. *126*, 578–586.
105. Baker, D. P., Van Lenten, B. J., Fogelman, A. M., Edwards, P. A., Kean, C. & Berliner, J. A. (1983) Atherosclerosis *4*, 248–255.
106. Van Lenten, B. J., Fogelman, A. M., Hokom, M. M., Benson, L., Haberland, M. E. & Edwards, P. A. (1983) J. Biol. Chem. *258*, 5151–5157.
107. Dresel, H. A., Friedrich, E., Via, D. P., Schettler, J. & Sinn, H. (1985) EMBO J. *4*, 1157–1162.
108. Gordon, T., Castelli, W. P., Hjortland, M. C., Kannel, W. B. & Dawber, T. R. (1977) Am. J. Med. *62*, 707–714.
109. Miller, G. J. & Miller, N. E. (1975) Lancet *97*, 16–19.
110. Eisenberg, S. (1984) J. Lipid Res. *25*, 1017–1058.
111. Kovanen, P. T., Schneider, W. J., Hillman, G. M., Goldstein, J. L. & Brown, M. S. (1979) J. Biol. Chem. *254*, 5498–5506.
112. Glass, C., Pittman, R. C., Weinstein, D. B. & Steinberg, D. (1983) Proc. Natl. Acad. Sci. USA *80*, 5435–5439.
113. Fidge, N. H., Nestel, P. J. & Suzuki, N. (1982) Biochim. Biophys. Acta *753*, 14–21.
114. Fidge, N., Leonard-Kanevsky, M. & Nestel, P. (1984) Biochim. Biophys. Acta *793*, 180–186.
115. Fidge, N. H. & Nestel, P. J. (1985) J. Biol. Chem. *260*, 3570–3575.
116. Miller, N. E., La Ville, A & Crook, D. (1985) Nature *314*, 109–111.
117. Stein, Y., Glangeaud, M. C., Fainaru, M. & Stein, O. (1975) Biochim. Biophys. Acta *380*, 106–118.
118. Biesbroeck, R., Oram, J. F., Albers, J. J. & Bierman, E. L. (1983) J. Clin. Invest. *71*, 525–539.
119. Oram, J. F., Brinton, E. A. & Bierman, E. L. (1983) J. Clin. Invest. *72*, 1611–1621.
120. Brinton, E. A., Kenagy, R. D., Oram, J. F. & Bierman, E. L. (1984) Arteriosclerosis *5*, 329–335.
121. Brinton, E. A., Oram, J. F., Chen, C.-H., Albers, J. J. & Bierman, E. L. (1986) J. Biol. Chem. *261*, 495–503.
122. Hoeg, J. M., Demosky, S. J., Jr. Edge, S. B., Gregg, R. E., Osborne, J. C., Jr. & Brewer, H. B. Jr. (1985) Arteriosclerosis *5*, 228–237.
123. Chako, G. K. (1985) J. Lipid Res. *26*, 745–754.
124. Tabas, I. & Tall, A. R. (1984) J. Biol. Chem. *259*, 13897–13905.
125. Havekes, L., Schouten, D., van Hinsbergh, V. & de Wit, E. (1984) Biochem. Biophys. Res. Commun. *122*, 785–790.
126. Ho, Y. K., Brown, M. S. & Goldstein, J. L. (1980) J. Lipid Res. *21*, 391–398.
127. Werb, Z. & Cohn, Z. A. (1972) J. Exp. Med. *135*, 21–44.
128. Innerarity, T., Pitas, R. E. & Mahley, R. W. (1982) Arteriosclerosis *2*, 114–124.
129. Schmitz, G., Robenek, H., Lohmann, U. & Assmann, G. EMBO J. *4*, 613–622.
130. Schmitz, Niemann, R., Brennhausen, B., Krause, R. & Assmann, G. (1985) EMBO J. *4*, 2773–2779.
131. Schmitz, G. Assmann, G., Robenek, H. & Brennhausen, B. (1985) Proc. Natl. Acad. Sci. USA *82*, 6305–6309.
132. Phillips, M. C., McLean, L. R., Stoudt, G. W. & Rothblat, G. H. (1980) Arteriosclerosis *35*, 409–422.
133. St. Clair, R. W. & Leight, M. A. (1983) J. Lipid Res. *24*, 183–191.
134. Rifici, V. A., Eder, H. A. (1984) J. Biol. Chem. *259*, 13814–13818.
135. O'Malley, J. P., Soltys, P. A. & Portman, O. W. (1981) J. Lipid Res. *22*, 1214–1224.
136. Bachorik, P. S., Franklin, F., Virgil, D. G. & Kwiterovich, P. O. Jr. (1982), Biochemistry *21*, 5675–5684.

137. Dashti, N., Wolfbauer, G. & Alaupovic, P. (1985) Biochim. Biophys. Acta *833*, 100–110.
138. Van Tol, A., Dallinga-Thie, G. M., Van Gent, T. & Van 't Hooft, F. M. (1986) Biochim. Biophys. Acta *876*, 340–351.
139. Van 't Hooft, F. M. & Van Tol, A. (1985) Biochem. J. *226*, 715–721.
140. Kagami, A., Fidge, N., Suzuki, B. & Nestel, P. (1984) Biochim. Biophys. Acta *795*, 179–190.
141. Gal, D., MacDonald, P. C., Porter, J. C. & Simpson, E. R. (1982) Arch. Biochem. Biophys. *214*, 726–733.
142. Chacko, G. K. (1984) Biochim. Biophys. Acta *795*, 417–426.

Chapter 6
Intravascular Lipoprotein Remodelling: Lipolytic Enzymes

M. J. Tikkanen

1 Introduction

In his classic study in 1953 Nikkilä (1) injected heparin intravenously to human subjects. He found that the "clearing factor" released into the circulation by heparin caused a shift of lipids from β-globulins to α-globulins (i. e., from lower density lipoproteins to high density lipoproteins), and suggested "that the redistribution of lipids elicited by the heparin injection is opposite to that found in many pathological states, i. a. in atherosclerosis." In line with this basic finding, we now have experimental evidence showing that the clearing factor, or lipoprotein lipase (LPL), is the rate-limiting enzyme of chylomicron and very low density lipoprotein (VLDL) catabolism, and that during this lipolytic process high density lipoproteins (HDL) serve as acceptors for lipid constituents liberated from triglyceride-rich lipoproteins (2). This also implies that inactivation of LPL-induced lipolysis results in an increase of VLDL and a decrease of HDL, a pattern that may be associated with atherosclerosis. Some of the mechanisms involved in the regulation of the plasma lipoprotein pattern are discussed in the present chapter. The emphasis will be on the two lipolytic enzymes, lipoprotein lipase and hepatic lipase, which together with the lipid transfer proteins (LTP) and lecithin:cholesterol acyl transferase (LCAT), contribute to a continuous remodelling of lipoproteins in the plasma compartment.

2 Lipoprotein Lipase

2.1 Physiology of lipoprotein lipase (LPL)

Lipoprotein lipase is an enzyme (or a group of related enzymes) with glycerol ester hydrolase activity attached to the endothelial surfaces of extrahepatic capillaries. The enzyme is releasable into the circulation by heparin injection and its activity can be assayed in ("postheparin") plasma. It is thought to be synthesized by parenchymal cells and subsequently transported to the capillary endothelium where it becomes active (3). The physiologic function of LPL is to hydrolyze chylomicron and VLDL triglycerides, which results in transfer of liberated fatty acids to tissues, such as adipose tissue (storage) or skeletal muscle (oxidation) (4). LPL is activated by the presence of apoprotein CII, a specific cofactor contained on the surface of triglyceride-rich lipoproteins. The degradation of chylomicrons and VLDL by LPL is a prerequisite for the subsequent removal of these particles from the circulation (5–8).

2.2 The inverse relationship between VLDL and HDL levels

Starting with Gofman's early work (9) a number of studies (10–14) have demonstrated an inverse relation between VLDL and HDL or HDL_2 levels in various populations. In addition, lowering of VLDL by various interventions was shown to increase HDL (α-lipoprotein) cholesterol (15–17), while different types of hypertriglyceridemia (13, 18) or VLDL triglyceride elevations induced by diet interventions (15, 19) were associated with lowering of HDL cholesterol. Most of the reciprocal relationship between plasma VLDL (triglyceride) and HDL (cholesterol) levels can be explained by two major regulatory systems, LPL-induced lipolysis and lipid transfer protein (LTP) activity (see also Chapter 7). Some relevant experimental and clinical studies will be reviewed below.

2.2.1 **The role of LPL-induced lipolysis**

Activation of LPL can be conveniently brought about by an intravenous heparin injection. In such experiments, heparin-induced lipolysis caused an increase in HDL (1, 19, 20–22) with surface components of VLDL (phospholipid, free cholesterol and C apoprotein) being transferred to HDL (23). In hypertriglyceridemic subjects heparin-induced lipolysis was followed by increases in phospholipid, cholesterol and protein in HDL_2 while the level of the smaller and less dense HDL_3 lipoprotein decreased (24). The "reverse" intervention, inactivation of lipolysis by antibody-induced inhibition of lipoprotein lipase (in an animal model) resulted in increased VLDL levels with decreases in HDL cholesterol, phospholipid and protein (25). The surface components of lipolyzed chylomicrons and VLDL may either be taken up by HDL_3 particles serving as acceptor molecules (26), or they may form discoidal structures that are converted to spherical HDL particles after acquisition of apolipoprotein AI and formation of cholesteryl esters by LCAT (27, 28). Either type of mechanism would result in net transfer of lipid and apoprotein constituents from TG-rich lipoproteins to the HDL density range. Since more lipid than protein is added, the average density of HDL particles moves toward that of HDL_2 (26). However, the fusion alone of "surface remnants" and HDL_3 does not produce true HDL_2 particles; in addition, the cholesteryl ester content must be doubled or tripled by the LCAT reaction (29).

The plasma triglyceride concentration is determined more by the rate of VLDL production than by its clearance, i.e., VLDL (triglyceride) levels are only weakly regulated by LPL activity (30). The contrary may be true for HDL, the level of which appears to be directly proportional to the efficiency of LPL-induced lipolysis (30, 31). This was first suggested by Nikkilä's studies which revealed a positive correlation between HDL cholesterol and LPL activity in postheparin plasma and adipose tissue (32–36). On the basis of these and subsequent studies it was postulated that the rate of VLDL (and chylomicron) lipolysis, regulated by LPL activity, was a factor determining the plasma HDL level (36, 37). This theory has been strengthened by reports demonstrating a consistent parallelism between HDL (HDL_2) and LPL activity or triglyceride clearance rate in a variety of experimental situations and disease conditions. For example, elevation of HDL or HDL_2 occurred when lipolysis was stimulated by the following interventions; oral (38, 39) and intravenous (40) administration of fat, increased physical activity (34, 41, 42), regular alcohol intake (43–45), or insulin therapy (32, 46, 47). Conversely, low HDL (HDL_2) levels were associated with conditions characterized by low *in vivo* LPL activity, such as ingestion of a low fat diet (39, 48), caloric restriction (49), insulin deficiency or insulin resistance in diabetic subjects (32, 46, 50). These and other relevant studies have been summarized in a recent review (30).

2.2.2 The role of the lipid transfer proteins

These specialized proteins, whose actions are described in detail in Chapter 7, mediate the transfer and exchange of core-lipids (cholesteryl esters and triglycerides) and phospholipids between the lipoproteins (51–54). During lipolysis, lipid transfer proteins probably facilitate transfer of phospholipids from TG-rich lipoproteins to HDL (55). Production of phospholipid-rich HDL particles may be followed by movement of free cholesterol from other lipoproteins and cells into HDL (28), providing substrate (phospholipid and cholesterol) for LCAT, the enzyme which catalyses esterification of cholesterol in the plasma compartment (56). LCAT-derived cholesteryl esters may either reside in the lipophilic cores of HDL particles, or they may be transferred to the TG-rich lipoproteins by lipid transfer proteins (LTP) in exchange for triglycerides (57, 58). LTP activity (i. e. the rate of cholesteryl ester-triglyceride exchange) is directly proportional to the ratio between TG-rich lipoproteins (chylomicrons and VLDL) and HDL in the plasma compartment (29). Thus, LTP activity is enhanced in hypertriglyceridemia and postprandial lipemia (55) causing, in these situations, reciprocal lowering of HDL (cholesterol) by accelerated transfer of cholesteryl esters to TG-rich lipoproteins.

Very little is known about other factors regulating lipid transfer activity. Enhanced cholesteryl ester transfer from HDL to VLDL has been demonstrated *in vitro* during lipolysis induced by triglyceride lipase or phospholipase (59). This enhancement was probably due to preferred binding of LTP to lipolyzed (as compared to non-lipolyzed) lipoprotein particles (59). Unfortunately, it is not known whether the activity of the LTP system is related to lipase activity *in vivo*.

The incorporation of cholesteryl esters into TG-rich lipoproteins may provide a part of a "reverse cholesterol transport" system by which cholesterol derived from various sources (lipoproteins or cells) arrives in apolipoprotein B containing lipoproteins. Such lipoprotein particles can be cleared from the circulation by LDL receptor-mediated uptake (Chapter 5).

2.2.3 **HDL in hypertriglyceridemia**

The lowering of HDL levels in hypertriglyceridemic patients can probably be brought about by several mechanisms. Limited availability of "surface remnants" derived from TG-rich lipoproteins for HDL_2 formation may result in low HDL_2 (and total HDL) concentrations (29). This occurs when elevation of (VLDL) triglycerides is caused by a removal defect (produced by a relative deficiency of LPL), or when incomplete conversion of hypertriglyceridemic VLDL to LDL takes place. In the latter process incompletely lipolyzed VLDL "remnant" particles are rapidly removed from the circulation and part of the surface material is diverted from HDL. Such metabolic diversions of hypertriglyceridemic VLDL have been discussed in recent reviews (60, 61). In addition to impaired formation of HDL_2, the presence of an increased mass of circulating VLDL and chylomicrons promotes plasma lipid transfer activity with loss of cholesteryl ester from HDL and production of triglyceride-enriched HDL particles. This TG-rich HDL is susceptible to lipolysis by hepatic lipase resulting in further HDL degradation and the generation of smaller HDL_3 particles (62, 63).

3 Hepatic Lipase

3.1 Physiology of hepatic lipase

Hepatic lipase is an enzyme that contains triglyceride lipase, monoglyceride lipase and phospholipase activity (64). The enzyme is probably synthesized by parenchymal liver cells and subsequently transferred to the endothelial cells of liver sinusoids where it becomes active. An enzyme resembling hepatic lipase has been found in the ovaries and adrenals in some animal species (65, 66). Several lines of evidence suggest that hepatic lipase catalyzes the degradation HDL particles, more specifically HDL_2. In addition, experimental data suggest that this enzyme may also be involved in the degradation of triglyceride-rich lipoproteins. Some of the evidence for these functional roles will be summarized below.

3.2 Regulation of HDL(HDL_2) by hepatic lipase

The role for hepatic lipase (HL) in the degradation of HDL was suggested in studies demonstrating an inverse relationship between HL activity in plasma after heparin injection (postheparin plasma) and HDL or HDL_2. Such inverse correlations were observed in normal men (67), in postmenopausal women prior to starting hormone therapy (68–71) and during stimulation of HL activity by progestin administration (68, 71) but not during suppression of the enzyme by estrogen (69–71). Experimental and clinical studies indicated that HDL_2 is preferred substrate for hepatic lipase, and that this enzyme regulates plasma total HDL levels by its effects on the HDL_2 fraction.

3.2.1 Experimental studies

Studies *in vitro* have shown that hepatic lipase binds to HDL more avidly than to triglyceride-rich lipoproteins (72). Inhibition of this enzyme by injecting anti-HL antibody into rats caused an elevation in plasma HDL levels (73–76). The HL-mediated effects on plasma HDL levels are probably due to alterations in the level of HDL_2, which is selectively degraded by hepatic lipase (77–80).

3.2.2 Clinical studies

Since postheparin plasma HL activity can effectively be suppressed by estrogen and enhanced by androgens (81, 82), such perturbations have been employed in order to clarify the role of this enzyme. A consistent reciprocal relationship between (postheparin plasma) HL activity and HDL(HDL_2) levels has been found in these studies. For example, administration of any type of estrogen, such as conjugated equine (83), alkylated (84), or natural estrogen (69, 70) suppresses HL activity in postheparin plasma, while HDL(HDL_2) cholesterol levels increase (82). Conversely, up-regulation of HL activity by androgenic steroids such as oxandrolone (85), stanozolol (86) and the progestational agent levonorgestrel (87) is associated with low HDL(HDL_2) levels (81, 82). Pooling of the results from several studies (88, 89) showed that sex hormone-induced *changes* in HL activity were correlated with *changes* in HDL_2 cholesterol, supporting the role of hepatic lipase as a key regulator of HDL_2 catabolism.

3.3 Catabolism of triglyceride-rich lipoproteins

The role for hepatic lipase in the degradation of triglyceride-rich lipoproteins is controversial. Some investigators propose that hepatic lipase functions in removing triglycerides from VLDL or other lower density lipoproteins (90–93). Most of the evidence supporting this theory comes from experimental or clinical conditions associated with lack of HL activity. Inhibition of HL activity by specific antibody in cynomolgus monkeys caused elevations of VLDL, IDL and "light LDL" (93), while similar inhibition studies in the rat caused increases in VLDL or LDL, or both (73–76). Breckenridge et al. reported increases in triglyceride-rich lipoproteins and in the triglyceride content of LDL in subjects with HL deficiency (94). All these studies were characterized by an almost complete absence of HL activity.

There is, however, no evidence indicating that HL activity determines the overall rate of catabolism of one or more of the TG-rich lipoproteins. No correlation has been found between the fractional catabolic rates of radioiodinated VLDL or VLDL remnants and hepatic lipase in postheparin plasma (95). Suppression of HL activity by estrogen in other turnover studies employing radiolabeled VLDL triglycerides (96), or VLDL apolipoprotein B (97) did not impair the removal of these tracers. The evidence available suggests that "physiologic" fluctuations in HL activity, such as occur during the normal menstrual cycle (98) or during estrogen therapy (96) do not influence the removal rate of TG-rich lipoproteins. Therefore, hepatic lipase is probably not a major factor determining the catabolic rate of TG-rich lipoproteins during physiologic circumstances. Nevertheless, complete loss of this enzyme activity may cause deterioration in the catabolism of these lipoproteins.

4 Dual Regulation of Plasma HDL_2 by Lipoprotein Lipase and Hepatic Lipase

Variation in total plasma HDL concentrations usually reflects variation in the HDL_2 subfraction (99, 100). The HDL_2 level is regulated by two major opposing forces. LPL-induced lipolysis in conjunction with the LCAT reaction promotes formation of HDL_2 particles (24, 30), while the LTP activity coupled with hepatic lipase-induced lipolysis of HDL_2 triglycerides causes degradation of these particles (29, 101). In a continuous process of remodelling, these opposing forces regulate the interconversions between HDL_2 and HDL_3 (Fig. 1). This "HDL cycle" may constitute a central part of the "reverse cholesterol transport" system mediating cholesterol transport from cells to the liver. In addition to the processes discussed above, HDL_2 and HDL_3 levels are influenced by many other factors beyond the scope of the present chapter, such as the synthetic and catabolic rates of apolipoproteins AI and AII, and production of nascent HDL particles in the splanchnic area.

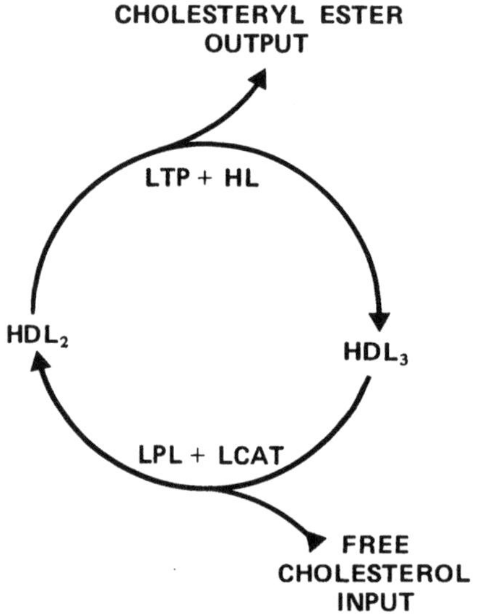

Fig. 1. Schematic outline of cholesterol transport in high density lipoprotein "cycle". LTP = lipid transfer protein, HL = hepatic lipase, LPL = lipoprotein lipase, LCAT = lecithin:cholesterol acyl transferase.

Acknowledgement

The writing of this review was supported by The Academy of Finland, The Sigrid Juselius and Paavo Nurmi Foundations, and the University of Helsinki.

References

1. Nikkilä, E. A. (1953) Scand. J. Clin. Lab. Invest. 5, suppl. 8, 1–101.
2. Nikkilä, E. A. (1983) in: The Metabolic Basis of Inherited Disease (Stanbury, J. B., Wyngaarden, J. B., Fredrickson, D. S., Goldstein, J. L. & Brown, M. S., eds.) pp. 622–642, McGraw-Hill, New York.
3. Nilsson-Ehle, P., Garfinkel A. S. & Schotz, M. C. (1980) Ann. Rev. Biochem. *49*, 667–693.
4. Quinn, D., Shirai, K. & Jackson, R. L. (1982) Prog. Lipid Res. *22*, 35–78.
5. Kompiang, I. P., Bensadoun A. & Yan, M. W. W. (1976) J. Lipid Res. *17*, 498–505.
6. Bensadoun A. & Kompiang, I. P. (1979) Fed. Proc. *38*, 2622–2626.
7. Scow, R. O., Hamosh, M., Blanchette-Mackie, E. J. & Evans, A. J. (1972) Lipids *7*, 497.
8. Fielding, C. J. & Higgins, J. M. (1974) Biochemistry *13*, 4324–4330.
9. Gofman, J. W., Delalla, O., Glazier, F., Freeman, N. K., Lindgren, F. T., Nichols, A. V., Strisower, B. & Tamplin, A. R. (1954) Plasma *2*, 413–484.
10. Nichols, A. V. (1967) Adv. Biol. Med. Physics *11*, 109–158.
11. Hoffmann, A. A., Nelson, W. R. & Goss, F. A. (1967) Am. J. Cardiol. *20*, 516–524.
12. Miller, G. J. & Miller, N. E. (1975) Lancet *1*, 16–19.
13. Olsson, A. G. & Carlson, L. A. (1975) Acta Med. Scand., suppl. 580.
14. Gordon, T., Castelli, W. P., Hjortland, M. C., Kannel, W. B. & Dawber, T. R. (1977) Am. J. Med. *62*, 717–714.
15. Wilson, D. E. & Lees, R. S. (1972) J. Clin. Invest. *51*, 1051–1057.
16. Altekruse, E. B. & Wilmore, J. H. (1973) J. Occupat. Med. *15*, 110–113.
17. Nichols, A. V., Coggiola, E. L., Jensen, L. C. & Yokoyama, E. H. (1968) Biochem. Biophys. Acta *168*, 87–94.
18. Fredrickson, D. S., Levy, R. I. & Lindgren, F. T. (1968) J. Clin. Invest. *47*, 2446–2457.
19. Levy, R. I., Lees, R. S. & Fredrickson, D. S. (1966) J. Clin. Invest. *45*, 63–77.
20. Lindgren, F. T., Nichols, A. V. & Freeman, N. K. (1955) J. Phys. Chem. *59*, 930–938.
21. Furman, R. H., Howard, R. P. & Alaupovic, P. (1962) Metabolism *11*, 879–892.
22. LaRosa, J. C., Levy, R. I., Brown, V. W. & Fredrickson, D. S. (1971) Am. J. Physiol. *220*, 785–791.
23. Eisenberg, S., Bilheimer, D. W., Levy, R. I. & Lindgren, F. T. (1973) Biochim. Biophys. Acta *326*, 361–377.
24. Forte, T. M., Krauss, R. M., Lindgren, F. T. & Nichols, A. V. (1979) Proc. Natl. Acad. Sci. USA *76*, 5934–5938.
25. Behr, S. R., Patsch, J. R., Forte, T. & Bensadoun, A. (1981) J. Lipid Res. *22*, 443–451.
26. Patsch, J. R., Gotto, Jr., A. M., Olivecrona, T. & Eisenberg, S. (1978) Proc. Natl. Acad. Sci. USA *75*, 4519–4523.
27. Redgrave, T. G. & Small, D. M. (1979) J. Clin. Invest. *64*, 162–171.
28. Tall, A. R., Green, P. H. R., Glickman, R. M. & Riley J. W. (1979) J. Clin. Invest. *64*, 977–989.
29. Eisenberg, S. (1984) J. Lipid Res. *25*, 1017–1058.
30. Nikkilä, E. (1984) in: Clinical and Metabolic Aspects of High-Density Lipoproteins (Miller, N. E. & Miller, G. J., eds.) pp. 217–245), Elsevier, Amsterdam, New York, Oxford.
31. Nikkilä, E. A., Taskinen, M.-R. & Sane, T. (1987) Am. Heart. J. *113*, 543–548.
32. Nikkilä, E. A., Huttunen, J. K. & Ehnholm, C. (1977) Diabetes *26*, 11–21.
33. Nikkilä, E. A., Taskinen, M.-R. & Kekki, M. (1978) Atherosclerosis *29*, 497–501.
34. Nikkilä, E. A., Taskinen, M.-R., Rehunen, S. & Härkönen, M. (1978) Metabolism *27*, 1661–1671.
35. Nikkilä, E. A. (1978) in: High Density Lipoproteins and Atherosclerosis (Gotto, Jr. A. M., Miller, N. E. and Oliver, M. F., eds.) pp. 177–192, Elsevier/North Holland, Amsterdam.

36. Nikkilä, E. A. (1978) Eur. J. Clin. Invest. *8*, 111–113.
37. Nikkilä, E. A., Kuusi, T. & Taskinen, M.-R. (1982) in: Metabolic Risk Factors in Ischemic Cardiovascular Disease (Carlson, L. A. & Pernow, B., eds.) pp. 205–215, Raven Press, New York.
38. Havel, R. J. (1958) Am. J. Clin. Nutr. *6*, 662–668.
39. Lithell, H., Jacobs, I., Vessby, B., Hellsing, K. & Karlsson, J. (1982) Metabolism *31*, 994–998.
40. Taskinen, M.-R., Nikkilä, E. A., Kuusi, T. & Tulikoura, I. (1983) Arteriosclerosis *3*, 607–615.
41. Krauss, R. M. (1982) Med. Clin. N. Am. *66*, 403–430.
42. Peltonen, P., Marniemi, J., Hietanen, E., Vuori, I. & Ehnholm, C. (1981) Metabolism *30*, 518–525.
43. Belfrage, P., Berg, B., Hägerstrand, I., Nilsson-Ehle, P., Törnqvist, H. & Wiebe, T. (1977) Eur. J. Clin. Invest. *7*, 127–131.
44. Taskinen, M.-R., Välimäki, M., Nikkilä, E. A., Kuusi, T. & Ylikahri, R. (1985) Metabolism *34*, 112.
45. Taskinen, M.-R., Nikkilä, E. A., Välimäki, M., Sane, T., Kuusi, T., Kesäniemi, Y. A. & Ylikahri, R. (1987) Am. Heart. J. *113*, 458–464.
46. Taskinen, M.-R. & Nikkilä, E. A. (1979) Diabetologia *17*, 351–356.
47. Nikkilä, E. A. & Hormila, P. (1978) Diabetes *27*, 1078–1086.
48. Mancini, M., Mattock, M., Rabaya, E., Chait, A. & Lewis, B. (1973) Atherosclerosis *17*, 445–454.
49. Taskinen, M.-R. & Nikkilä, E. A. (1979) Atherosclerosis *32*, 289–299.
50. Taskinen, M.-R., Nikkilä, E. A., Kuusi, T. & Harno, K. (1982) Diabetologia *22*, 46–50.
51. Morton, R. E. & Zilversmit, D. B. (1982) J. Lipid Res. *23*, 1058–1067.
52. Ihm, J., Ellsworth, J. L., Chataing, B. & Harmony, J. A. K. (1982) J. Biol. Chem. *257*, 4818–4827.
53. Albers, J. J., Tollefson, J. H., Chen, C.-H. & Steinmetz, A. (1984) Arteriosclerosis *4*, 49–58.
54. Abbey, M., Calvert, G. D. & Barter, P. J. (1984) Biochim. Biophys. Acta *793*, 471–480.
55. Tall, A. R. (1986) J. Lipid Res. *27*, 361–367.
56. Glomset, J. A. (1968) J. Lipid Res. *9*, 155–167.
57. Rajaram, O. V. & Barter, P. J. (1985) Biochim. Biophys. Acta *835*, 41–49.
58. Deckelbaum, R., Eisenberg, S., Granot, E., Oschry, Y. & Olivecrona, T. (1982) Atherosclerosis *2*, 437a.
59. Sammett, D. & Tall, A. R. (1985) J. Biol. Chem. *260*, 6687–6697.
60. Eisenberg, S. (1987) Am. Heart J. *113*, 555–561.
61. Shepherd, J. & Packard, C. (1987) Am. Heart J. *113*, 503–508.
62. Eisenberg, S., Gavish, D., Oschry, Y., Fainaru, M. & Deckelbaum, R. (1984) J. Clin. Invest. *74*, 470–482.
63. Deckelbaum, R. J., Granot, E., Oschry, Y., Rose, L. & Eisenberg, S. (1984) Arteriosclerosis *4*, 225–231.
64. Kuusi, T., Nikkilä, E. A., Taskinen, M.-R., Somerharju, P. & Ehnholm, C. (1982) Clin. Chim. Acta *122*, 39–45.
65. Jansen, H. & De Greef, W. J. (1981) Biochem. J. *196*, 739–745.
66. Cordle, S. R., Yeaman, S. J. & Clegg, R. A. (1983) Biochim. Biophys. Acta *753*, 213–219.
67. Kuusi, T., Saarinen, P. & Nikkilä, E. A. (1980) Atherosclerosis *36*, 589–593.
68. Tikkanen, M. J., Nikkilä, E. A., Kuusi, T. & Sipinen, S. (1981) Clin. Chim. Acta *115*, 63–71.
69. Tikkanen, M. J., Kuusi, T., Vartiainen, E. & Nikkilä, E. A. (1979) Acta Obstet. Gynecol. Scand., *Suppl. 88*, 83–88.
70. Tikkanen, M. J., Nikkilä, E. A., Kuusi, T. & Sipinen, S. (1982) J. Clin. Endocr. Metab. *54*, 1113–1117.
71. Tikkanen, M. J., Nikkilä, E. A., Kuusi, T. & Sipinen, S. (1982) Acta Endocrinol. *99*, 630–635.
72. Bengtsson, G., Olivecrona, T. (1980) FEBS Lett. *119*, 290–292.
73. Kuusi, T., Kinnunen, P. K. J. & Nikkilä, E. A. (1979) FEBS Lett. *104*, 384–388.
74. Jansen, H., van Tol, A. & Hülsmann, W. C. (1980) Biochem. Biophys. Res. Commun. *92*, 53–59.
75. Grosser, J., Schrecker, O. & Greten, H. (1981) J. Lipid Res. *22*, 437–442.
76. Murase, T. & Itakura, H. (1981) Atherosclerosis *39*, 293–300.
77. Groot, P. H. E., Jansen, H. & Van Tol, A. (1981) FEBS Lett. *129*, 269–272.
78. Groot, P. H. E., Scheek, L. M. & Jansen, H. (1983) Biochim. Biophys. Acta *751*, 393–400.
79. Van Tol, A., Van Gent, T. & Jansen, H. (1980) Biochem. Biophys. Res. Commun. *94*, 101–108.

80. Shirai, K., Barnhart, R. L. & Jackson, R. L. (1981) Biochem. Biophys. Res. Commun. *100*, 591 – 598.
81. Tikkanen, M. J., Kuusi, T., Nikkilä, E. A. & Sipinen, S. (1986) Maturitas *8*, 7 – 17.
82. Tikkanen, M. J. & Nikkilä, E. A. (1987) Am. Heart J. *113*, 562 – 567.
83. Glueck, C. J., Gartside, P., Fallat, R. W. & Mendoza, S. (1976) Metabolism *25*, 625 – 632.
84. Applebaum, D. M., Goldberg, A. P., Pykälistö, O. J., Brunzell, J. D. & Hazzard, W. R. (1977) J. Clin. Invest. *59*, 601 – 608.
85. Ehnholm, C., Huttunen, J. K., Kinnunen, P. J., Miettinen, T. A. & Nikkilä, E. A. (1975) N. Engl. J. Med. *292*, 1314 – 1317.
86. Taggart, H. M., Applebaum-Bowden, D., Haffner, S., Warnick, G. R., Cheung, M. C., Albers, J. J., Chestnut, III C. H. & Hazzard, W. R. (1982) Metabolism *31*, 1147 – 1152.
87. Tikkanen, M. J., Nikkilä, E. A., Kuusi, T. & Sipinen, S. (1981) Atherosclerosis *40*, 365 – 369.
88. Kuusi, T., Nikkilä, E. A., Tikkanen, M. J., Taskinen, M.-R. & Ehnholm, C. (1983) in: Atherosclerosis VI (Schettler, F. G., Gotto, A. M., Middelhoff, G., Habenicht, A. J. R. & Jurutka, K. R., eds.) pp. 628 – 632, Springer, Berlin, New York.
89. Tikkanen, M. J. & Nikkilä, E. A. (1986) J. Reprod. Med. *31*, 898 – 905.
90. Applebaum-Bowden, D., Haffner, S. M., Wahl, P. W., Hoover, J. J., Warwick, G. R., Albers, J. J. & Hazzard, W. (1985) Arteriosclerosis *5*, 273 – 282.
91. Krauss, R. M., Levy, R. I. & Fredrickson, D. S. (1974) J. Clin. Invest. *54*, 1107 – 1124.
92. Musliner, T. A., Herbert, P. N., Kingston, M. J. (1979) Biochim. Biophys. Acta *575*, 277 – 288.
93. Goldberg, I. J., Le, N.-A., Paterniti, J. R., Ginsberg, H. N. & Lindgren, F. T. (1982) J. Clin. Invest. *70*, 1184 – 1192.
94. Breckenridge, W. C., Little, J. A., Alaupovic, P., Wang, C. S., Kuksis, A., Kakis, G., Lindgren, F. & Gardiner, G. (1982) Atherosclerosis *45*, 161 – 179.
95. Reardon, M. F., Sakai, H. & Steiner, G. (1982) Arteriosclerosis *2*, 396 – 402.
96. Tikkanen, M. J., Kuusi, T., Nikkilä, E. A. & Sane, T. (1985) FEBS Lett. *181*, 160 – 164.
97. Schaefer, E. J., Foster, D. M., Zech, L. A., Lindgren, F. T., Brewer, H. B., Jr. & Levy, R. I. (1983) J. Clin. Endocrinol. Metab. *57*, 262 – 267.
98. Tikkanen, M. J., Kuusi, T., Nikkilä, E. A. & Stenman, U.-H. (1986) Metabolism *35*, 99 – 104.
99. Anderson, D. W., Nichols, A. V., Pan, S. S. & Lindgren, F. T. (1978) Atherosclerosis *29*, 161 – 179.
100. Shepherd, J., Packard, C. J., Stewart, J. M., Vallance, B. D., Lawrie, T. D. V. & Morgan, H. G. (1980) Clin. Chim. Acta *101*, 57 – 62.
101. Patsch, J. R., Prasad, S., Gotto, Jr., A. M. & Bengtsson-Olivecrona, G. (1984) J. Clin. Invest. *74*, 2017 – 2023.

Chapter 7
Intravascular Lipoprotein Remodelling: Neutral Lipid Transfer Proteins

S. Eisenberg and R. Deckelbaum

1 Introduction

Lipid transfer reactions in human plasma have been described more than three decades ago. Phospholipid (PL) and unesterified cholesterol (UC) exchange were defined already during the mid-1950's (1, 2). A decade later, in 1965, neutral lipid (triglycerides and cholesteryl esters (CE)) transfers were described (3). The observations by Nichols and Smith differed from the earlier reports in two important respects: firstly, the lipids that moved between lipoproteins were the very hydrophobic TG and CE molecules present predominantly in the hydrophobic core of the particles, and secondly the reaction caused a pronounced change of the lipid composition of the lipoproteins (3). During the 24 hours incubation, Nichols and Smith observed accumulation of cholesteryl ester within the TG-rich lipoprotein, very low density lipoprotein (VLDL), whereas the CE-rich low and high density lipoproteins (LDL and HDL) acquired TG molecules. These pioneering studies remained almost unnoticed for another decade. Renewed interest in the reaction began when Zilversmit et al. reported in 1975 that a protein, or proteins, present in rabbit lipoprotein deficient plasma ($\rho > 1.25\,kg/l$) promote exchange of labeled CE between VLDL and LDL from hypercholesterolemic rabbit plasma (4).

In the next few years, numerous investigations on the nature of the reaction have been published. Through this scientific effort, the protein responsible for lipid transfer has been purified and identified and the mechanism(s) of the reaction have been elucidated. But, again, almost a decade has passed before it has been realized that lipid transfer reactions are responsible for major structural alterations of the plasma lipoproteins (5). That such structural and compositional alterations of lipoproteins have profound metabolic implications, has been realized only very recently, when reports demonstrating abnormal metabolism of lipoproteins excessively affected by lipid transfers have been published. This chapter describes the known effects of lipid transfer reactions on lipoprotein metabolism and speculates on how such effects modify the process of plasma fat transport in health and disease. To achieve this aim, we will begin with a brief description of the reaction, continue with construction of *in vitro* models and finally describe the metabolic implications of lipid transfer processes.

2 The Lipid Transfer Reaction

2.1 Lipid transfer protein (LTP)

The properties of LTP, its distribution among lipoproteins and its mechanism of action have recently been reviewed (6) and are described here only in brief. Several investigators have reported procedures to isolate LTP from human (7–13) and rat (12) plasma. Initially, identity between human LTP and apolipoprotein D was reported (7), and it was speculated that particles containing LCAT, LTP and Apo AI exist in plasma (14). Soon, however, it became apparent that the major LTP in human plasma is distinctly different from all other known apolipoproteins (10–13). The protein has a molecular mass of approximately 60 000 to 70 000 daltons and facilitates exchange of CE, TG and PL molecules (9, 10, 13). Whenever tested, the three lipid transfer activities co-purify during the different purification steps and are most probably mediated by the same protein. Phospholipid transfer, however, has also been reported to be mediated by a protein discernible from the classic LTP (6, 13, 15). In the rat, an animal species devoid of CE and TG transfer activity (16, 17), protein-mediated PL exchange is readily demonstrable (18, 19). In human plasma, LTP is associated mainly with HDL (20), but after centrifugation, distributes equally between HDL and lipoprotein-deficient plasma ($\rho > 1.25$ g/l). LTP binding to VLDL and LDL has been recently demonstrated, but whereas VLDL- and LDL-LTP complexes rapidly dissociate, HDL-LTP complexes remain stable (21). HDL-LTP complexes are stable at pH of 7.4 even in the presence of 3.5 mol/l NaCl or NaBr but dissociation occurs when the pH is lowered to 4.5 (20). LTP activity is not affected by heating the plasma at 58 °C for one hour, and is insensitive to sulfhydryl blocking substances (22, 23), two treatments that abolish LCAT activity. LTP activity accumulates in rabbit liver perfusates (24, 25) indicating that the liver is an important source of the protein. LTP is also synthesized and secreted by human monocyte-derived macrophages (26). Of interest, an LTP inhibitor has been described in human plasma (27, 28). The inhibitor is a sialoglycoprotein that possibly interacts with the substrate rather than with the transfer protein. Whether the inhibitor is active *in vivo* is, however, not known.

LTP activity can be identified in the plasma of some, but not all animal species (29). High activities are found in rabbits, possum and trout, while values are low or undetectable in rats, sheep, cows, pigs and dogs. Humans have intermediate activity.

2.2 Mode of action

Lipid transfer reactions cause either an exchange phenomenon (replacement of a molecule in a lipoprotein by the *same* molecule from another lipoprotein, e.g., CE exchange between LDL and HDL), or a transfer phenomenon, i.e., replacement of a molecule in a lipoprotein by a *different* molecule of another lipoprotein, e.g., exchange of VLDL-TG for HDL-CE. These two reactions, shown schematically in Figure 1, have been termed *homoexchange* and *heteroexchange*, respectively (21) and appear to be mediated by the same mechanism (30). TG and CE transfers are

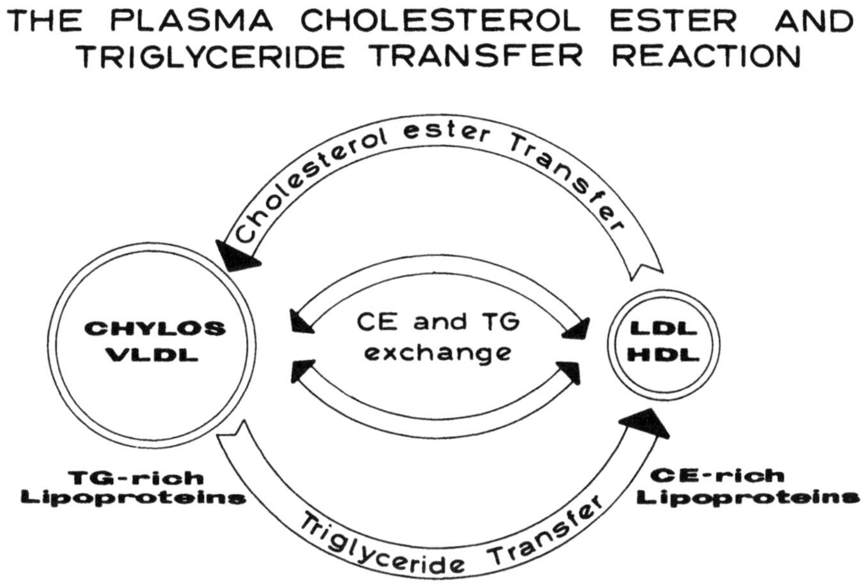

Fig. 1. Schematic representation of the plasma lipid transfer reaction. The figure shows the reactions that take place between TG-rich lipoproteins (chylomicrons and VLDL) and CE-rich lipoproteins (LDL + HDL). Homo-exchange of CE or TG between lipoproteins of *dissimilar* core-lipid composition (inner arrows) is of small magnitude; hetero-exchange of CE for TG (outer arrows), predominates. Homo-exchange is of course the predominant reaction between lipoproteins of *similar* core-lipid composition, e. g., LDL and HDL (CE exchange) or chylomicrons and VLDL (TG exchange).

inter-related and the sum of TG and CE transfer is independent of donor and acceptor lipoprotein concentrations and of their lipid composition (30). It has therefore been suggested that heteroexchange of CE and TG proceeds by a reciprocal exchange of these molecules without net transfer of core lipids between lipoproteins (30). Yet, in heteroexchange reactions, the core lipid composition of both donor and acceptor lipoproteins changes: TG-rich lipoproteins lose TG molecules and acquire equimolar amounts of CE, while CE-rich lipoproteins lose CE and acquire TG molecules. It is this heteroexchange phenomenon that causes alterations of the structure, composition and metabolism of the lipoproteins. In a recent study, we investigated the capacities of three VLDL density subfractions – VLDL-I, VLDL-II and VLDL-III – to serve as acceptors of CE from HDL, as compared to LDL (31). The study demonstrated homoexchange of CE with all four lipoprotein fractions that was proportional to the surface area of the lipoproteins. Heteroexchange (measured by increments of CE content in the VLDL's and LDL), however, was best correlated to the TG/CE ratios of the lipoproteins (31). Morton and Zilversmit reached similar conclusions while studying native and reconstituted lipoproteins (30). In our study, heteroexchange between VLDL-I and HDL accounted for 80% of total CE transfer (homo- and heteroexchange), but only 5–10% of total CE transfer to LDL (31); VLDL-II and VLDL-III exhibited intermediate behavior. Figure 2 presents graphically these data and demonstrates

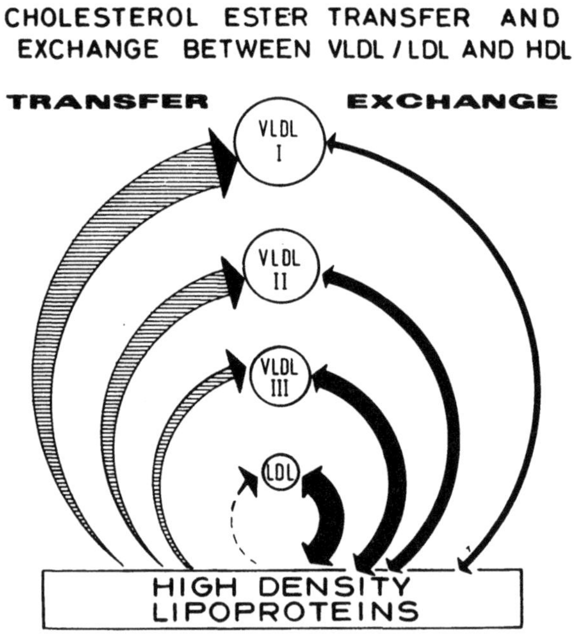

Fig. 2. Lipid transfer and exchange between VLDL subfractions, LDL and HDL. The scheme summarizes data published recently on CE transfer and exchange between VLDL-I, II, III and LDL and HDL (31). In VLDL-I, transfer predominates while exchange is minimal. The opposite is found for LDL. The behavior of VLDL-II and VLDL-III is intermediate. As discussed in the text, the findings support the notion that the core-lipid composition (TG and CE) of the lipoproteins determines the relative magnitude of lipid transfer compared to lipid exchange.

the effects of core lipid composition on the relative magnitude of homo- and heteroexchange in the four lipoprotein preparations.

Two mechanisms have been proposed for the mode of action of LTP. The first assumes that LTP acts as a carrier of lipid molecules between lipoproteins, i. e., a ping-pong mechanism (32). Ihm et al., however, provided evidence that LTP facilitates lipid transfer by enhancing formation of a ternary collision complex consisting of LTP, donor lipoprotein and acceptor lipoprotein (33). This latter model is currently regarded as the favored mechanism of action of LTP's (6).

3 In vitro Remodelling of Lipoproteins

3.1 Chylomicrons and VLDL

In transfer reactions with LDL or HDL, chylomicrons and VLDL serve as donors of TG and acceptors of CE. Marcel et al. demonstrated *in vitro* transfer of CE from HDL to either ascitic fluid chylomicrons or plasma chylomicrons when LTP was added to the incubation mixture (34). CE transfer was unrelated to Apo E transfer and appeared to be more pronounced with HDL fractions devoid of Apo E. Because chylomicrons have very high TG/CE ratios, it is to be expected that they serve as excellent acceptors of CE (see below). That, however, has not been studied.

Many investigators reported that VLDL actively participates in CE transfer reactions (3, 4, 31, 35–42). Back-transfer of TG from VLDL to HDL and LDL has clearly been demonstrated (37, 38). There are both homo- and heteroexchange processes, and the VLDL-CE mass increases. Barter et al. demonstrated 7–11 times greater capacity of VLDL to serve as an acceptor of LCAT-derived CE from HDL, as compared to LDL (40, 41). Noel et al. compared the abilities of nascent rat hepatic perfusate VLDL, rat serum VLDL and human plasma VLDL to accept CE from human HDL (42). Mass CE transfer was greatest to rat hepatic perfusate VLDL, lower to rat serum VLDL and lowest to human plasma VLDL. Mass CE transfer thus clearly was related to the TG/CE ratio in the acceptor lipoproteins. A similar conclusion was reached by us while studying the capacity of VLDL density subfractions to serve as acceptors for HDL-CE (31). Tall and colleagues (43, 44) demonstrated that VLDL lipolysis by bovine milk lipoprotein lipase enhances transfer of CE from HDL to VLDL, and speculated that accumulation of fatty acids in the VLDL (and HDL) increases binding of LTP to the lipoproteins. That, in turn, enhances CE transfer. More recently, similar conclusions were derived for chylomicrons (45, 46).

CE transfer to VLDL and chylomicrons takes place also *in vivo*. Nestel et al. calculated that the apparent flux of CE from HDL to VLDL is similar to reported values for total cholesterol esterification in human plasma *in vivo* (47). Thus, most or even all of the LCAT-derived CE in human plasma is metabolized after being transferred to TG-rich lipoproteins.

3.2 Low density lipoproteins

LTP-facilitated homoexchange of CE between LDL and HDL, without mass transfer, has been demonstrated in virtually all studies on lipid transfer reactions. Whether CE mass transfer to LDL (from HDL) occurs when the LCAT reaction is active, is uncertain. Fielding and Fielding reported increased CE content of LDL during plasma incubations (39). Other investigators, however, found only minimal CE mass transfer under similar, albeit not identical conditions (31, 30, 41).

Heteroexchange of CE and TG between LDL and VLDL has been studied in depth by us (48–50). This reaction appears to predispose to marked compositional, structural and metabolic consequences. More than half of the CE molecules in LDL are replaced by TG after 18 h incubation of LDL with VLDL and an LTP source (48, 50). This change of core lipid composition is associated with a change of the surface constituent composition of the LDL, and the LTP modified LDL (M-LDL) acquires Apo AI, Apo C and Apo E from the VLDL and/or the lipoprotein-deficient plasma fraction. The M-LDL also is larger (by electron microscopy) and less dense than the native LDL, presumably because of the larger volume and weight of TG molecules as compared to CE (48). Next, we attempted to deplete the M-LDL of transferred TG molecules by adding lipoprotein lipase to the incubation mixture. More than 80% of M-LDL triglycerides could be hydrolyzed. This resulted in changed composition of M-LDL towards the native CE-rich lipoprotein with loss of the added apoproteins. The post-lipolysis M-LDL however, was significantly smaller than either M-LDL or the native LDL. The cycle of transfer and lipolysis, therefore, resulted in net loss of core lipids from the LDL. Undoubtedly, as discussed below, such activities provide a basis for many metabolic events occurring *in vivo*. Two other observations are worth mentioning. Firstly, the artificial TG lipid emulsion, Intralipid, is as active in similar experiments as VLDL (51), indicating that none of the VLDL specific constituents (e.g., apoproteins) is necessary. Secondly, CE-depleted, TG-enriched LDL is less effective than native LDL in regulating cellular cholesterol synthesis and LDL-receptor activity (50). This last observation has been the first clue that the lipid transfer reaction may affect lipoprotein- and cellular-lipid metabolism.

3.3 High density lipoproteins

Remodelling of HDL particles by amounts of core lipids – predominantly CE – is a process responsible for the regulation of HDL subpopulation (HDL_1, HDL_2, HDL_3, and HDL_4) distribution (52, 53). In this complex cycle, supply of free cholesterol and phospholipids (from lipolysis and cell membranes) together with LCAT is responsible for conversion of the small, dense HDL_3 to HDL_2 (53, 54), while heteroexchange of CE and TG in HDL_2 followed by hydrolysis of transferred triglycerides causes reverse-conversion of HDL_2 to HDL_3 (52, 53). The events occurring during the reverse-conversion process were recently reported by us (55). Using systems similar to those described above for LDL, human HDL_2 was incubated with human VLDL in the presence of LTP sources (human lipoprotein-deficient plasma or purified LTP). As with LDL, HDL_2 lost CE to, and acquired TG from, VLDL. The modified HDL_2 then lost most of the acquired TG molecules after incubation with lipases, either hepatic lipase or bovine milk lipoprotein lipase. After repetitive *in vitro* cycles of HDL_2 incubation with VLDL and LTP followed by M-HDL_2 lipolysis, HDL_2 changed toward HDL_3 as judged by increasing density and decreasing CE content (by 2/3) and particle size (55). Although in these cycles both lipoprotein lipase and hepatic lipase were active against HDL-TG, *in vivo*, it is the latter enzyme that more probably is responsible for the reverse-conversion process (53, 56).

3.4 Lipoprotein remodelling: structural implications

Figures 3 and 4 summarize some of the consequences of CE and TG heteroexchange reactions. The main feature of the reaction is replacement of molecules that cannot be metabolized in the circulation, CE, by molecules that are avidly metabolized, TG. CE molecules transferred to the TG-rich chylomicrons and VLDL constitute the major source of CE, and remain associated with the particles throughout their intravascular metabolism. As for LDL and HDL, lipid transfer followed by TG

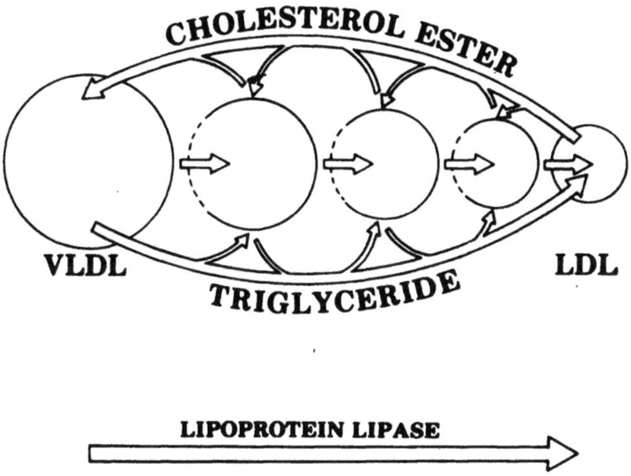

Fig. 3. Pathways for removing CE from LDL. While CE-TG hetero-exchange between LDL and VLDL (or chylomicrons) takes place, most of the acquired TG is hydrolyzed by lipases resulting in a smaller, CE-depleted, LDL.

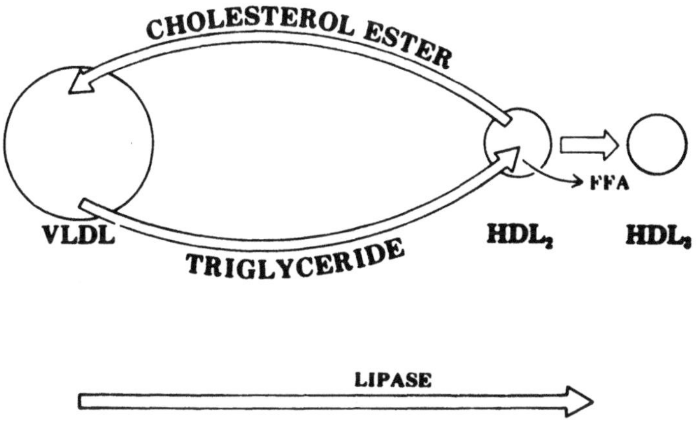

Fig. 4. Pathways for removing CE from HDL. Subsequent to CE-TG hetero-exchange between HDL and VLDL (or chylomicrons), acquired TG is hydrolyzed by lipases resulting in a smaller, CE-depleted, HDL. These processes provide a pathway whereby HDL_2 can be modified towards HDL_3.

hydrolysis provides a mechanism by which core lipid molecules are deleted. This mechanism appears to be an important determinant of the size, density, CE content, and subpopulation distribution of the two lipoproteins. Lipoprotein remodelling, moreover, appears to have profound effects on the metabolic fate of VLDL, LDL and possibly HDL, as discussed below.

4 Effects of LTP on the Structure, Composition and Metabolism of in vivo Lipoproteins

Until recently, the metabolic implications of lipid transfer processes, if any, were obscure. Yet, it was clear that transfer of CE and TG molecules between lipoproteins must affect the structure, composition and metabolism of the particles. For example, addition or deletion of cholesteryl ester molecules is a process that must affect the function of lipoproteins, e. g., their capacity to regulate cellular metabolic pathways. Elucidation of such metabolic consequences of lipid transfer processes, has, surprisingly, begun only in the last few years, and is still poorly understood. Nevertheless, the available data allow formulation of schemes that describe the potential role of lipid transfer reactions in metabolic cycles. These data are based mainly on studies of three models: plasma lipoproteins in rats, an animal species devoid of lipid transfer; human patients with abetalipoproteinemia, whose plasma lacks acceptors for cholesteryl esters; and patients with hypertriglyceridemia (HTG), a condition with presumably excessive lipid transfer. In the following text, lessons learnt from all three models are examined although it is realized that important differences exist between the three.

4.1 Chylomicrons and very low density lipoproteins

4.1.1 Structure and composition

Data on the structure and composition of chylomicrons and VLDL (predominantly the latter), that can be related to the LTP reaction are available in rats and in humans with hypertriglyceridemia (HTG).

In rats, VLDL contains only a few CE molecules, (1.5–2.5% of the total lipoprotein mass (17)) and the fatty acid pattern of these esters indicates their hepatic origin. Whether paucity of CE contributes to the rapid degradation of VLDL remnants in rats and to their low conversion to LDL (only 5–10% of total VLDL-Apo B becomes LDL (57, 58)), has not been investigated so far.

The structure and composition of VLDL from subjects with HTG before and after initiation of lipid lowering therapy (with bezafibrate) was recently reported (59). As compared to N-VLDL, HTG-VLDL contains less protein, more free and esterified cholesterol and less TG, but no change is found for total phospholipids. All four abnormalities are strongly and significantly related to plasma TG levels and revert towards normal when TG lowering therapy is instituted. In a subsequent study (60), the VLDL (N and HTG) was fractionated to three populations, I, II and III, according to their size and density. Of the three, VLDL-I, the fraction containing the largest and least dense particles, was found to be the most abnormal. For example, CE to protein ratios in HTG-VLDL-I were about 2.5–3.0-fold higher than in N-VLDL-I. HTG-VLDL-III, in contrast, was almost normal, while HTG-VLDL-II was moderately abnormal. It thus appears that abnormalities of the large-sized VLDL particles are responsible for most of the abnormalities of

HTG-VLDL. The metabolic processes that are responsible for enrichment of HTG-VLDL with cholesteryl esters are presumably related to an excessive lipid transfer process. Experiments carried out with VLDL and VLDL subfractions *in vitro* (31) demonstrated that there was a high capacity of VLDL-I to serve as an acceptor of CE and donor of TG in lipid transfer reactions. VLDL-II had moderate capacity to participate in the reaction, while VLDL-III, had least. LDL did not acquire CE (from HDL) although an avid exchange process was demonstrated. It is therefore possible that during their prolonged circulating lives, large VLDL particles become excessively enriched with CE molecules *in vivo*.

4.1.2 **Metabolism**

Abnormal metabolism of VLDL in HTG states has been demonstrated in many studies (61 – 64). Two abnormalities are common to all patients, regardless of disease entity, experimental procedure or method of analysis. One is a slow delipidation rate and the other, a decreased fractional conversion of VLDL-Apo B to LDL-Apo B. The first abnormality reflects most probably the pathogenesis of HTG-states: absolute, or relative (to high synthetic rates) decreases in the rates of VLDL triglyceride transport (65, 66). The metabolic basis of the reduced VLDL → LDL conversion is unknown. One possibility is that the incomplete conversion of VLDL to LDL in HTG states is due to the excessive content of CE in HTG-VLDL (59, 60). To test this hypothesis, we determined the structure and composition of "in vitro LDL" produced by nearly complete TG hydrolysis of three VLDL subfractions (I, II and III) obtained from N and HTG human subjects (60). LDL-like particles (by density, diameter and lipid and protein composition) were produced from all three N-VLDL fractions. In fact, the LDL-like particles produced from N-VLDL-I contained considerably less CE per Apo B than N-LDL (weight ratios of 1.27 and 2.0, respectively). Apparently, N-VLDL-I becomes *enriched* with CE on its conversion to LDL. In HTG, in contrast, LDL-like particles were found only after lipolysis of VLDL-III. Neither VLDL-I nor VLDL-II (especially the former) produced LDL-like particles, in spite of almost complete TG hydrolysis ($>95\%$). Post-lipolysis VLDL-I and VLDL-II in such patients were enriched with CE, less dense and much larger than LDL. For some HTG-VLDL-I preparations, the diameter of the post-lipolysis particles exceeded 30 nm, and the CE to protein weight ratios were between 4.0 and 5.0. This data suggested that the bulk of HTG VLDL-I and VLDL-II cannot form LDL and is degraded independent of the LDL pathway. Hence, in HTG states, only VLDL-III is the source of LDL. This hypothesis, shown schematically in Figure 5, explains not only the reduced VLDL → LDL conversion in HTG but also the increase of LDL levels observed when TG levels are reduced by therapy (59, 63). We suggest that with therapy, VLDL delipidation is accelerated, less CE molecules are transferred to the VLDL (especially VLDL-I and VLDL-II), and more VLDL particles complete the VLDL → LDL conversion pathway.

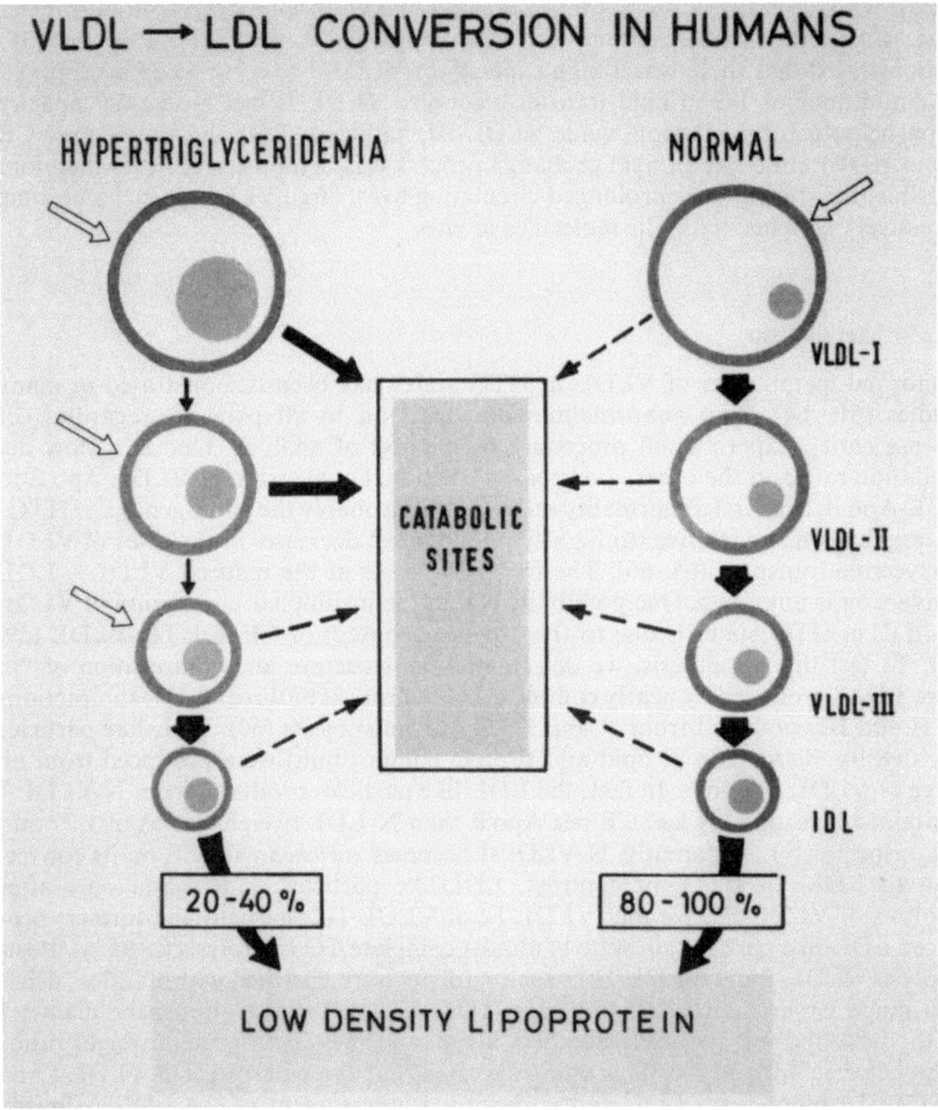

Fig. 5. Effects of excessive lipid transfer on VLDL metabolism in hypertriglyceridemic (HTG) human subjects. In normals, almost complete conversion of VLDL-I to LDL is achieved and only a few particles are cleared from the plasma along the cascade. In HTG, the situation is different. VLDL-I and VLDL-II become excessively enriched with CE molecules, cannot complete the conversion process and are cleared from the plasma before IDL or LDL are formed. That accounts for the low conversion of VLDL to LDL (20–60%). VLDL-III is apparently secreted independently of VLDL-I or VLDL-II, and is the predominant precursor of IDL and LDL.

4.2 Intermediate density lipoproteins

4.2.1 **Structure and composition**

The structure and composition of IDL, N or HTG, has rarely been reported. In our study, IDL preparations obtained by zonal ultracentrifugation from the plasma of normotriglyceridemic subjects or patients with HTG (moderate and severe) are of similar structure and composition (60). On average, IDL from either source contains 17–20% protein, 20–25% TG, 25–30% CE, 8–10% free cholesterol and 18–22% phospholipids. The molar ratio of lecithin to sphingomyelin is 2.2 to 2.4, and the average diameter of the particles, about 27 nm. More recently, we determined the apoprotein composition of N- and HTG-IDL (unpublished). Again, preparations were similar, with Apo B contributing 60–70% of total protein, Apo E 8–15%, and Apo C 20–30%. These similarities support the view that IDL in either normals or patients is produced from the same precursor, i.e., VLDL-III. It thus appears that IDL is not extensively remodelled by lipid transfer reactions, the main exception being found in type III hyperlipoproteinemia. In this disorder, IDL (and VLDL) circulate for exceptionally long periods, and both are makedly enriched with CE (67). We suggest that this phenomenon reflects unusual enrichment of the particles (VLDL and IDL) with cholesteryl esters by the lipid transfer reactions.

4.2.2 **Metabolism**

It is generally accepted that, with the exception of type III hyperlipoproteinemia, most or even all IDL particles become converted in the circulation to LDL (61). Because IDL interacts specifically with the LDL receptor, some IDL particles, however, must be degraded before forming LDL. This pathway is an important regulator of LDL synthesis (from IDL) and appears to explain the differences of LDL synthetic rates between species (57, 58). Whether HTG-IDL behave metabolically differently from N-IDL, is not known.

4.3 Low density lipoproteins

4.3.1 **Structure and composition**

In rats, LDL can be isolated free of other lipoproteins by zonal centrifugation techniques. The LDL thus obtained is of normal diameter (22 nm) and density and contains Apo B as the predominant apoprotein (17). The main difference between rat and human LDL is their core-lipid composition. In the rat, TG is a major core-lipid, contributing about 20% of the total lipoprotein mass, while CE contributes 27% (17). In humans, the corresponding values are 5% and 40%, respectively. This core-lipid composition of LDL in rats reflects the paucity of CE in the precursor lipoprotein, VLDL, and the inability to enrich either VLDL, IDL or LDL with CE in an animal devoid of LTP activity.

Abnormalities of LDL in HTG states have been reported by several investigators (5, 59, 68, 69). In our study (59), LDL from 16 patients with HTG (plasma triglycerides 2.8–28 mmol/l) was analyzed before, and during triglyceride lowering therapy. In a zonal ultracentrifugation system, HTG-LDL separates into two or even three populations. The main LDL (corresponding to N-LDL) was further characterized. HTG-LDL is denser than N-LDL, relatively enriched with protein (predominantly Apo B) and TG and contains significantly less free and esterified cholesterol. In addition HTG-LDL is smaller (by electron microscopy) than N-LDL, and diameters less than 17 nm (N-LDL = 21 – 22 nm) are found. These six abnormalities are strongly and significantly related to the degree of triglyceridemia, and tend to revert towards normal when plasma TG are reduced. Molecular analysis of HTG-LDL indicates that the lipoprotein contains more TG and less cholesterol (molecules/particle), but similar amounts of protein. Of the six abnormalities, we focused on the cholesterol content of the lipoprotein. On the basis of regression analysis of the change of LDL core-volume and CE to protein (Apo B) ratio we were able to demonstrate a reduced content of cholesterol molecules in HTG-LDL that was curvilinearly related to plasma TG levels (70). We believe that the abnormal density, composition and size of HTG-LDL are best explained by excessive lipid transfers between a large pool of slowly turning-over TG-rich lipoproteins in a system with normal, or even low rates of LDL production.

4.3.2 **Metabolism**

The compositional analysis predicts that the catabolism of HTG-LDL is abnormal, and that this abnormality reflects loss of free and esterified cholesterol from the lipoprotein. This is to be expected even on theoretical grounds alone. It is well established that the regulation of cellular cholesterol synthesis and of LDL receptor activity – the major pathway of LDL catabolism – depends on the amount of cholesterol that enters cells with LDL (71). As HTG-LDL is a cholesterol-poor lipoprotein, it should down regulate the receptor activity less efficiently than N-LDL and should be cleared from plasma (by the receptor pathway) at an accelerated rate. Indeed, in studies where LDL-Apo B plasma decay rates were determined in subjects with HTG, they were found to be faster than in normals (63, 72, 73). Recently, we initiated a series of investigations on the ability of HTG-LDL to regulate LDL receptor activity and cholesterol synthesis in cultured human skin fibroblasts (74). HTG-LDL exhibited lower affinity towards the LDL receptor and was a less efficient regulator of cellular sterol synthesis and cellular receptor activity than N-LDL. These abnormalities were strongly and significantly correlated with the cholesterol content of the LDL and reverted towards normal when bezafibrate therapy was instituted. Analysis of the data revealed that the inefficient regulation of cellular cholesterol synthesis and of LDL receptor activity were due to the lower cholesterol content of HTG-LDL and that the cholesterol content of LDL was the best predictor for the capacity of the lipoprotein to regulate cellular metabolic activities, regardless of source or type of subject. More recently, we determined the actual metabolism of HTG-LDL in regulated cells (unpublished observations). The cellular metabolism of HTG-LDL in cells regulated with HTG-LDL was higher than N-LDL in cells regulated with N-LDL. It thus appears that cells exposed to HTG-LDL contain more receptors than cells exposed to N-LDL and degrade considerable amounts of LDL, in spite of its reduced affinity for the receptor.

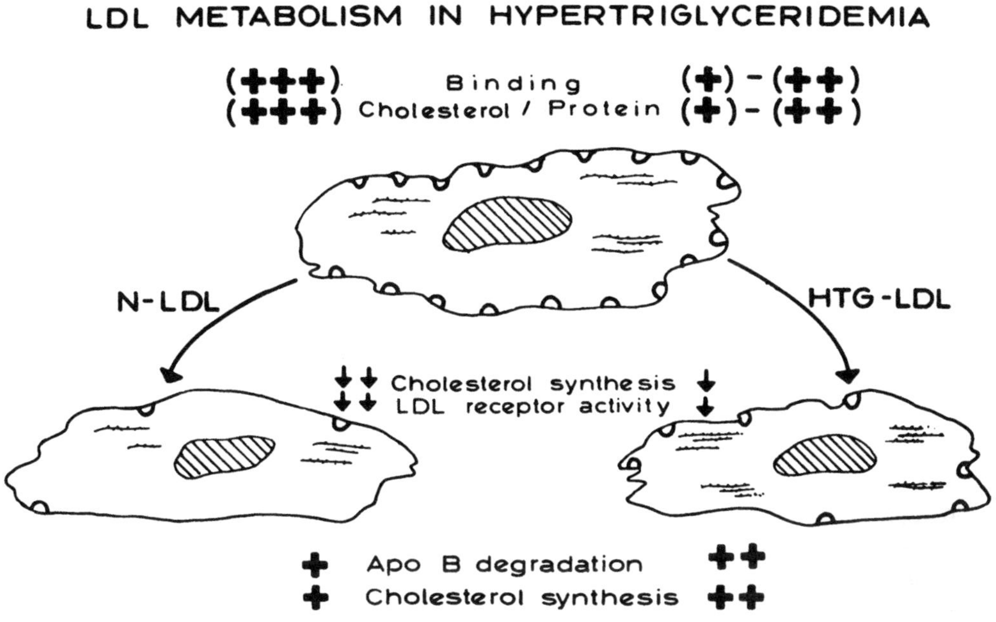

Fig. 6. Effects of excessive lipid transfer on LDL metabolism in hypertriglyceridemic (HTG) human subjects. In normals, LDL is efficiently taken up by the cells, delivers relatively large amounts of cholesterol and down-regulates cellular cholesterol synthesis and LDL receptor activity. HTG-LDL is CE-depleted, and does not interact well with the LDL receptor protein. Therefore, the amount of cholesterol that enters the cells is relatively small, and both cholesterol synthesis and LDL receptor activity are high.

The molecular basis for defective affinity of HTG-LDL towards the receptor is as yet unknown. This phenomenon must reflect reversible conformational changes of the Apo B moiety in HTG-LDL. That indeed was demonstrated by us in experiments using monoclonal antibodies specific for the receptor binding domain of Apo B (75). The abnormality of HTG-LDL responsible for the change in Apo B, however, remains to be elucidated. Regardless of mechanism, these investigations demonstrate a defect of cellular metabolism of HTG-LDL, as shown in Figure 6.

4.4 High density lipoproteins

4.4.1 Structure and composition

Data on the structure and composition of HDL related to either absence or excessive lipid transfers are available in humans with abetalipoproteinemia (52), in rats (17) and in human subjects with HTG states (5, 59). Lessons learnt from all three models have established that core-lipid transfer reactions are the principal modulators of HDL subpopulation distribution (53). In abetalipoproteinemia, a condition where chylomicrons, VLDL and LDL are absent, HDL levels are low, but most of the HDL is associated with abnormally large-sized, light density and CE-rich HDL_2, the ABL-HDL_2 (52). ABL-HDL_3 is also CE-enriched (52). Relatively large quantities of Apo E-HDL have been detected (76). All these deviations from normal are compatible with accretion of LCAT-derived CE molecules in the HDL system, and conversion of small-size particles to larger ones. Rat HDL is composed of two populations: HDL_2 and HDL_1 (17). The dense, small-size HDL_3 is not identified. Again, these findings are best explained by absence of LTP activity. Of particular relevance to the notion that the LTP reaction determines HDL structure and composition are reports of Japanese patients with hyperalphalipoproteinemia and markedly reduced, or absent LTP activity (77, 78). The main HDL in these subjects is CE-rich HDL_2 while VLDL is CE-poor (1.7% of total mass) and LDL is TG-rich. A converse situation is found in HTG where excessive transfer between an expanded pool of TG-rich lipoproteins and HDL takes place. HTG-HDL is exclusively in the dense HDL_3 fraction, also referred to as HDL-3D (79). In our studies we showed dense, TG-rich and CE-poor HDL_3 in groups of HTG patients (5, 59). These abnormalities are strongly related to plasma TG levels and tend to revert towards normal when TG lowering therapy (with bezafibrate) is instituted. These findings in HTG are best explained by the mechanism discussed above (Section 2, Fig. 4), i.e., heteroexchange of HDL-CE and VLDL-TG, followed by hydrolysis of the transferred TG molecules.

4.4.2 **Metabolism**

Limited information is available on LTP-related metabolism of HDL populations in the rat. Taking advantage of the absence of LTP activity in the rat, we have determined the *in vivo* metabolism of rat plasma lipoproteins labeled in their CE moiety. Using [^{3}H]-CE labeled rat plasma HDL_2, we showed conversion of some HDL_2 particles to HDL_1 (80). More recently, we used [^{3}H]-CE labeled human HDL_3 in a similar experimental system (81). Conversions of the HDL_3 first to HDL_2 and then to Apo E-HDL_1 were documented. Next, the experiments were repeated in rats pre-injected with human LTP. LTP had only a minimal effect on $HDL_3 \rightarrow HDL_2$ conversion, probably because of the rapidity (30–60 min) of this process. Formation of HDL_1, however, was markedly affected and the amount of HDL_1 formed during 18 hours of circulation was reduced to about one half. Loss of HDL_1 in LTP injected rats was also reported by Barter and associates, but the mechanism by which LTP acts has not been elucidated (82, 83). Extrapolation of these *in vivo* investigations in the rat to the human predicts that HDL subpopulation distribution indeed reflects processes of CE enrichment and CE removal along a size and density cascade. Figure 7 illustrates the situation in abetalipoproteinemia and, as discussed elsewhere, the same pathways appear to be responsible for the regulation of HDL subpopulations in general (53).

Whether LTP affects the HDL apoprotein metabolism is unknown. Yet, it is of interest to note that Apo AI metabolism in subjects with hyperalphalipoproteinemia or high HDL from other causes is slower (84, 85), while Apo AI metabolism in HTG is usually somewhat faster (84, 86) than in normals. Whether these observations reflect differences between HDL_2 and HDL_3 metabolism, or other LTP related effects remains to be established.

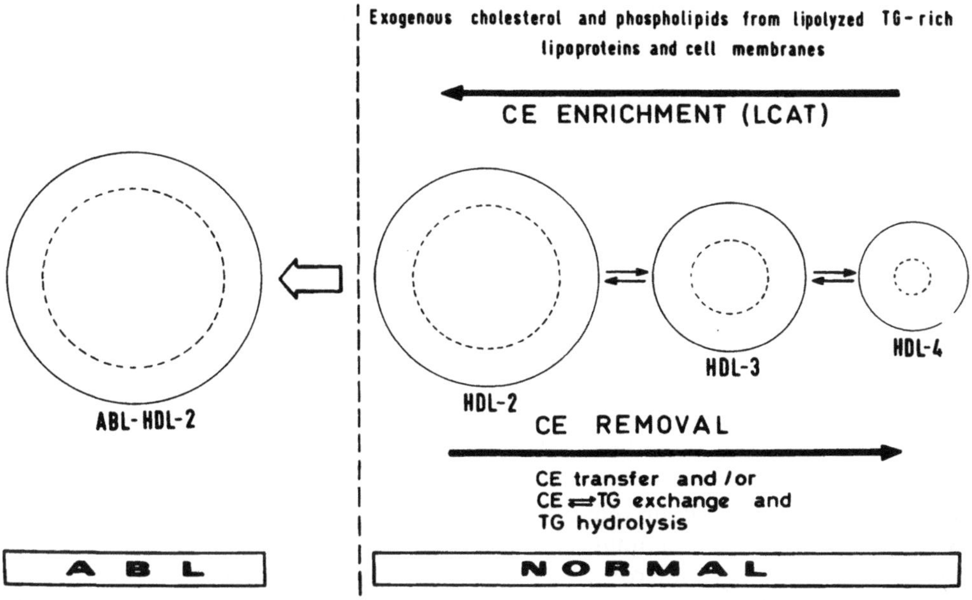

Fig. 7. Regulation of HDL subpopulation distribution in normolipidemic and abetalipoproteinemic (ABL) human subjects. The scheme predicts that HDL subpopulation distribution is determined by a balance of processes responsible for enrichment and removal of CE from HDL. It is proposed that supply of free cholesterol and phospholipids followed by LCAT activity cause conversion of small HDL to larger size particles, while CE transfer and CE-TG heteroexchange followed by hydrolysis of acquired TG, results in the reverse conversion of large HDL to smaller size particles. (Reproduced by permission from the Journal of Lipid Research (*23*, 1280 (1982) and *25*, 1017 (1984).)

4.5 LTP and reverse cholesterol transport

Reverse cholesterol transport is a chain of metabolic reactions that causes transport of cholesterol from peripheral cells to the liver. The pathway may in part involve LTP mediated processes. Fielding and associates suggested that LTP activity facilitates egress of cholesterol from cells to the plasma lipoproteins (87, 88). The hypothesis assumes that transfer of CE out of a subpopulation of HDL is necessary for removal of cholesterol from cells. The following chain of events has been proposed: 1. LCAT dependent net transport of cholesterol from cells to the plasma; 2. LTP facilitated esterification of such cholesterol in a lipoprotein complex containing LCAT, Apo AI and LTP which catalyzes removal of CE from the complex; and 3. accumulation of CE in acceptor lipoproteins. These investigators, moreover, demonstrated that plasma samples may differ in their capacity to remove cholesterol from cells and that plasma from HTG patients with coronary artery disease have the lowest capacity to accept cholesterol from cells (88). Another possible mode of action of LTP in reverse cholesterol removal processes is CE efflux from extracellular and cellular elements of the arterial wall (89). In that study, LTP and human lipoprotein deficient plasma, but not rat lipoprotein deficient plasma, were shown to increase by 3–12 fold the efflux of CE, or its ether analogue, CLE, from a variety of cell types, including fibroblasts, smooth muscle cells and macrophages. It thus appears that LTP associated reactions may both indirectly and directly affect cholesterol and cholesteryl ester removal from cellular and extracellular elements. Whether these processes provide protection against atherosclerosis, however, is uncertain.

5 Lipoprotein Remodelling, Cholesteryl Ester Distribution and Lipoprotein Metabolism

The data and concepts described in this chapter deal predominantly with individual lipoproteins. Plasma lipid transport in lipoproteins, however, must be viewed as one system where each component is one part of a whole. The LTP reaction is an excellent example of the mandatory relationships that exist between all lipoprotein classes. In plasma, chylomicrons, VLDL, IDL, LDL and HDL and their subclasses are all simultaneously exposed to and affected by lipid transfer processes. LTP's facilitate transfer of triglycerides, phospholipid and cholesteryl ester molecules. Metabolically, heteroexchange of TG and CE and the resulting triglyceride and cholesteryl ester distributions, appear to be of greatest importance. Initially, TG molecules are secreted with chylomicrons and VLDL, while CE is generated in HDL. In human plasma, redistributions of TG and CE begin as soon as the nascent lipoproteins enter the plasma compartment and continue throughout their circulating lives. TG redistribution causes mainly structural and compositional alterations of all lipoproteins but may also affect metabolic routes, e. g., LDL metabolism. CE redistributions is followed by immediate metabolic implications, that are possibly related to atheroma formation and development of atherosclerotic diseases. The best example of this is the lipoprotein system in HTG states where excessive TG and CE transfers occur. Figure 8 illustrates the effects of CE redistribution on plasma cholesterol clearance in N- and HTG-subjects. Let us begin with LDL and HDL. In normotriglyceridemic subjects, an amount of cholesterol and cholesteryl ester (estimated to be 1.5 to 2.5 g/d) is cleared from the plasma with irreversible degradation of LDL and HDL. In HTG, however, considerably less cholesterol is cleared from the plasma via these two lipoproteins, even if their degradation is accelerated. That is inferred by the consideration that if normal amounts of cholesterol enter cells (with a larger number of HTG-LDL particles), then the LDL-receptor activity would be down regulated to normal levels. Hence, even in the face of accelerated LDL catabolism, the equilibrium reached is bound to reflect sub-normal clearance of plasma cholesterol with the lipoprotein. The same considerations lead to the conclusion that cholesterol clearance with HDL in HTG subjects is also sub-normal. Therefore, in HTG, an additional pathway for removal of cholesterol from plasma must exist, and the amounts of cholesterol cleared by that pathway must be quite substantial. Our own studies indicate that direct catabolism of cholesterol and cholesteryl ester rich VLDL or VLDL-remnants in HTG provides

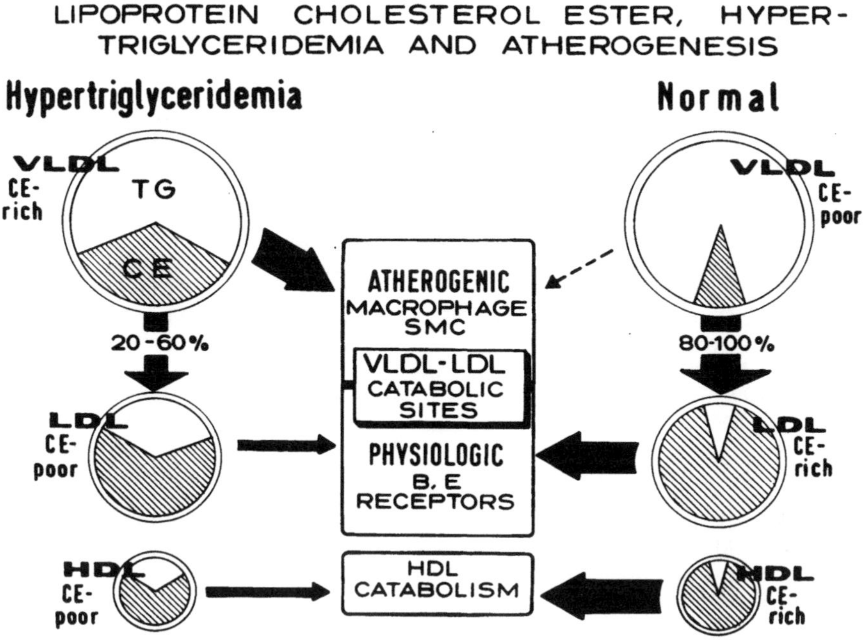

Fig. 8. Cholesterol redistribution, lipoprotein metabolism and atherogenesis. The figure compares a situation of excessive lipid transfer (hypertriglyceridemia) to the normolipemic state. See text for details.

an "escape" mechanism responsible for the clearance from plasma of cholesterol that normally is catabolized with LDL and HDL. Indeed, both fibroblasts and macrophages were shown to degrade large-sized VLDL particles (VLDL-I and VLDL-II) isolated from HTG but not normal human subjects (90). Degradation of Type III β-VLDL in macrophages (91) may represent the extreme situation of inefficient VLDL → LDL conversion, and LDL-independent VLDL catabolism. Figure 8 suggests that that pathway is atherogenic. That suggestion, however, remains to be tested.

References

1. Eder, H. A. (1957) Am. J. Med. *23*, 269–282.
2. Bell, F. P. (1978) Prog. Lipid Res. *17*, 207–243.
3. Nichols, A. V. & Smith, L. (1965) J. Lipid Res. *6*, 206–210.
4. Zilversmit, D. B., Hughes, L. B. & Balmer, J. (1975) Biochim. Biophys. Acta *409*, 393–398.
5. Deckelbaum, R. J., Granot, E., Oschry, Y., Rose, L. & Eisenberg, S. (1984) Arteriosclerosis *4*, 226–231.
6. Tall, A. R. (1986) J. Lipid Res. *27*, 359–365.
7. Chajek, T. & Fielding, C. J. (1978) Proc. Natl. Acad. Sci. USA *75*, 3445–3449.
8. Pattnaik, N. M., Montes, A., Hughes, L. B. & Zilversmit, D. B. (1978) Biochim. Biophys. Acta *530*, 428–438.
9. Ihm, J., Harmony, J. A. K., Ellsworth, J. & Jackson, R. L. (1980) Biochem. Biophys. Res. Commun. *93*, 1114–1120.
10. Morton, R. E. & Zilversmit, D. B. (1981) Biochim. Biophys. Acta *663*, 350–355.

11. Morton, R. E. & Zilversmit, D. B. (1982) J. Lipid Res. *23*, 1058–1067.
12. Ihm, J., Ellsworth, J. L., Chataing, B. & Harmony, J. A. K. (1982) J. Biol. Chem. *257*, 4818–4827.
13. Albers, J. J., Tollefson, J. H., Chen, C. H. & Steinmetz, A. (1984) Arteriosclerosis *4*, 49–58.
14. Fielding, P. E. & Fielding, C. J. (1980) Proc. Natl. Acad. Sci. USA *77*, 3327–3330.
15. Tall, A. R., Abreu, E. & Shuman, J. (1983) J. Biol. Chem. *258*, 2174–2180.
16. Barter, P. J., Gooden, J. M. & Rajaram, O. V. (1979) Atherosclerosis *33*, 165–169.
17. Oschry, Y. & Eisenberg, S. (1982) J. Lipid Res. *23*, 1099–1106.
18. Eisenberg, S. (1978) J. Lipid Res. *19*, 229–236.
19. Brewster, M. E., Ihm, J., Brainard, J. R. & Harmony, J. A. K. (1978) Biochim. Biophys. Acta *529*, 147–159.
20. Pattnaik, N. M. & Zilversmit, D. B. (1979) J. Biol. Chem. *254*, 2782–2786.
21. Morton, R. E. (1985) J. Biol. Chem. *260*, 12593–12599.
22. Sniderman, A., Teng, B., Vezina, C. & Marcel, Y. L. (1978) Atherosclerosis *31*, 327–333.
23. Barter, P. J., Gorjatschko, L. & Hopkins, G. J. (1982) Biochim. Biophys. Acta *710*, 349–358.
24. Abbey, M., Savage, J. K., Mackinnon, A. M., Barter, P. J. & Calvert, G. D. (1984) Biochim. Biophys. Acta *793*, 481–484.
25. De Parscau, L. & Fielding, P. E. (1984) J. Lipid Res. *25*, 721–728.
26. Tollefson, J. H., Faust, R., Albers, J. J. & Chait, A. (1985) J. Biol. Chem. *260*, 5887–5890.
27. Morton, R. E. & Zilversmit, D. B. (1981) J. Biol. Chem. *256*, 11992–11995.
28. Son, Y. S. C. & Zilversmit, D. B. (1984) Biochim. Biophys. Acta *795*, 473–480.
29. Ha, Y. C. & Barter, P. J. (1982) Comp. Biochem. Physiol. *71B*, 265–269.
30. Morton, R. E. & Zilversmit, D. B. (1983) J. Biol. Chem. *258*, 11751–11757.
31. Eisenberg, S. (1985) J. Lipid Res. *26*, 487–494.
32. Barter, P. J. & Jones, M. E. (1980) J. Lipid Res. *21*, 238–249.
33. Ihm, J., Quinn, D. M., Busch, S. J., Chataing, B. & Harmony, J. A. K. (1982) J. Lipid Res. *23*, 1328–1341.
34. Marcel, Y. L., Vezina, C. & Milne, R. W. (1983) Biochim. Biophys. Acta *750*, 411–417.
35. Marcel, Y. L., Vezina, C., Teng, B. & Sniderman, A. (1980) Atherosclerosis *35*, 127–133.
36. Marcel, Y. L., Vezina, C., Emond, D., Verdery, R. B. & Milne, R. W. (1981) J. Lipid Res. *22*, 1198–1205.
37. Hopkins, G. J. & Barter, P. J. (1980) Metabolism *29*, 546–550.
38. Eisenberg, S. & Perret, B. (1981) Arteriosclerosis *1*, 363a.
39. Fielding, C. J. & Fielding, P. E. (1981) J. Biol. Chem. *256*, 2102–2104.
40. Barter, P. J., Hopkins, G. J. & Calvert, G. D. (1982) Biochim. Biophys. Acta *713*, 136–148.
41. Barter, P. J. & Hopkins, G. J. (1983) Biochim. Biophys. Acta *751*, 33–40.
42. Noel, S. P., Dupras, R., Vezina, C. & Marcel, Y. L. (1984) Biochim. Biophys. Acta *796*, 277–284.
43. Tall, A. R., Sammett, D., Vita, G. M., Deckelbaum, R. & Olivecrona, T. (1984) J. Biol. Chem. *259*, 9587–9594.
44. Sammett, D. & Tall, A. R. (1985) J. Biol. Chem. *260*, 6687–6697.
45. Castro, G. R. & Fielding, C. J. (1985) J. Clin. Invest. *75*, 874–882.
46. Tall, A., Sammett, D., Granot, E. (1986) J. Clin. Invest. *77*, 1163–1172.
47. Nestel, P. J., Reardon, M. and Billington, T. (1979) Biochim. Biophys. Acta *573*, 403–407.
48. Deckelbaum, R. J., Eisenberg, S., Oschry, Y., Butbul, E., Sharon, I. & Olivecrona, T. (1982) J. Biol. Chem. *257*, 6509–6517.
49. Deckelbaum, R. J., Olivecrona, T. & Eisenberg, S. (1984) in: Treatment of Hyperlipoproteinemia (Carlson, L. A. & Olsson, A. G., eds.) pp. 85–93, Raven Press, New York.
50. Chait, A., Eisenberg, S., Steinmetz, A., Albers, J. & Bierman, E. L. (1984) Biochim. Biophys. Acta *795*, 314–325.
51. Granot, E., Deckelbaum, R. J., Eisenberg, S., Oschry, Y. & Bengtsson-Olivecrona, G. (1985) Biochim. Biophys. Acta *833*, 308–315.
52. Deckelbaum, R. J., Eisenberg, S., Oschry, Y., Cooper, M. & Blum, C. (1982) J. Lipid Res. *23*, 1274–1282.
53. Eisenberg, S. (1984) J. Lipid Res. *25*, 1017–1058.
54. Patsch, J. R., Gotto, A. M., Olivecrona, T. & Eisenberg, S. (1978) Proc. Natl. Acad. Sci. USA *75*, 4519–4523.
55. Deckelbaum, R. J., Eisenberg, S., Oschry, Y., Granot, E., Sharon, I. & Bengtsson-

Olivecrona, G. (1986) J. Biol. Chem. *261*, 5201–5208.
56. Patsch, J. R., Prasad, S., Gotto, A. M. & Bengtsson-Olivecrona, G. (1984) J. Clin. Invest. *74*, 2017–2023.
57. Eisenberg, S. & Rachmilewitz, D. (1973) Biochim. Biophys. Acta *326*, 378–390.
58. Eisenberg, S. & Rachmilewitz, D. (1973) Biochim. Biophys. Acta *326*, 391–405.
59. Eisenberg, S., Gavish, D., Oschry, Y., Fainaru, M. & Deckelbaum, R. J. (1984) J. Clin. Invest. *74*, 470–482.
60. Oschry, Y., Olivecrona, T., Deckelbaum, R. J. & Eisenberg, S. (1985) J. Lipid Res. *26*, 158–167.
61. Berman, M., Eisenberg, S., Hall, M., Levy, R. I., Bilheimer, D. W., Phair, R. D. & Goebel, R. H. (1978) J. Lipid Res. *19*, 38–56.
62. Reardon, M. F., Fidge, N. H. & Nestel, P. J. (1978) J. Clin. Invest. *61*, 850–860.
63. Beltz, W. F., Kesaniemi, A., Howard, B. V. & Grundy, S. M. (1985) J. Clin. Invest. *76*, 575–585.
64. Ginsberg, H. N., Le, N. A. & Gibson, J. C. (1985) J. Clin. Invest. *75*, 614–623.
65. Grundy, S. M., Mok, H. Y. I., Zech, L., Steinberg, D. & Berman, M. (1979) J. Clin. Invest. *63*, 1274–1283.
66. Kekki, M. (1980) Ann. Clin. Res. *12*, 64–76.
67. Mahley, R. W. & Angelin, B. (1984) Adv. Int. Med. *29*, 385–411.
68. Vakakis, N., Redgrave, T. G. & Small, D. M. (1983) Biochim. Biophys. Acta *751*, 280–285.
69. Nelson, C. A. & Morris, M. D. (1983) Lipids *18*, 553–557.
70. Eisenberg, S. (1985) Adv. Exptl. Med. Biol. *183*, 73–84.
71. Brown, M. S. & Goldstein, J. L. (1986) Science *232*, 34–47.
72. Sigurdsson, G., Nicoll, A. & Lewis, B. (1976) Eur. J. Clin. Invest. *6*, 151–158.
73. Janus, E. D., Nicoll, A. M., Turner, P. R., Magill, P. & Lewis, B. (1980) Eur. J. Clin. Invest. *10*, 161–172.
74. Kleinman, Y., Eisenberg, S., Oschry, Y., Gavish, D., Stein, O. & Stein, Y. (1985) J. Clin. Invest. *75*, 1796–1803.
75. Kleinman, Y., Schonfeld, G., Gavish, D., Oschry, Y. & Eisenberg, S. (1987) J. Lipid Res. *28*, 540–548.
76. Blum, C. B., Deckelbaum, R. J., Witte, L. D., Tall, A. R. & Cornicelli, J. (1982) J. Clin. Invest. *70*, 1157–1169.
77. Kurasawa, T., Yokoyama, S., Miyake, Y., Yamamura, T. & Yamamoto, A. (1985) J. Biochem. *98*, 1499–1508.
78. Koizumi, J., Mabuchi, H., Yoshimura, A., Michishita, I., Takeda, M., Itoh, H., Sakai, Y., Sakai, T., Ueda, K. & Takeda, R. (1985) Atherosclerosis *58*, 175–186.
79. Patsch, J. R. & Gotto, A. M. (1979) in: Reports of the High Density Lipoprotein Methodology Workshop, pp. 310–324, Dept. Health, Education and Welfare – NIH Publ 79-1661, NIH, Washington DC.
80. Eisenberg, S., Oschry. & Zimmerman, J. (1984) J. Lipid Res. *25*, 122–128.
81. Gavish, D., Oschry, Y. & Eisenberg, S. (1987) J. Lipid Res. *28*, 257–267.
82. Ha, Y. C., Chang, L. B. F. & Barter, P. J. (1985) Biochim. Biophys. Acta *833*, 203–210.
83. Ha, Y. C. & Barter, P. J. (1986) Comp. Biochem. Physiol. *83B*, 463–466.
84. Fidge, N., Nestel, P., Ishikawa, T., Reardon, M. & Billington, T. (1980) Metabolism *29*, 643–653.
85. Herbert, P. N., Bernier, D. N., Cullinane, E. M., Edelstein, L., Kantor, M. A. & Thompson, P. D. (1984) J. Am. Med. Assoc. *252*, 1034–1037.
86. Schaefer, E. J., Zech, L. A., Jenkins, L. L., Bronzert, T. J., Rubalcaba, E. A., Lindgren, F. T., Aamodt, R. L. & Brewer, H. B. Jr. (1982) J. Lipid Res. *23*, 850–862.
87. Fielding, C. J. & Fielding, P. E. (1982) Med. Clin. N. America *66*, 363–374.
88. Fielding, P. E., Fielding, C. J., Havel, R. J., Kane, J. P. & Tun, P. (1983) J. Clin. Invest. *71*, 449–460.
89. Stein, O., Halperin, G. & Stein, Y. (1986) Arteriosclerosis *6*, 70–78.
90. Gianturco, S. H., Gotto, A. M. Jr., Hwang, S. L. C., Karlin, J. B., Lin, A. H. Y., Prasad, S. C. & Bradley, W. A. (1983) J. Biol. Chem. *258*, 4526–4533.
91. Fainaru, M., Mahley, R. W., Hamilton, R. L. & Innerarity, T. L. (1982) J. Lipid Res. *23*, 702–714.

Chapter 8
The Dyslipoproteinemias of Diabetes Mellitus

P. Alaupovic and P. R. Blackett

1 Introduction

Atherosclerosis has become a major cause of morbidity and mortality in patients with diabetes mellitus in Western countries (1–3). The incidence of atherosclerotic vascular disease, including coronary heart disease, cerebrovascular disease and peripheral vascular disease, is higher in patients with diabetes than in the general population (2, 4–7). Although nondiabetic and, particularly, premenopausal women have less disease than nondiabetic men, diabetes eliminates this relative protection and makes diabetic women as vulnerable to various forms of vascular disease as diabetic men (6, 8, 9). In fact, the incidence of cardiovascular disease (6) or the relative risk of death from ischemic heart disease (9) is higher for diabetic women than diabetic men. Macrovascular disease is the major cause of death in 75–80% of diabetics (1, 10). It is not at all clear, however, why the rates are so high in diabetics, nor why some diabetics escape macrovascular disease. There may be several inherent and/or environmental factors contributing to its development. The WHO Multinational Study of Vascular Disease in Diabetes (3, 11–13) shows marked variation in the prevalence between diabetics from various geographical regions. The rates of atherosclerosis and its clinical manifestations are substantially higher in European and American diabetic populations than in those from Japan and Hong Kong (1). This variation in prevalence is similar to that in nondiabetic patients from the same geographical regions suggesting that factors other than diabetes may be responsible for the accelerated rate of atherosclerosis demonstrated in some but not all diabetic populations. Thus, it appears, that environmental factors play a major role in determining the prevalence rates, although in some populations the genetic factors may be more important than environmental (13). The significance of various risk factors for the pathogenic mechanisms of atherosclerosis in general and in diabetic patients in particular has been discussed extensively in several recent reviews (3, 14–16). Among these factors, genetic and acquired abnormalities in lipid metabolism and transport may play an important, if not crucial, role in the development of excessive and premature vascular disease in nondiabetic as well as diabetic patients.

It is well established that diabetes is very frequently associated with increased concentrations of plasma lipids, particularly in untreated or poorly controlled patients (17–23). Hypertriglyceridemia occurs more frequently than hypercholesterolemia suggesting that an impairment in metabolism of triglyceride-rich lipoproteins may be the main defect of lipid transport in diabetes. However, the plasma levels of triglyceride and triglyceride-rich lipoproteins vary from normal to very high values depending on the type of diabetes, degree of control and other factors and conditions known to influence the composition and concentration of plasma lipoproteins such as obesity, renal disease, diet, and the use of hypolipidemic, diuretic and beta-adrenergic blocking drugs.

Estimates on the prevalence of hyperlipoproteinemia in diabetic populations vary greatly depending most probably on the selection of patients, dietary regimen, diabetic control and criteria selected for hyperlipidemic states (20). However, it has been shown in more recently conducted studies using cutoff points for triglyceride ranging between 2.1 – 2.3 mmol/l, that hyperlipoproteinemia occurs in 30 – 40% of diabetic patients (24 – 28). In contrast to the clinical setting of these studies, a recent survey of diabetics and nondiabetics in a community-based population showed no significant difference in the mean cholesterol level and the prevalence of hypercholesterolemia between the diabetics and nondiabetics who were 50 years of age and older (29). However, the mean triglyceride level and the prevalence of hypertriglyceridemia were significantly higher in diabetics than nondiabetics; in diabetics, hypertriglyceridemia (> 2.15 mmol/l) was found to occur in 29% of nonobese and 25% of obese men, and in 10% of nonobese and 21% of obese women.

Prevalence studies indicate clearly that at least one-third of the diabetic population has an increased concentration of plasma triglyceride or, in terms of lipoproteins, an increased concentration of chylomicrons and VLDL. It should be recognized, however, that expressing a lipoprotein abnormality in terms of a single plasma lipid or lipoprotein reveals very little about the possible changes in the composition and concentration of other constituents of triglyceride-rich lipoproteins or their interaction with other plasma lipoproteins.

The lack of a consistent and uniform occurrence of hyperlipoproteinemia in all diabetic patients represents one of the most intriguing questions regarding the lipid transport abnormality in this disease. Is an underlying structural and/or metabolic abnormality only present in a certain percentage of diabetic patients or is it expressed in an overt form in some and in a latent form in others? Is there a single or are there multiple metabolic derangements of lipid transport? Is it possible to identify particular lipid, apolipoprotein or lipoprotein profiles reflecting the underlying structural and metabolic aberrations associated with diabetes or independent of diabetes? Before any attempts are made to answer such questions on the basis of available evidence about the chemistry and metabolism of plasma lipoproteins in diabetic patients, it seems only appropriate to consider briefly the rather complex relationship between hyperlipoproteinemia and clinical categories of diabetes. According to the new classification of diabetes mellitus (30), there are three major clinical categories which include type I or insulin-dependent diabetes mellitus (IDDM), type II or noninsulin-dependent diabetes mellitus (NIDDM), and various subclasses of impaired glucose tolerance (IGT). Reaven and Greenfield have suggested that distinct pathogenetic mechanisms of hypertriglyceridemia in these three clinical categories result from changes in the concentrations of plasma insulin and/or free fatty acids (18). A number of studies have shown (for reviews see 17, 22, 31, 32) that insulin deficiency as well as hyperinsulinemia exert profound effects on the synthesis, secretion and removal of VLDL-triglycerides. Insulin plays several roles in the regulation of triglyceride metabolism. It suppresses the activity of hormone-sensitive lipase, an adipose tissue enzyme responsible for the lipolysis and mobilization of free fatty acids, and it maintains the normal activity of adipose tissue lipoprotein lipase, an enzyme catalyzing the lipolytic degradation of plasma VLDL-triglycerides. Furthermore, insulin stimulates the synthesis and secretion of hepatic VLDL-triglycerides.

In insulin deficiency (type I diabetes), the enhanced release of free fatty acids from adipose tissue stimulates the hepatic synthesis of triglycerides and their secretion in the form of VLDL. However, in severe hypoinsulinemia the rates of VLDL-triglyceride secretion may be subnormal relative to the concentration of available free fatty acids (18, 22). The hypertriglyceridemia in type I diabetes may be further aggravated by decreased activity of lipoprotein lipase in adipose tissue and skeletal muscle (33). Thus, in the insulin-deficient diabetic patients, the increased concentrations of triglyceride-rich lipoproteins may result from varying degrees of increased secretion and decreased removal. Insulin treatment normalizes VLDL production and secretion, raises the activity of lipoprotein lipase and reciprocally inhibits hormone-sensitive lipase with concomitant lowering of free fatty acids for plasma triglyceride production (22, 34).

In type II diabetes, the hypertriglyceridemia results mainly from an increased secretion of VLDL-triglycerides (18, 22, 34 – 37). Due to peripheral insulin resistance (38), there is less glucose uptake and less inhibition of lipolysis in adipose tissue. Since the plasma insulin levels are within the normal range, the liver is capable of utilizing the elevated concentrations of free fatty acids resulting in an increased synthesis and secretion of VLDL-triglycerides (18, 23, 34). Although in most patients with type II diabetes hypertriglyceridemia is due to an increase in VLDL production, in some patients reduced removal of VLDL represents the main defect leading to hypertriglyceridemia. In this respect, Pima Indians constitute a unique group of type II diabetics whose mild hypertriglyceridemia is caused exclusively by decreased clearance of VLDL (39). Hypertriglyceridemia may be further enhanced by the high prevalence of obesity among patients with type II diabetes, because obesity itself is associated with insulin resistance and increased production of VLDL-triglycerides (17, 22, 40, 41). In contrast to the predictive effect of insulin treatment on hyperlipoproteinemia in type I diabetes, the correction of lipoprotein abnormalities in type II diabetes will depend on the use of a particular hypoglycemic therapy. While caloric restriction and weight loss are accompanied by reduction in VLDL-triglyceride levels and increased concentrations of HDL-cholesterol (42 – 44), the commonly used oral hypoglycemic drugs seem to have very little effect on plasma triglyceride levels (45 – 50). It has been shown recently, however, that oral hypoglycemic drugs can decrease total and VLDL-triglyceride if patients' weight and diet composition are rigorously controlled (51). Change in VLDL-triglyceride is mainly due to a reduced rate of VLDL production (52). According to some reports, oral drugs seem to increase (53), to decrease (45, 49) or to have no effect on the concentration of HDL-cholesterol (46 – 48, 51, 54, 55). Treatment with insulin decreases plasma triglycerides (45, 46, 49, 56, 57) by diminishing the rate of VLDL synthesis (37 – 58) or by increasing lipoprotein lipase activity (56, 59). This may be reflected by an elevation of HDL-cholesterol levels (45, 46, 49, 57, 60 – 62). In addition to the effect of insulin on plasma lipoproteins, the evidence indicates a possible link between hyperinsulinemia and the pathogenesis of atherosclerosis (63 – 65). This would suggest that insulin treatment should be more extensively evaluated in future studies.

Impaired glucose tolerance is characterized by slightly abnormal glucose levels during a glucose tolerance test (30). It is thought (18) to be associated with insulin resistance and a compensatory increase in plasma insulin levels; the hyperinsulinemia results in an increased rate of synthesis and secretion of VLDL-triglycerides and, in the absence of adequate removal capacity, in hypertriglyceridemia. Impaired glucose tolerance may also be associated with an increased risk of developing coronary artery disease (66 – 68).

Type I and type II diabetes are genetically distinct diseases (30, 69). It is evident that each of these two types of diabetes is characterized by a unique pathophysiological mechanism contributing to aberrations in lipid transport. In this brief review of more recent advances in the chemistry and metabolism of plasma lipoproteins, types I and II diabetes will be treated as separate entities.

The vast literature on plasma lipids and lipoproteins in diabetes has been covered and discussed in a number of excellent reviews (17, 19 – 23, 31, 32) and only salient facts based on a few selected studies will be presented in this review. In reviewing the literature, one is impressed by an overwhelming amount of available data on the concentrations of plasma lipids, particularly cholesterol and triglyceride, and, more recently, cholesterol levels in major lipoprotein density classes. In contrast, the information about the concentration of plasma apolipoproteins and, particularly, the apolipoprotein composition of major lipoprotein density classes, is rather sketchy or, in the case of some minor apolipoproteins, unavailable. The recent developments in the chemistry of plasma lipoproteins indicate that lipoprotein density classes cannot be regarded as the fundamental physicochemical and metabolic entities of the lipid transport system, because each density class consists of several distinct lipoprotein species which have similar hydrated densities but different apolipoprotein compositions (70 – 72). We have suggested that lipoproteins be characterized and classified on the basis of their apolipoprotein composition; accordingly, lipoprotein particles containing a single apolipoprotein are called simple lipoproteins, while lipoprotein particles containing two or more apolipoproteins are referred to as complex lipoproteins (72). Although there are already several reports on simple and complex lipoprotein particles isolated from normolipidemic subjects (72), studies on lipoprotein particles in diabetes have not been reported.

2 Insulin-Dependent Diabetes Mellitus (Type I)

2.1 Untreated and partially treated diabetes

Patients with partially treated but nonketotic type I diabetes have increased plasma concentrations of triglyceride ranging between 1.2 and 5.5 mmol/l (73 – 83). Slightly increased plasma cholesterol concentrations (3.8 – 7.5 mmol/l) are reported in some (73, 74, 76, 79, 82) but not in all of these studies. However, in some cases of severe

Table 1. Plasma lipid concentrations (mmol/l) in ketoacidosis: effect of insulin treatment[a]

Days[b]	Triglyceride	Total cholesterol	Cholesterol		
			VLDL	LDL	HDL
1	164.3	28.2	26.8	0.5	0.05
2	7.1	13.0	8.4	3.7	0.3
4	3.8	10.8	3.9	6.5	0.4
7	1.4	7.9	0.4	6.7	0.5
37	0.8	4.8	0.2	2.9	1.1

[a] This case of severe ketoacidosis and hyperlipoproteinemia and the methodology for measuring plasma lipids are described in ref. 87.

[b] During the first two days, blood glucose was lowered with short-acting insulin given by intravenous infusion. Following the correction of dehydration, acidosis and electrolyte imbalance, the insulin was administered by continuous subcutaneous infusion with boluses of insulin before meals.

ketoacidosis (84–87), concentrations of plasma triglycerides and cholesterol may reach extremely high values (Table 1). Plasma phospholipids are in the upper range of reference values (2.0–2.12 mmol/l) (78).

Patients with partially treated type I diabetes have varying concentrations of VLDL-, LDL- and HDL-cholesterol (74–83). On the average, the increase of VLDL-cholesterol (range 0.2–1.68 mmol/l) is more significant than the increase of LDL-cholesterol (range 2.6–3.9 mmol/l). On the other hand, the levels of HDL-cholesterol (range 0.74–1.4 mmol/l) are low in comparison with age-matched nondiabetics or better controlled diabetics. In newly diagnosed patients, the levels of HDL_2-cholesterol are significantly lower than in control subjects or in the same patients after insulin treatment (88); there is no change in the concentration of HDL_3-cholesterol. It seems that the reduction in HDL_2-cholesterol levels may be a sex-related phenomenon, because low levels of HDL_2-cholesterol were found in partially controlled men but not women (89); the HDL_3-cholesterol concentrations of male and female diabetics were not different from those of nondiabetics.

In contrast to a number of studies on plasma lipid concentrations, there are very few reports on the plasma apolipoprotein levels in inadequately controlled, nonketotic children and adults with type I diabetes. In children, the levels of Apo B were found to be increased significantly, but the levels of Apo AI and Apo AII were unchanged (82). In two studies on inadequately controlled adult diabetics, the levels of Apo AI were found to be significantly lower and the levels of Apo B significantly higher than in well controlled patients (83, 90). However, in two other studies, no significant changes were detected in the levels of these two apolipoproteins between poorly and well controlled patients (77, 91). Poor control of diabetes had no effect on the concentration of Apo AII (83, 91) or Apo D (83). There are no reports on the possible effect of poor control on the levels of Apo C-peptides and Apo E. However, in ketoacidosis the levels of Apo C-peptides and Apo E may reach very high values (Table 2).

Although numerous studies on the determination of lipoprotein-cholesterol concentrations indicate that inadequately controlled, nonketotic patients with type

Table 2. Concentrations (mg/dl) of plasma apolipoproteins in ketoacidosis: effect of insulin treatment[a]

Days[b]	Apolipoproteins							Apo CIII ratio[c]
	AI	AII	B	CII	CIII	D	E	
1	105	52	88	21	195	23	72	0.01
2	46	7	98	7	12	15	26	1.58
4	75	20	177	6	13	10	22	0.41
7	79	26	192	4	10	9	9	0.65
37	115	57	75	2	7	2	7	2.8

[a] This case of severe ketoacidosis and hyperlipoproteinemia and the procedures used for the quantitative determination of plasma apolipoproteins are described in ref. 87.

[b] During the first two days, blood glucose was lowered with short-acting insulin given by intravenous infusion. Following the correction of dehydration, acidosis and electrolyte imbalance, the insulin was administered by continuous subcutaneous infusion with boluses of insulin before meals.

[c] Apo CIII ratio represents the ratio of apolipoprotein CIII measured in the supernatant fraction and precipitate obtained by heparin-Mn^{2+} precipitation of patient's plasma samples. This ratio is considered to be a measure of the efficiency of lipolytic degradation of triglyceride-rich lipoproteins (102); the higher the ratio, the more efficient the lipolysis. The normal value for Apo CIII ratio is 2.44 ± 1.5 (mean $\pm$ S. D., $n = 128$).

I diabetes have increased levels of VLDL-cholesterol as compared to nondiabetics, very little is known about the composition and distribution of lipid and apolipoprotein constituents in the major lipoprotein density classes. It has been suggested that the composition of VLDL particles accumulating in insulin deficiency differs very little, if at all, from that of nondiabetic controls (34, 92). There is indication, however, that some subtle changes do occur in the composition of LDL and HDL. In these patients, the relative content of triglyceride in both LDL ($\varrho = 1.006-1.063$ kg/l) (92) and LDL_2 (90) was found to be higher than in corresponding lipoprotein fractions from nondiabetic subjects. In LDL (93) but not in LDL_2 (90), the ratio of cholesterol/Apo B was also found to be higher in diabetics than nondiabetics. It has been suggested that the increased cholesterol/Apo B ratio results from the accumulation of LDL_1 (the intermediate density lipoprotein subclass of LDL) and that this compositional abnormality might lead to accelerated atherogenesis (93). However, it is necessary to determine the concentration and composition of Apo C-peptides and Apo E so that possible mechanisms causing the accumulation of these putative, atherogenic lipoprotein particles may be inferred and further investigated. Lopes-Virella et al. (90) showed that LDL_2 isolated from diabetic patients, before establishing metabolic control, were internalized and degraded by normal human fibroblasts to a lesser degree than LDL_2 isolated from well controlled diabetics or nondiabetic controls. The authors suggested that the higher content of triglyceride in LDL_2 from the less well controlled patients might have been responsible for the altered interaction of lipoprotein with its receptor, although the possible effect of nonenzymatic glycation of Apo B on the metabolic behavior of LDL (94) was not excluded. The composition of HDL particles also seems to be characterized by a higher than normal content of triglyceride (92).

There is very little information on the distribution of lipids and apolipoproteins among major lipoprotein density classes. The inverse correlation between HDL and VLDL (23, 95) implies, however, that the distribution of lipids and apolipoproteins, particularly apolipoproteins B, C and E, shifts from higher density to lower density lipoproteins. Since such changes in apolipoprotein distribution may result in the formation of potentially atherogenic lipoprotein particles of very low and low hydrated densities, the nature and extent of lipid and apolipoprotein rearrangements caused by poor or inadequate control of diabetes should be explored in future studies.

The increased concentrations of plasma triglyceride and VLDL in patients with type I diabetes are due primarily to increased rates of their production (22, 34, 42, 96). Although the average fractional catabolic rate appears to be normal, in some cases impaired removal may be an important contributing factor responsible for the accumulation of triglyceride-rich lipoproteins (96). The finding of decreased activities of lipoprotein lipase in adipose tissue, skeletal muscle and postheparin plasma (33, 97) provide further evidence that varying degrees of hypertriglyceridemia seen in individual patients may result from altered anabolic and/or catabolic processes. However, the available experimental evidence indicates that overproduction may be a more important factor than impaired removal (22).

Although the proper treatment of insulin deficiency ameliorates and normalizes the concentration and composition of plasma lipoproteins, occasional fluctuations in diabetic control may variably affect the lipoprotein system and contribute to its atherogenic potential. These fluctuations in control vary among individuals some of whom may progressively undergo metabolic decompensation and develop ketoacidosis.

Diabetic ketoacidosis is a condition commonly associated with absolute or relative insulin deficiency and an ensuing hormonal stress response (98). Catecholamines, glucagon, growth hormone, and glucocorticoids mobilize fatty acids from fat cell triglycerides. The fatty acids are preferentially oxidized to form ketone bodies in addition to being utilized as a substrate for triglyceride synthesis and triglyceride-rich lipoprotein formation (99). It follows that both the insulin deficiency and excessive fatty acid mobilization may contribute towards the moderate to high degree of hyperlipoproteinemia in ketoacidosis. There are, however, few studies on the changes in lipids and apolipoproteins during ketoacidosis due to the priority of required medical attention and difficulty in obtaining blood samples for investigation. Gofman et al. (95, 100) were the first to study by analytical ultracentrifugation the concentration and distribution of the major lipoprotein density classes in cases of diabetes mellitus with severe ketoacidosis or diabetic coma. Most of the cases displayed an abnormal distribution of lipoproteins characterized by a very high proportion of chylomicrons and VLDL (70 – 90% of total lipoproteins) and low percentages of LDL (3 – 15%) and HDL (3 – 10%) when compared to the distribution of corresponding lipoproteins in nondiabetic controls (15,45 and 40%, respectively). However, the clinical severity of the diabetes apparently has no relationship to the abnormal distribution of lipoproteins or the degree of hyperlipoproteinemia, as indicated by normal distribution and concentration of major density classes in some cases of ketoacidosis. In the majority of cases studied, the concentration of chylomicrons and VLDL was well above the control value (1.05 g/l) ranging from 2.5 to 52 g/l; however, a value as low as 0.3 g/l was also recorded. The concentrations of LDL and HDL were also variable. The levels of LDL ranged from a low value of 0.83 to a value as high as 5.5 g/l, in comparison with a mean value of 3.8 g/l for nondiabetic controls. The HDL concentrations ranged between 1.03 and a normal value of 3 g/l. The variable relationship of hyperlipoproteinemia to the severity of ketoacidosis would indicate that, besides insulin deficiency, other factors are operative during the pathogenesis of the ketoacidosis such as dietary intake or stress.

The large variability of plasma lipid levels has been observed in several studies (81, 84–86, 101). The reported range of plasma triglyceride was between 0.7 and 32 mmol/l (81, 85). In some cases, triglyceride concentrations were higher than 45 mmol/l (84, 87). Similarly, the reported plasma cholesterol concentrations ranged between 1.45 and 13.5 mmol/l (81, 83, 101), but values exceeding 30 mmol/l were also reported in some individual patients (84, 87).

Few attempts have been made to characterize hyperlipoproteinemia in ketoacidosis by measuring the concentrations of plasma apolipoproteins (85–87). The levels of Apo AI tend to be lower and the levels of Apo B higher than in adequately controlled diabetics or nondiabetic subjects (Table 3). Apo AI levels were found to be lowest in the more hypertriglyceridemic patients (85). In one case of severe ketoacidosis accompanied by a massive hypertriglyceridemia (87), the concentration of Apo CII was 207 mg/dl and those of Apo CIII and Apo E were 195 mg/dl and 716 mg/dl, respectively. These concentrations were among the highest recorded in the literature for these apolipoproteins. In the same study, the distribution of Apo CIII between heparin-Mn^{2+} supernate ("HDL") and heparin-Mn^{2+} precipitate ("VLDL + LDL") was abnormal; in contrast to a normal value of 2.5 for the Apo CIII ratio in heparin-Mn^{2+} supernate/heparin-Mn^{2+} precipitate, in the ketoacidotic patient, this ratio was 0.01 indicating a marked, if not total, block in the lipolytic degradation of triglyceride-rich lipoproteins and concomitant transfer of Apo CIII from VLDL to HDL (102). This abnormal lipoprotein pattern was restored to normal during treatment with insulin.

The marked hypertriglyceridemia in ketoacidosis is mainly due to an increased production of triglyceride-rich lipoproteins (96). However, in some cases of severe ketoacidosis, the impaired removal of triglyceride-rich lipoproteins is an equally important factor contributing to the massive hypertriglyceridemia (96). Bagdade et al. (103) assessed the effect of acute insulin deprivation on plasma triglyceride removal. Triglyceride increased and postheparin plasma lipoprotein lipase activity decreased in all patients studied, two of whom became ketoacidotic during a 48-hour period of insulin withdrawal. The lowest responses in postheparin plasma lipoprotein lipase activity after insulin withdrawal were observed in the ketoacidotic patients.

Since extremes of hyperlipoproteinemia occur in insulin-deficient states, the question of whether insulin therapy reverses the hyperlipoproteinemia was addressed by Weidman et al. (85) who observed rapid decreases in triglyceride-rich lipoproteins (chylomicrons and VLDL) with all subjects studied attaining triglyceride levels below 2.3 mmol/l after 24 h of insulin therapy. A decrease in plasma Apo B levels corresponded with the decrease of Apo B in VLDL but not in LDL_1 or LDL_2. A transient decrease in Apo AI was thought to be due to either a reduction in Apo AI

Table 3. Plasma concentrations (mg/dl) of apolipoproteins AI, AII and B in patients with type I diabetes

Sex	Age (years)	Apolipoproteins AI	AII	B	Method[a]	Reference
Ketoacidosis						
M + F	16–50	119	–	101	EIA	85
M + F	38–73	122	–	154	RID	86
Adequate Control						
M + F	11–14	131	42	64	EIA	114
M + F	9–21	132	–	91	EIA	115
M	18± 2	92↓[b]	35↓	92↓	RIA	112
F	18± 2	111↓	37	106	RIA	112
M	7–22	236[c]		93	RID	82
F	7–22	225		92	RID	82
F	21–30	126	59	90	EIA	116
F	27± 2	135↓	32↓	97	EIA	91
F	31± 7	140↑	31↓	–	RID	108
M	32± 9	131↑	32↓	–	RID	108
M	36± 2	137↓	34↓	108↑	EIA	91
M	30–60	149↑	34	113↑	EIA	83
M + F	35–55	–	–	139↑	EIA	77
M	62± 5	152↓	41↓	–	RIA	117
F	64± 6	165↓	50↓	–	RIA	117

[a] EIA = electroimmunoassay; RIA = radioimmunoassay; RID = radial immunodiffusion assay
[b] Direction of arrow indicates whether the values for apolipoproteins in diabetic patients are lower or higher than the corresponding values in normolipidemic controls.
[c] Values for combined apolipoproteins AI and AII

secretion or an increase in its catabolism brought about by insulin (85). More recently, we have studied a 13-year old boy with severe ketoacidosis associated with an initial triglyceride value of 165 mmol/l and a cholesterol value of 28.2 mmol/l (87). As expected, the levels of VLDL-cholesterol were extremely high (26.8 mmol/l) and those of LDL-cholesterol and HDL-cholesterol very low (0.52 and 0.05 mmol/l, respectively) (Table 1). The concentration of Apo AI was relatively low (105 mg/dl) while the levels of Apo AII (52 mg/dl) and Apo B (88 mg/dl) were within the normal range (Table 2). In contrast, the levels of Apo CII (20 mg/dl), Apo CIII (195 mg/dl) and Apo E (71 mg/dl) were extremely high. The changes in the concentrations of lipids and apolipoproteins were evaluated over a 37-day period during which the diabetes was initially corrected with intravenous insulin followed after the first two days by continuous subcutaneous infusion with a pump (Table 1 and 2). Plasma triglyceride and cholesterol levels declined rapidly and this was due largely to the reduction in VLDL levels. After the first week of insulin therapy, the concentration of VLDL-cholesterol reached normal values. Similarly, the levels of Apo CII, Apo CIII and Apo E were within the normal range. As a result of an increased catabolic conversion of VLDL to LDL, there was a transient elevation of LDL-cholesterol and Apo B levels; after 7 days, the levels of LDL-cholesterol and Apo B declined and reached normal values possibly due to an increased uptake of LDL by receptors which are known to be increased in number by insulin (104). At the same time intervals, HDL-cholesterol underwent a steady increase from 0.05 to 1.1 mmol/l suggesting an improved lipolytic degradation of chylomicrons and VLDL. Further evidence of improved lipolysis during insulin treatment was provided by an increase in the ratio of Apo CIII in the heparin-Mn^{2+} supernate to heparin-Mn^{2+} precipitate indicating increased transfer of Apo CIII from Apo B-containing, triglyceride-rich lipoproteins to HDL (102). In agreement with an earlier study (85), there was a transient decline in the levels of Apo AI and Apo AII. It is not clear whether this phenomenon is caused by an enhanced catabolism or reduced production of major HDL apolipoproteins. After 37 days of insulin administration the levels of all lipids and apolipoproteins were normalized.

It appears that in most ketotic patients, the composition of VLDL particles is not changed during insulin-deficiency (22, 34). Weidman et al. (85) found that the mean Apo E/Apo C and Apo CII/Apo CII + Apo CIII ratios did not change with insulin treatment; the ratio of Apo B to soluble apolipoproteins also remained constant. However, in cases of severe hypertriglyceridemia (87), the Apo E/Apo CIII and Apo CII/Apo CII + Apo CIII ratios were significantly lower before than after insulin administration (Table 2) suggesting an increased secretion of Apo CIII-containing lipoprotein particles. The chemical nature of these lipoprotein particles including their possible similarity to lipoproteins accumulating in nondiabetic hypertriglyceridemia is not known.

Ketoacidosis occuring in type I diabetics is characterized by a wide spectrum of lipid and apolipoprotein profiles ranging from almost normal to extreme hyperlipidemia and hyperapolipoproteinemia. Insulin deficiency appears to be the major determinant of hyperlipoproteinemia, since all cases with obvious type I diabetes respond to insulin treatment during the correction of ketoacidosis and hyperglycemia by normalizing the levels of lipids and apolipoproteins. However, factors such as counterhormonal reponses and preceding dietary fat intake may be responsible for the variability in hyperlipoproteinemia between individual patients with ketoacidosis. In addition, some coexisting familial forms of hypertriglyceridemia in the diabetic may be exacerbated by ketoacidosis (22, 23).

2.2 Adequately controlled diabetes

It is well-established that proper glycemic control in patients with type I diabetes results in normalization of plasma lipid and apolipoprotein profiles. During conventional treatment of diabetic children with insulin, the concentrations of plasma cholesterol (3.8 – 5.15 mmol/l) and triglyceride (0.7 – 1.36 mmol/l) are within the normal range (75, 76, 78, 79, 80 – 82, 105 – 107). In adequately controlled adult diabetics, the comparable concentrations of plasma cholesterol (4.6 – 6.5 mmol/l) and triglyceride (1.14 – 1.38 mmol/l) are also within the normal range (77, 83, 91, 108 – 110). No significant change was found in the plasma phospholipid concentrations between untreated and treated diabetic children (78).

When adequately controlled, children and adult diabetics have normal concentrations of VLDL-cholesterol (0.17 – 0.6 mmol/l), LDL-cholesterol (2.1 – 4.9 mmol/l) and HDL-cholesterol (0.97 – 1.7 mmol/l) (74 – 78, 80 – 83, 88, 105, 107 – 112). In fact, diabetic patients under strict glycemic control achieved by either intensified conventional insulin treatment or by continuous, subcutaneous insulin infusion may have plasma concentrations of cholesterol, triglyceride, VLDL-cholesterol and LDL-cholesterol in the lower ranges of normal values (90, 118, 119); in contrast, HDL-cholesterol levels may be in the upper ranges (90, 115, 118 – 120) of normal values. The increase in HDL-cholesterol levels in response to improved glycemic control requires a longer period of treatment in contrast to other lipids (22, 115, 118, 119). In most studies, the concentrations of plasma cholesterol, triglyceride, VLDL-cholesterol and LDL-cholesterol correlated significantly with the degree of glycemic control (74, 76, 80, 107, 110, 112, 121 – 123). In some studies, however, no significant correlation was found between the levels of LDL-cholesterol and glycated hemoglobin (107, 110, 114) or between the concentrations of plasma cholesterol, triglyceride, and VLDL-cholesterol and glycemic control (75, 114, 124). Despite some variability in these findings, influenced most probably by factors such as type of treatment, diet, physical activity, body mass index, smoking and alcohol consumption, the concentrations of plasma triglyceride and triglyceride-rich lipoproteins appear to represent a reliable measure of metabolic control. On the other hand, the concentrations of HDL-cholesterol show little or no correlation with the degree of glycemic control (22, 74, 80, 89, 107 – 110, 112, 114, 122, 123, 125); the only exception is ketoacidosis which is characterized, as previously described, by very low levels of HDL-cholesterol (22). The already mentioned slow response of HDL-cholesterol levels to restoration of euglycemia is indicative of the well-documented lack of association between the diabetic control and concentrations of this lipoprotein class. Nikkilä has suggested (22, 126) that one of the possible causes for the relatively high concentrations of HDL in well controlled diabetic patients is related to the increased concentration of insulin in peripheral tissues with a consequent stimulation of tissue lipoprotein lipase activity. The increased lipolytic degradation of triglyceride-rich lipoproteins leads to decreased levels of VLDL and increased levels of HDL. Experimental evidence supporting this hypothesis is based on the observations that in adequately controlled diabetics, the increased activity of postheparin plasma lipoprotein lipase (97, 109) is positively correlated with the levels of HDL-cholesterol (109) and that the elevated cholesterol concentration of HDL_2, a subfraction primarily affected by intravascular lipolysis of triglyceride-rich lipoproteins, is negatively associated with the levels of VLDL-triglyceride (127). It has been suggested that the Apo AI/Apo AII ratios may be used as a measure of the relative concentration of HDL_2 and HDL_3 subfractions, the higher values indicating a rise in the former and lower values, an elevation in the latter subfraction (108, 128). The finding of increased Apo AI/Apo AII ratios in controlled diabetics (83, 108) lends further support to the suggestion that the normalization of VLDL catabolism may be one of the main contributing factors to the increased levels of HDL. There are, however, reports indicating that HDL_3 rather than HDL_2 is the main contributor to elevated levels of HDL-cholesterol (89, 93). Although the reason for this discrepancy is not clear, the already mentioned slow response of HDL levels to adequate diabetic control suggests that mechanisms other than an increased rate of VLDL lipolysis may also be responsible for high concentrations of HDL. One of such possible mechanisms may be the direct effect of insulin on the hepatic production of protein and lipid constituents of HDL (22).

In adequately controlled children and adults with type I diabetes, the concentrations of apolipoproteins AI, AII and B are within the normal ranges (Table 3). In general, however, the values for any apolipoprotein will depend on the duration and form of treatment, age, sex and the methodology used for the quantification of apolipoproteins. The values for Apo AI are between 125 and 140 mg/dl; however, lower (92 mg/dl) as well as higher values (165 mg/dl) for Apo AI have also been reported. As in normolipidemic subjects, female diabetics have slightly higher levels of Apo AI than diabetic males. The concentrations of Apo AII range between 31 and 59 mg/dl. There is no discernible sex difference in the levels of this apolipoprotein. The levels of Apo AI and Apo AII were found to be either slightly higher (83, 108) or lower (91, 112, 117) than those of the corresponding normolipidemic controls. This discrepancy is most probably due to the type and duration of treatment. Wilson et al. (115) have shown that, after intensified insulin delivery during a six-month period, concentrations of both Apo AI and HDL-cholesterol increased significantly without a corresponding change in glycated hemoglobin, cholesterol, triglyceride and Apo B. A similar increase in the levels of Apo AI and Apo AII without a significant change in other apolipoproteins was observed after treatment of type I diabetics with an insulin pump for 2 weeks (129). The reported Apo B levels in adequately controlled diabetic patients are within the normal range (64 – 113 mg/dl) (Table 3). The slightly elevated Apo B concentration reported in one study (77) was found to be statistically insignificant in comparison with the control values. Owing to the effect of improved control on decreasing VLDL and increasing HDL concentrations, one would expect the levels of Apo C-peptides to approach normal values in well controlled diabetic patients. Although the concentrations of Apo CII and Apo CIII may be highly elevated in poorly controlled patients (Table 2), the levels of these apolipoproteins reach normal values shortly after the initiation of proper treatment. The reported values for Apo CII and Apo CIII range between 2 – 5 and 7 – 18 mg/dl, respectively (87, 116, 117, 124). Values for Apo CI have not been reported. Based on one study, it appears that Apo D levels may not be affected by glycemic control in type I diabetics (83). In our study (87) of a 13-year old boy, the concentrations of Apo E and Apo D (Table 2) were reduced to the normal range with good control. In two other studies involving 6 adequately controlled adult women (116) and 18 adult women and men (130), the mean value for Apo E was 9 mg/dl.

The available data indicate that the proper treatment of type I diabetes normalizes the concentration, distribution and composition of major lipoprotein density classes within a relatively short period of time (95, 98). This includes an increase in the distribution of lipids in LDL_2 and HDL and a corresponding decrease in VLDL and LDL_1 (98). One of the important compositional changes reported is a decreased percentage of triglyceride and an increased percentage of cholesterol in LDL_2 (79). It has recently been reported (93) that in insulin-treated patients, the LDL cholesterol/plasma Apo B ratio was significantly higher than in nondiabetic controls with the implication that this change may have resulted from increased levels of intermediate density lipoproteins (LDL_1 fraction of LDL). Because of the potential atherogenicity of these lipoproteins (131), this finding deserves to be confirmed and followed up by additional studies focusing on the concentration and composition of intermediate density lipoproteins in various stages of glycemic control. As these isolated cases show, there is still a paucity of data regarding the composition of lipids and apolipoproteins in major lipoprotein density classes of patients with moderately to well controlled hyperglycemia. For this reason, it is not yet possible to ascertain the degree of control necessary for adequate normalization of plasma lipoproteins.

The mean and median concentrations of Lp(a), a complex lipoprotein of relatively high hydrated density (132) shown to correlate with the presence of coronary artery disease in nondiabetic patients (133), were found not to differ statistically between diabetic patients and control subjects (91, 129). However, Lp(a) levels higher than 20 mg/dl were found in 14% of diabetics and in only 5% of the control subjects (91). Additional studies with larger diabetic populations are needed in order to confirm these findings and further explore the significance and role of Lp(a) in type I diabetics.

In diabetics with good control, the production rates of VLDL triglycerides are within the normal range (96, 134, 135). However, in one study of conventionally treated diabetics, increased production rates of VLDL triglyceride were still observed in some patients (96). In contrast, treatment of type I diabetes with continuous subcutaneous insulin infusion resulted in a significant reduction in mean VLDL-triglyceride transport rates to levels lower than those seen in the nondiabetic controls (135). Since there was no change in the fractional catabolic rates, the results of the latter study suggested that the normalization of triglyceride levels in euglycemic patients was due to suppression of hepatic VLDL-triglyceride synthesis rather than to an increased catabolic rate of triglyceride-rich lipoproteins.

Type I diabetics in good control have normal or slightly elevated levels of postheparin plasma lipoprotein lipase activity (97, 109). The markedly reduced levels of lipoprotein lipase activities in adipose tissue and skeletal muscle of untreated diabetics increase with insulin administration but the restoration of normal activities appears to be a rather slow process (33). Of these two tissue activities, the skeletal muscle lipoprotein lipase is less sensitive to insulin than is the lipoprotein lipase in adipose tissue. Since the slow restoration of normal lipoprotein lipase activities parallels equally slow normalization of HDL-cholesterol levels, Taskinen and Nikkilä (33) have suggested that these findings are consistent with the view that concentrations of HDL-cholesterol are dependent on the efficiency of catabolic processes responsible for the removal of triglyceride-rich lipoproteins from the circulation.

In insulin-treated patients with type I diabetes, the activity of lecithin:cholesterol acyltransferase (LCAT), an enzyme catalyzing cholesterol esterification in plasma, was found to be either within the normal range (91, 136) or higher than in nondiabetic controls (137).

3 Noninsulin-Dependent Diabetes Mellitus (Type II)

3.1 Untreated diabetes

Patients with type II diabetes comprise 85–90% of all diabetics (13). This large subclass of ketosis-resistant diabetes represents a heterogeneous disorder of carbohydrate metabolism frequently associated with obesity and highly variable degrees of hyperglycemia and hypertriglyceridemia. Newly diagnosed diabetics may present with normal or only slightly elevated concentrations of triglyceride-rich lipoproteins in association with extreme hyperglycemia at one end of this wide spectrum or with high levels of chylomicrons and VLDL and mild elevations in blood glucose at the other (16, 22). The estimates of the prevalence of hypertriglyceridemia in type II diabetes vary between 20 and 70% (24–29, 138–140) depending on the criteria used for defining hypertriglyceridemia and the setting (clinical vs. community populations) selected for the conduct of the survey. It should be emphasized that the concentration and composition of lipid and apolipoprotein constituents of plasma lipoproteins are influenced by and depend not only on the heterogeneous nature of the underlying metabolic disorder but also on a number of other genetic and environmental factors known to affect lipid transport in nondiabetic subjects. Of these factors, obesity is frequently present (30) and may be associated with both hyperlipidemia and severity of the diabetes.

In newly diagnosed or untreated type II diabetics, the concentrations of plasma triglycerides range between 1.5 and 4.9 mmol/l (27, 51, 52, 55–57, 83, 91, 141–149). With one exception (147), the mean triglyceride levels in all these studies were found to be significantly higher in diabetic patients than controls. Triglyceride levels are positively correlated with the body mass index (25, 55, 57, 138, 141, 151), and the obese diabetics have higher levels of triglycerides than the nonobese patients (25, 150). In most studies of newly diagnosed diabetics, men seem to have higher levels of triglyceride than women (55, 91, 143, 145, 150). In both diabetic men and women, triglyceride concentrations increase with age up to the sixth decade when they start to decline (25). In addition to moderate hypertriglyceridemia, newly diagnosed diabetic patients have normal or slightly elevated levels (4.3–6.8 mmol/l) of plasma cholesterol (27, 51, 52, 55–57, 83, 91, 141–149). Significantly increased cholesterol concentrations were only detected in a few untreated diabetic populations (27, 137, 141). In a community-based population, the cholesterol levels and the prevalence of hypercholesterolemia were not significantly different in diabetics and nondiabetics (29). In a selected group of patients referred to a hospital clinic, the prevalence of hypertriglyceridemia was 38% among diabetic women and 26% among men, while increased levels of cholesterol were found in only 11% of women and 4.5% of men (25). In a recently conducted study on the lipid and apolipoprotein levels in Indian diabetics of Oklahoma (152), 30% of all patients had increased levels of triglyceride (cut-off point $\geqq$ 2.5 mmol/l), but only 11% had increased concentrations of cholesterol (cut-off point $\geqq$ 6.25 mmol/l). The concentrations of plasma phospholipids were within the normal range (1.5–3.5 g/l) (46, 47, 153).

Untreated type II diabetics have increased concentrations of VLDL-cholesterol ranging from 0.4 to 1.3 mmol/l (47, 83, 146, 147, 154). In these same studies, the corresponding concentrations of LDL-cholesterol were normal (3.4–3.5 mmol/l) or slightly elevated (3.9 – 4.7 mmol/l) in comparison with nondiabetic controls. It should be emphasized that these values represent mean concentrations of VLDL- and LDL-cholesterol; in hypertriglyceridemic and/or hypercholesterolemic diabetics, the levels of VLDL- and LDL-cholesterol are significantly higher than the mean values for the entire diabetic population studied (92, 150, 154). In untreated diabetic populations, the levels of HDL-cholesterol are found to be either normal (1.25 – 1.45 mmol/l) (89, 147) or slightly but significantly reduced (0.77 – 1.17 mmol/l) (47, 51, 83, 154, 155). Since hypertriglyceridemia and obesity are inversely related to levels of HDL-cholesterol (126), the variability in the mean HDL-cholesterol values among various diabetic populations may be due, at least partially, to the selection of patients and corresponding nondiabetic controls (22). However, in two studies the HDL-cholesterol levels were found to be significantly lower in male and female diabetic patients than nondiabetic controls after adjustment for obesity, age, alcohol consumption, cigarette smoking and estrogen use in women (143, 156). The levels of both HDL_2- and HDL_3-cholesterol seem to be lower in diabetic patients than in nondiabetic controls (51, 157). However, noninsulin-dependent diabetics who had suffered a previous myocardial infarction had significantly lower levels of HDL_2-cholesterol than diabetics without any symptoms of coronary heart disease; on the other hand, there was no difference between these two groups of patients in the levels of HDL_3-cholesterol (158). This difference in HDL_2-cholesterol was evident in men and women and persisted even after adjustment for the effect of obesity, glycemic control and other confounding factors. In studies where LDL- and HDL-cholesterol were measured, the LDL/HDL cholesterol ratio was significantly higher in diabetics than nondiabetic controls (51, 91, 137, 142, 159).

As a consequence of hypertriglyceridemia, the concentrations of VLDL-, LDL- and HDL-triglyceride are increased in type II diabetics when compared to nondiabetic controls (26, 91, 92, 137, 154, 159). Only the triglyceride enrichment of LDL was found to be statistically significant in all of these studies.

Most of the available information on the levels of plasma apolipoproteins in newly diagnosed diabetic patients relates to Apo AI (Table 4). Although there are differences in the reported Apo AI values between laboratories, the consensus is that in newly diagnosed or uncontrolled diabetics the concentrations of plasma Apo AI are significantly lower than in nondiabetic controls. As in nondiabetics, the diabetic women have a slightly higher level of Apo AI than the diabetic men. The Apo AII concentrations are either normal or slightly reduced. In contrast, the levels of plasma Apo B seem to be higher in diabetics than the corresponding controls. Thus, similarly to the LDL/HDL cholesterol ratio, the Apo B/Apo AI ratio is also increased in newly diagnosed noninsulin-dependent diabetics. There is a scarcity of data on the levels of minor apolipoproteins. In a study of five men whose diabetes had been managed by diet only, the concentrations of plasma Apo CII (8 ± 2 mg/dl) and Apo CIII (15 ± 2 mg/dl) were significantly higher than in the control subjects (162). The Apo CII/Apo CIII ratio was slightly higher in diabetics than nondiabetics. No difference was found in the levels of plasma Apo D between untreated diabetics and controls. In one study, the mean concentration of patients' Apo D was 17 ± 4 mg/dl (83) and in the other it was 5 ± 1 mg/dl (160). The concentration of apolipoprotein E (11 – 14 mg/dl) was significantly higher in untreated type II diabetics than in the controls (130, 160). As this survey shows,

Table 4. Plasma concentrations (mg/dl) of apolipoproteins AI, AII and B in patients with type II diabetes

Sex	Age (years)	Control	Apolipoproteins			Method[a]	Reference
			AI	AII	B		
M	51	UT[b]	114[c]	–	204	EIA	55
M	51±10	UT	113↓	–	–	EIA	143
M	57± 3	UT	87↓	–	–	EIA	155
F	52	UT	116	–	203	EIA	55
F	52± 9	UT	116↓	–	–	EIA	143
F	58± 4	UT	95↑	–	–	EIA	155
M + F	31± 3	UT	114↓	–	–	RID	51
M + F	52± 5	UT	92↓	29	138↑	RID	160
M + F	59	UT	126	30	–	RID	161
M	49± 9	D	147↓	21↓	–	RIA	162
M + F	56±13	D	–	–	110↑	RIA	92
M	14–49	OA	87↓	–	123↑	RIA	142
M	30–60	OA	135↑	44↑	110↑	EIA	83
M	61.4± 2	OA	125↓	43↓	115↓	EIA	91
F	14–49	OA	091↓	–	118↑	RIA	142
F	64± 4	OA	126↓	43↓	120↑	EIA	91
M	63± 5	M	170↓	42↓	–	RIA	117
F	64± 6	M	170↓	45↓	–	RIA	117
M + F	56±12	M	124↓	–	105	EIA	152

[a] EIA = electroimmunoassay; RIA = radioimmunoassay; RID = radial immunodiffusion assay
[b] UT = untreated patients; D = dietary treatment; OA = treatment with oral agents; M = miscellaneous treatments (diet; insulin; oral agents)
[c] Direction of arrow indicates whether the values for apolipoproteins in diabetic patients are lower or higher than corresponding values in normolipidemic controls.

there is an obvious need for additional studies on the concentrations of all plasma apolipoproteins, but particularly Apo C-peptides and Apo E.

There are very few reports on the concentration of the total lipid and protein mass of major lipoprotein density classes in untreated or inadequately treated type II diabetics. In an older study (163), based on the analytical ultracentrifugation of lipoproteins, uncontrolled female patients with nonketotic diabetes had significantly increased concentrations of the VLDL subfraction, S_f 100–400, and LDL_2 than when treated with insulin or compared to nondiabetic controls. The levels of the VLDL subfraction, S_f 20–100, LDL_1, HDL_1, HDL_2 and HDL_3 did not differ from controls. More recently, Lisch and Sailer (153) found no significant difference in the concentrations of VLDL, LDL_1, LDL_2, HDL_2 and HDL_3 between control subjects and diet-treated diabetics. These results show that measuring the total mass of lipoprotein density classes provides a more realistic approach to the evaluation of lipoprotein concentrations than measurement of a single lipoprotein constituent. Its biochemical significance should be further explored and evaluated in future studies.

The variability in the reported results on the lipid and apolipoprotein composition of major lipoprotein density classes is another reflection of the heterogeneous nature of type II diabetes. A handful of studies have shown that the lipoprotein composition depends on the severity of disease, the extent of hypertriglyceridemia and the mode of treatment. In mildly hypertriglyceridemic patients who had not been treated by insulin or hypoglycemic drugs, the lipid and protein composition of VLDL, LDL and HDL was not different from the composition of corresponding lipoproteins in normolipidemic controls (153, 159). However, in untreated hypertriglyceridemic patients, LDL_2, but not VLDL, had a significantly higher content of triglyceride and a lower content of cholesterol esters than LDL_2 from control subjects (164); furthermore, both VLDL and LDL_2 had a higher ratio of free cholesterol/phospholipid than controls. In a study of diabetic patients treated either by insulin or hypoglycemic drugs, Schonfeld et al. (92) found a significant elevation in the triglyceride content and triglyceride/cholesterol ratio of LDL and HDL irrespective of the patients' plasma levels of triglyceride or cholesterol; however, the relative increase in triglyceride content appeared to be greater in hypertriglyceridemic than in normolipidemic diabetic patients. In a more recent study of a mixed group of diabetic patients treated either by oral hypoglycemic agents or insulin, Hiramatsu et al. (165) reported no difference in the lipid composition of LDL_2 between normolipidemic diabetics and nondiabetic controls; however, LDL_2 isolated from hypertriglyceridemic patients had a significantly higher mass ratio of triglyceride/protein and a lower ratio of total cholesterol/protein as compared with the control subjects. A free cholesterol enrichment of LDL_2 was not confirmed in this report. An increased ratio of triglyceride/protein was detected in both HDL_2 and HDL_3 of untreated diabetic patients with normal or slightly elevated levels of plasma triglyceride (51). The information on the apolipoprotein composition of lipoprotein density classes is scarce. Measurement of apolipoproteins by scanning densitometry of polyacrylamide gel electrophoretic patterns revealed small but significant increases in the relative contents of Apo CII and Apo CIII in diabetic VLDL as compared with control VLDL (166). Determination of insoluble (Apo B) and soluble (Apo C and Apo E) apolipoproteins showed that the percentage of the former was lower and that of the latter greater in VLDL of poorly controlled, normolipidemic type II diabetics than in normals or well controlled patients (167). This, as yet, unconfirmed finding suggests the existence of a latent defect in the turnover of VLDL that may be characteristic of normolipidemic and hyperlipidemic diabetes. No difference was found between hypertriglyceridemic diabetic and nondiabetic VLDL in the Apo CII/Apo CIII ratio and percent distribution of Apo CIII isomorphic forms (168), thus excluding an abnormal composition of Apo C-peptides as a factor contributing to the pathogenesis of hypertriglyceridemia in type II diabetes. On the other hand, the analysis of Apo E isomorphic forms in VLDL of type II diabetic patients showed a decrease in the ratio of Apo E_3 to Apo E_2 (169, 170). A change towards more acidic isomorphs of Apo E, caused by an increase in sialic acid content, may result in decreased receptor binding of such VLDL particles (171) and their increased accumulation in plasma. However, adequate control of hyperglycemia seems to improve the Apo E-3/Apo E-2 ratio (170). There is apparently no difference between the normal and diabetic subjects in the apolipoprotein composition of LDL (164, 166). The apolipoprotein composition of diabetic HDL is characterized by a slight but significant reduction in the relative content of Apo AI and significant increases in the relative contents of apolipoproteins D, CI, CII and CIII (166).

Results of a recent study on the composition of HDL from untreated diabetic patients (172) suggest that, despite a decreased content of cholesterol, the actual number of HDL particles is not reduced. This may be due to replacement of cholesterol esters with triglyceride in the lipid core of HDL particles as evidenced by increased ratios of triglyceride/cholesterol and Apo AI/cholesterol. Thus, a compositional change in HDL particles may not be necessarily accompanied by a change in their number and mass. It has been suggested (172) that triglyceride enrichment of HDL particles might impair their putative antiatherosclerotic role by decreasing their potential as acceptors of cholesterol from peripheral tissue sites.

Although the concentration of Lp(a) in patients with type II diabetes seems to be slightly increased in comparison with the control group, this increase is not statistically significant (91, 129). Further studies are needed to confirm and expand these preliminary observations.

Posttranslational nonenzymatic glycation of apolipoproteins represents a qualitatively important compositional feature of lipoproteins in the diabetic state. *In vitro* glycation of VLDL, LDL and HDL, through a covalent linkage of glucose to free amino group of lysine residues, was shown to be a linear function of time and glucose concentration (94, 173–175). Recent studies (176) with monoclonal antibodies to glucitollysine, the reduced glycated lysine, have confirmed the initial observation (173) that several apolipoproteins are glycated including Apo AI, Apo AII, Apo B, Apo CI and Apo E. Furthermore, these studies have demonstrated that glycation of apolipoproteins occurs *in vivo* in both normal, euglycemic subjects and in diabetic patients; however, the glycation of lipoproteins from normal subjects appears to be less than 8% of that observed in diabetics (176). In hyperglycemic and hypertriglyceridemic patients the majority of glycated proteins are present in triglyceride-rich lipoproteins, while in euglycemic subjects the majority of glycated proteins occurs in HDL (176); it is of some interest that glycated Apo B was detected in LDL_1 and LDL_2 but not in chylomicrons and VLDL. Although the *in vitro* reductive glycation of LDL inhibits its binding and degradation by the high affinity receptor pathway (94, 174, 175), no difference was found between LDL_2 isolated from moderately hyperglycemic type II diabetic patients and nondiabetic controls in their binding and degradation by cultured human fibroblasts (177). In an independent study, Hiramatsu et al. (165) have also shown unimpaired binding by cultured human fibroblasts of LDL_2 from diabetic patients with normal or slightly elevated triglyceride levels. However, the high affinity binding of LDL_2 from hypertriglyceridemic patients with and without diabetes has been significantly reduced, suggesting that the altered metabolism of LDL_2 is associated with hypertriglyceridemia rather than diabetes. Kissebah et al. (178) have established that the fractional catabolic rate of LDL-Apo B is higher in mild diabetics but lower in moderately severe diabetics than in nondiabetic controls implying that the glycation of Apo B might be of some significance only in the moderately severe form of diabetes. However, the lower fractional catabolic rate might also be associated with triglyceride enrichment of LDL_2 particles. In a recently reported study (179) no difference was found between the fractional catabolic rates of LDL isolated from insulin-deficient diabetic patients and sex, age and weight matched nondiabetic controls. The exact pathophysiological role of nonenzymatic glycation of plasma apolipoproteins in diabetes remains to be clarified.

Metabolic and hormonal heterogeneity of type II diabetes resulting in a variable degree of insulin resistance and plasma insulin levels is also reflected in the complex kinetics of VLDL transport (18, 180). There is general agreement that the mild to severe hypertriglyceridemia in untreated diabetic patients is due to the overproduction of VLDL-triglyceride (36, 37, 88, 96, 134, 180). Kissebah et al. (180) have demonstrated a similarity in the turnover rates of VLDL-triglyceride and VLDL-Apo B suggesting that the production of these two VLDL constituents is integrated at their synthetic sites. This finding in conjunction with results of lipid and protein analyses indicates that varying degrees of hypertriglyceridemia are due to overproduction of VLDL particles of normal composition (180). The production rates of VLDL-triglyceride and Apo B are also significantly higher in normolipidemic diabetics when compared to nondiabetic controls (180). However, the increased production rates are compensated for in these patients by increased fractional catabolic rates resulting in normal or only slightly elevated triglycerides. In most patients with mild hypertriglyceridemia, the fractional catabolic rates of VLDL-triglyceride are within the normal range (96, 180). However, in moderately to severely hypertriglyceridemic patients, the catabolic rates are significantly lower than in controls (58, 96, 108). As a consequence, the conversion of VLDL-Apo B to LDL is greatest in the normolipidemic mildly diabetic patients but decreases with increasing concentrations of triglyceride (180); the amount of VLDL-Apo B catabolized via the non-LDL pathway is estimated to be three to five-fold greater in moderately to severely hypertriglyceridemic diabetics than in control subjects. Decreased catabolism of VLDL particles seems to be paralleled by decreased lipoprotein lipase activities in postheparin plasma and in adipose tissue of diabetic patients. Whereas in normolipidemic patients the postheparin plasma lipoprotein lipase activity is normal (22, 47, 97), in hypertriglyceridemic patients both the plasma and adipose tissue lipoprotein lipase activities are significantly reduced (38, 56, 59, 97, 181). Thus, in some patients with type II diabetes, both overproduction and reduced removal capacity may lead to a marked accumulation of triglyceride-rich lipoproteins including chylomicrons (181). In contrast, Abrams et al. (37) have found no difference in the postheparin plasma lipoprotein lipase activity between a mildly to severely hypertriglyceridemic group of diabetic patients and nondiabetic controls; the activity of postheparin plasma hepatic triglyceride lipase was also found to be normal in agreement with similar results by other investigators (37, 56, 97). These divergent results indicate the necessity for further studies to evaluate postheparin plasma lipoprotein lipase activities in hypertriglyceridemic diabetic patients. Since lipoprotein lipase activities are underestimated in the presence of triglyceride-rich lipoproteins, we suggest that these activities be estimated by procedures which eliminate triglyceride-rich lipoproteins as a potential source of lipase inhibitors (182).

In Pima Indian diabetics, the moderately increased levels of VLDL-triglycerides were found to be a result of decreased clearance capacity rather than overproduction of VLDL (39). One of the possible explanations for this unexpected finding is that levels of plasma-free fatty acids are not elevated in this diabetic population.

In most type II diabetic populations studied, the LDL concentrations are within the normal range. However, in moderately severe diabetics, the LDL levels are increased (178) and LDL-cholesterol concentrations correlate positively with diabetic control (183). A study on the transport kinetics of LDL_2 (178) has shown that in mild diabetics the production rate of LDL_2-Apo B is increased in comparison with normal subjects. However, the plasma levels of LDL_2 are not increased because of a simultaneous increase in their fractional catabolic rate. In moderately severe diabetics, the production rate of LDL_2-Apo B is normal, but a reduced fractional catabolic rate results in the elevation of LDL levels. It is quite possible that structural changes in LDL particles such as the replacement of cholesterol esters with triglycerides and/or excessive glycation of Apo B may be the underlying cause for their impaired removal.

Apart from the effect of obesity and other factors known to affect the levels of HDL-cholesterol in nondiabetic populations, the kinetic and biochemical causes of slightly reduced concentrations of HDL in untreated type II diabetics remain to be investigated.

The LCAT activities in type II diabetic patients are found to be either normal (91) or significantly increased (137) in comparison with nondiabetic controls. However, Fielding et al. (162) have observed in a group of untreated patients with type II diabetes a slight but significant reduction of the esterification rate in comparison with normal subjects. Furthermore, there was a marked inhibition of cholesterol net transport between cultured fibroblasts and plasma coupled with a reduced cholesterol ester transfer from HDL to VLDL and LDL. In a subsequent report (164), these changes were considered to be due to an increased content of free cholesterol in VLDL and LDL. The authors have suggested that the free cholesterol enrichment of these lipoproteins prevents the uptake of cholesterol esters and causes free cholesterol to be transported back to cells. These potentially atherogenic abnormalities could be reversed by adequate control of hyperglycemia or by removal of an Apo E-containing lipoprotein (182) thought to mediate the excessive influx of free cholesterol from diabetic plasma to cells. Since free cholesterol enrichment of VLDL and LDL in diabetic patients has not been observed in all reported studies (23, 153, 159, 165), these intriguing and potentially important findings need to be confirmed.

3.2 The chylomicronemia syndrome

Severe hypertriglyceridemia has often been associated with clinical manifestations of the chylomicronemia syndrome (184) consisting of abdominal pain, mental confusion and memory loss, symptoms which are also present in nondiabetic subjects with extreme hypertriglyceridemia. In addition to lipemia retinalis, eruptive xanthomata may be observed which have been found to contain lipids of similar composition to chylomicrons (185).

When the triglyceride rises to levels above 25 mmol/l, Chait and Brunzell suggest that the hypertriglyceridemia results from an interaction of familial and secondary factors such as diabetes mellitus (186). Evidence for this hypothesis comes from studies on families of extremely hypertriglyceridemic, diabetic subjects. Relatives of these index patients had high triglyceride levels but much lower than when diabetes was present (186). Lipid-lowering therapy, including treatment of the diabetes with oral hypoglycemic agents or insulin, seldom resulted in reduction of the plasma triglyceride levels to within the normal range suggesting the presence of a residual disorder (186). Similar observations were made by Fallat and Glueck who also observed marked but incomplete reductions in the triglyceride level in type V patients after treatment of various secondary hyperlipidemias including diabetes (187). These diabetic subjects were mostly noninsulin-dependent and obese (186, 187).

When the type II diabetes is associated with an inherited type V hyperlipoproteinemia, two possible mechanisms may account for the extreme hypertriglyceridemia. Overproduction of VLDL is a feature of both familial hypertriglyceridemia (188) and type II diabetes (58, 96), in addition to which reduced adipose tissue lipoprotein lipase activity occurs (189). Furthermore, it is likely that when production is increased or fat intake is excessive, the triglyceride removal mechanism becomes saturated followed by further accumulation of plasma triglyceride-rich lipoproteins (190).

Although studies have been conducted on the relationship of the chylomicronemia syndrome to genetic and acquired causes, there is a virtual absence of studies on the distribution of lipids in the density classes other than identification of the presence of chylomicrons in subjects with triglyceride levels greater than 25 mmol/l (191). We have observed a lipid and apolipoprotein distribution resembling type V hyperlipoproteinemia with low levels of LDL and HDL in addition to relative triglyceride enrichment of these density classes (Blackett and Alaupovic, unpublished data).

Apolipoprotein levels could potentially be useful in characterizing the hyperlipidemia and identifying apolipoprotein profiles in nondiabetic hypertriglyceridemic relatives so helping to define the genetic disorder in the index case with diabetes. Assays of apolipoproteins and studies on their association with each other may help to define the lipoprotein composition and study relationships to atherosclerotic risk in these subjects when they present with extreme hypertriglyceridemia and during therapeutic intervention.

3.3 Treatment of noninsulin-dependent diabetes

In contrast to the uniform treatment of hyperglycemia in type I diabetes, the glycemic control in patients with type II diabetes may be achieved by several treatment modalities including diet, oral hypoglycemic agents and insulin. In contrast to a relatively uniform response to insulin administration in insulin-dependent diabetes, each of the treatment modalities used in type II (noninsulin-dependent) diabetes exerts a variable effect on lipids and lipoproteins. Frequently, adequate control of hyperglycemia is not accompanied by comparable normalization of deranged composition and metabolism of plasma lipoproteins. In addition to the relative ineffectiveness of a particular treatment, failure to reverse the hyperlipidemia may be the result of a number of genetic and acquired factors such as obesity, familial forms of hyperlipoproteinemia, use of hypotensive and contraceptive drugs, smoking and alcohol consumption. The important topic of glycemic control and its effect on lipids and lipoprotein metabolism has been covered in several recent reviews (19, 20, 22, 23) and will only be presented in a summary form.

Because patients with type II diabetes are frequently obese, weight reduction by hypocaloric diets is the preferred initial treatment for controlling hyperglycemia, particularly in overweight patients with mild diabetes (20, 22, 24). Treatment with hypocaloric diets during 3 – 12 month periods leads to improved glycemic control and significantly reduces plasma triglyceride levels (42 – 44, 52, 55, 145, 146), VLDL-cholesterol and VLDL-triglyceride (42, 43, 52, 145); the decrease in plasma triglyceride correlates with the percentage weight loss (44). The effect of caloric restriction on plasma cholesterol and LDL-cholesterol is minimal, but the HDL-cholesterol levels are significantly increased (44, 52, 55). However, in another study (43), a significant increase in HDL-cholesterol levels was only achieved when caloric restriction was combined with gastric bypass surgery. After one year of caloric restriction, there was a slight but statistically insignificant rise in the levels of Apo AI and a decrease in Apo B levels in male and female diabetics (55); a significant increase in the concentration of Apo AI in females and an equally significant decrease in Apo B in male patients was only reached when caloric restriction was combined with the use of an oral hypoglycemic agent (Metformin) (55). A hypocaloric regimen caused a significant reduction in VLDL-Apo B and VLDL-Apo E, but no change in Apo C-peptides and the Apo CII/Apo $CIII_1$ ratio (146); these findings were interpreted as preferential removal of remnant particles containing Apo B and Apo E from VLDL. In general, studies on the hypocaloric treatment of patients with type II diabetes indicate that the most significant observation is a reduction in VLDL particles associated with their decreased synthesis rather than increased catabolism (42, 52). Changes in lipid and apolipoprotein levels are less marked when a lipid-lowering dietary regimen is used without accompanying caloric restriction (49, 155).

Studies on the effect of oral hypoglycemic agents on plasma lipids and lipoproteins have produced variable and inconsistent results. The use of sulphonylureas resulted in either decreased (51, 53, 54, 56, 145, 192) or increased (25, 27, 45, 49, 153) levels of plasma triglycerides. Other investigators found no significant change in the concentration of triglyceride (46, 47, 55, 155). Accordingly, the levels of VLDL-cholesterol were found to be either decreased (28, 51), increased (49, 154) or without change (47). Phenformin, a biguanide hypoglycemic drug, lowered plasma triglyceride levels by decreasing the production and increasing the degradation of triglyceride-rich lipoproteins (50). In two hypertriglyceridemic patients treated with the sulphonylurea, chlorpropamide, a decrease in the concentration of VLDL-triglyceride was also mediated through a reduced production and increased catabolism of VLDL particles (58). The effect of sulphonylureas on the levels of plasma cholesterol is variable; plasma cholesterol was decreased in some (45, 47, 56, 145, 153) but not in other studies (27, 46, 155). In a few reported studies, the levels of LDL-cholesterol were found to be either significantly decreased (47, 51) or without a significant change (49, 154). The published data on the levels of HDL-cholesterol have created considerable controversy because of the possible association between the adverse effect of oral hypoglycemic agents on cardiovascular mortality

(22). Several studies have shown that treatment of diabetic patients with oral hypoglycemic drugs lowers the levels of HDL-cholesterol (49, 60, 125, 153, 154, 192, 193). In contrast, results of prospective studies have indicated that the HDL-cholesterol levels are either increased (53) or not changed (46, 47, 51, 55) by sulphonylureas. The concentration of HDL_2-cholesterol was found to be unchanged (46, 47, 51), while the concentration of HDL_3-cholesterol was either slightly but significantly decreased (51) or unchanged (47). On the basis of these results, Taskinen et al. (47) have suggested that low levels of HDL-cholesterol in type II diabetes are a result of a basic abnormality in HDL metabolism rather than the consequence of sulphonylurea treatment. Oral hypoglycemic agents seem to exact very little effect on the levels of Apo AI (51, 55, 155) and Apo B (55) with the possible exception of a group of female diabetics whose Apo AI levels were significantly increased after one year of treatment with either caloric restriction and Metformin or carbohydrate restriction and Glibenclamide, a second generation sulphonylurea (55). Data on the concentration of minor apolipoproteins are not available. The lipid and protein composition of major lipoprotein density classes is not affected by sulphonylurea treatment (153) nor are the activities of adipose tissue and postheparin plasma lipoprotein lipase (47).

Insulin treatment of type II diabetics significantly lowers the levels of plasma triglyceride and VLDL-cholesterol (25, 27, 37, 46, 49, 57, 58, 163, 194, 195). It has little effect on the concentration of LDL-cholesterol (37, 48) but significantly increases the levels of HDL-cholesterol (37, 45, 46, 49, 53, 61, 154). Determination of the total lipoprotein mass has shown that the increased concentration of HDL is due to the elevation of HDL_2 rather than HDL_3 particles (46). The lipid and protein composition of VLDL, LDL_1, LDL_2, HDL_2 and HDL_3 is not significantly influenced by insulin treatment (153). With the exception of a report showing decreased levels of Apo B in total plasma and LDL (45), there are no studies on the effect of insulin on either the concentration or distribution of apolipoproteins. The lowering of VLDL and elevation of HDL seems to coincide with the observed increases of postheparin plasma lipolytic activity (46) or adipose tissue lipoprotein lipase (59). Turnover studies have shown that the reduction of VLDL by insulin treatment is due to both their decreased production and increased removal (37, 58).

4 Conclusions

Although evidence for an increased incidence of atherosclerotic vascular disease in diabetic patients appears conclusive, the full biological significance of this association is still unclear. Abnormalities in the lipid transport system may be an important, if not crucial, contributing factor. Hyperlipidemia which is present in at least a third of the diabetic population is characterized by hypertriglyceridemia and only occasionally by hypercholesterolemia. Control of the diabetes, environmental factors and inherent genetic potential may all affect its expression. Hyperlipidemia is accounted for by predominantly excess hepatic production and, to a lesser degree, by defective disposal of triglyceride-rich lipoproteins and is often reversible by insulin in type I diabetes and partially by oral hypoglycemic agents and insulin in type II diabetes. Lipid levels during treatment will depend on the control of the blood glucose levels and free fatty acid flux which are important variables in both type I and type II diabetes. Decreased levels of HDL-cholesterol, inversely related to the levels of triglyceride-rich lipoproteins, show little or no correlation with the degree of glycemic control. However, insulin may significantly influence the HDL-cholesterol levels coinciding with its effect on adipose tissue lipoprotein lipase.

Untreated type I diabetes and ketoacidosis is characterized by moderate to very high levels of plasma triglycerides and slight to moderate increases in plasma cholesterol. Increased levels of Apo B and, especially, Apo C-peptides and Apo E are accompanied by decreased concentrations of Apo AI and Apo AII. Hypertriglyceridemia is reflected by significantly elevated levels of VLDL including chylomicrons. In contrast, the LDL and HDL levels are decreased. This lipoprotein profile results from the hepatic overproduction of triglyceride-rich lipoproteins on one hand and their partially impaired degradation on the other. The main compositional change is the triglyceride enrichment of LDL and HDL and, in severe cases of hypertriglyceridemia, the predominant Apo CIII enrichment of VLDL. Good control of hyperglycemia by insulin administration reverses this lipoprotein profile and normalizes the levels of major lipoproteins.

Patients with untreated type II diabetes have increased levels of triglycerides but normal or only slightly elevated levels of cholesterol. Levels of Apo C-peptides and Apo E are slightly to moderately elevated depending on the severity of hypertriglyceridemia: the levels of Apo B are usually within the normal range, while the levels of apolipoproteins AI and AII tend to be slightly reduced. The characteristic lipoprotein abnormality is the increased concentration of triglyceride-rich lipoproteins, mainly VLDL and only rarely chylomicrons, brought about by their increased production. The LDL levels are mostly within the normal range, except in cases of moderately severe hypertriglyceridemia when an altered removal mechanism leads to their accumulation. The HDL levels are normal or slightly reduced. Both LDL and HDL have increased contents of triglyceride. In contrast to type I diabetes, the lipid and lipoprotein abnormalities are only partially normalized by the administration of oral hypoglycemic agents and insulin. Caloric restriction appears to be the most effective therapeutic regimen.

Measurement of plasma lipids and lipoprotein density classes shows that lipid, apolipoprotein and lipoprotein profiles of type I and type II diabetes differ very little from similar profiles characteristic of other dyslipoproteinemic states. The differences between diabetic lipoprotein patterns are of a quantitative rather than qualitative nature. However, there are several recent findings indicating the possible presence of structural and functional abnormalities that may, indeed, be characteristic of the lipid transport system in diabetes. These include the well-characterized glycation of several major and minor apolipoproteins, free cholesterol enrichment of VLDL and LDL, abnormal ratios of Apo E isomorphs, abnormal Apo B/cholesterol ratio in VLDL and LDL_2 subfractions, and the occurrence of Apo E-containing particles thought to mediate the abnormal transport of free cholesterol from plasma to peripheral cells. In future studies, the separation of lipoprotein particles from all major density classes on the basis of their apolipoprotein composition may allow the identification of these and other subtle compositional changes resulting from or reflecting functional abnormalities characteristic of diabetes. Furthermore, such developments may reveal additional potential risk factors for the accelerated macrovascular disease in diabetic patients and contribute to the modification of present and/or the introduction of new measures for normalizing both hyperglycemia and abnormal lipid transport.

Acknowledgements

We wish to thank Ms. Anne Harris for her help with the preparation and typing of the manuscript.

This work was supported in part by Grant HL23181 from the U.S. Public Health Service and by the resources of the Oklahoma Medical Research Foundation.

References

1. West, K. M. (1978) Epidemiology of Diabetes and its Vascular Lesions. Elsevier, New York, Oxford
2. Ruderman, N. B. & Hauderschild, C. (1984) Diabetes as an atherogenic factor. Progr. Cardiovascul. Dis. *26*, 373–412
3. West, K. M., Ahija, M. M., Bennett, P. H., Czyzyk, A., Mateo de Acosta, O., Fuller, J. H., Grab, B., Grabauskas, V., Jarrett, R. J., Kosaka, K., Keen, H., Krolewski, A. S., Miki, E., Schliack, V., Teuschev, A., Watkins, P. J. & Stober, J. A. (1983) The role of circulating glucose and triglyceride concentrations and their interactions with other "risk factors" as determinants of arterial disease in nine diabetic population samples from the WHO multinational study. Diabetes Care *6*, 361–369
4. Roberston, W. B. & Strong, J. P. (1968) Atherosclerosis in persons with hypertension and diabetes mellitus. Lab. Invest. *18*, 538–551
5. Garcia. M. J., McNamara, P. M., Gordon, T. & Kannel, W. B. (1974) Morbidity and mortality in diabetics in the Framingham population. Sixteen year follow-up study. Diabetes *23*, 105–111
6. Kannel, W. B. & McGee, D. L. (1979) Diabetes and cardiovascular risk factors: the Framingham study. Circulation *59*, 8–13
7. Jarret, R. J., McCartney, P. & Keen, H. (1982) The Bedford survey: ten-year mortality rates in newly diagnosed diabetics, borderline diabetics and normoglycaemic controls and risk indices for coronary heart disease in borderline diabetics. Diabetologia *22*, 79–84
8. Heyden, S., Heiss, G., Bartel, A. G. & Hames, C. G. (1980) Sex differences in coronary mortality among diabetics in Evans County, Georgia. J. Chron. Dis. *33*, 265–273
9. Barrett-Connor, E. & Wingard, D. L. (1983) Sex differential in ischemic heart disease mortality in diabetics: a prospective population-based study. Am. J. Epidemiol. *118*, 489–496
10. Marble, A. (1976) Late complications of diabetes: a continuing challenge. Diabetologia *12*, 193–199
11. Keen, H. & Jarrett, R. J. (1979) The WHO multinational study of vascular disease in diabetes: 2. Macrovascular disease prevalence. Diabetes Care *2*, 187–197
12. The World Health Organization Multinational Study of Vascular Disease in Diabetics (Diabetes Drafting Group) (1985) Prevalence of small vessel and large vessel disease in diabetic patients from 14 centers. Diabetologia *28*, 615–640
13. Zimmet, P. (1982) Type 2 (non-insulin-dependent) diabetes – an epidemiological overview. Diabetologia *22*, 399–411

14. Ganda, O. P. (1980) Pathogenesis of macrovascular disease in the human diabetic. Diabetes *29*, 931–942
15. Colwell, J. A., Winocour, P. D., Lopes-Virella, M. & Halushka, P. V. (1983) New concepts about the pathogenesis of atherosclerosis in diabetes mellitus. Am. J. Med. *74*, 67–80
16. Chait, A., Bierman, E. L. & Brunzell, J. D. (1985) in: The Diabetes Annual/1 (Alberti, K. G. M. M. & Krall, L. P., eds.) pp. 323–348, Elsevier, Amsterdam, New York, Oxford
17. Stout, R. W., Bierman, E. L. & Brunzell, J. D. (1975) in: Diabetes, Its Physiological and Biochemical Basis (Vallance-Owen, J., ed.) pp. 125–169, University Park Press, Baltimore
18. Reaven, G. M. & Greenfield, M. S. (1981) Diabetic hypertriglyceridemia: evidence for three clinical syndromes. Diabetes *30* (Suppl. 2), 66–75
19. Goldberg, R. B. (1981) Lipid disorders in diabetes. Diabetes Care *4*, 561–572
20. Dunn, F. L. (1982) Hyperlipidemia and diabetes. Med. Clin. North Am. *77*, 1347–1360
21. Eder, H. A. & Bergman, M. (1983) in: Diabetes Mellitus, Theory and Practice (Ellenberg, M. & Rifkin, H., eds.) pp. 61–76, 3rd Ed., Medical Examination Publishing, New Hyde Park, New York
22. Nikkilä, E. A. (1984) in: Diabetes and Heart Disease (Jarrett, R. J., ed.) pp. 133–167, Elsevier, Amsterdam, New York, Oxford
23. Brunzell, J. D., Chait, A. & Bierman, E. L. (1985) in: Diabetes Annual/1 (Alberti, K. G. M. M. & Krall, L. P., eds.) pp. 463–479, Elsevier, Amsterdam, New York, Oxford
24. Albrink, M. J. (1974) Dietary and drug treatment of hyperlipidemia in diabetes. Diabetes *23*, 913–918
25. Wahl, P., Hasslacher, C. & Vollmar, J. (1974) Diabetes und Hyperlipoproteinämien. Dtsch. Med. Wochenschr. *99*, 2158–2165
26. Mancini, M., Rivellese, A., Rubba, P. & Riccardi, G. (1980) Plasma lipoproteins in maturity onset diabetes. Nutr. Metab. (Suppl. 1), 65–73
27. Retnam, V. J., Nerurkar, S. V., Gupte, M. H., Bhandarkar, S. D. & Taskar, S. P. (1983) Plasma lipids and lipoproteins in treated diabetics. J. Postgrad. Med. *29*, 193–200
28. Hughes, T. A., Kramer, J. O. & Segrest, J. P. (1985) Effects of glyburide therapy on lipoproteins in non-insulin-dependent diabetes mellitus. Am. J. Med. *79* (Suppl. 3B) 86–91
29. Barrett-Connor, E., Grundy, S. M. & Holdbrook, M. J. (1982) Plasma lipids and diabetes mellitus in an adult community. Am. J. Epidemiol. *115*, 657–663
30. National Diabetes Data Group. (1979) Classification and Diagnosis of Diabetes Mellitus and Other Categories of Glucose Intolerance. Diabetes *28*, 1039–1057
31. Nikkilä, E. A. (1969) Control of plasma and liver triglyceride kinetics by carbohydrate metabolism and insulin. Adv. Lipid Res. *7*, 63–134
32. Nikkilä, E. A. (1973) Triglyceride metabolism in diabetes mellitus. Progr. Biochem. Pharmacol. *8*, 271–299
33. Taskinen, M. R. & Nikkilä, E. A. (1979) Lipoprotein lipase activity of adipose tissue and skeletal muscle in insulin-deficient human diabetes. Diabetologia *17*, 351–356
34. Nikkilä, E. A. (1985) in: Dyslipoproteinaemias and Diabetes (James, R. W. & Pometta, D., eds.) pp. 44–52, Karger, Basel
35. Olefsky, J. M., Farquhar, J. W. & Reaven, G. M. (1974) Reappraisal of the role of insulin in hypertriglyceridemia. Am. J. Med. *57*, 551–560
36. Kissebah, A. H., Alfarsi, S., Adams, P. W. & Wynn, V. (1976) The metabolic fate of plasma lipoproteins in normal subjects and in patients with insulin resistance and endogenous hypertriglyceridemia. Diabetologia *12*, 501–509
37. Abrams, J., Ginsberg, H. & Grundy, S. M. (1982) Metabolism of cholesterol and plasma triglycerides in nonketotic diabetes mellitus. Diabetes *31*, 903–910
38. Taskinen, M. R., Nikkilä, E. A., Kuusi, T. & Harno, K. (1982) Lipoprotein lipase activity and serum lipoproteins in untreated type 2 (insulin-dependent) diabetes associated with obesity. Diabetologia *22*, 46–50
39. Howard, B. V., Reitman, J. S., Vasquez, B. & Zech, L. (1983) Very-low-density lipoprotein triglyceride metabolism in non-insulin-dependent diabetes mellitus. Diabetes *32*, 271–276
40. Olefsky, J. M. & Kolterman, O. G. (1981) Mechanisms of insulin resistance in obes-

ity and non-insulin dependent (type II) diabetes. Am. J. Med. *70*, 151–168

41. Grundy, S. M., Mok, H. Y. I., Zech, L., Steinberg, D. & Berman, M. (1979) Transport of very low density lipoprotein triglyceride in varying degrees of obesity and hypertriglyceridemia. J. Clin. Invest. *63*, 1274–1283
42. Ginsberg, H. & Grundy, S. M. (1982) Very low density lipoprotein metabolism in non-ketotic diabetes mellitus: effect of dietary restriction. Diabetologia *23*, 421–425
43. Hughes, T. A., Gwynne, J. T., Switzer, B. R., Herbst, C. & White, G. (1984) Effects of caloric restriction and weight loss on glycomic control, insulin release and resistance, and atherosclerotic risk in obese patients with type II diabetes mellitus. Am. J. Med. *77*, 7–17
44. Kennedy, L., Walshe, K., Hadden, D. R., Weaver, J. A. & Buchanan, K. D. (1982) The effect of intensive dietary therapy on serum high density lipoprotein cholesterol in patients with type 2 (non-insulin-dependent) diabetes mellitus: a prospective study. Diabetologia *23*, 24–27
45. Tamai, T., Nakai, T., Yamada, S., Kobayashi, T., Hayashi, T., Kutsumi, Y., Oida, K. & Takeda, R. (1981) The effects of glibenclamide and insulin on plasma high density lipoprotein in diabetics. Artery *9*, 477–493
46. Breier, C., Lisch, H.-J. & Sailer, S. (1982) Effect of treatment on the concentration of lipoproteins and the postheparin lipolytic activity in the plasma of noninsulin-dependent diabetics. Klin. Wochenschr. *60*, 551–554
47. Taskinen, M.-R., Harno, K. & Nikkilä, E. A. (1984) Serum lipids and lipoproteins in type 2 diabetes. Acta Endocrinol. (Suppl. 262) 95–99
48. Osei, K., O'Dorisio, T. M. & Falko, J. M. (1984) Concomitant insulin and sulfonylurea therapy in patients with type II diabetes. Am. J. Med. *77*, 1002–1009
49. Walden, C. E., Knopp, R. H., Wahl, P. W., Beach, K. W. & Standness, E., Jr. (1984) Sex differences in the effect of diabetes mellitus on lipoprotein triglyceride and cholesterol concentrations. N. Engl. J. Med. *311*, 953–959
50. Kissebah, A. H., Adams, P. W. & Wynn, V. (1974) Inter-relationship between insulin secretion and plasma free fatty acid and triglyceride transport kinetics in maturity onset diabetes and the effect of phenethylbiguanide (Phenformin). Diabetologia *10*, 119–130
51. Howard, B. V., Xiaoren, P., Harper, I., Foley, J. E., Cheung, M. C. & Taskinen, M.-R. (1985) Effect of sulfonylurea therapy on plasma lipids and high-density lipoprotein composition in non-insulin-dependent diabetes mellitus. Am. J. Med. *79* (Suppl. 3B) 78–85
52. Vessby, B., Selinus, I. & Lithell, H. (1985) Serum lipoprotein and lipoprotein lipase in overweight, type II diabetics during and after supplemental fasting. Arteriosclerosis *5*, 93–100
53. Paisey, R., Elkeles, R. S., Hambley, J. & Maggill, P. (1978) The effects of chlorpropamide and insulin on serum lipids, lipoproteins and fractional triglyceride removal. Diabetologia *15*, 81–85
54. Greenfield, M. S., Doberne, L., Rosenthal, M., Vreman, H. J. & Reaven, G. M. (1982) Lipid metabolism in non-insulin-dependent diabetes mellitus. Effect of glipizide therapy. Arch. Intern. Med. *142*, 1498–1500
55. Taylor, K. G., John, W. G., Matthews, K. A. & Wright, A. D. (1982) A prospective study of the effect of 12 months treatment on serum lipids and apolipoproteins A-I and B in type 2 (non-insulin-dependent) diabetes. Diabetologia *23*, 507–510
56. Pfeifer, M. A., Brunzell, J. D., Best, J. D., Judzewitsch, R. G., Halter, J. B. & Porte, D., Jr. (1983) The response of plasma triglyceride, cholesterol, and lipoprotein lipase to treatment in non-insulin-dependent diabetic subjects without familial hypertriglyceridemia. Diabetes *32*, 525–531
57. Rabkin, S. W., Boyko, E. & Streja, D. A. (1983) Changes in high density lipoprotein cholesterol after initiation of insulin therapy in non-insulin-dependent diabetes mellitus: relationship to changes in body weight. Am. J. Med. Sci. *285*, 14–20
58. Dunn, F. L., Raskin, P., Bilheimer, D. W. & Grundy, S. M. (1984) The effect of diabetic control on very low-density lipoprotein – triglyceride metabolism in patients with type II diabetes mellitus and marked hypertriglyceridemia. Metabolism *33*, 117–123
59. Taylor, K. G., Galton, D. J. & Holdsworth, G. (1979) Insulin-independent diabetes: a defect in the activity of lipoprotein lipase in adipose tissue. Diabetologia *16*, 313–317

60. Calvert, G. D., Mannik, T., Graham, J. J., Wise, P. H. & Yeates, R. A. (1978) Effects of therapy on plasma-high-density-lipoprotein-cholesterol concentration in diabetes mellitus. Lancet *2*, 66–68
61. Sommariva, D., Scotti, L., Massaroli, C., Zanaboni, L., Bonfiglioli, D., Negrati, M., Micossi, P., Raggi, U. & Fasoli, A. Serum lipoproteins in diabetes: relation of alpha-lipoprotein level to therapy. Acta Diabet. Lat. *17*, 237–245
62. Masarei, J. R. L., Constable, I. J., Stanton, K. & Davis, R. E. (1982) Cholesterol and lipoprotein-cholesterol levels in western Australian rural diabetics. Aus. N.Z. J. Med. *12*, 241–247
63. Stout, R.W. (1977) Diabetes and atherosclerosis – the role of insulin. Diabetologia *16*, 141–150
64. Orchard, T. J., Becker, D. J., Bates, M., Kuller, L. H. & Drash, A. L. (1983) Plasma insulin and lipoprotein concentrations: an atherogenic association? Am. J. Epidemiol. *118*, 326–337
65. Steiner, G. & Vranic, M. (1982) Hyperinsulinemia and hypertriglyceridemia, a vicious cycle with atherogenic potential. Inter. J. Obesity *6* (Suppl. 1), 117–124
66. Keen, H. (1976) Glucose intolerance, diabetes mellitus and atherosclerosis; prospects for prevention. Postgrad. Med. J. *52*, 445–451
67. Wilson, P.W. F., Kannel, W. B. & Anderson, K. A. (1985) in: Dyslipoproteinemias and Diabetes (James, R.W. & Pometta, D., eds.) pp. 1–11, Karger, Basel
68. Epstein, F. H. (1985) in: Dyslipoproteinemias and Diabetes (James, R.W. & Pometta, D., eds.) pp. 92–97, Karger, Basel
69. Rotter, J. I. & Rimoin, D. L. (1981) The genetics of the glucose intolerance disorders. Am. J. Med. *70*, 116–126
70. Alaupovic, P. (1972) Conceptual development of the classification systems of plasma lipoproteins. Protides Biol. Fluids Proc. Colloq. *19*, 9–19
71. Osborne, J. C., Jr. & Brewer, H. B., Jr. (1977) The plasma lipoproteins. Adv. Protein Chem. *31*, 253–337
72. Alaupovic, P. (1984) in: Clinical and Metabolic Aspects of High-Density Lipoproteins (Miller, N. E. & Miller, G. J., eds.) pp. 1–45, Elsevier, Amsterdam, New York, Oxford
73. Salt, H. B., Wolff, O. H., Nestadt, A. & Lloyd, J. K. (1960) Control of lipemia in children with diabetes mellitus. The role of insulin and the effects of a diet rich in unsaturated fatty acids. Lancet *1*, 71–75
74. Sosenko, J. M., Breslow, J. L., Miettinen, O. S. & Gabbay, K. H. (1980) Hyperglycemia and plasma lipid levels. A prospective study of young insulin-dependent diabetic patients. N. Engl. J. Med. *302*, 650–654
75. Pollak, A., Widhalm, K., Havelec, L., Frisch, H. & Scholar, E. (1980) Glycosylated hemoglobin (HbA_{1c}) and plasma lipoproteins in juvenile onset diabetes mellitus. Acta Paediatr. Scand. *69*, 475–479
76. Lopes-Virella, M. F., Wohltmann, H. J., Loadholt, C. B. & Buse, M. G. (1981) Plasma lipids and lipoproteins in young insulin-dependent diabetic patients: relationship with control. Diabetologia *21*, 216–223
77. Rini, G. B., Averna, M. R., Di Fede, G., Montalto, G. & Notarbartolo, A. (1982) Apo-lipoproteic pattern in type I complicated diabetes (I.D.D.) with microangiopathy and with no clinico-instrumental signs of macroangiopathy – note II. Boll. Soc. Ital. Biol. Sper. *58*, 919–924
78. Soltész, G., Molnár, D., Klujber, L. & Kardos, M. (1982) Relationship between metabolic control and plasma lipoprotein level in diabetic children. Acta Paediatr. Acad. Sci. Hung. *23*, 75–83
79. Lopes-Virella, M. F., Sherer, G. K., Lees, A. M., Wohltmann, H., Mayfield, R., Sagel, J., Le Roy, E. C. & Colwell, J. A. (1982) Surface binding, internalization and degradation by cultured human fibroblasts of low density lipoproteins isolated from type I (insulin-dependent) diabetic patients: changes with metabolic control. Diabetologia *22*, 430–436
80. Peja, M., Madácsy, L., Kassay, L. & Bognár, M. (1983) Metabolic relations of serum lipids and lipoproteins in diabetic children. Acta Paediatr. Hung. *24*, 217–220
81. Carvajal, F., Quesada, X. & Gonzalez, P. (1983) High density lipoprotein cholesterol in insulin-dependent diabetic children. Acta Diabetol. Lat. *20*, 289–295
82. Richard, L., Delaunay, J., Dorleac, E. & Gillet, P. (1983) Apolipoprotéines et hémoglobine glycosylée chez le jeune diabétique insulino-dépendant. Arch. Fr. Pediatr. *40*, 11–14
83. Bittolo-Bon, G., Cazzolato, G. & Avogaro, P. (1984) Lipids, lipoproteins and apolipoproteins in type 1 (insulin-depend-

ent) and type 2 (non-insulin-dependent) diabetes mellitus. Acta Diabetol. Lat. *21*, 315–324
84. Ginies, J. L., Berthelot, J. & Limal, J. M. (1980) Diabète sucré avec hypertriglycéridémie majeure et xanthomatose éruptive lors de l'acidocétose. Arch. Fr. Pediatr. *37*, 55–57
85. Weidman, S. W., Ragland, J. B., Fisher, J. N., Jr., Kitabchi, A. E. & Sabesin, S. B. (1982) Effects of insulin on plasma lipoproteins in diabetic ketoacidosis: evidence for a change in high density lipoprotein composition during treatment. J. Lipid Res. *23*, 171–182
86. Joven, J., Rubiés-Prat, J., De la Figuera, M., Valdés, M., Villabona, C. M., Chacon, P. & Masdeu, S. (1985) High density lipoprotein changes during treatment of diabetic ketoacidosis. Diabete Metab. *11*, 102–105
87. Blackett, P. R., Holcombe, J. H., Alaupovic, P. & Fesmire, J. (1986) Case Report: plasma lipids and apolipoproteins in a 13-year old boy with diabetic ketoacidosis and extreme hyperlipidemia. Am. J. Med. Sci. *291*, 342–346
88. Pacifico, A., Cherchi, G. M., Baggi, M. A., Deiana, M. G. & Maioli, M. (1983) Modificazioni delle subfrazioni delle HDL HDL nei pazienti con diabete mellito di tipo I prima e dopo compenso metabolico. Boll. Soc. Ital. Biol. Sper. *59*, 1618–1624
89. Durrington, P. N. (1982) Serum high density lipoprotein cholesterol subfractions in type I (insulin-dependent) diabetes mellitus. Clin. Chim. Acta *120*, 21–28
90. Lopes-Virella, M. F., Wohltmann, H. J., Mayfield, R. K., Loadholt, C. B. & Colwell, J. A. (1983) Effect of metabolic control on lipid, lipoprotein, and apolipoprotein levels in 55 insulin-dependent diabetic patients. A longitudinal study. Diabetes *32*, 20–25
91. Schernthaner, G., Kostner, G. M., Dieplinger, H., Prager, R. & Mühlhauser, I. (1983) Apolipoproteins (A-I, A-II, B), Lp(a) lipoprotein and lecithin: cholesterol acyltransferase activity in diabetes mellitus. Atherosclerosis *49*, 277–293
92. Schonfeld, G., Birge, C., Miller, J. P., Kessler, G. & Santiago, J. (1974) Apolipoprotein B levels and altered lipoprotein composition in diabetes. Diabetes *23*, 827–834
93. Winocour, P. H., Durrington, P. N., Ishola, M. & Anderson, D. C. (1986) Lipoprotein abnormalities in insulin-dependent diabetes mellitus. Lancet *1*, 1176–1178
94. Witztum, J. L., Mahoney, E. M., Branks, M. J., Fisher, M., Elam, R. & Steinberg, D. (1982) Nonenzymatic glucosylation of low density lipoprotein alters its biologic activity. Diabetes *31*, 283–291
95. Kolb, F. O., De Lalla, O. F. & Gofman, J. W. (1955) The hyperlipemias in disorders of carbohydrate metabolism: serial lipoprotein studies in diabetic acidosis with xanthomatosis and in glycogen storage disease. Metabolism *4*, 310–317
96. Nikkilä, E. A. & Kekki, M. (1973) Plasma triglyceride transport kinetics in diabetes mellitus. Metabolism *22*, 1–22
97. Nikkilä, E. A., Huttunen, J. K. & Ehnholm, C. (1977) Postheparin plasma lipoprotein lipase and hepatic lipase in diabetes mellitus. Relationship to plasma triglyceride metabolism. Diabetes *26*, 11–21
98. Schade, D. S. & Eaton, R. P. (1979) Pathogenesis of diabetic ketoacidosis: a reappraisal. Diabetes Care *2*, 296–306
99. Foster, D. W. & McGarry, J. D. (1983) The metabolic derangements and treatment of diabetic ketoacidosis. N. Engl. J. Med. *309*, 159–169
100. Gofman, J. W., De Lalla, O., Glazier, F., Freeman, N. K., Lindgren, F. T., Nichols, A. V., Strisower, B. & Tamplin, A. R. (1954) The serum lipoprotein transport system in health, metabolic disorders, atherosclerosis, and coronary heart disease. Plasma *2*, 413–484
101. Harris, L. V. D., Albrink, M. J., Van Eck, W. F., Man, E. B. & Peters, J. P. (1953) Serum lipids in diabetic acidosis. Metabolism *2*, 120–132
102. Alaupovic, P. (1981) David Rubenstein Memorial Lecture: the biochemical and clinical significance of the interrelationship between very low density and high density lipoproteins. Can. J. Biochem. *59*, 565–579
103. Bagdade, J. D., Porte, D., Jr. & Bierman, E. L. (1968) Acute insulin withdrawal and the regulation of plasma triglyceride removal in diabetic subjects. Diabetes *17*, 128–132
104. Chait, A., Bierman, E. L. & Albers, J. J. (1979) Low-density lipoprotein receptor activity in cultural human skin fibroblasts – mechanism of insulin-induced stimulation. J. Clin. Invest. *64*, 1309–1319

105. Ewald, U., Tuvemo, T., Vessby, B. & Wålinder, O. (1982) Serum apolipoproteins A-I, A-II and B in diabetic children and matched healthy controls. Acta Paediatr. Scand. *71*, 15–18

106. Chase, H. P. & Glasgow, A. M. (1976) Juvenile diabetes mellitus and serum lipids and lipoprotein levels. Am. J. Dis. Child. *130*, 1113–1117

107. Anderson, G. E., Christiansen, J. S., Mortensen, H. B., Christiansen, K. M., Pedersen-Bjergaard, L., Kastrup, K.W. & Vestermark, S. (1983) Serum lipids and lipoproteins in 157 insulin dependent diabetic children and adolescents in relation to metabolic regulation, obesity and genetic hyperlipoproteinemia. Acta Paediatr. Scand. *72*, 361–365

108. Eckel, R. H., Albers, J. J., Cheung, M. C., Wahl, P.W., Lindgren, F.T. & Bierman, E. L. (1981) High density lipoprotein composition in insulin-dependent diabetes mellitus. Diabetes *30*, 132–138

109. Nikkilä, E. A. & Hormila, P. (1978) Serum lipids and lipoproteins in insulin-treated diabetes. Demonstration of increased high density lipoprotein concentrations. Diabetes *27*, 1078–1086

110. Fedele, D., Fellin, R., Lapolla, A., Baggio, G., De Romedi, F., Frigato, F. & Crepaldi, G. (1982) Serum lipid and lipoprotein levels and metabolic control in insulin-treated diabetics. Acta Diabetol. Lat. *19*, 151–159

111. Cruickshanks, K. J., Orchard, T. J. & Becker, D. J. (1985) The cardiovascular risk profile of adolescents with insulin-dependent diabetes mellitus. Diabetes Care *8*, 118–124

112. Gonen, B., White, N., Schonfeld, G., Skor, D., Miller, P. & Santiago, J. (1985) Plasma levels of apoprotein B in patients with diabetes mellitus: the effect of glycemic control. Metabolism *34*, 675–679

113. Goldberg, R. B., Reeves, M. L., Seigler, D. E., Ryan, E. A., Miller, N., Hsia, S. L. & Skyler, J. S. (1985) Lack of a persistent reduction in serum lipid and apoprotein levels in insulin-dependent diabetic patients in receiving intensified insulin treatment. Acta Diabetol. Lat. *22*, 93–101

114. Strobl, W., Widhalm, K., Schober, E., Frisch, H., Pollak, A. & Westphal, G. (1985) Apolipoproteins and lipoproteins in children with type I diabetes: relation to glycosylated serum protein and HbA_1. Acta Paediatr. Scand. *74*, 966–971

115. Wilson, D. P., Fesmire, J. D., Endres, R. K. & Blackett, P. R. (1985) Increased levels of HDL-cholesterol and apolipoprotein A-I after intensified insulin therapy for diabetes. Southern Med. J. *78*, 636–638

116. Hollenbeck, C. B., Connor, W. E., Riddle, M. C., Alaupovic, P. & Leklem, J. E. (1985) The effects of a high-carbohydrate low-fat cholesterol-restricted diet on plasma lipid, lipoprotein, and apoprotein concentrations in insulin-dependent (type I) diabetes mellitus. Metabolism *34*, 559–566

117. Briones, E. R., Mao, S. J. T., Palumbo, P. J., O'Fallon, W. M., Chenoweth, W. & Kottke, B. A. (1984) Analysis of plasma lipids and apolipoproteins in insulin-dependent and non-insulin-dependent diabetics. Metabolism *33*, 42–49

118. Dunn, F. L., Pietri, A. & Raskin, P. (1981) Plasma lipid and lipoprotein levels with continuous subcutaneous insulin infusion in type I diabetes mellitus. Ann. Intern. Med. *95*, 426–431

119. Pietri, A., Dunn, F. L. & Raskin, P. (1980) The effect of improved diabetic control on plasma lipid and lipoprotein levels. A comparison of conventional therapy and continuous subcutaneous insulin infusion. Diabetes *29*, 1001–1005

120. Falko, J. M., O'Dorisio, T. M. & Cataland, S. (1982) Improvement of high-density lipoprotein-cholesterol levels. Ambulatory type I diabetics treated with the subcutaneous insulin pump. JAMA *247*, 37–39

121. Glasgow, A., August, G. P. & Hung, W. (1981) Relationship between control and serum lipids in juvenile-onset diabetes. Diabetes Care *4*, 76–80

122. Sosenko, J. M., Breslow, J., Miettinen, O. S. & Gabbay, K. H. (1982) Hyperglycemia and plasma lipid levels: covariations in insulin-dependent diabetes. Diabetes Care *5*, 40–43

123. Vialettes, B., Cartouzon, G., Berge-Lefranc, J.-L., Moulin, J.-P. & Vague, P. (1981) Fractions lipidiques plasmatiques du diabétique insulino dépendant. Effet du contrôle glycémique à court time. Nouv. Press Med. *10*, 2185–2192

124. Ewald, U., Gustafson, S., Tuvemo, T. & Vessby, B. (1984) Increased high density lipoproteins in diabetic children. Eur. J. Pediatr. *142*, 154–156

125. Kennedy, A. L., Lappin, T. R. J., Lavery, T. D., Hadden, D. R., Weaver, J. A. & Montgomery, D. A. D. (1978) Relation of high-density lipoprotein cholesterol concentration to type of diabetes and its control. Br. Med. J. *2*, 1191 – 1194
126. Nikkilä, E. A. (1981) High density lipoproteins in diabetes. Diabetes *30*, Suppl. 2, 82 – 87
127. Mattock, M. B., Salter, A. M., Fuller, J. H., Omer, T., El-Gohari, R., Redmond, S. D. & Keen, H. (1982) High density lipoprotein subfractions in insulin-dependent diabetic and normal subjects. Atherosclerosis *45*, 67 – 79
128. Cheung, M. C. & Albers, J. J. (1979) Distribution of cholesterol and apolipoprotein A-I and A-II in human high density lipoprotein subfractions separated by NaCl equilibrium gradient centrifugation: evidence for HDL subpopulations with differing A-I/A-II molar ratios. J. Lipid. Res. *20*, 200 – 207
129. Kostner, G. M. & Schernthaner, G. (1985) in: Dyslipoproteinemias and Diabetes (James, R. W. & Pometta, D., eds.) pp. 12 – 24, Karger, Basel
130. Weisweiler, P., Merk, W. & Schwandt, P. (1985) in: Dyslipoproteinemias and Diabetes (James, R. W. & Pometta, D., eds.) pp. 104 – 110, Karger, Basel
131. Mahley, R. W. (1985) Atherogenic lipoproteins and coronary artery disease: concepts derived from recent advances in cellular and molecular biology. Circulation *72*, 943 – 948
132. Berg, K. (1979) in: The Biochemistry of Atherosclerosis (Scanu, A. M., Wissler, R. W. & Getz, G. S., eds.) pp. 418 – 490, Dekker, New York
133. Kostner, G. M., Avogaro, P., Cazzolato, G., Marth, E., Bittolo-Bon, G. & Quinci, G. B. (1981) Lipoprotein Lp(a) and the risk for myocardial infarction. Atherosclerosis *38*, 51 – 61
134. Greenfield, M., Kolterman, O., Olefsky, J. & Reaven, G. M. (1980) Mechanism of hypertriglyceridemia in diabetic patients with fasting hyperglycaemia. Diabetologia *18*, 441 – 446
135. Pietri, A. O., Dunn, F. L., Grundy, S. M. & Raskin, P. (1983) The effect of continuous subcutaneous insulin infusion on very-low-density lipoprotein triglyceride metabolism in type I diabetes mellitus. Diabetes *32*, 75 – 81
136. Misra, D. P., Staddon, G., Powell, N., Misra, J. & Cook, D. (1974) Lecithin: cholesterol acyltransferase activity in diabetes mellitus and the effect of insulin in these cases. Clin. Chim. Acta *56*, 83 – 89
137. Mattock, M. B., Fuller, J. H., Maude, P. S. & Keen, H. (1979) Lipoproteins and plasma cholesterol esterification in normal and diabetic subjects. Atherosclerosis *34*, 437 – 449
138. Kudo, H. (1969) Serum triglyceride levels of untreated diabetics in relation to vascular lesions and obesity. Tohoku J. Exp. Med. *97*, 47 – 56
139. Wilson, D. E., Schreibman, P. H., Day, V. D. & Arky, R. A. (1970) Hyperlipidemia in an adult diabetic population. J. Chron. Dis. *23*, 501 – 506
140. Hayes, T. M. (1972) Plasma lipoproteins in adult diabetes. Clin. Endocrinol. *1*, 247 – 251
141. Beylot, M., Garcia, I., Temori, S. A., Orgiazzi, J., Revol, A. & Noel, G. (1983) Lipides sériques dans le diabète sucré. Diabete Metab. *9*, 199 – 203
142. Jialal, I., Deppe, W. & Joubert, S. M. (1985) Lipid and lipoprotein aberrations in Indian patients with non-insulin-dependent diabetes in the young. SAMJ *67*, 1001 – 1004
143. Taylor, K. G., Wright, A. D., Carter, T. J. N., Valente, A. J., Betts, S. W. & Matthews, K. A. (1981) High-density lipoprotein cholesterol and apolipoprotein A-I levels at diagnosis in patients with non-insulin dependent diabetes. Diabetologia *20*, 535 – 539
144. Zimmerman, B. R., Palumbo, P. J., O'Fallon, W. M., Ellefson, R. D., Osmundson, P. J. & Kazmier, F. J. (1981) A prospective study of peripheral occlusive arterial disease in diabetes. III. Initial lipid and lipoprotein findings. Mayo Clin. Proc. *56*, 233 – 242
145. Simpson, R. W., Mann, J. I., Hockday, T. D. R., Hockday, J. M., Turner, R. C. & Jelfs, R. (1979) Lipid abnormalities in untreated maturity-onset diabetics and the effect of treatment. Diabetologia *16*, 101 – 106
146. Weisweiler, P., Drosner, M. & Schwandt, P. (1982) Dietary effect on very low-density lipoproteins in type 2 (non-insulin-dependent) diabetes mellitus. Diabetologia *23*, 101 – 103
147. Ballantyne, D., White, C., Strevens, E. A., Lawrie, T. D. V., Lorimer, A. R., Mander-

son, W. G. & Morgan, H. G. (1977) Lipoprotein concentrations in untreated adult onset diabetes mellitus and the relationship of the fasting plasma triglyceride concentration to insulin secretion. Clin. Chim. Acta *78*, 323–329

148. Reckless, J. P. D., Betteridge, D. J., Wu, P., Payne, B. & Galton, D. J. (1978) High-density and low-density lipoproteins and prevalence of vascular disease in diabetes mellitus. Br. Med. J. *1*, 883–886
149. Uusitupa, M., Siitonen, O., Pyörälä, K., Aro, A., Hersio, K., Pentillä, I. & Voutilainen, E. (1985) The relationship of cardiovascular risk factors to the prevalence of coronary heart disease in newly diagnosed type 2 (non-insulin-dependent) diabetes. Diabetologia *28*, 653–659
150. Cohen, A. M., Fidel, J., Cohen, B., Furst, A. & Eisenberg, S. (1979) Diabetes, blood lipids, lipoproteins, and change of environment: re-study of the "new immigrant Yemenites" in Israel. Metabolism *28*, 716–728
151. Sailer, S., Sandhofer, F. & Braunsteiner, H. (1966) Overweight and triglyceride level in normal persons and patients with diabetes mellitus. Metabolism *15*, 135–137
152. Alaupovic, P., Knight, C., Lee, E. T. & West, K. M. (1982) Lipid and apolipoprotein profiles of patients with insulin-independent diabetes. Fed. Proc. *41*, 1228
153. Lisch, H.-J. & Sailer, S. (1981) Lipoprotein patterns in diet, sulphonylurea, and insulin treated diabetics. Diabetologia *20*, 118–122
154. Nerurkar, S. V., Retnam, V. J., Taskar, S. P. & Bhandarkar, S. D. (1983) Lipid composition of plasma lipoproteins in treated diabetics. J. Postgrad. Med. *29*, 201–211
155. Billingham, M. S., Leatherdale, B. A., Hall, R. A. & B Bailey, C. J. (1982) High density lipoprotein cholesterol and apolipoprotein A-I concentrations in non-insulin dependent diabetics treated by diet and chlorpropamide. Diabete Metabol. *8*, 229–233
156. Barrett-Connor, E., Witztum, J. L. & Holdbrook, M. (1983) A community study of high density lipoproteins in adult noninsulin-dependent diabetics. Am. J. Epidemiol. *117*, 186–192
157. Bergman, M., Gidez, L. I. & Eder, H. A. (1982) HDL and HDL subclasses in diabetes. Diabetes *31*, Suppl. 2, 14A
158. Laakso, M., Voutilainen, E., Pyöräla, K. & Sarlund, H. (1985) Association of low HDL and HDL_2 cholesterol with coronary heart disease in noninsulin-dependent diabetics. Arteriosclerosis *5*, 653–658
159. Howard, B. V., Savage, P. J., Bennion, L. J. & Bennett, P. H. (1978) Lipoprotein composition in diabetes mellitus. Atherosclerosis *30*, 153–162
160. Fielding, C. J., Reaven, G. M. & Fielding, P. E. (1982) Human noninsulin-dependent diabetes: identification of a defect in plasma cholesterol transport normalized *in vivo* by insulin and *in vitro* by selective immunoadsorption of apolipoprotein E. Proc. Natl. Acad. Sci. USA *79*, 6365–6369
161. Nakano, E., Ozawa, K., Kumosaka, K., Kawano, K., Tsuchiya, T. & Watanabe, K. (1983) in: Electrophoresis '83 (Hirai, H., ed.) pp. 689–694, de Gruyter, Berlin, New York
162. Briones, E. R., Palumbo, J. P., Mao, S. J. T., Gerich, J. E. & Kottke, B. A. (1984) Acute effects of a glucose load on plasma apolipoproteins A-I, A-II, C-II and C-III in normal and non-insulin-dependent diabetic men. Mayo Clin. Proc. *59*, 399–403
163. Strisower, E. H., Weed, R. O., Gofman, J. W., Strisower, B. & De Lalla, O. F. (1958) The effect of insulin on serum lipoproteins of diabetic patients. J. Clin. Endocrinol. Metab. *18*, 1418–1427
164. Fielding, C. J., Reaven, C. M., Liu, G. & Fielding, P. E. (1984) Increased free cholesterol in plasma low and very low density lipoproteins in non-insulin-dependent diabetes mellitus: its role in the inhibition of cholesteryl ester transfer. Proc. Natl. Acad. Sci. USA *81*, 2512–2516
165. Hiramatsu, K., Bierman, E. L. & Chait, A. (1985) Metabolism of low-density lipoprotein from patients with diabetic hypertriglyceridemia by cultured human skin fibroblasts. Diabetes *34*, 8–14
166. Gabor, J., Spain, M. & Kalant, N. (1980) Composition of serum very-low-density and high-density lipoproteins in diabetes. Clin. Chem. *26*, 1261–1265
167. Maruhara, Y., Saito, F., Hikichi, I., Hashimoto, T., Takahashi, K. & Kaito, I. (1982) Apoprotein-B-poor very low density lipoprotein in diabetic patients without hyperlipoproteinemia. Tohoku J. Exp. Med. *138*, 419–425

168. Stalenhoef, A. F. H., Demacker, P. N. M., Lutterman, J. A. & van't Laar, A. (1982) Apolipoprotein C in type 2 (non-insulin-dependent) diabetic patients with hypertriglyceridemia. Diabetologia *22*, 489–491
169. Weisweiler, P., Jungst, D. & Schwandt, P. (1983) Quantitation of apolipoprotein E-isoforms in diabetes mellitus. Horm. Metab. Res. *15*, 201
170. Sano, R., Abe, R., Oikawa, S.-I., Fujii, Y. & Toyota, T. (1985) ApoE-2/E-3 ratio of very low density lipoprotein in diabetes mellitus. Tohoku, J. Exp. Med. *146*, 131–136
171. Havel, R. J., Chao, Y. S., Windler, E. E., Lotite, L. & Guo, L. S. S. (1980) Isoprotein specificity in the hepatic uptake of apolipoprotein E and the pathogenesis of familial dysbetalipoproteinemia. Proc. Natl. Acad. Sci. USA *77*, 4349–4353
172. Biesbroeck, R. C., Albers, J. J., Wahl, P. W., Weinberg, C. R., Bassett, M. L. & Bierman, E. L. (1982) Abnormal composition of high density lipoproteins in non-insulin-dependent diabetics. Diabetes *31*, 126–131
173. Schleicher, E., Duefel, T. & Wieland, O. H. (1981) Non-enzymatic glycosylation of human serum lipoproteins. Elevated ε-lysine glycosylated low density lipoprotein in diabetic patients. FEBS Lett. *129*, 1–4
174. Gonen, B., Baenzinger, J., Schonfeld, G., Jacobsen, D. & Farrar, P. (1981) Nonenzymatic glycosylation of low density lipoproteins *in vivo*. Effects on cell-interactive properties. Diabetes *30*, 875–878
175. Sasaki, J. & Cottam, G. L. (1982) Glycosylation of LDL decreases its ability to interact with high-affinity receptors of human fibroblasts *in vitro* and decreases its clearance from rabbit plasma *in vivo*. Biochim. Biophys. Acta *713*, 199–207
176. Curtiss, L. K. & Witztum, J. L. (1985) Plasma apolipoproteins AI, AII, B, CI, and E are glycosylated in hyperglycemic diabetic subjects. Diabetes *34*, 452–461
177. Kraemer, F. B., Chen, Y.-D. I., Cheung, R. M. C. & Reaven, G. M. (1982) Are the binding and degradation of low density lipoprotein altered in type 2 (non-insulin-dependent) diabetes mellitus? Diabetologia *23*, 28–33
178. Kissebah, A. H., Alfarsi, S., Evans, D. J. & Adams, P. W. (1983) Plasma low density lipoprotein transport kinetics in noninsulin-dependent diabetes mellitus. J. Clin. Invest. *71*, 655–667
179. Kesäniemi, Y. A. (1985) in: Dyslipoproteinaemias and Diabetes (James, R. W. & Pometta, D., eds.) pp. 63–73, Karger, Basel
180. Kissebah, A. H., Alfarsi, S., Evans, D. J. & Adams, P. W. (1982) Integrated regulation of very low density lipoprotein triglyceride and apolipoprotein-B kinetics in non-insulin-dependent diabetes mellitus. Diabetes *31*, 217–225
181. Haffner, S. M., Foster, D. M., Kushawa, R. S. & Hazzard, W. R. (1984) Retarded chylomicron apolipoprotein B catabolism in type 2 (non-insulin-dependent) diabetic patients with lipaemia. Diabetologia *26*, 349–354
182. Wang, C.-S., Bass, H. B., Downs, D. & Whitmer, R. K. (1981) Modified heparin-sepharose procedure for determination of plasma lipolytic activities of normolipidemic and hyperlipidemic subjects after injection of heparin. Clin. Chem. *27*, 663–668
183. Schmitt, J. K., Poole, J. R., Lewis, S. B., Shore, V. G., Maman, A., Baer, R. M. & Forsham, P. H. (1982) Hemoglobin A_1 correlates with the ratio of low- to high-density-lipoprotein cholesterol in normal weight type II diabetes. Metabolism *31*, 1084–1089
184. Chait, A., Robertson, H. T. & Brunzell, J. D. (1981) Chylomicronemia syndrome in diabetes mellitus. Diabetes Care *4*, 343–348
185. Parker, F., Bagdade, J. D., Orland, G. F. & Bierman, E. L. (1970) Evidence for the chylomicron origin of lipids accumulating in diabetic eruptive xanthomas: a correlative lipid, biochemical, histochemical, and electron microscopic study. J. Clin. Invest. *49*, 2172–2187
186. Chait, A. & Brunzell, J. D. (1983) Severe hypertriglyceridemia: role of familial and acquired disorders. Metabolism *32*, 209–214
187. Fallat, R. W. & Glueck, C. J. (1976) Familial and acquired type V hyperlipoproteinemia. Atherosclerosis *23*, 41–62
188. Chait, A., Albers, J. J. & Brunzell, J. D. (1980) Very low density lipoprotein overproduction in genetic forms of hypertriglyceridemia. Eur. J. Clin. Invest. *10*, 17–22
189. Pykälistö, O. J., Smith, P. H. & Brunzell, J. D. (1975) Determinants of human adi-

pose tissue lipoprotein lipase. Effect of diabetes and obesity on basal and diet-induced activity. J. Clin. Invest. *56*, 1108–1117

190. Brunzell, J. D., Hazzard, W. R., Porte, D., Jr. & Bierman, E. L. (1973) Evidence for a common, saturable, triglyceride removal mechanism for chylomicrons and very low density lipoproteins in man. J. Clin. Invest. *52*, 1533–1543
191. Brunzell, J. D. & Bierman, E. L. (1982) Chylomicronemia syndrome: interaction of genetic and acquired hypertriglyceridemia. Med. Clin. N. Am. *66*, 455–468
192. Shoukry, M. & Jayyab, A. K. A. (1983) Plasma HDL in non-insulin-dependent diabetes and the effect of various types of treatment. Atherosclerosis *49*, 333–338
193. Berry, E. M. & Bar-On, H. (1981) Comparison of sulphonylurea and insulin treatment on lipid levels in maturity-onset diabetic men and women. Isr. J. Med. Sci. *17*, 384–387
194. Agardh, C.-D., Nilsson-Ehle, P. & Schersten, B. (1982) Improvement of the plasma lipoprotein pattern after institution of insulin treatment in diabetes mellitus. Diabetes Care *5*, 322–325

Chapter 9
Dyslipoproteinemia in Renal Disease

D. Robert, R. C. Mordasini and W. Riesen

1 Dyslipoproteinemia in the Nephrotic Syndrome

Under physiological conditions, the kidneys are not involved in the regulation of lipid and lipoprotein metabolism. However, recent studies have shown that high density lipoproteins (HDL) in particular undergo glomerular filtration and proximal nephron reabsorption (1) and that apoproteins derived from HDL have been detected in normal urine (2). The physiological relevance of these observations still remains unknown.

Hyperlipidemia is a frequent consequence of renal disease and was first described by Bright over a century ago (3). Impairment of lipid metabolism in renal diseases arises in two different types of pathological events: in response to an abnormal loss of protein in urine (such as in the nephrotic syndrome) or in response to impaired renal function. However, the distinction is often arbitrary since the nephrotic syndrome may be complicated by renal insufficiency.

Hyperlipidemia and lipiduria are common features of the renal nephrotic syndrome. An increased concentration of cholesterol and less frequently an increased concentration of triglycerides and phospholipids is the rule. Dyslipoproteinemia of types II b and II a are observed with equal frequency (4); types V and IV are less frequent (5).

The alterations of lipid and lipoprotein metabolism in the nephrotic syndrome depend on several factors: 1. an increased production of lipids and lipoproteins in the liver, 2. loss of lipoproteins and apoproteins in urine, 3. defective catabolism of lipoproteins.

Clinical and experimental studies have clearly shown that the liver responds to an abnormal urinary protein loss by increasing its synthesis of lipoproteins. In experimental nephrosis, the increased production of nascent VLDL and HDL has been confirmed in liver perfusions (6–8) or in humans by triglyceride turnover studies (9). These studies have confirmed that the expanded VLDL pool in these patients is primarily a result of raised production (10). However, the mechanism whereby the liver responds with an increased lipid and lipoprotein production remains uncertain. Hypoalbuminemia (11), decreased oncotic pressure (12) or changes in plasma viscosity (13) have been proposed to explain the increased hepatic lipoprotein synthesis. It is interesting to note that depletion of plasma proteins by plasmapheresis induces a similar effect to a smaller extent (14). The degree of hyperlipidemia in the nephrotic syndrome is correlated with the degree of hypoalbuminemia or with the decrease of oncotic pressure (14, 15). Another strong argument for the role of hypoalbuminemia in the pathogenesis of hyperlipidemia is the significant decrease in serum lipids observed after infusion of albumin or dextran (16).

During the course of the nephrotic syndrome, analysis of the different apoproteins shows that Apo B is elevated during the active phase and decreases progressively after disappearance of the proteinuria (17). Apo CI, Apo CII and Apo CIII are also increased in the acute phase and return to normal levels within 4 weeks of normalisation of the urine (17). A decreased Apo CII concentration has also been reported (18), whereas the Apo AI and Apo AII concentration have mostly been found within the reference range (19, 20).

Increased hepatic lipoprotein synthesis leads to an elevated serum level of VLDL and LDL (21). On the other hand, the HDL concentration has been found mostly to be in the reference range (20, 21) or slightly decreased (22). The discussion about the HDL concentration in the nephrotic syndrome and its possible protective effect against an increased risk for cardiovascular complications (23–25) may reflect the fact that the nephrotic syndrome includes a very heterogenous group of diseases with different histological and/or functional manifestations (26). Minimal change nephropathy is characterised by the absence of morphological changes in light microscopy and a proteinuria of high selectivity (over 90% albumin) while other glomerulopathies such as membranous or membrano-proliferative glomerulonephritis show proteinuria of variable selectivity. In glomerular diseases associated with a proteinuria of moderate selectivity, lipoproteins such as VLDL, LDL, IDL and HDL have been detected in the urine (27). Moreover, HDL which are relatively small molecules are quantitatively the most important lipoproteins in urine whereas VLDL or LDL do not appear in significant amounts in urine despite their higher concentration in serum. The urinary loss of HDL could be, at least in part, responsible for the relatively high LDL/HDL ratio observed in this category of nephrotic patients (28). In contrast, in glomerular diseases associated with a proteinuria of high selectivity, no HDL or Apo AI has been detected in the urine (20). These patients show a normal or increased HDL concentration.

Several other causes have been raised to explain the hyperlipidemia of the nephrotic syndrome such as defective catabolism of triglyceride rich lipoprotein. Studies using a fat tolerance test to examine triglyceride removal have shown that catabolism is only slightly disturbed (29). Furthermore, several studies report a decreased plasma lipolytic activity in patients with the nephrotic syndrome affecting either lipoprotein lipase activity (LPL) (18, 30) or hepatic triglyceride lipase (HTGL) (19). Which of these is more important is difficult to assess at present, since few studies of this kind have been performed in patients with the nephrotic syndrome without renal insufficiencies. It may be possible that the increased concentration of triglyceride rich remnants reported by several authors reflects a decrease in glomerular filtration rate and not the nephrotic syndrome. Finally, mechanisms such as impaired metabolism of renal mevalonate, a precursor of cholesterol synthesis, may contribute to the lipoprotein alterations during the course of the nephrotic syndrome (31).

2 Dyslipoproteinemia in Chronic Renal Insufficiency

Hyperlipidemia is a frequent finding in patients with chronic renal insufficiency. Interest in this finding started in the early seventies with a paper of Lindner (32) reporting a very high incidence of cardiovascular complications in patients treated by hemodialysis for end stage renal disease, possibly associated with alterations of lipid metabolism.

Hyperlipidemia in chronic renal insufficiency is not related to a specific underlying pathogenetic mechanism and can be reproduced in experimental uremia in animals (33, 34). Disturbances of lipid metabolism are found even in patients with only moderate renal failure (35). They cannot be reversed by hemodialysis and persist after successful renal transplantation. However, the pattern of hyperlipidemia varies widely from patient to patient (Fig. 1).

Hypertriglyceridemia is a common finding in conservatively treated uremic patients whether undergoing hemodialysis or peritoneal dialysis (36–38). The incidence is approximatively 20 to 30 percent of the patients treated conservatively and 30 to 60 percent on chronic hemodialysis (39–41).

Total cholesterol is generally normal or only slightly increased (37). Uremic patients usually show a type IV pattern. The disturbances of lipid and lipoprotein metabolism in renal failure are very complex and affect composition, structure and metabolic regulation of all circulating molecules responsible for the transport of triglycerides, cholesterol and phospholipids (42–44). Ritz has shown that modifications in the composition and structure of lipoproteins in chronic uremia are independent of the presence of hypertriglyceridemia (45).

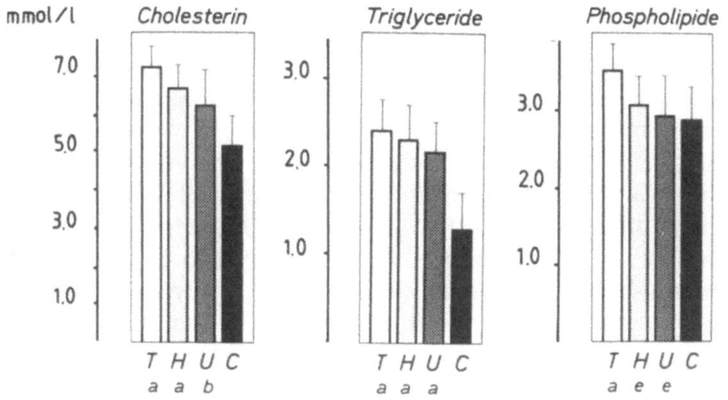

Fig. 1. Cholesterol (CHOL) and triglyceride (TG) content of LDL, HDL and VLDL of patients with renal disease and normal controls. T = patients after renal transplantation; H = patients on maintenance hemodialysis; U = conservatively treated patients with chronic uremia; C = controls; a = $p < 0.0005$; b = $p < 0.005$; c = $p < 0.01$; d = $p < 0.025$; e = difference not significant. (from Riesen, W. F. and Mordasini, R. (1984) Contr. Nephrol. *41*, 312–320).

Analysis of the different lipoprotein classes using a variety of methods (ultracentrifugation, precipitation, electrophoresis, use of specific antibodies against various apoproteins or a combination of these) provided a better perspective of the pathogenetic mechanisms underlying the lipoprotein alterations in chronic renal insufficiency.

VLDL are almost always elevated. Their lipid content shows an enrichment in triglycerides and to a lesser extent in cholesterol (46). The apoprotein composition of VLDL shows several abnormalities. They contain Apo B48 (an apoprotein usually found only in chylomicrons) and Apo AIV (47). The Apo CII content is decreased (48) whereas Apo CIII is increased (49) and shows an abnormal enrichment in sialic acid (5). Apo E is also elevated.

In uremic patients, the intermediate density lipoproteins (IDL) are elevated (51) and enriched in triglycerides. Lipoprotein particles with Apo CIII as their main protein have recently been reported in such individuals (52). The accumulation of IDL in chronic uremia is due to incomplete catabolism of VLDL in the plasma.

The density of LDL in uremic individuals is usually normal, but the particles show several abnormalities in their composition in that they are enriched in triglycerides but have a normal cholesterol content, and contain Apo CII, Apo AIV and Apo E (47). Their Apo B content is normal or slightly increased.

HDL are constantly decreased in uremia (46, 48, 53) and exhibit a low cholesterol content and enrichment in triglycerides (54). The apoprotein composition of HDL shows different abnormalities in comparison with non uremic persons. Apo AI, the main apoprotein of the HDL is normal or decreased (4, 54).

Racial or sexual differences seem to be important. If anephric patients are studied, it appears that in comparison with patients with their defective kidneys still *in situ*, the Apo AI concentration is decreased in men whereas it remains normal in females (Fig. 2). The Apo CII level of HDL was found to be decreased and the Apo CIII as well as the Apo E level increased (48). The enrichment of the different lipoprotein classes in triglycerides results from an impairment of the catabolism of triglyceride rich lipoprotein. This defect in lipolytic activity has been confirmed using a variety of methods including intravenous fat tolerance tests with the determination of the fractional clearance rate or K2 (56, 57), turnover studies using tritiated glycerol (58) or measurement of post-heparin lipolytic activity in plasma or in different tissues. The last problem is particularly relevant in view of the pathogenesis of hyperlipidemia in chronic uremia. The triglyceride rich lipoproteins, chylomicrons and VLDL are catabolized in plasma by two enzymes: lipoprotein lipase (LPL) from muscle and adipose tissue and hepatic triglyceride lipase (HTGL) bound to the hepatocyte membrane. Both enzymes, LPL and HTGL, are released into plasma after heparin administration. Immuno-chemical methods made selective measurement of the activity of both enzymes possible. Post heparin lipolytic activity (PHLA) is found to be decreased in chronic uremia but it is still uncertain which one of the two key enzymes is primilarily reduced (59–62) although recent studies have shown that the activity of both enzymes is decreased in chronic uremia. LPL in adipose tissue of hypertriglyceridemic patients is reduced even postprandially (63). In uremic animals LPL activity is also significantly decreased (64). On the other hand, there are studies indicating that the reduced lipolytic activity measured

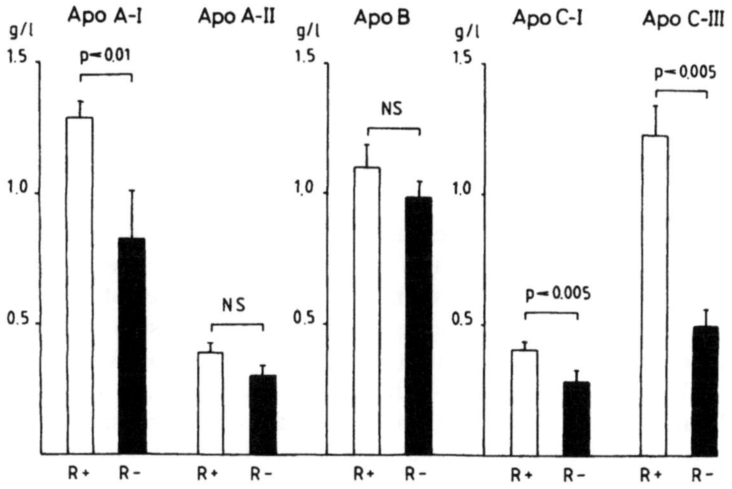

Fig. 2. (a) Apolipoprotein levels in anephric patients and in hemodialysed patients with remaining kidneys. R− = anephric patients (n = 11); R+ = hemodialysed patients (n = 20); Values are means + 1 SEM, statistics were done by the student's t-test.

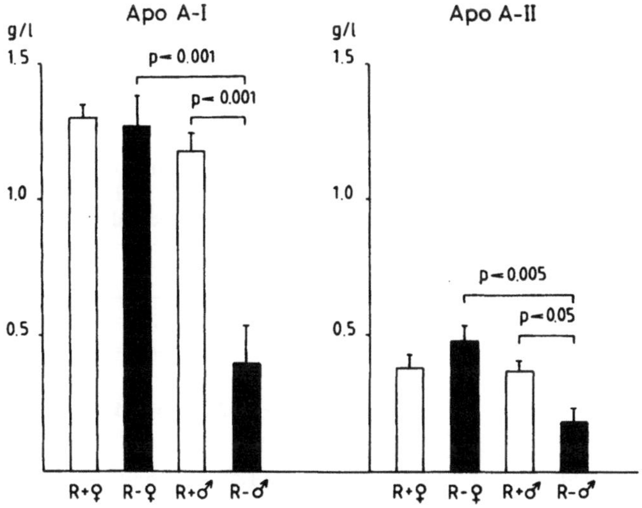

Fig. 2. (b) Sex-related differences of apolipoproteins AI and AII in anephric patients and in patients with remaining kidneys. R + ♀ = women with remaining kidneys (n = 10); R + ♂ = men with remaining kidneys (n = 10); R − ♀ = anephric women (n = 5); R − ♂ = anephric men (n = 6).

in plasma is mainly due to a decrease in HTGL activity which represents about 70 percent of the total lipolytic activity measured in post heparin plasma (61).

The reduction in LPL activity leads to defective catabolism of chylomicrons and VLDL (65) and interferes with the normal transformation of VLDL to LDL (66).

Several studies in animals and in humans have shown that the preferential substrates of HTGL are chylomicrons and VLDL remnants (67, 68). The decreased HTGL activity in chronic uremia probably results in the accumulation of IDL (61, 69) and the triglyceride enrichment of LDL (70). The reduced lipolytic activity also has effects on HDL. A low HDL-cholesterol is associated with a decreased LPL activity (71); and disturbed conversion from HDL_2 into HDL_3 has been linked to a reduction in HTGL activity (72).

A variety of different factors may explain the decreased lipolytic activity, including 1. hormonal influences particularly insulin (73) which regulates LPL secretion, 2. reduced activity of activators such as a decreased level of Apo CII, the main activator of LPL (74), 3. the presence of inhibitors such as an increased level of Apo CIII, a modulator of LPL activity (75) or changes in Apo AI and Apo AII which can modify HTGL activity (76), 4. the uremic toxins (77), 5. a reduced pool of releasable LPL and HTGL which might follow repeated heparinisation necessary for hemodialysis treatment (57). A decrease in PHLA has been noted in patients receiving long term heparin treatment (78). If hemodialyzed patients are switched from conventional heparin treatment to treatment with low molecular weight heparin (LMWH, Fragmine R) which releases LPL from the vascular wall to a lesser extent than heparin, no changes in the concentration of plasma cholesterol and triglycerides are noted, even after 6 weeks of treatment. However, after the third day of treatment with LMWH, a significant enrichment of triglycerides in HDL is observed (Table 1). In uremia lecithin:cholesterol acyl transferase (LCAT) a third key enzyme of lipid metabolism has also been found to be reduced (79, 80).

Table 1. In hemodialyzed patients, treatment with low molecular weight lepocrin (LMWH) induces changes in lipoprotein lipase (LPL), hepatic triglyceride lipase (HTGL) activity and changes in the concentration of HDL cholesterol and HDL triglycerides.

	Day				
	0	1	5	21	50
Lipoprotein lipase activity (μmol/ml FFA)	9.58 ± 1.27	7.65 ± 1.88	6.34 ± 2.02[a]	4.61 ± 1.02[b]	5.02 ± 0.12[b]
Hepatic triglyceride lipase activity (μmol/ml FFA)	4.96 ± 2.61	3.91 ± 2.17	5.0 ± 1.56	3.16 ± 0.53[c]	3.03 ± 0.81[c]
HDL cholesterol (mmol/l)	0.86 ± 0.25	0.91 ± 0.30	0.89 ± 0.24	0.83 ± 0.15	0.91 ± 0.14
HDL triglycerides (mmol/l)	0.25 ± 0.05	0.28 ± 0.12	0.37 ± 0.14[a]	0.44 ± 0.23	0.39 ± 0.14

LPL and HTGL activities and HDL lipid levels were measured 10 minutes after 60 i U heparin/kg body weight on day 0 or 60 i U low molecular weight heparin/kg body weight on days 1, 5, 21 and 50.

[a] $\varrho < 0{,}002$ in comparison with Day 0
[b] $\varrho < 0{,}001$ in comparison with Day 0
[c] $\varrho < 0{,}05$ in comparison with Day 5
[d] $\varrho < 0{,}02$ in comparison with Day 0

n = 10.

This enzyme plays an important role in the regulation of cholesterol transport. The decreased LCAT activity leads to defective redistribution of cholesterol ester from HDL to VLDL and LDL, a step necessary in the regulation of cholesterol homeostasis by the liver (81–83). Several factors may explain the decreased LCAT activity in uremic patients, such as a low level of Apo AI, the main activator of LCAT (84), the presence of a possibly dialysable uremic toxin (85) and finally decreased hepatic synthesis of LCAT (80).

The pathogenetic mechanisms leading to hypertriglyceridemia in chronic uremia remain unknown. The role of the different factors proposed to explain the disturbances of lipid metabolism are still uncertain. Impaired glucose metabolism associated with insulin resistance is observed in uremia and may induce an increased hepatic triglyceride production (86, 87); however this hypothesis was not confirmed in a study using isolated hepatocytes of uremic rats (88). Other hormonal factors have been proposed, such as hyperglucagonemia (40) thyroid dysfunction (89), and secondary hyperparathyroidism (90, 91). With regard to the latter, it is interesting to note that there is a decrease in triglyceride after parathyroidectomy (42). Uremic toxins, poorly dialysable, have been incriminated in this process (78) as has the loss of carnitine induced by the dialysis treatment. Carnitine, a natural quarternary amine, allows the transfer of long chain free fatty acids across the mitochondrial membrane (57). Oral or intravenous supplementation of carnitine in hemodialysis patients was followed in some studies by a normalisation of lipid metabolism with a decrease in triglycerides and an increase in HDL-cholesterol (92), but other studies have not confirmed these observations (93).

Exogenous factors also play a significant role in the pathogenesis of hyperlipidemia in chronic renal failure (94). The diet which contains a high amount of carbohydrates and of fat could promote the hypertriglyceridemia of hemodialyzed patients. Dietetic changes either with restriction of carbohydrates (95) or of fat (96) reduce elevated triglycerides. The composition of the dialysis fluid may also influence the degree of hyperlipidemia: plasma lipoprotein levels are increased in patients treated by hemodialysis with acetate buffer. Acetate is a precursor of fatty acid synthesis, labeled acetate clearly appearing in tissue and in plasma lipids (97). On the other hand, patients treated with bicarbonate buffer showed a decrease in their plasma triglycerides (98) whereas HDL-cholesterol was little influenced (99). In patients treated by continuous ambulatory peritoneal dialysis (CAPD), the carbohydrate load is greater than in hemodialysis patients because of the glucose content of dialysis fluid. In CAPD patients, a rise in total cholesterol and in triglycerides (100) has been reported. On the other hand some other parameters of lipoprotein metabolism may be positively influenced by CAPD. For example the concentration of esterified cholesterol was found to be within the normal range in contrast to values observed in hemodialyzed patients (101).

Since Lindner's report (32), numerous studies have confirmed that cardiovascular complications are frequent in patients with chronic renal insufficiency. Recent data collected from the EDTA Registry show that cardiovascular complications account for 52 percent of the causes of death, myocardial infarction being diagnosed in 12.5 percent of men and 7 percent of females (102). However, it is important to realize that various factors in chronic uremia contribute to this high rate of fatal cardiovascular diseases, especially hypertension, smoking habits or other metabolic abnormalities such as an impaired carbohydrate or calcium metabolism (103, 104). Among these different risk factors, the relevance of hyperlipoproteinemia is difficult to define (105). Hypertriglyceridemia, for example, is considered to be only weakly associated with an increased risk of coronary heart disease in non uremic (106, 107) as well as in uremic patients (108). However, there are also important findings to suggest that there might be a strong relationship between increased triglycerides and the degree of coronary atherosclerosis (109) although in some studies of uremic patients, no correlations were found between atherosclerotic alterations and triglyceride levels (110). The most reliable link between the impairment of lipid metabolism and atherogenesis in uremia is the accumulation of abnormal VLDL (β VLDL) and lipoprotein remnants whose atherogenetic role has been confirmed by several studies in animals (111, 112) and in humans (113, 114). The concomitant decrease which is seen in the HDL cholesterol may be considered a promoting factor (112). Several epidemiologic studies have confirmed an inverse correlation between the incidence of coronary heart disease and the HDL concentration (115) as occurs in hemodialyzed patients (116). In this regard, Hahn et al. (116) demonstrated that a high VLDL cholesterol and a low HDL cholesterol are associated with an increased risk for coronary heart disease.

3 Dyslipoproteinemia after Renal Transplantation

Renal transplantation succeeds in reversing most uremic alterations. However, even a successful transplantation cannot reverse the alterations in lipid metabolism. In contrast to the predominant hypertriglyceridemia of hemodialyzed patients, hypercholesterolemia is the most frequent finding after renal transplantation (110). In transplanted patients type IIa and type IIb hyperlipoproteinemia are observed in the majority of cases (4, 118, 119). Increased cholesterol is observed in 90 percent and elevated triglycerides in about 65 percent of the patients (139). The distribution of cholesterol and triglycerides among the different lipoprotein classes shows an enrichment in triglyceride in VLDL and LDL whereas the triglyceride content of HDL is decreased in comparison either with control subjects or uremic patients (40). The cholesterol content of VLDL and LDL is increased while the HDL-cholesterol content was found in some cases to be decreased (37, 53) or in other studies, to be within the reference range (121, 122). The apoprotein composition of the different lipoprotein classes is not altered to the same extent as in patients with chronic renal insufficiency. Apo B is increased in VLDL and more significantly in LDL. The Apo AI and Apo AII concentrations are reduced in transplanted adults (4) whereas they were found to be normal in transplanted children (123) as compared to normal controls.

Post heparin lipolytic activity after successful renal transplantation does not differ from that of normal subjects (61), whereas transplant recipients with impaired renal function have reduced lipolytic activity which is comparable to that of uremic patients (39). It is interesting to note that the intravenous fat tolerance test remains significantly reduced after renal transplantation (57).

Different factors play an important pathogenetic role in the hyperlipidemia which follows renal transplantation, and corticosteroid therapy should not be forgotten in this context. This drug increases hepatic VLDL synthesis and produces peripheral insulin resistance (124).

Some authors have found a correlation between the corticosteroid dosage and the frequency and degree of hyperlipidemia (128). In other studies this correlation could not be confirmed (126). Low doses of corticosteroids or alternate day application improved the hyperlipidemia of transplant recipients (127). In a recent study the pathogenetic role of corticosteroids in generating the hyperlipidemia of patients receiving a renal graft was questioned. Indeed patients treated by Cyclosporin A as monotherapy and patients under conventional therapy with corticosteroids and azathioprin showed no difference in their incidence or severity of hyperlipidemia (128). Other drugs such as diuretics and beta blockers also influence the hyperlipidemia of transplant recipients (124). Dietetic treatment limiting carbohydrate and fat intake may be effective in reducing hyperlipidemia after renal transplantation (120).

References

1. Peterson, D. R., Hjelle, J. T., Carone, F. A. & Moore, P. A. (1984) Kidney Int. *26*, 411–421.
2. Segal, P., Gidez, L. I., Vega, G. L., Edelstein, D., Eder, H. A. & Roheim, P. S. (1979) J. Lipid Res. *20*, 784–788.
3. Bright, R. (1827) Reports of medical cases selected with a view of illustrating the symptoms and cures of diseases by reference to morbid anatomy. Longman, Rus, Orme, Brown and Green, London.
4. Riesen, W. F. & Mordasini, R. (1984) Contr. Nephrol. *41*, 312–320.
5. Newmark, S. R., Anderson, C. F., Donadio, J. V. et al. (1975) Mayo Clin. Proc. *50*, 359–364.
6. Marsh, J. B. Drabkin, D. L. (1960) Metab. Clin. Exp. *9*, 946–955.
7. Marsh, J. B. (1974) J. Lipid. Res. *15*, 544–550.
8. Marsh, J. B. & Sparks, C. E. (1979) J. Clin. Invest. *64*, 1229–1237.
9. McKenzie, I. F. C. & Nestel, P. J. (1968) J. Clin. Invest. *47*, 1685–1695.
10. Kikki, M. & Nikkilä, E. A. (1971) Eur. J. Clin. Invest. *1*, 345–351.
11. Rosenman, R. H., Friedman, M. & Byers, S. O. (1956) J. Clin. Invest. *5*, 522–532.
12. Appel, J. B., Blum, C. B., Chien, S. et al. (1985) N. Engl. J. Med. *312*, 1544–1548.
13. Yedgar, S., Eilam, O. & Shafrir, E. (1985) Am. J. Physiol. *248*, E 10–E 14.
14. Shafrir, E. & Brenner, T. (1979) Lipids *14*, 695–702.
15. Baxter, J. H., Goodman, H. & Havel, R. J. (1960) J. Clin. Invest. *38*, 455–462.
16. Allen, J. C., Baxter, J. H. & Goodman, H. C. (1961) J. Clin. Invest. *40*, 499–508.
17. Ohta, T. & Matsuda, I. (1981) Clin. Chim. Acta *117*, 133–143.
18. Kashyap. M. L., Srivastava, L. S., Hynd, B. A. et al. (1980) Atherosclerosis *35*, 29–40.
19. Breier, C. H., Lisch, H. J., Drexel, H. & Braunsteiner, H. (1983) Schweiz. med. Wochenschr. *113*, 909–913.
20. Lopes-Virella, M., Virella, G., Debeukelaer, M. et al. (1979) Clin. Chim. Acta *94*, 73–81.
21. Oetliker, O. H., Mordasini, R., Lutsdig, J. & Riesen, W. (1980) Pediatr. Res. *14*, 64–66.
22. Gherardi, E., Rota, E., Calandra, S. et al. (1977) Eur. J. Clin. Invest. *7*, 563–570.
23. Mallick, N. P. & Short, C. D. (1981) Nephron. *27*, 54–57.
24. Wass, V. & Cameron, J. S. (1981) Nephron. *27*, 58–61.
25. Wass, V. J., Chilvers, C., Jarrett, R. J. & Cameron, J. S. (1979) Lancet II, 664–667.

26. Glassock, R. J., Cohen, A. H., Bennet, C. M. & Martinez-Malonado, M. (1981) Primary glomerular disease in Brenner & Rector, pp. 1351–1492. Saunders, Philadelphia, London, Toronto.
27. Shore, V. G., Forte, T., Licht, H. & Lewis, S. B. (1982) Metabolism *31*, 258–268.
28. Short, C. D., Durrington, P. N., Mallick, N. P. et al. (1986) Kidney Intern. *29*, 1224–1228.
29. Chan, M. K., Persaud, J. W., Ramdial, L. et al. (1981) Clin. Chim. Acta *117*, 317–323.
30. Yamada, M. & Matsuda, I. (1970) Clin. Chim. Acta *30*, 787.
31. Golper, T. A. & Swartz, S. H. (1982) Metabolism *31*, 471–476.
32. Lindner, A., Charra, B., Sherrard, D. J. & Scribner, D. H. (1974) N. Engl. J. Med. *290*, 697–701.
33. Felts, J. M., Zacherle, B. & Childress, G. (1979) Clin. Chim. Acta *93*, 127–134.
34. Bagdade, J., Yee, E., Wilson, D. E. & Shafrir, E. (1978) J. Lab. Clin. Med. *91*, 176–186.
35. Grützmacher, P., Radtke, H. W., Schifferdecker, E. et al. (1984) Contr. Nephrol. *41*, 332–336.
36. Bagdade, J. D. & Porte, D. (1968) N. Engl. J. Med. *279*, 181–185.
37. Mordasini, R. & Riesen, W. (1982) Schweiz. med. Wochenschr. *112*, 1335–1340.
38. Ramos, J. M., Gokal, R., Siamopoulos, K. et al. (1983) Quart. J. Med. *206*, 165–186.
39. Brons, M., Christensen, N. C. & Horder, M. (1972) Acta Med. Scand. *192*, 119–123.
40. Bagdade, J., Casaretto, A. & Albers, J. (1976) J. Lab. Clin. Med. *87*, 37–48.
41. Chan, M. K., Varghese, Z. & Moorhead, J. F. (1981) Kidney Intern. *19*, 625–637.
42. Drüeke, T., Lacour, B., Roullet, J. B. & Funck-Brentano, J. L. (1983) Kidney Intern. *24*, Suppl. 16, 134–138.
43. Daubresse, J. C., Lerson, G. & Plomteux, G. (1976) Europ. J. Clin. Invest. *6*, 159–166.
44. Norbeck, H. E. & Carlson, L. A. (1980) Eur. J. Clin. Invest. *10*, 423–426.
45. Ritz, E., Augustin, J., Bommer, J. et al. (1985) Kidney Intern. *28*, S17, 84–87.
46. Norbeck, H. E. (1981) Acta med. Scand. (Suppl.) *649*, 1–49.
47. Nestel, P. J., Fidge, N. H. & Tan, M. H. (1982) N. Engl. J. Med. *307*, 329–333.
48. Rapoport, J., Aviram, M., Chaimovitz, C. & Brook, G. (1978) N. Engl. J. Med. *299*, 1326–1329.
49. Staprans, I., Felts, J. M. & Zacherle, B. (1979) Clin. Chim. Acta *93*, 135–143.
50. Holdsworth, G., Stocks, J., Dodson, P. & Galton, D. J. (1982) J. Clin. Invest. *69*, 932–939.
51. Ron, D., Oren, I., Aviram, M. et al. (1983) Atherosclerosis *46*, 67–75.
52. Attman, P. O., Alaupovic, P., Knight, C. et al. (1984) Contr. Nephrol. *41*, 328–331.
53. Bagdade, J. W. & Albers, J. J. (1977) N. Engl. J. Med. *296*, 1436–1439.
54. Brunzell, J. D., Albers, J. J., Haas, L. B. et al. (1977) Metabolism *26*, 903–910.
55. Goldberg, A. P., Harter, H. R., Patsch, W. et al. (1983) N. Engl. J. Med. *308*, 1245–1252.
56. Attman, P. O. & Gustafson, A. (1979) Eur. J. Clin. Invest. *9*, 285–291.
57. Ibels, L. S., Reardon, M. F. & Nestel, P. J. (1976) J. Lab. Clin. Med. *87*, 648–658.
58. Cattran, D. C., Fenton, S. S. A., Wilson, D. R. & Steiner, G. (1976) Ann. Intern. Med. *85*, 29–33.
59. Crawford, G. A., Savdie, E. & Stewart, J. H. (1979) Clin. Science *57*, 155–165.
60. Chan, M. K., Persaud, J., Varghese, Z. & Moorhead, J. F. (1984) Kidney Intern. *25*, 812–818.
61. Mordasini, R., Frey, F., Flury, W. et al. (1977) N. Engl. J. Med. *25*, 1362–1366.
62. Applebaum-Bowden, D., Goldberg, A. P., Hazzard, W. R. et al. (1979) Metabolism *28*, 917–924.
63. Goldberg, A., Sherrard, D. J. & Brunzell, J. D. (1978) J. Clin. Endocrin. Metabolism *47*, 1173–1182.
64. Drüeke, T., Roullet, J. B., Coutard, M. & Funck-Brentano, J. L. (1984) Contr. Nephrol. *41*, 321–327.
65. Havel, R. J. (1982) Med. Clin. N. Am. *66*, 319–334.
66. Reardon, M. F., Sakai, H. & Steiner, G. (1982) Arteriosclerosis *2*, 396–402.
67. Nicoll, A. & Lewis, B. (1980) Eur. J. Clin. Invest. *10*, 487–495.
68. Rubinstein, A., Gibson, J. C. & Paternini, J. R. (1985) J. Clin. Invest. *75*, 710–721.
69. Murase, T. & Itakura, H. (1981) Atherosclerosis *39*, 293–300.
70. Bolzano, K., Krempler, F. & Sandhofer, F. (1978) Eur. J. Clin. Invest. *8*, 289–293.
71. Kekki, M. (1980) Atherosclerosis *37*, 143–150.

72. Nozaki, S., Kubo, M., Sudo, H. et al. (1986) Metabolism *35*, 53–58.
73. Persson, B. (1973) Acta Med. Scan. *193*, 457–462.
74. LaRosa, J. E., Levy, R. I., Herbert, P. et al. (1970) Biochem. Biophys. Res. Commun. *41*, 57–62.
75. Wang, S. C., McConathy, W. J., Kloer, H. U. & Alaupovic, P. (1985) J. Clin. Invest. *75*, 384–390.
76. Kubo, M., Matsuzawa, Y., Tajima, S. et al. (1981) Biochem. Biophys. Res. Commun. *100*, 261–266.
77. Murase, T., Cattran, D. C., Rubenstein, B. & Steiner, G. (1975) Metabolism *24*, 1279–1286.
78. Huttunen, J. K., Pasternack, A., Vänttinen, T. et al. (1978) Acta Med. Scand. *204*, 211–218.
79. Guarnieri, G. F., Moracchiello, M., Campanacci, L. et al. (1978) Kidney Intern. *13*, Suppl. 8, 26–30.
80. McLeod, R., Reeve, C. E. & Frohlich, J. (1984) Kidney Intern. *25*, 683–688.
81. Glomset, J. A. (1968) J. Lipid Res. *9*, 165–167.
82. Hsia, S. L., Perez, G. O. & Mendez, A. J. (1985) J. Lab. Clin. Med. *106*, 53–61.
83. Chajek, T. & Fielding, C. J. (1978) Proc. Natl. Acad. Sci. USA *75*, 3345–3349.
84. Fielding, C. J., Shores, V. G. & Fielding, P. E. (1972) Biochem. Biophys. Res. Commun *46*, 1493–1498.
85. Bories, P. C., Subbaiah, P. V. & Bagdade, J. D. (1982) Nephron *32*, 22–27.
86. Maloff, B. L., McCaleb, M. L. & Lockwood, D. H. (1983) Am. J. Physiol. *245*, E 178–184.
87. McCaleb, M. L., Izzo, M. S. & Lockwood, D. H. (1985) J. Clin. Invest. *75*, 391–396.
88. Caro, J. F. & Lanza-Jacoby, S. (1983) J. Clin. Invest. *72*, 882–892.
89. Akmal, M., Massry, S. G., Goldstein, D. A. et al. (1985) J. Clin. Invest. *75*, 1037–1044.
90. Lacour, B., Basile, C., Drüeke, T. & Funck-Brentano, J.-L. (1982) Mineral Electrolyte Metab. *7*, 157–165.
91. Bohmer, T., Bergrem, H. & Eiklid, K. (1978) Lancet *1*, 126–128.
92. Vacha, G. M., Giorcelli, G., Siliprandi, N. & Corsi, M. (1983) Am. J. Clin. Nutr. *38*, 532–540.
93. Chan, M. K., Varghese, Z., Persuad, J. W. & Moorhead, J. F. (1980) Lancet *2*, 1028–1029.
94. Golper, T. A. (1984) Nephron *38*, 217–225.
95. Sanfelippo, M. L., Swenson, R. S. & Reaven, G. M. (1977) Kidney Intern. *11*, 54–61.
96. Gokal, R., Mann, J. I., Oliver, D. O. & Ledingham, J. G. G. (1978) Am. J. Clin. Nutr. *31*, 1915–1918.
97. Davidson, W. D., Rorke, S. J., Guo, L. S. S. & Morin, R. J. (1978) Am. J. Clin. Nutr. *31*, 1897–1902.
98. Van Stone, J. C. & Cook, J. (1978) Proc. Clin. Dial. Transplant. Forum *8*, 103–107.
99. Heuck, C. C., Ritz, E., Kluge, R., Wildberger, D. & Wirth, A. (1979) N. Engl. J. Med. *300*, Letter 1055.
100. Ramos, J. M., Heaton, A., McGurk et al. (1983) Nephron *35*, 20–23.
101. Dieplinger, H., Schoenfeld, P. Y. & Fielding, C. J. (1986) J. Clin. Invest. *77*, 1071–1083.
102. Wing, A. J., Brunner, F. P., Brynger, H. et al. (1984) Contr. Nephrol. *41*, 306–311.
103. Haire, H., Sherrard, D., Scardapane, D. et al. (1978) Cardiovasc. Med. *3*, 1163–1168.
104. Scharf, S., Wexler, J., Longnecker, R. E. & Blaufox, M. D. (1980) Prog. Cardiovasc. Dis. *22*, 343–356.
105. Lacour, B., Roulet, J. B., Beyne, P. & Drüeke, T. (1985) Blood Purification *3*, 42–51.
106. Hulley, S. B., Rosenman, R. H., Bawol, R. D. & Brand, R. J. (1980) N. Engl. J. Med. *302*, 1383–1389.
107. Kannel, W. B., Castelli, W. P. & Gordon, T. (1979) Ann. Intern. Med. *90*, 85–91.
108. Rostand, R. G., Kirk, K. A. & Rutski, E. A. (1984) Kidney Intern. *25*, 653–659.
109. Fielding, C. (1981) Diabetes *30*, Suppl. 2, 19–23.
110. Vicenti, F., Amend, W. J., Abela, J., Fedusko, N. J. et Salucticra O. (1980) An. J. Med. *68*, 363–369.
111. Fielding, C. (1981) Diabetes *30*, Suppl. 2, 19–23.
112. Mahley, R. W. (1982) Med. Clin. North Am. *66*, 375–402.
113. Zilversmit, D. B. (1979) Circulation *60*, 473–485.
114. Tatami, R., Mabuchi, H., Ueda, K. et al. (1981) Circulation *64*, 1174–1184.
115. Miller, N. E. (1983) in: Atherosclerosis: Mechanisms and Approaches to Therapy (Miller, N. E., ed.), pp. 158–163, Raven Press, New York.

116. Hahn, R., Oette, K., Mondorf, H. et al. (1983) Atherosclerosis *48*, 279–288.
117. Gokel, R., Mann, J. J., Moore, R. A., er Morris, P. J., (1979) Quat. 3. Med. 192, 507–517.
118. Casaretto, A., Marchioro, T. L., Goldsmith, R. & Bagdade, J. D. (1974) Lancet *I*, 481–484.
119. Ponticelli, C., Barbi, G. L., Cantaluppi, A. et al. (1978) Nephron *20*, 189–195.
120. Shen, S. Y., Lukens, C. W., Alongi, S. V. et al. (1983) Kidney Intern. *24*, Suppl. 16, 147–152.
121. Ibels, L. S., Simons, L. A. & King, J. U. (1975) Quart. J. Med. *44*, 601–614.
122. Curtis, J. J., Galla, J. H., Woodford, S. Y. et al. (1978) Transplantation *26*, 364–366.
123. Goldstein, S., Duhamel, G., Laudat, M. H. et al. (1984) Nephron *38*, 87–92.
124. Cattran, D. C., Steiner, G., Wilson, D. R. & Fenton, S. S. A. (1979) Ann. Intern. Med. *91*, 554–559.
125. Chan, M. K., Varghese, Z., Persaud, J. W. et al. (1981) Clin. Nephrol. *15*, 309–313.
126. Gosh, P., Evans, D. B., Tomlinson, S. A. et al. (1973) Transplantation *15*, 521–523.
127. Curtis, J. J., Galla, J. H., Woodford, S. Y. et al. (1982) Kidney Intern. *22*, 42–47.
128. Hodel, K., Mordasini, R. C., Brunner, F. P. & Thiel, G. (1986) Schweiz. Med. Wochenschr. *116*, 885–888.

Chapter 10
Dyslipoproteinemias and Liver Disease

W. F. Riesen and H. Kloer

1 Introduction

Since the last century it has been well known that obstructive jaundice is associated with elevated cholesterol levels (1). Subsequently it was discovered that the levels of plasma cholesteryl esters are decreased in patients with either biliary obstruction or hepatic parenchymal disease (2, 3). In recent years there has been considerable increase in knowledge concerning the metabolism of lipoproteins in hepatic disease.

Besides the intestine, the liver is the major organ involved in the biosynthesis of plasma lipoproteins. It plays a key role in the synthesis of very low density lipoproteins (VLDL) and also secretes a nascent high density lipoprotein (HDL). It is the site of synthesis of several enzymes and apolipoproteins which regulate lipoprotein metabolism in the plasma and peripheral tissues. In addition, the liver is also involved in the uptake of intermediate and end products of lipoprotein catabolism and in the final degradation of these remnant lipoproteins. Finally, the liver regulates total body cholesterol stores by excreting cholesterol and other lipids into the bile.

Hepatic involvement in lipoprotein metabolism is very complex. Basically, the lipoprotein abnormalities that arise may result from the impairment of hepatic parenchymal function (4) or they may be cholestatic in origin (5). However, this distinction may be somewhat arbitrary, since parenchymal injury is frequently accompanied by intrahepatic cholestasis. Chronic biliary obstruction, on the other hand, may also lead to parenchymal dysfunction.

Since the lipoprotein alterations which occur secondary to hepatic injury disappear with clinical recovery from the disease, they may serve as valuable parameters in the follow-up of the disease. In addition, such an "experiment of nature" may help us to understand the mechanism involved in lipid transport and metabolism.

2 The Role of the Liver in Lipoprotein Metabolism

2.1 Apolipoproteins

As far as apolipoproteins are concerned the liver has been shown to synthesize most major species. Studies with both isolated perfused rat livers and with rat hepatocytes in culture have demonstrated the production of Apo AI, B, CII, CIII and E (7, 8, 9, 10). The most abundant apolipoprotein produced and secreted is Apo B, followed by Apo E. Apo AI and the C peptides seem to be made in relatively small quantities (9, 10). In terms of the relative contribution to the circulating levels of Apo AI, the small intestine actually seems to be the major source (11). In man, organ cultures of human fetal liver were shown to produce Apo AI, AII, B, CII, CIII and E (12). A well differentiated human hepatoma cell line (Hep-G2) released mostly Apo B, while minor amounts of Apo E, AI and of the C-apoproteins were also found in the culture medium (13). The presence of Apo AI and Apo B could also be demonstrated by immunofluorescence (14) in adult liver from normal subjects and from patients with primary hyperlipidemia. It was suggested recently that human liver secretes several isoforms of Apo AI and Apo E which may be subsequently modified in the plasma compartment (12).

2.2 Nascent lipoproteins

After assembly of lipid and apoprotein components in the endoplasmatic reticulum and the Golgi apparatus, nascent lipoproteins are secreted into the perisinusoidal space of Disse (15, 16). The main lipoprotein secretion products of the liver seem to be particles in the density range of VLDL and HDL, using the isolated perfused rat liver as a model system (6, 7). Both in recirculating perfusion systems and in primary cultures of rat hepatocytes, lipoprotein particles have also been isolated in the LDL density range (17–19). It remains unclear, however, whether these particles represent true de novo synthesis of LDL by the liver or whether there is metabolic conversion of VLDL to LDL particles in the media of the respective systems due to the release of hepatic lipase by the organ in *vitro* (17).

Nascent VLDL is quite different in composition from circulating plasma VLDL. It shows a high content of triglycerides and Apo B, the one apoprotein that is essential for the formation and secretion of triglyceride-rich VLDL and chylomicrons (20). It contains little Apo C and Apo E. To what extent the cholesterol and the phospholipids of circulating VLDL represent those of nascent VLDL is unknown. Apo C and Apo E are mainly acquired through exchange processes during the conversion of nascent to circulating VLDL in the plasma compartment (21).

The other immature lipoprotein, nascent HDL, is characterized by a high content of phospholipids and unesterified cholesterol (6), and a high content of Apo E, Apo AII and Apo C in its protein part. Under the electron microscope these nascent HDL particles show a discoidal structure. A similar nascent HDL accumulates in the plasma of patients with severe alcoholic hepatitis. This is associated with an almost complete absence of lecithin: cholesterol acyl transferase (LCAT) and cholesteryl esters in plasma (6, 22).

In humans, most of the LDL, the major plasma lipoprotein, is not directly synthesized by the liver, but it is formed in plasma by metabolic conversion from VLDL (23). LDL contains almost entirely Apo B and its principal lipid component is esterified cholesterol that may be acquired by transfer from HDL.

2.3 Lecithin: cholesterol acyl transferase (LCAT) and lipoprotein metabolism

The hepatic synthesis and secretion of LCAT is very important in lipoprotein metabolism. LCAT transfers an acyl group from lecithin to cholesterol to form cholesteryl ester (24). It seems to be responsible for the formation of most of the cholesteryl esters found in plasma under normal conditions (25). The enzyme is of hepatic origin, since LCAT is essentially absent in hepatectomized animals. It can be recovered in isolated liver perfusates (26) and its activity is low in many types of liver disease (27–30).

Nascent HDL is apparently the principal substrate for LCAT (6). As soon as the nascent HDL particles have entered the plasma compartment, they interact with the enzyme. The free cholesterol of HDL is esterified by the LCAT reaction, becoming hydrophobic in the process and moving to the core of the HDL particle. This converts the HDL from a discoidal to a spherical structure. The change in structure and composition is associated with a change in protein composition. Apo E and Apo C are substituted by Apo AI. This apolipoprotein which is of hepatic and intestinal origin is a cofactor for LCAT (24).

Apo C and Apo E which are released from HDL become available to nascent VLDL (31). By means of a cholesteryl ester transfer protein (32, 33), the esterified cholesterol generated by the LCAT reaction is transported from HDL to LDL and VLDL and thus enters the metabolic pathway which leads ultimately to cholesterol elimination via the liver LDL receptor pathway.

2.4 Post heparin lipolytic activity and lipoprotein metabolism

Two other enzymes, which constitute the enzymes of so-called "post heparin lipolytic activity" are involved in the degradation of the VLDL. One of them is hepatic triglyceride lipase (HTGL), which is attached to the plasma membrane of liver cells, and which may be important in the final removal of triglycerides from the remnant particles during their catabolism by the liver (34, 35). The other enzyme, lipoprotein lipase (LPL), is also a key-enzyme involved in plasma lipoprotein metabolism since it hydrolyzes the chylomicron and VLDL triglycerides at the endothelial surface of capillaries in muscle and adipose tissue (36).

LPL needs Apo CII for activation (37). The degradation of the triglyceride-rich lipoproteins by LPL and HTGL is associated with a reflux of Apo C and Apo E into the HDL density class. Finally, the mature HDL, which is rich in Apo AI, is taken up and degraded by the liver.

3 Lipid and Lipoprotein Alterations in Liver Disease

Changes in lipoprotein metabolism due to liver dysfunction may result from disorders which are caused by genetic factors affecting hepatic function or they may be due to acquired liver dysfunction. The latter is by far more frequent. In this review only changes in lipoprotein metabolism arising from an acquired alteration of liver function other than that induced by alcohol will be discussed.

In contrast to the hyperlipidemias of diabetes, in chronic renal failure or in hypothyroidism, the clinical importance of lipid disorders in liver diseases is rather limited. Usually, the lipoprotein alterations are reversible and tend to disappear during convalescence. In many cases, however, valuable diagnostic information regarding liver function may be obtained by analysis of lipoprotein patterns. In addition, lipoprotein alterations in liver disease may provide important physiological and pathophysiological clues to our understanding of lipoprotein metabolism due to the central role played by the liver in these processes.

In principle, two mechanisms of acquired disorders associated with hepatic dysfunctions may be discerned, i. e. cholestatic liver disease and parenchymal liver disease. Cholestatic liver disease is due to defective excretion of bile and its constituents such as cholesterol or bile acids. Parenchymal liver disease results from hepatocellular injury. There are several mechanisms by which liver diseases can alter lipoprotein metabolism. These include (1) failure of lipoprotein synthesis and/or secretion; (2) production of abnormal lipoproteins; (3) impaired synthesis or secretion of apoprotein co-factors; (4) impaired synthesis or secretion of the enzymes LCAT and HTGL; (5) impaired lipoprotein catabolism or clearance.

3.1 Lipid and lipoprotein disturbances in parenchymal liver disease

The impact of hepacellular injury on lipoprotein metabolism is manifold and may vary according to whether the injury is acute or chronic. Generally, lipoprotein abnormalities are more pronounced in the acute than in the chronic type of hepatitis. In acute hepatitis and during an active phase of chronic hepatitis an increase of plasma triglycerides is a general rule (38, 39). This increase reaches its maximum in the second or third week after onset of acute hepatitis and normalizes concomitantly with serum transaminases. The course of the triglycerides during acute viral hepatitis is shown in Figure 1, the course of cholesterol in Figure 2.

This elevation is accounted for by an abnormal triglyceride-rich LDL particle (38, 39). The characteristic triglyceride-rich LDL associated with acute liver injury could represent the accumulation of lipoprotein remnants secondary to impaired hepatic clearance. Evidence in support of this hypothesis was obtained by isolation of the abnormal triglyceride-rich LDL, which was characterized by electrophoretic mobility in 1% agar gel and reactivity with antibodies to apoproteins B and C (38, 39).

Besides liver disease there are other secondary and primary dyslipoproteinemic states where a significantly higher proportion of TG in the LDL_2 density range has been described. Most of these conditions are associated with a moderate to severe increase of VLDL-TG as well. For example, in primary hyperlipoproteinemias type I and V the relative triglyceride content of the LDL_2 density range is increased to 20–30% of total mass as compared to a normal range of 4–8%. The relative proportions of the other lipoprotein constituents are normal except for a commensurate decrease in cholesteryl esters and free cholesterol. The most important difference between these primary hypertriglyceridemias and liver disease is that the

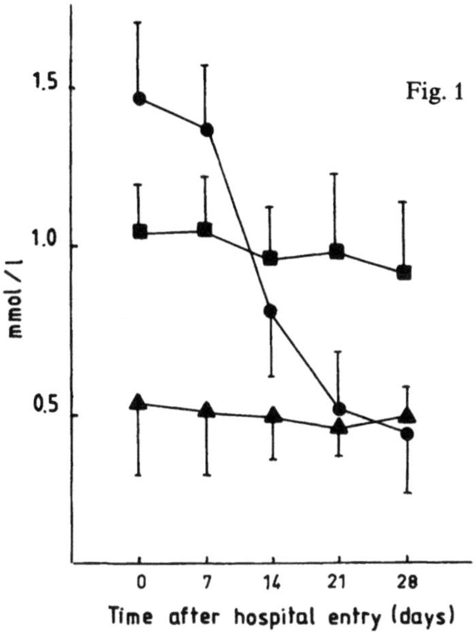

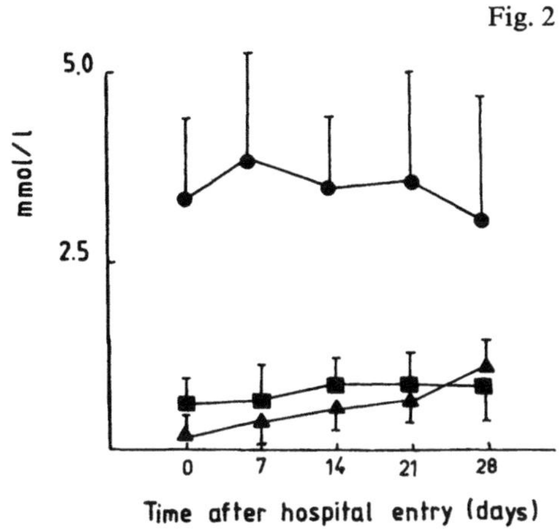

Fig. 1. Changes in the level of triglycerides in the different lipoprotein density classes during acute viral hepatitis (■ VLDL triglycerides, ● LDL triglycerides, ▲ HDL triglycerides).

Fig. 2. Changes in the level of cholesterol in the different lipoprotein density classes during acute viral hepatitis (■ VLDL cholesterol, ● LDL cholesterol, ▲ HDL cholesterol).

VLDL-TG is significantly elevated in the former while this is usually not the case in the latter. In the primary hepatic lipase deficiency syndrome recently described by Breckenridge and collaborators (40) the appearance of TG-rich LDL_2 was also associated with an increase in VLDL. The secondary hyperlipidemia associated with chronic renal failure is also a disease state characterized by simultaneous accumulation of VLDL and of TG-rich LDL. It has also been postulated that a decrease in hepatic lipase activity in postheparin plasma may be the reason for these lipoprotein changes (41).

Other dyslipoproteinemias, which may be categorized under "HDL-deficiency syndromes" are also associated with the appearance of TG-rich LDL. However, they are usually not characterized by increased VLDL in the fasting state, with the exception of some cases of familial LCAT deficiency (42). In Tangier disease (43) and in LpA-deficiency (44), the TG content of LDL_2 has been reported to be in the range of 25–30% total mass with a concomitant reduction of cholesteryl ester. In A-I-Milano disease (45) and in fish eye disease (46) no complete compositional analysis of LDL_2 was reported, but the molar ratio of triglyceride/total cholesterol was 3–4 fold higher than normal. In AI/CIII deficiency (47) and in AI absence (48), however, the increase in triglyceride relative to cholesteryl ester was not as remarkable as in the previously mentioned diseases characterized by a lack of plasma HDL.

The common denominator of this last group of lipoprotein abnormalities is the lack of a pool of normal, cholesteryl ester-rich HDL which could serve as a partner for the exchange of triglyceride from the TG-rich LDL for cholesteryl ester from the normal HDL. This exchange process which is mediated by cholesteryl ester transfer protein would then achieve the conversion of a "final VLDL and/or chylomicron remnant" (i. e. the TG-rich LDL) into the normal, cholesteryl ester-rich LpB particle of LDL_2 density. The functional deficiency of this conversion process could, in addition to deficient triglyceride-hydrolysis by hepatic lipase, furnish one explanation for the accumulation of triglyceride-rich LDL in liver disease.

Another most characteristic alteration in patients with hepatocellular injuries is a particularly abnormal pattern on lipoprotein electrophoresis. In the acute phase the α- and pre-β-bands disappear, while a high and broad β-band remains (49—51). The recurrence of the pre-β- and, especially of the α-band, is an early sign of a favourable prognostic course of hepatitis. Figure 3 shows a typical electrophoretic pattern during the course of acute vial hepatitis.

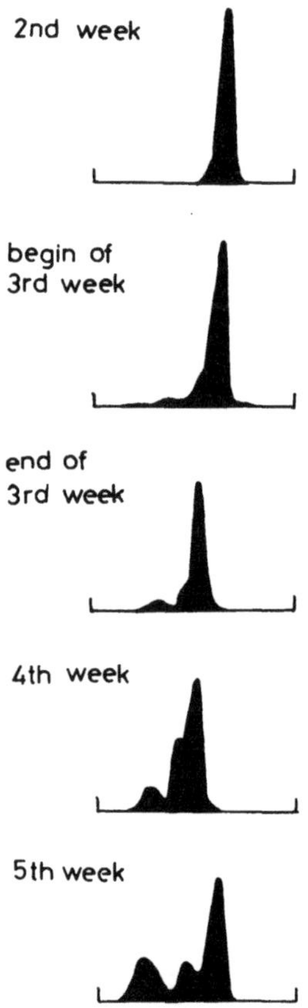

Fig. 3

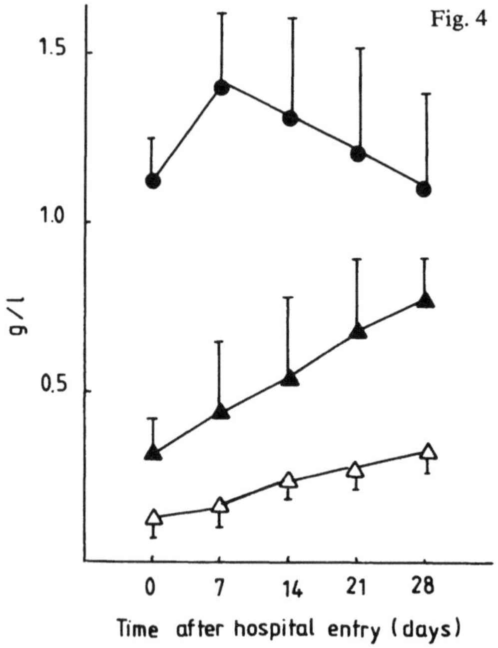

Fig. 4

Fig. 3. Changes in lipoprotein electrophoretic pattern in the course of acute hepatitis. Densitometric evaluation of lipoprotein electrophoresis on agarose with lipid staining.

Fig. 4. Changes of apolipoprotein levels in the course of acute viral hepatitis (● apolipoprotein B, ▲ apolipoprotein AI, △ apolipoprotein AII).

The normal relationship between electrophoretic mobility as α-lipoproteins and HDL isolated by ultracentrifugation does not necessarily hold true in liver disease (see Chapter 1). The same applies to β-lipoprotein and LDL and pre-β and VLDL. Patients with acute hepatitis, in whom α-lipoproteins are absent on electrophoresis have HDL of abnormal composition and morphology that migrate as a component of the broad β-band (30, 51). Instead of triglycerides, these HDL bind bile acids, which explains why they do not pick up typical lipid stains. The changes in the composition of HDL do not only affect the lipid part of the molecule but also the protein part, which contains mostly apoprotein E and lacks Apo AI and AII, the major apolipoproteins of normal HDL. Accordingly, the surface charge of the HDL molecules is altered, leading to abnormal mobility. During recovery, the levels of Apo AI and Apo AII return to normal. Because the synthetic capacity of the liver is mostly unimpaired in acute hepatitis, the low level of the A apolipoproteins is probably due to increased catabolism of these apolipoproteins after reduction of LCAT activity described by several authors in acute hepatitis (27, 53, 54). A typical course of the levels of the apolipoproteins AI, AII and B during acute viral hepatitis is shown in Figure 4.

Apart from hypertriglyceridemia impaired cholesterol esterification is characteristic of parenchymal liver disease. This kind of alteration of lipoprotein metabolism in hepatic disease has been recognized even earlier than the hypertriglyceridemia and abnormal electrophoretic patterns (1, 2). Low levels of LCAT account for many changes seen in plasma lipoproteins in liver disease. This includes reduced cholesteryl esters and elevated free cholesterol, while the total cholesterol is frequently normal. The reduced cholesteryl esters correlate with a reduced LCAT activity (30, 55). The cause of apparent LCAT deficiency is probably diminished hepatic synthesis and secretion of the enzyme, rather than the reduced synthesis of Apo AI which is a cofactor for LCAT (56), although this has to be proven by the quantitation of the enzyme protein instead of its activity. In addition, the decrease in LCAT activity is associated with discoidal HDL due to the high amount of free cholesterol. However, the abnormal lipoproteins which are typical of LCAT deficiency (28) are not detectable. This is probably due to the presence of traces of LCAT activity. Severe LCAT deficiency indicates a poor prognosis (29, 30).

The lipid and lipoprotein alterations occurring in liver dysfunction vary with the stage and the type of the disease and according to whether it is chronic or acute. The most pronounced alterations are observed at the onset of acute hepatitis. These consist of mild hypertriglyceridemia, the absence of α- and the pre-β-band on electrophoresis, and the presence of β_2-lipoproteins, i. e. LDL rich in triglycerides. In addition, a decrease of the level of the major apoproteins of HDL, Apo AI and AII, to less than 50% of the reference values, a decrease of LCAT activity and an abnormally low concentration of cholesteryl esters are frequently observed. All these changes are reversible and disappear during recovery from the disease.

In contrast, in chronic liver disease triglyceride levels were found to be within the normal range (38, 57, 58). In addition, a clear trend towards low levels of total cholesterol and Apo A, reflecting the severity of liver injury, may be observed. Patients with decompensated cirrhosis tend to have lower levels of cholesterol and triglycerides than patients with compensated cirrhosis. These show lower cholesterol and triglyceride concentrations than patients with chronic active hepatitis. In hepatic coma very low concentrations of total cholesterol and Apo A and the absence of α- and pre-β-lipoproteins were found (57, 59).

On the other hand, primary biliary cirrhosis is associated with high levels of total cholesterol and triglyceride and with the presence of Lp-X. Vergani and coworkers (57) found Apo A levels nearly normal in 5 out of 8 patients with primary biliary cirrhosis.

3.2 Lipoprotein alterations in cholestatic liver disease

The presence of elevated serum levels of predominantly unesterified cholesterol in patients with obstructive jaundice was reported first by Flint more than a century ago (1). The discovery of an abnormal serum lipoprotein (Lp-X) which constitutes the circulating lipoprotein form of the increased phospholipid and free cholesterol found in obstructive jaundice by Russ et al. (60), Switzer (61) and Seidel et al. (62) provided the explanation for this observation at the molecular level.

The protein-lipid composition of isolated Lp-X is very typical and is characterized by a high percentage of phospholipids (about 65%), a low percentage of cholesteryl esters (about 2%), about 25% of free cholesterol, traces of proteins (5%) and of triglycerides (5%) Bile acids may be bound in various amounts to Lp-X and have an impact on its physico-chemical properties, notably on its structural stability and hydrated density. The major protein part of Lp-X consists of albumin (40%), which is essentially hidden in the inner part of the vesicle together with the triglycerides. Small quantities of Apo AI, of the C apolipoproteins, of Apo E and of bile acids are also found (63–65). Lp-X can be isolated in the density range of LDL. It may be subfractionated into 3 subfractions, Lp-X1, Lp-X2 and Lp-X3 (63), which differ slightly with respect to chemical composition, hydrated density and molecular mass. Lp-X2 and Lp-X3 seem to differ from Lp-X1 by their Apo AI and Apo E content. Apart from albumin and the known apolipoproteins, Lp-X may associate with several enzymes such as alkaline phosphatase and γ-glutamyltransferase (66).

An important physicochemical property of Lp-X is its electrophoretic mobility towards the cathode in agar gel electrophoresis, which has permitted its quantitation in plasma (67). The differentiation between intrahepatic and extrahepatic cholestasis on the basis of Lp-X quantitation seems to be impossible, although this was suspected for some time (68, 69). Although patients with extrahepatic biliary obstruction tend to have higher levels of Lp-X, which suggests that the quantitation of Lp-X might permit a differentiation between extrahepatic obstruction and intrahepatic cholestasis (69), considerable overlap between plasma values in both instances precludes the use of this test in a discriminatory fashion (70).

Lp-X is not the only abnormal protein within the LDL density range found in obstructive jaundice. Kostner and co-workers (63) have characterized 2 additional abnormal LDL apart from Lp-X, i. e. Lp-Y, a triglyceride-rich LDL, which showed in addition to apoprotein C the presence of Apo B and "lipoprotein B", a lipoprotein particle with higher triglyceride and free cholesterol content than that of normal individuals and an unusually high content of Apo C. They found that Lp-X, Lp-Y and lipoprotein B varied from patient to patient.

While there are numerous abnormalities observed within the LDL-class in cholestasis, this is not the case for HDL. HDL of patients with cholestasis are typically discoidal in shape, similar to nascent HDL secreted by perfused rat-liver (54, 71). However, normal spherical HDL are also observed, which suggests that patients with cholestasis still possess some LCAT activity which may esterify limited quantities of cholesterol. On the other hand, the disturbance in the metabolism of HDL in cholestasis is also manifested by the reduction of the serum concentration of Apo AI (72).

The mechanism of formation of Lp-X is still not entirely clear. Since the concentration of triglycerides, free cholesterol, cholesteryl esters and phospholipids of bile are similar to that of Lp-X, it has been postulated that Lp-X represents bile that has extravasated into the plasma compartment, secondary to cholestasis (73). Indeed, it has been possible to make Lp-X artificially by incubation of serum or albumin with bile (73). Apparently, the formation of characteristic Lp-X vesicles follows pure physico-chemical principles and is probably independent of enzymatic or energy producing mechanisms.

Lp-X is also found in familial LCAT deficiency. It was suspected that the disturbed esterification of cholesterol leads to an increase of free cholesterol and of lecithin, which might provoke the formation of LCAT. However, like in cholestasis, Lp-X in LCAT deficiency seems to be the consequence of special physico-chemical conditions in the plasma. The question whether Lp-X or its lipids serve as a substrate for LCAT has been recently resolved (74). According to these data, Lp-X is not a substrate for LCAT and it is not degraded by post-heparin-lipolytic activity (LPL + HTGL). Studies in animals by Walli and Seidel (75) have indicated that, similarly to other macromolecules but in contrast to normal lipoproteins, Lp-X is taken up and degraded by the reticulo-endothelial system.

Lp-X might also have some importance in the diagnosis of biliary obstructions due to malignant tumors. Havel et al. (76) found levels of Lp-X exceeding 4 g/l in biliary cirrhosis and in hepatic cancer. Such concentrations are rarely seen in non-malignant liver diseases.

Efforts to reduce the concentration of Lp-X in cholestasis so far have had little success. Only plasmaphereses and plasma exchange have effectively lowered cholesterol levels and have led to a reduction of the xanthomas and the lymphopathic symptoms in these patients (76).

4 Pathogenesis of Lipoprotein Abnormalities in Liver Disease

Deficiency of hepatic function may lead to a number of lipoprotein alterations. These abnormalities are due in part to disturbed biosynthesis. More important, however, is abnormal intraplasmatic transformation and/or catabolism.

Apart from apolipoproteins and lipolytic enzymes, the liver synthesizes LCAT, which appears to be a key enzyme for normal lipoprotein metabolism. Virtually all cholesteryl esters in human plasma are considered to be the result of the LCAT reaction, for which nascent HDL is the principal substrate and Apo AI the major activator (24). Part of Apo AI is synthesized by the liver. In liver disease, the activity of LCAT is decreased as may be seen in patients with LCAT deficiency. LCAT deficiency results in decreased incorporation of free cholesterol into the core of newly synthesized HDL, leading to decreased production of normal HDL (77). Cholesteryl esters, which are normally present in great amounts in HDL, are therefore decreased and free cholesterol and phospholipids, which constitute substrates for the LCAT reaction, may increase. Nestel et al. (78) found that the catabolic rate of Apo AI in patients with alcoholic hepatitis is about 2 to 4 times higher than in normal individuals. This indicates that the reduced concentration of Apo AI in such patients is probably due to increased catabolism rather than to decreased synthesis. The abnormal HDL transfers a smaller amount of cholesteryl esters and apoproteins to intermediates in the catabolism of VLDL and chylomicrons via the cholesteryl ester transfer pathway. This leads to an accumulation of remnant lipoproteins in the LDL fraction, since the core cannot be enriched in cholesteryl esters by the cholesteryl ester transfer process. As a consequence, hypertriglyceridemia occurs, and, notably, an abnormal increase of triglycerides, normally not seen in the LDL fraction. The decrease of LCAT activity in liver disease, however, is not the only cause of the observed changes in plasma lipoproteins associated with liver dysfunction, because it was shown that in cholestasis Lp-X may be detected, even when LCAT activity shows normal or even high levels. This is supported by quantitative determination of LCAT and Lp-X in patients with liver disease and in familial LCAT deficiency by Ritland et al. (68), who where not able to show a negative statistical correlation between those two parameters. Lp-X has a great affinity for apolipoproteins, notably Apo C and Apo D, and it may, therefore, inhibit the exchange of apolipoproteins between the various lipoproteins during lipolysis.

Impaired cholesteryl esterification, abnormal electrophoretic pattern and changes in the composition of lipoproteins all reflect abnormalities which have to be regarded as direct consequences of an impairment of the normal liver functions. These abnormalities are sensitive indicators of metabolic defects that are related to deficiencies of LCAT, of HTGL and LPL, of activator apolipoproteins and of the hepatic clearance of remnant lipoproteins.

References

1. Flint, A., jr. (1862) Am. J. Med. Sci. *44*, 305–365
2. Epstein, E. Z. & Peters, J. P. (1945) Arch. Intern. Med. *50*, 203–222
3. Man, E. B., Karlin, B. L., Durchlacher, S. H. & Peters, J. P. (1945) Clin. Invest. *24*, 623–643
4. McIntyre, N., Calandra, S. & Pearson, A. J. G. (1974) Scand. J. Clin. Lab. Invest. *33* (137), 115–1250
5. McIntyre, N., Harry, D. S. & Pearson, A. J. G. (1975) Gut *16*, 379–391
6. Hamilton, R. L., Williams, M. C., Fielding, C. J. & Havel, R. J. (1976) J. Clin. Invest. *58*, 667
7. Marsh, J. B. (1976) J. Lipid Res. *17*, 85
8. Dolphin, P. J. & Rubinstein, D. (1977) Can. J. Biochem. *55*, 83
9. Felker, T. E., Fainaru, M. & Hamilton, R. I. (1977) J. Lipid Res. *18*, 465

10. Dashti, N., McConathy, W. & Ontko, J. A. (1980) Biochim. Biophys. Acta *618*, 347
11. Green, P. & Glickman, J. (1979) J. Clin. Invest. *64*, 233
12. Zannis, V., Kurnit, D. & Breslow, J. (1982) J. Biol. Chem. *257*, 536
13. Rash, P. (1981) Biochim. Biophys. Acta *666*, 294
14. Hopf, R. & Assmann, G. (1979) Gut *20*, 219
15. Jones, A., Ruderman, N. & Herrera, M. (1967) J. Lipid Res. *8*, 429
16. Alexander, C., Hamilton, R. & Havel, R. (1976) J. Cell. Biol *69*, 241
17. Bell-Quint, J. & Forte, T. (1981) Biochim. Biophys. Acta *663*, 83
18. Marsh, J. (1974) J. Lipid Res. *15*, 544
19. Ide, T. & Ontko, J. (1981) J. Biol. Chem. *256*, 10247
20. Sabesin, S. M. & Isselbacher, K. J. (1965) Science *147*, 1149–1151
21. Schaefer, E. J., Eisenberg, S. & Levy, R. I. (1975) J. Lipid Res. *19*, 667–687
22. Ragland, J. B., Bertram, P. D. & Sabesin, S. M. (1978) Biochem. Biophys. Res. Commun *80*, 81
23. Eisenberg, S. & Rachmilewitz, D. (1973) Biochem. Biophys. Acta *326*, 391
24. Glomset, J. A. (1968) J. Lipid Res. *9*, 155–167
25. Glomset, J. A. (1972) Blood Lipids and Lipoproteins: Quantitation, Composition and Metabolism, pp. 445–487 Wiley, New York
26. Osuga, R. & Portman, O. W. (1971) Am. J. Physiol. *220*, 735–741
27. Gjone, E., Blomhoff, J. P. & Wiebecke, I. (1971) Scand. J. Gastronenterol. *6*, 161–168
28. Blomhoff, J. P., Skrede, S. & Ritland, S. (1974) Clin. Chim. Acta *53*, 197–207
29. Simon, J. B. (1974) Scand. J. Clin. Lab. Invest. *33* (137), 107–113
30. Sabesin, S. M., Hawkins, H. L., Kuiken, L. & Ragland, J. B. (1977) Gastroenterology *72*, 510–518
31. Havel, R. J., Kane, J. P. & Kashyap, M. L. (1979) J. Clin. Invest. *64*, 233–242
32. Chajek, T. & Fielding, P. E. (1978) Proc. Natl. Acad. Sci. USA *75*, 3444–3449
33. Pattnaik, N. M., Montes, A., Hughes, L. B. & Zilversmit, D. B. (1978) Biochim. Biophys. Acta *530*, 428–438
34. Krauss, R. M., Levy, R. I. & Fredrickson, D. S. (1974) J. Clin. Invest. *54*, 1107
35. Felts, J. M., Itakura, H. & Crane, R. I. (1975) Biochem. Biophys. Res. Commun. *66*, 1467
36. Fielding, P. E., Shore, V. G. & Fielding, C. J. (1974) Biochemistry *13*, 4318
37. La Rosa, J. C., Levy, R. I., Herbert, P. N., Lux, S. E. & Fredrickson, D. S. (1970) Biochem. Biophys. Res. Commun. *41*, 57–62
38. Kloer, H. U., Ditschuneit, H. H., Mico, L. & Ditschuneit, H. (1972) Z. Gastroenterol. *10*, 489
39. Müller, P., Fellin, R., Lanbracht, J., Agostini, H., Wieland, H., Rose, W. & Seidel, D. (1974) Eur. J. Clin. Invest. *4*, 419–428
40. Breckenridge, J. & Little, A. (1982) Atherosclerosis *45*, 161
41. Mordasini, R., Frey, F., Flury, W., Klose, G. & Greten, H. (1977) New Engl. J. Med. *297*, 1362–1366
42. Glomset, J. A., Norum, K. R., Nichols, A. V., Forte, T., King, W. C., Albers, J., Mitchell, C. D., Applegate, K. R. & Gjone, E. (1974) Scand. J. Lab. Clin. Invest. *33*, 165–171
43. Heinen, J. & Levy, R. (1978) J. Clin. Invest. *52*, 1078
44. Gustavsson, A. (1979) Scand. J. Clin. Lab. Invest. *39*, 377
45. Franceschini, R. & Sirtori, C. (1980) J. Clin. Invest. *66*, 892
46. Carlson, L. A. (1982) Eur. J. Clin. Invest. *12*, 41
47. Norum, J. (1982) N. Engl. J. Med. *306*, 1513
48. Schaefer, E. & Brewer, B. (1982) Arteriosclerosis *2*, 16
49. Papadopoulos, N. M. & Charles, M. A. (1970) Proc. Soc. Exp. Biol. Med. *134*, 797–799
50. Seidel, D., Greten, H., Geisen, H. P., Wengeler, H. & Wieland, H. (1972) Eur. J. Clin. Invest. *2*, 359–364
51. Thallasinos, N., Hatzioannou, J., Scliros, P., Kanaghinis, F., Anastasion, C., Crocos, P., Thomopoulos, D. & Gardikas, C. (1975) Amer. J. Dig. Dis. *20*, 148–155
52. Ragland, J. B., Bertram, P. D. & Sabesin, S. M. (1978) Biochem. Biophys. Res. Commun. *80*, 81
53. Calandra, S., Martin, M. J. & McIntyre, N. (1971) Eur. J. Clin. Invest. *1*, 352–360
54. Simon, J. B. & Scheig, R. (1976) New Engl. J. Med. *283*, 841–846
55. Blomhoff, J. P. (1976) Scand. J. Gastroenterol. *11*, 753–756
56. Fielding, C. J., Shore, V. G. & Fielding, P. E. (1972) Biochem. Biophys. Res. Commun. *46*, 1493–1498

57. Vergani, C., Trovato, G., Delu, A., Pietrogrande, M. & Dioguardi, N. (1978) J. Clin. Pathol. *31*, 772–778
58. Cordova, C., Musca, A. & Violi, F. (1984) Clin. Chem. Acta *137*, 61–66
59. Kloer, H. U., Staudacher, U., Paulini, K. & Ditschuneit, H. (1973) Digestion *8*, 434
60. Russ, E. M., Rayment, J. & Barr, D. P. (1956) J. Clin. Invest. *35*, 133–144
61. Switzer, S. (1967) J. Clin. Invest. *46*, 1855, 1866
62. Seidel, D., Agostini, B. & Müller, P. (1972) Biochim. Biophys. Acta *260*, 146–154
63. Kostner, G. M., Laggner, P., Prexe, H. J. & Holasek, A. (1976) Biochem. J. *157*, 401–407
64. Patsch, J. R., Aune, K. C., Gotto, A. M. & Morrisett, J. D. (1977) J. Biol. Chem. *252*, 2113–2120
65. Schwartz, C. J., Halloran, L. G., Vlahcevic, Z. R., Gregory, D. H. & Swell, L. (1978) Science *200*, 62–64
66. Huseby, N. E. (1982) Clin. Chim. Acta *124*, 103–112
67. Ritland, S. (1974) Clin. Chim. Acta *55*, 359–366
68. Ritland, S., Blomhoff, J. P. & Gjone, R. J. (1973) Clin. Chim. Acta *49*, 251–259
69. Magnani, H. N. & Alaupovic, P. (1976) Gastroenterology *71*, 87–93
70. Simon, J. B. & Poon, R. W. M. (1978) Gastroenterology *75*, 177–180
71. Jones, A. L., Ruderman, N. B. & Herrera, M. G. (1967) J. Lipid Res. *8*, 429
72. Yamamoto, K., Koya, S. & Ibayashi, H. (1978) Clin. Chim. Acta *87*, 85–92
73. Manzato, E., Fellin, R., Baggio, G., Walch, S., Neubeck, W. & Seidel, D. (1976) J. Clin. Invest. *57*, 1248–1260
74. Bartholomé, M., Niedmann, D., Wieland, H. & Seidel, D. (1981) Biochim. Biophys. Acta *664*, 327–334
75. Walli, A. & Seidel, D. (1980) 3rd Eur. Symp. Metabolism, Padova
76. Havel, R. J., Goldstein, J. L. & Brown, M. S. (1980) Metabolic Control and Diseases. (Bondy, P. K. & Rosenberg, L. B., eds.), pp. 393–394, W. B. Saunders Co., Philadelphia
77. Glomset, J. A., Norum, K. R. & King, W. (1970) J. Clin. Invest. *49*, 1827–1837
78. Nestel, P. J., Tada, N. & Fidge, N. H. (1980) Metabolism *29*, 101–104

Chapter 11
Dyslipoproteinemias in Thyroid Disease

H. N. Ginsberg and I. I. Goldberg

1 Introduction

Alterations in the concentrations of the plasma lipids, cholesterol and triglycerides, in states of excessive or deficient thyroid function have been clearly documented during the previous thirty years. The association between levels of circulating thyroid hormones and cholesterol and triglyceride levels is not surprising when one considers the multitude of effects that thyroid hormone has upon intermediary metabolism. As we will discuss in the following pages, T_4 (thyroxine) and T_3 (triiodothyronine) have important effects on essentially all aspects of lipid and apolipoprotein synthesis, lipoprotein secretion from the liver, intraplasmic metabolism of lipoprotein lipids, and removal of lipoproteins from the plasma. The clinically significant consequences of thyroid hormone deficiency include increased plasma concentrations of very low density (VLDL), intermediate density (IDL) and low density lipoproteins (LDL). In addition, altered plasma levels of high density lipoproteins (HDL) are present frequently in patients with thyroid dysfunction and may also be clinically relevant.

In order to fully describe the abnormalities in lipid and lipoprotein metabolism that have been observed in hyper and hypothyroidism we must first review the normal actions of the thyroid hormones at various regulatory points in these pathways.

2 Metabolic Effects of Thyroid Hormones

2.1 Effect of thyroid hormone on hepatic triglyceride synthesis

Thyroid hormone has been shown to have multiple effects on the metabolism of fatty acids and glycerol. A recent review by Heimberg (1) outlined these effects in detail and they will only be summarized here. Thyroid hormone stimulates oxidation of fatty acids and formation of ketones. This occurs mainly as a result of the stimulation of the enzyme carnitine palmitoyltransferase I on the outer surface of the mitochondrial membrane. There is a parallel stimulation of fatty acid synthesis. The simultaneous stimulation of fatty acid oxidation and synthesis is a state that is unique to the action of thyroid hormones. However, the overall effect of thyroid hormone on hepatic fatty acid metabolism is that less fatty acid is available for triglyceride synthesis. The hepatic content of glycerol-3-phosphate, a major precursor of triglyceride, is significantly reduced by thyroid hormone excess. The sum of these effects on hepatic fatty acid and glycerol metabolism should be a reduction in the availability of substrates for triglyceride synthesis, and this conclusion is supported by most of the data generated from in vitro studies with liver perfusion systems and isolated hepatocytes. Lack of thyroid hormone reverses the activities described above with reductions in fatty acid oxidation and ketone formation, and increased hepatocyte levels of glycerol-3-phosphate prevailing. These conditions favor the synthesis of triglycerides, and such increases in triglyceride production by isolated perfused livers have been demonstrated.

However, *in vivo*, a major modulator of triglyceride synthesis is the flux of plasma free fatty acids into the liver. Thyroid hormone, by stimulating adipose tissue hormone sensitive lipase, causes increased free fatty acid release into the circulation from adipocytes. As hepatic extraction of these fatty acids is unaffected by thyroid status, the increased plasma levels result in increased hepatic uptake of free fatty acids. This elevated rate of delivery of fatty acids to the hepatocyte may overcome the effects of thyroid hormone described above and result in increased hepatic triglyceride synthesis and secretion. In a similar fashion, in states of deficient thyroid hormone there will be reduced fatty acid delivery to the liver. This may offset the hepatic effects of thyroid hormone deficiency which are to increase triglyceride synthesis. The complexity of this system, obvious from this brief summary, will be apparent as we discuss triglyceride metabolism in humans and laboratory animals with altered thyroid function.

2.2 Effect of thyroid hormone on hepatic cholesterol metabolism

Because the studies of the effect of thyroid hormone on cholesterol metabolism in animals and humans are limited, we will address these data as they relate to hypo and hyperthyroidism (see below).

2.3 Effect of thyroid hormone on apolipoprotein synthesis

Very little literature exists describing the regulation of apolipoprotein synthesis, particularly synthesis of apolipoprotein B (Apo B), in states of deficient or excess thyroid hormone. The few studies that have been completed are compatible with the view that hepatic synthesis of Apo B and Apo E are not significantly affected by alterations in circulating thyroid hormone levels. Hence, in hypothyroid rats, Apo B secretion in VLDL appears to be normal, while triglyceride secretion in VLDL is elevated (2, 3). However, in hypothyroid, cholesterol fed rats, secretion of VLDL Apo B and Apo E is increased (3). Hepatic Apo AI secretion has also been reported to increase in hyperthyroid rats (4). From these limited studies it would appear that the changes in plasma apolipoproteins observed in thyroid dysfunction result from altered plasma metabolism of the lipoproteins.

2.4 Effect of thyroid hormone on plasma lipoprotein catabolism

Circulating lipoproteins interact with several enzymes and cell surface receptors as they undergo metabolism in plasma. Two enzymes that play major roles in the catabolism of plasma lipoproteins are lipoprotein lipase (LPL) and hepatic triglyceride lipase (HTGL). These two triglyceride hydrolytic enzymes are also referred to as the post-heparin lipolytic enzymes since they are both released into the circulation of humans and most animals after the intravenous injection of heparin. LPL is well established to be the major enzyme responsible for hydrolysis of triglyceride in chylomicrons and newly secreted VLDL. As hydrolysis of these triglyceride-rich particles is associated with the transfer of surface components (phospholipids, free cholesterol and apolipoproteins) to HDL, LPL activity is also intimately involved in the metabolism of HDL.

The roles of HTGL are less clearly defined. Inherited deficiencies of HTGL have been reported in humans (5), and immunologically induced HTGL deficiency has been produced in primates (6) and in rodents (7, 8). These HTGL deficient states have been associated with elevations in plasma triglyceride levels with increases in small VLDL and IDL. Additionally, increases in HDL cholesterol and phospholipid, associated with rises in HDL_2, have been observed during inhibition of HTGL activity. Hence it has been hypothesized that HTGL plays a role in regulating catabolism of circulating HDL and in the conversion of HDL_2 to HDL_3. This hypothesis is supported by the inverse correlation between plasma HDL levels and measurements of post-heparin HTGL activity (9).

The activities of both LPL and HTGL have been shown to be regulated, in part, by thyroid hormones. The sites of regulation, which are likely to include gene transcription, messenger RNA translation and protein turnover, have not been elucidated. The relevance of these regulatory actions of thyroid hormone is clearly demonstrated by the defect in chylomicron clearance that is demonstrable in hypothyroid states (10, 11), and the increase in chylomicron and VLDL triglyceride hydrolysis that occurs in thyroid hormone treated humans (11, 12).

Probably the major effect of thyroid hormone on plasma lipoprotein metabolism occurs via this hormone's effects on the LDL receptor (13). Thyroid hormone appears necessary for the maintenance of normal numbers of LDL receptors on cell surfaces. Whether this is a specific activity of the hormone, related to basic cellular metabolic regulation, or a consequence of thyroid hormone action on protein synthesis in general is unknown. Regardless, stimulation of receptor function by thyroid hormone may lead to increased removal from plasma of VLDL remnants and IDL, and certainly causes increased rates of clearance from plasma of LDL (14, 15) with resultant reductions in plasma LDL cholesterol levels.

It should be clear that thyroid hormones are intimately involved in essentially all aspects of lipid and lipoprotein metabolism. The various dyslipoproteinemias associated with either reduced or increased levels of circulating T3 and T4, to be described in the following sections, are the logical outcomes of those involvements.

3 Hypothyroidism

The association between elevations in plasma lipid levels and hypothyroidism spans more than 50 years. While there has been general agreement concerning the hypercholesterolemia of hypothyroidism, hypertriglyceridemia as a concomitant of hypothyroidism has been reported only intermittently. This latter is most likely the result of the complex interrelationship between the effects of thyroid hormones on fatty acid and glycerol metabolism in the liver; on triglyceride metabolism in adipocytes; and on triglyceride rich lipoprotein catabolism in plasma. The situation is further complicated by the effects of obesity on triglyceride metabolism and on the likely inclusion in all studies of some patients with coexistent primary hypertriglyceridemia.

3.1 Triglyceride metabolism in hypothyroidism

The metabolism of exogenously derived triglyceride in chylomicrons is altered in hyothyroid states in humans. In the early 1970s Nikkila et al. (10), Tulloch et al. (11), and Rossner et al. (16) all demonstrated that the removal of intravenously injected triglyceride-rich liposomes or emulsions was reduced in hypothyroid subjects. Rossner actually found that triglyceride removal in these subjects was slower than in individuals with primary hypertriglyceridemia (16). Tulloch reported that the intravenous fat tolerance test improved after thyroid hormone replacement therapy (11) and this response to treatment was subsequently confirmed by others (17, 18). Abrams et al. (12) did not find significantly reduced chylomicron clearance in most of their patients with hypothyroidism and hypertriglyceridemia, although they did observe that the rate of chylomicron removal from plasma did increase in their subjects after hormone replacement. It is important to note that fasting chylomicronemia is rarely present in hypothyroid patients, and if present, is an indication of underlying, severe primary hyperlipidemia.

Plasma levels of endogenously derived, VLDL triglyceride have been described as normal or moderately elevated in hypothyroidism. As noted above, the combined effects of thyroid hormone on the hepatocyte and the adipocyte should result in normal or actually reduced triglyceride synthesis by the liver in hypothyroidism (1). Nikkila and Kekki (1), in 1972, published the first report of VLDL triglyceride turnover in hypothyroid patients. Using tritiated glycerol as a labeled precursor of triglyceride, they found that production of VLDL triglyceride was normal in nonobese hypothyroid patients and increased in obese hypothyroid subjects. The fractional catabolic rates (FCR) of VLDL triglyceride in their patients were reduced compared to normal subjects and this appeared to be the major reason for the modest hypertriglyceridemia that occurred. The obese patients, in whom both elevated VLDL triglyceride production and reduced *fractional clearance* were present, had the most significant triglyceride elevations. Abrams, Grundy and Ginsberg (12) reported similar findings in their group of hypothyroid patients studied with tritiated glycerol. In the obese hypothyroid subjects in that study, VLDL triglyceride levels were elevated and this was associated with both increased rates of VLDL production and reduced FCRs. Interestingly, the VLDL production rates were not greater than in weight matched controls. However, the obese euthyroid controls had much higher FCRs than their obese hypothyroid counterparts. In the report by Abrams et al. (12) the tritiated glycerol studies were repeated after thyroid hormone replacement. In many patients with normal or only modest hypertriglyceridemia there was actually an increase in VLDL triglyceride production after treatment. This finding is consistent with the increase in free fatty acid delivery to the liver that would take place after thyroid hormone replacement. A surprising finding was that the FCRs for triglyceride did not increase after therapy. The basis for this finding is unclear, especially in view of the increase in the rate of chylomicron clearance that was demonstrated (12).

3.2 Lipoprotein compositional changes in hypothyroidism

Although consistent changes in triglyceride concentration may not occur in hypothyroid states, there is a more constant pattern of compositional change in VLDL, IDL, LDL and HDL. First, VLDL appear to become cholesterol enriched in the presence of thyroid deficiency. Wahlqvist et al. (19), Ballantyne et al. (20) and Aviram et al. (21) have demonstrated the presence of increased VLDL cholesterol during hypothyroid states. Ballantyne also found elevated levels of IDL in her hypothyroid subjects. Hormone replacement therapy reduced the cholesterol content in these lipoproteins in all studies. Dory et al. (22) noted the appearance of β-VLDL in rat made hypothyroid with propylthiouracil, even though VLDL triglyceride levels fell in those animals. Finally, Hazzard et al. (23) reported a case of coexistent hypothyroidism and Type III hyperlipoproteinemia. This patient presented with marked hyperlipidemia and high concentrations of β-VLDL. After thyroid hormone replacement total plasma cholesterol and triglyceride levels fell to the normal ranges but VLDL remained enriched with cholesterol.

A second compositional finding has been the enrichment of LDL and HDL with triglyceride. Wahlqvist et al. (19), Lithell et al. (17) and Rossner et al. (16) described this triglyceride enrichment of LDL and HDL and noted that the composition of these lipoproteins normalized with thyroid hormone replenishment. Both these compositional changes in LDL and HDL, and the cholesterol enrichment of VLDL described above are compatible with the reduced activity of HTGL that has been reported in hypothyroid individuals. Alterations of both lipolytic enzymes, LPL and HTGL, in hypothyroidism will be discussed in detail in the following section.

3.3 Plasma lipases in hypothyroidism

Lipoprotein lipase and hepatic triglyceride lipase have been studied in detail in hypothyroid subjects. Studies by Porte et al. initially described a deficiency of total post-heparin lipolytic activity in patients with hypothyroidism (24). This measurement included a summation of both LPL and HTGL, which are released into the plasma after intravenous injection of heparin. Subsequently, Krauss et al. demonstrated that in human post-heparin plasma the HTGL activity could be specifically measured by inhibition of LPL with protamine sulfate and that only the HTGL was significantly reduced in patients with hypothyroidism (25). These findings were confirmed by Abrams et al. (12) who also reported that the low levels of post-heparin HTGL activity increased after treatment of hypothyroid patients with thyroid hormone. Similar decreases in HTGL have been found in thyroidectomized rats (26). While little or no change in post-heparin plasma LPL activity has been observed in hypothyroidism, measurements of LPL activity in tissue samples are not concordant with those results. Lithell et al. reported that both adipose tissue and skeletal muscle LPL are reduced in hypothyroidism and that LPL activity increases with treatment (17). Pykalisto et al. noted decreased levels of heparin releasable LPL activity in adipose tissue biopsies from hypothyroid patients (27), but no significant difference was noted in total adipose tissue LPL activity. In summary, post-heparin plasma HTGL activity is reduced in hypothyroidism and the reduced level of this enzyme activity may relate to the compositional changes that occur in VLDL, IDL and LDL in the hypothyroid state. The effects of hypothyroidism on LPL activity are less certain, but the changes demonstrated in heparin releasable adipose tissue activity may play a role in the reduced removal of triglyceride rich lipoproteins from plasma that has been demonstrated in some studies.

3.4 Cholesterol metabolism in hypothyroidism

Because of the association between altered thyroid function and increased risk for developing cardiovascular disease (28), there has been intense interest in the hypercholesterolemia of hypothyroidism for several decades. More detailed studies of plasma cholesterol metabolism began to appear in the period 1950–1960, when Kurland and coworkers (29) reported reduced plasma clearance of radiolabeled cholesterol in hypothyroid subjects. These findings were expanded when Chobanian et al. (30) and Rabinowitz et al. (31) used similar methodologies to describe the plasma turnover of cholesterol in hypothyroid patients. In both of these studies the fractional removal of cholesterol from plasma was significantly prolonged compared to normal, and in the study by Rabinowitz (31), thyroid hormone treatment resulted in an increase in that rate. At about this same time, Furman et al. (32) and Cornwell et al. (33) demonstrated increased concentrations of cholesterol in ultracentrifugally separated LDL. In the latter study, hormone replacement therapy was associated with reductions in LDL cholesterol (33). The next advance was achieved by Walton et al. (14) who, in 1965, reported that the clearance of iodinated LDL from plasma was slow in hypothyroid patients and increased after treatment. This finding was confirmed in 1981 by Thompson et al. who demonstrated that the FCR for iodinated LDL in a patient with severe hypothyroidism was essentially identical to that seen in subjects with homozygous familial hypercholesterolemia (15). This group went on to show that the reduced LDL removal was associated with decreased receptor and nonreceptor mediated catabolism of LDL and that both pathways increased their activity after thyroid replacement (15). All of this information is quite compatible with the demonstration by Chait et al. (13) that thyroid hormone is necessary for normal LDL receptor function in cultured cells.

At about the same time that work related to plasma cholesterol metabolism was accelerating, several groups were examining the role of thyroid hormone in whole body sterol balance. Friedman, Byers and Rosenman carried out studies indicating that there was both reduced synthesis of cholesterol (34) and decreased conversion of cholesterol to bile acids in hypothyroid states (35). Their studies were followed by a number of investigations in humans and rats which demonstrated reduced synthesis of cholesterol in the hypothyroid state (36). However, reduced synthesis of cholesterol is an unlikely candidate for the cause of elevated plasma cholesterol concentrations. Decreased conversion of cholesterol to bile acids, with concomitantly increased secretion of hepatic cholesterol in lipoproteins would be a more likely cause of this hypercholesterolemia. Unfortunately, more recent data do not support this mechanism. In the study by Abrams and Grundy (37), bile acid synthesis was normal during the hypothyroid state and unchanged by hormone replacement therapy. In contrast, cholesterol synthesis was only slightly reduced in nonobese hypothyroid individuals but increased significantly after treatment. In obese, hypothyroid patients cholesterol synthesis was elevated pretreatment but increased even further post-therapy. Hence these authors concluded that factors regulating plasma clearance of cholesterol must dominate in the development of hypercholesterolemia of hypothyroidism.

All of these physiologic studies were complemented by detailed investigations of the effects of thyroid hormones on plasma cholesterol concentrations. The sum effect of these many studies was to clearly demonstrate that total plasma cholesterol and LDL cholesterol are very frequently elevated in patients with hypothyroidism (16–20, 38–41) and that thyroid hormone treatment reduces the cholesterol levels to normal or near normal (16–20, 38–40). Several studies focused on the correlation between the severity of thyroid dysfunction and the levels of cholesterol. The consensus from these reports is that total and LDL cholesterol concentrations are elevated only in more severe cases of hypothyroidism, that is those cases where the thyroid stimulating hormone (TSH) level is $>$ 40 units/ml and where the T4 and T3 levels have begun to fall below the normal range (42–44). This is in contrast to some earlier reports which suggested that hypercholesterolemia was a marker for preclinical myxedema (45, 46). These earlier studies probably suffered from patient selection difficulties. This is not to suggest, however, that hypothyroidism is not an important cause of hypercholesterolemia. In fact, it is probably the most common cause of secondary hyperlipidemia, accounting for approximately one percent of all hyperlipidemia (43).

3.5 High density lipoproteins in hypothyroidism

During the past 10 years, in parallel with the increasing interest in HDL cholesterol as an indicator of reduced risk for development of cardiovascular disease, investigators have closely examined the effect of deficient thyroid hormone action on HDL metabolism. However, in 1961, both Furman et al. (32) and Cornwell et al. (33) presaged the present activity with reports in this area. Furman reported that

HDL cholesterol was variable in hypothyroid patients and could be significantly reduced in such individuals who were also hypertriglyceridemic, while Cornwell presented four patients with hypothyroidism and increased HDL cholesterol concentrations. He also noted that the HDL levels were reduced by hormone replacement therapy in three of the four. Sachs et al. (48), in 1968, reported that alpha cholesterol (the older terminology for HDL) was elevated in hypothyroid patients and decreased after hormone treatment.

Little more about HDL and hypothyroidism was published until 1977 when Scottolini et al. (49) reported a study of 41 males and 200 females. The females had increased HDL cholesterol levels that were normalized by hormone treatment. Elevated HDL in hypothyroidism has been confirmed by several groups (18, 21, 38, 41, 50), although unchanged (51) or even reduced (52) HDL cholesterol levels have been reported in similar patient groups. Although some of this variability can be attributed to the effects of obesity on HDL levels and to the coexistence of hypertriglyceridemia, not all of the conflict can be resolved in this manner. Thus, in the paper of Abrams and Grundy (37), although HDL cholesterol was reduced in their obese, often hypertriglyceridemic hypothyroid subjects, it was only elevated in some (but not all) of their nonobese hypothyroid patients. Similar variability is present in studies of rats made hypothyroid with propylthiouracil (4, 22).

The ability of thyroid replacement therapy to reduce HDL levels, first reported by Cornwell (33) and Sachs (48), has been confirmed by many groups (18, 21, 38, 48, 49). But even here, some controversy exists, as a few studies have failed to observe reductions in HDL cholesterol after treatment (37). We believe the confusion concerning this issue derives from the variability of the patient populations studied. As noted above, obesity and plasma triglyceride levels are two significant confounding variables in many of these reports. In addition, the severity of the thyroid deficiency must be taken into account. The importance of the latter variable is made clear by the studies of the Lipid Research Clinics in which HDL cholesterol concentrations varied inversely with the T4 concentrations (53).

More detailed investigations have revealed that HDL_2 appears to be the subfraction of HDL that is affected by lack of thyroid hormone (21, 38, 50). This conclusion, based on ultracentrifugal data, is supported by the report by Muls et al. (38) that Apo AI was increased while Apo AII was unchanged in a group of severely hypothyroid individuals. These latter findings, together with the observations that HDL triglycerides are also increased in hypothyroid states (17, 19) indicate that many if not all of the changes in HDL that have been reported may result from the reductions of HTGL demonstrated in thyroid deficiency.

4 Hyperthyroidism

4.1 Triglyceride metabolism in hyperthyroidism

Most studies of the effect of excessive thyroid hormone on VLDL triglyceride metabolism have demonstrated the expected; that is they have shown lower plasma triglyceride levels in hyperthyroid patients than in hypothyroid subjects. However, in no studies were the levels of plasma triglycerides markedly reduced compared to levels in control subjects. Similarly, there were not marked increases in the level of this lipid after the patients became euthyroid. In the few studies in which more detailed approaches were undertaken, the complex basis of the response of plasma triglyceride to thyroid hormone excess was apparent. Sandhofer et al. (54) in 1966 demonstrated increased rates of fatty acid flux and increased free fatty acid concentrations in subjects with hyperthyroidism. However, using radioactively labeled fatty acid, they reported decreased conversion of the fatty acids to triglyceride, leading to reduced VLDL triglyceride entry into plasma. They concluded that diversion of fatty acids away from triglyceride synthesis "protected" the liver from the increased delivery of plasma free fatty acids. Later authors criticized their methodology, and the criticisms were supported by the findings of Nikkila and Kekki (10) that VLDL production was slightly increased in hyperthyroid individuals. In this study, using radiolabeled glycerol as a precursor of VLDL triglyceride, the authors demonstrated that the increase in VLDL triglyceride production was almost balanced by an increased FCR for plasma triglyceride, and the subjects had, therefore, only slightly increased plasma levels of triglyceride. Increased capacity for triglyceride removal in their patients was also shown with Intralipid infusions. After treatment with antithyroid drugs, the triglyceride FCR fell, while production of VLDL triglyceride did not change. These post therapy changes were associated with increases in plasma triglyceride levels.

In support of the concept that more efficient removal of triglycerides is the predominant alteration associated with excess thyroid hormone, Tulloch et al. (11) demonstrated increased removal of exogenous fats during lipid infusions in hyperthyroid individuals. These high rates fell to normal after antithyroid treatment. More recently Abrams, Grundy and Ginsberg (12) quantitated the turnover of exogenous and endogenous triglyceride in hyperthyroid subjects and confirmed most of the findings of Nikkila and Kekki (10). Abrams et al. reported that the FCR for VLDL triglyceride was increased in hyperthyroidism and that production of this lipid was normal or slightly elevated. In contrast to the previous report, however, the FCR did not change after antithyroid therapy. Differences in the patient groups studied may account for these contrasting results. These authors concluded that increased removal of plasma triglycerides "prevented" the elevations of VLDL that might have otherwise resulted from increased VLDL triglyceride production in their hyperthyroid subjects (12).

4.2 Plasma lipases in hyperthyroidism

In the past decade only two reported studies have focused upon post-heparin plasma lipase activities in hyperthyroid patients before and after treatment (12, 55). Both groups of investigators reported that hyperthyroidism was associated with a significant increase in post-heparin plasma HTGL activity with no significant change in LPL. It was postulated that this increase in HTGL activity was causal in the decrease in HDL levels noted in hyperthyroid patients.

4.3 Cholesterol metabolism in hyperthyroidism

As was the case for thyroid hormone deficiency, plasma total and LDL cholesterol concentrations are significantly altered by excess hormone action. This association was noted early and has been fully characterized during the past 25 years. In 1961, Cornwell et al. (33) demonstrated that the S_f 3–9 LDL were reduced in hyperthyroid patients. This finding was confirmed a few years later by Walton et al. (40), who also showed that antithyroid treatment resulted in a rise in LDL cholesterol towards normal. At this same time, Rabinowitz and coworkers (31) observed that radiolabeled cholesterol had a shortened half-life in plasma in rats treated with D-T4. Walton et al. (14) injected radioiodinated LDL into hyperthyroid subjects and found a shortened half-life and increased FCR. These abnormalities were reversed to normal by therapy with antithyroid medications.

More recently, several authors have confirmed the inverse relationship between thyroid hormone activity and LDL cholesterol concentrations (41, 49, 51, 56). In addition, Muls et al. (57), Aviram and coworkers (21), and Boberg et al. (58) demonstrated normalization of LDL concentrations after antithyroid therapy. In the reports by Muls (57) and Boberg (58), Apo B levels were also low during hyperthyroidism and increased to normal after treatment.

Studies of whole body cholesterol metabolism in hyperthyroid states have been less revealing than those concerned with plasma cholesterol concentrations and flux. In a recent study by Abrams and Grundy (37), slight increases in cholesterol synthesis and in neutral steroid excretion were noted in a small group of hyperthyroid patients, with reversal of these trends after therapy. In contrast to earlier reports (35), these authors did not find evidence for increased conversion of cholesterol to bile acids during the hyperthyroid period. However, cholesterol absorption was somewhat reduced in their patients, and this alteration reverted to normal after antithyroid treatment (37).

4.4 High density lipoproteins in hyperthyroidism

The effects of excess thyroid hormones on plasma concentrations of HDL are in accord with the effects of these hormones on the other major lipoproteins: HDL cholesterol is reduced in most studies of thyrotoxic patients (21, 33, 49, 51, 52, 57, 58). This inverse relationship between HDL and T4 was first reported in 1961 by Cornwell et al. (33). Interestingly, little else appeared in the literature concerned with this question until the 1980s when this finding was confirmed by several groups. In addition, most of these studies demonstrated a rise in HDL cholesterol after treatment of the thyrotoxic patients with antithyroid medications. The study by Scottolini et al. (49) suggested some sex differences as males with hyperthyroidism had normal HDL cholesterol concentrations while female patients had reduced HDL levels. Only the females in that study showed a rise in HDL cholesterol levels after therapy (49). Aviram and coworkers (21) reported that HDL_2 was increased while HDL_3 concentration was reduced in thyrotoxic subjects. Antithyroid therapy, however, was associated with an increase in both fractions, although the rise was greater in HDL_2.

The changes in HDL cholesterol associated with hyperthyroidism are generally paralled by alterations in Apo AI levels. Boberg et al. (58) reported that Apo AI levels were reduced in untreated subjects and rose to normal levels after antithyroid treatment. Muls et al. (57) observed normal Apo AI concentrations in their hyperthyroid group, but the levels increased after therapy. In the latter study (57), the finding that Apo AII concentrations were normal in the untreated subjects and unchanged by treatment is supportive of the differences between HDL_2 and HDL_3 reported by Aviram et al. (21).

5 Summary

We have attempted, in this review, to present the multiple effects that thyroid hormone has upon the regulation of the synthesis, transport and catabolism of the plasma lipoproteins. While the complexity of the interaction between T4 and T3 and lipoprotein metabolism results in variability in the individual responses of hypo- and hyperthyroid subjects, several themes can be observed.

Overall, deficiency of thyroid hormones results in a strong predisposition to hypertriglyceridemia. This trend is initially driven by increased fatty acid delivery to the liver, stimulating triglyceride synthesis even in the face of reduced availability of glycerol-3-phosphate and the diversion of fatty acids away from incorporation into triglycerides; and is further supported by reduced activity of both LPL and HTGL, with resultant decreased fractional clearance of plasma VLDL triglyceride. In states of thyroid excess, all these circumstances are reversed, with increased lipase activities producing normal or reduced plasma triglyceride concentrations in the face of somewhat increased rates of synthesis.

A key point of regulation by thyroid hormones appears to be at the level of HTGL. Thus, in thyroid deficiency, reduced levels of HTGL activity are associated with cholesterol enrichment of VLDL and IDL, and triglyceride enrichment of LDL and HDL. All of these findings are compatible with previously documented effects of HTGL on lipoprotein metabolism. The elevated levels of HDL cholesterol present in hypothyroid patients (who do not have hypertriglyceridemia or obesity) are also consistant with current views of the action of HTGL.

The association between thyroid hormones and LDL cholesterol metabolism appears to be the most straightforward. It seems clear that the regulation of LDL receptor activity by T4 and T3 plays a significant physiologic role *in vivo*, as demonstrated by LDL turnover studies. The interaction of thyroid hormones and whole body cholesterol metabolism is less clear, although it appears that reduced gastrointestinal absorption of dietary cholesterol may contribute to the lowered plasma cholesterol concentrations in hyperthyroidism. Cholesterol synthesis appears to be reduced in thyroid deficiency and stimulated in states of thyroid hormone excess, but these are modest alterations and are overshadowed by the effects of the hormones on plasma LDL removal.

Finally, HDL levels are increased in hypothyroid patients, possibly as the result of reduced catabolism of this circulating lipoprotein. HDL appear to be removed from plasma by receptor mediated pathways and it is possible that these receptors are regulated by thyroid hormone. As noted above, the reduced level of HTGL activity in hypothyroidism may also play a significant role in elevating the plasma levels of HDL.

References

1. Heimberg, M., Olubadewo, J. O. & Wilcox, H. G. (1985) Plasma lipoproteins and regulation of hepatic metabolism of fatty acids in altered thyroid states. Endo. Rev. *6*, 590–607.
2. Dolphin, P. J. & Forsyth, S. J. (1983) Nascent hepatic lipoproteins in hypothyroid rats. J. Lipid Res. *24*, 541–551
3. Krul, E. S. & Dolphin, P. J. (1982) Secretion of nascent lipoproteins by isolated hepatocytes from hypothyroid and hypothyroid, hypercholesterolemic rats. Biochem. Biophys. Acta *213*, 609–621
4. Wilcox, H. G., Keyes, W. G., Hale, T. A., Frank, R., Morgen, A. & Heimberg, M. (1982) Effects of triodothryonine and propylthiouracil on plasma lipoproteins in male rats. J. Lipid Res. *23*, 1159–1166
5. Breckenrigde, W. C., Little, J. A., Alaupovic, P., Wang, C. S., Kuksis, A., Kakis, G., Lindgren, F. & Gardner, G. (1982) Lipoprotein abnormalities associated with a familial deficiency of hepatic lipase. Atherosclerosis *45*, 161–179
6. Goldberg, I. J., Le, N. A., Paterniti, J. R., Ginsberg, H. N., Lindgren, F. T. & Brown, W. V. (1982) Lipoprotein metabolism during acute inhibition of hepatic triglyceride lipase in the cynomolgus monkey. J. Clin. Invest. *70*, 1184–1192
7. Jansen, H., van Tol, A. & Hulsmann, W. C. (1980) On the metabolic function of heparin releasable liver lipase. Biochem. Biophys. Res. Comm. *92*, 53–59
8. Grosser, J., Schrecker, O. & Greten, H. (1981) Function of hepatic triglyceride lipase in lipoprotein metabolism. J. Lipid Res. *22*, 437–442
9. Kuusi, T., Saarinen, P. & Nikkila, E. A. (1980) Evidence for the role of hepatic endothelial lipase in the metabolism of plasma high density lipoprotein 2 in man. Artherosclerosis. *36*, 589–593
10. Nikkila, E. A. & Kekki, M. (1972) Plasma triglyceride metabolism in thyroid disease. J. Clin. Invest. *51*, 2103–2114
11. Tulloch, B. R., Lewis, B. & Fraser, T. R. (1973) Triglyceride metabolism in thyroid disease. Lancet. *1*, 391–394
12. Abrams, J. J., Grundy, S. M. & Ginsberg, H. (1981) Metabolism of plasma triglycerides in hypothyroidism and hyperthyroidism in man. J. Lipid Res. *22*, 307–322
13. Chait, A., Bierman, E. & Albers, J. (1979) Regulatory role of triiodothyronine in the degradation of low density lipoprotein by cultured skin fibroblasts. J. Clin. Endo. Metab. *48*, 887–891
14. Walton, K. W., Scott, P. J., Dykes, P. W. & Davies, J. W. L. (1965) The significance of alterations in serum lipids in thyroid dysfunction II. Alternations of the metabolism and turnover of ^{131}I-low density lipo-proteins in hypothyroidism and thyrotoxicosis. Clin. Sci. *29*, 217–238
15. Thompson, G. R., Soutar, A. K., Spengel, F. A., Jadhav, A., Gavigan, S. J. P. & Myant, M. B. (1981) Defects of receptor-mediated low density lipoprotein catabolism in homozygous familial hypercholesterolemia and hypothyroidism in vivo. Proc. Natl. Acad. Sci. USA *78*, 2591–2595
16. Rossner, S. & Rosenquist, U. (1974) Serum lipoproteins and the intraveneous fat tolerance test in hypothyroid patients before and during substitution therapy. Athero. *20*, 365–381
17. Lithell, H., Boberg, J., Hellsing, K., Ljunghall, S., Lundquist, G., Vessby, B. & Wide, L. (1981) Serum lipoprotein and apolipoprotein concentrations and tissue lipoprotein-lipase activity in overt and subclinical hypothyroidism: the effect of substitution therapy. Eur. J. Clin. Invest. *11*, 3–10
18. Valdemarsson, S., Hedner, P. & Nilsson-Ehle, P. (1982) Reversal of decreased hepatic lipase and lipoprotein lipase after treatment of hypothyroidism. Euro. J. Clin. Invest. *12*, 423–428
19. Wahlquist, M. L., Fidge, H. H. & Lomas, F. (1977) Lipoprotein composition in hy-

pothyroidism. Clin. Chem. Acta *77*, 269–274

20. Ballantyne, F. C., Epenetos, A. A., Caslake, M., Forsythe, S. & Ballantyne, D. (1979) The composition of low-density lipoprotein and very-low density lipoprotein subfractions in primary hypothyroidism and the effect of hormone-replacement therapy. Clin. Sci. *57*, 83–88
21. Avivam, M., Luboshitzky, R. & Brook, J. G. (1982) Lipid and lipoprotein pattern in thyroid dysfunction and the effect of therapy. Clin. Biochem. *15*, 62–66
22. Dory, L. & Roheim, P. S. (1981) Rat plasma lipoproteins and apolipoproteins in experimental hypothyroidism. J. Lipid Res. *22*, 287–296
23. Hazzard, W. & Bierman, E. L. (1972) Aggravation of broad beta disease (Type 3 Hyperlipoproteinemia) by hypothyroidism. Arch. Int. Med. *130*, 822–828
24. Porte, D., Jr., O'Hara, D. & Williams, R. H. (1966) The relationship between post heparin lipolytic activity and plasma triglyceride in myxedema. Metabolism *15*, 107–113
25. Krauss, R. M., Levy, R. I. & Fredrickson, D. S. (1974) Selective measurement of two lipase activities in postheparin plasma from normal subjects and patients with hyperlipoproteinemia. J. Clin. Invest. *54*, 1107–1124
26. Scottova, N., Wallinder, L. & Bergtsson, G. (1983) Activity of lipoprotein lipase in thyroidectomized rats. Biochim. Biophys. Acta *750*, 533–538
27. Pykalisto, O., Goldberg, A. P. & Brunzell, J. D. (1976) Reversal of decreased human adipose-tissue lipoprotein-lipase and hypertriglyceridemia after treatment of hypothyroidism. J. Clin. Endo. Metab. *43*, 591–600
28. Bastenie, P. A., Vanhaelst, L., Bonnyns, M., Neve, P. & Staquet, M. (1971) Preclinical hypothyroidism: A risk factor for coronary heart-disease. Lancet II, 203–204
29. Kurland, G. S., Lucas, J. L. & Freedberg, A. S. (1961) The metabolism of intravenously infused C^{14}-labeled cholesterol in euthyroidism and myxedema. J. Lab. Clin. Med. *57*, 574–585
30. Chobanian, A. V., Burrows, B. A. & Hollander, W. (1972) Body cholesterol metabolism in man. II. Measurement of the body cholesterol miscible pool and turnover rate. J. Clin. Invest. *41*, 1738–1744
31. Rabinowitz, J. L., Rodman, T. & Myerson, R. M. (1963) Effects of dextrothyroxine on metabolism of C^{14} labeled cholesterol and tripalmitin. JAMA *183*, 758–760
32. Furman, R. H., Palmer-Howard, R., Lakshi, K. & Norcia, L. N. (1961) The serum lipids and lipoproteins in normal and hyperlipidemic subjects as determined by preparative ultracentrifugation. Am. J. Clin. Nutr. *9*, 72–102
33. Cornwell, A. G., Kruger, F. A., Hamwi, G. J. & Brown, J. B. (1961) Studies on the characterization of human serum lipoproteins separated by ultracentrifugation in a density gradient. Am. J. Clin. Nutr. *9*, 24–40
34. Byers, S. O., Rosenman, R. H., Friedman, M. & Biggs, M. W. (1952) Rate of cholesterol synthesis in hypo- and hyperthyroid rats. J. Exp. Med. *96*m, 513–516
35. Rosenman, R. H., Byers, S. O. & Friedman, M. (1952) The mechanism responsible for the altered blood cholesterol content in deranged thyroid states. J. Clin. Endo. Metab. *12*, 1287–1299
36. Miettinnen, T. A. (1968) Mechanism of serum cholesterol reduction by thyroid hormons in hypothyroidism. J. Lab. Clin. Med. *71*, 537–547
37. Abrams, J. J. & Grundy, S. M. (1981) Cholesterol metabolism in hypothyroidism and hyperthyroidism in man. J. Lipid Res. *22*, 323–338
38. Muls, E., Rosseneu, M., Blaton, V., Lasaffre, E., Lamberiqts, G. & DeMoor, P. (1984) Serum lipids and apolipoproteins A-I, A-II and B in primary hypothyroidism before and during treatment. Euro. J. Clin. Invest. *14*, 12–15
39. Ballantyne, F. C., Epentos, Caslake, Forsythe, S. & Ballantyne, D. (1979) The composition of low-density lipoprotein and very-low density lipoprotein subfractions in primary hypothyroidism and the effect of hormone-replacement therapy. Clin. Sci. *57*, 83–88
40. Walton, K. W., Campbell, D. S. & Tonks, E. L. (1965) The significance of alterations in serum lipids in thyroid dysfunction. I. The relation between serum lipoproteins, carotenoids and vitamin A in hypothyroidism and thyrotoxicosis. Clin. Sci. *29*, 199–216
41. Raziel, A., Bosenznug, B., Botvinic, V., Beigel, I., Landau, B. & Blum, I. (1982) The influence of thyroid function on serum lipid profile. Athero. *41*, 321–326

42. Kutty, K. M., Bryant, D. G. & Farid, N. R. (1978) Serum lipids in hypothyroidism — a re-evaluation. J. Clin. Endo. Metab. *46*, 55—60
43. Nilsson, G., Nordlander, S. & Levin, K. (1976) Studies on subclinical hypothyroidism with special reference to the serum lipid pattern. Acta Med. Scand. *200*, 63—67
44. Evered, D. C., Ormston, B. J., Smith, P. A., Hall, R. & Bird, T. (1973) Grades of hypothyroidism. Br. Med. J. *1*, 657—662
45. Bastenie, P. A., Vanhaelst, L., Bonnyns, M., Neve, P. & Slaquet, M. (1971) Preclinical hypothyroidism: A risk factor for coronary heart-disease. Lancet II, 203—204
46. Fowler, P. B. S., Swale, J. & Andrews, H. (1970) Hypercholesterolemia in borderline hypothyroidism. Lancet II, 488—491
47. Koppers, L. E. & Palumbo, P. J. (1972) Lipid disturbances in endocrine disorder. Med. Clin. NA *56*, 1013—1020
48. Sachs, B. A., Wolfman, L. & Murthy, G. (1968) Lipid and clinical response to a new thyroid hormone combination. Am. J. Med. Sci. *256*, 232—238
49. Scottolini, A. G., Bhagavan, N. V., Oshiro, T. H. & Abe, S. Y. (1980) Serum high-density lipoprotein cholesterol concentrations in hypo- and hyperthyroidism. Clin. Chem. *26*, 584—587
50. Brook, J. G., Aviran, M., Luboshitsky, R. & Barzilai, D. (1981) High density lipoprotein and arteriosclerosis. Different patterns in hypothyroidism and type II a hyperlipoproteinemia. Isr. J. Med. Sci. *17*, 318—323
51. Valdemarsson, S., Hansson, P., Hedner, P. & Nilsson-Ehle, P. (1983) Relations between thyroid function, hepatic and lipoprotein lipase activities and plasma lipoprotein concentrations. Acta endocrinol. *104*, 50—56
52. Agdeppa, D., Macaron, C., Mallik, T. & Schnuda, N. D. (1979) Plasma high density lipoprotein cholesterol in thyroid disease. J. Clin. Endo. Metab. *49*, 726—729
53. (1980) Association of plasma high density lipoprotein cholesterol with clinical chemistry data: The LRC Program Prevalence Study. Circ. *62* (IV), 31—41
54. Sandhofer, F., Sailer, S. & Braunsteiner, W. (1966) Fettsäure- und Triglyceridumsatz bei Schilddrüsenüberfunktion. Klin. Wochenschr. *44*, 1389—1392
55. Valdemarsson, S., Hedner, P. & Nilsson-Ehle, P. (1984) Treatment of hyperthyroidism: Effects on hepatic lipase lipoprotein lipase, LCAT, and plasma lipoproteins. Scand. J. Clin. Lab. Invest. *44*, 183—189
56. Facly, J., Iqbal, S. J., Isles, R. E. & Baky, M. (1980) Total cholesterol/HDL cholesterol ratio in hyperthyroidism, hypothyroidism and subclinical hypothyroidism. Horm. Mct. Res. *12*, 560—561
57. Muls, E., Blaton, V., Rosseneu, M., Lasaffre, E., Lamberiqts, G. & DeMoor, P. (1982) Serum lipids and apolipoproteins AI, AII and B in hyperthyroidism before and after treatment. J. Clin. Endo. Metab. *55*, 459—464
58. Boberg, J., Dahlberg, P. A., Vessby, B. & Lithell, H. (1984) Serum lipoproteins and apolipoprotein concentrations in patients with hyperthyroidism and the effect of treatment with carbimazole. Acta Med. Scand. *215*, 453—459

Chapter 12
Ethanol and Dyslipoproteinemias

C. S. Lieber and J.-P. Pignon

1 Introduction

The interaction of ethanol with lipid metabolism is complex. When ethanol is present, it becomes a preferred fuel for the liver and displaces fat as a source of energy. This favors fat accumulation. In addition, the altered redox state secondary to the oxidation of ethanol promotes lipogenesis, for instance, through enhanced formation of acylglycerols. The depressed oxidative capacity of the mitochondria injured by chronic alcohol feeding also contributes to the development of the fatty liver. Accumulation of fat acts as a stimulus for the secretion of lipoproteins and the development of hyperlipemia. Hyperlipemia may also be facilitated by the proliferation of the endoplasmic reticulum after chronic ethanol consumption and the associated increase of enzymes involved in the production of triglycerides and lipoproteins. The propensity to enhance lipoprotein secretion is offset, at least in part, by a decrease in microtubules and an impairment of the secretory capacity of the liver. The level of blood lipids depends on the balance between these two opposite changes: At the early stages of alcohol abuse, when liver damage is still small, hyperlipemia will prevail, whereas the opposite occurs with severe liver injury. When hyperlipemia occurs, it involves all lipoprotein classes, including HDL. The latter have been incriminated in the lower incidence of coronary complications of moderate drinkers compared to teetotalers, but in fact, the subtype of HDL involved (HDL_3) differs from the major subtype associated with protection, namely HDL_2.

2 Interaction of the Metabolism of Ethanol and Lipids

Ethanol is oxidized almost exclusively in the liver by three pathways, each located in a different subcellular compartment of the hepatocyte: the alcohol dehydrogenase pathway of the cytosol, the microsomal ethanol oxidizing system (MEOS) in the endoplasmic reticulum, and catalase of the peroxisomes (1). The liver also plays a central role in the metabolism of lipids and therefore there are several interactions of ethanol with lipid metabolism. Most of the effects are due to the generation of NADH during cytoplasmic ethanol oxidation, or because of the interaction of ethanol with microsomal functions. Some of the effects are caused by a direct action of ethanol or its metabolites, acetaldehyde and acetate, and part of the effects are mediated by hormonal changes. Other changes observed depend upon the nutritional status of the subjects, the level of ethanol consumed, and the duration of ethanol intake. It has now been established that the effects observed following an acute dose of ethanol and those in subjects consuming ethanol for prolonged periods of time differ.

One difficulty in assessing the effect of ethanol on lipid metabolism is the fact that acetate-^{14}C, which is commonly used as a precursor for lipid studies, is heavily diluted by the unlabeled acetate that results from ethanol metabolism. Effects of ethanol on lipids are therefore easily overshadowed by isotopic dilution problems. This is one of the reasons why the literature is replete with conflicting assessments of the effects of ethanol on lipid metabolism. Despite the complexity of the possible mechanism involved, ethanol interference with lipid metabolism may lead not only to the well-known deposition of neutral lipids in the liver and some other tissues, but also to functional disturbances because of changes in the lipid environment of cellular membranes. Ethanol-induced lipid changes have been implicated in the pathogenesis of diseases such as pancreatitis, hemolytic anemia, cardiomyopathy and atherosclerosis.

2.1 Effects of excessive hepatic NADH generation

Fig. 1 shows an overview of ethanol effects on the major pathways of hepatic lipid metabolism. As shown in this figure, the oxidation of ethanol results in the transfer of reducing equivalents to NAD. The excess NADH produces changes in the flux of other substrates and alters the concentration of those metabolites that are dependent for reduction on the NADH/NAD ratio. This alteration is responsible for a number of metabolic abnormalities associated with alcohol abuse (2).

In the absence of ethanol, the reducing equivalents utilized in the respiratory chain are provided mainly by the oxidation of fatty acids. During the oxidation of ethanol, the activity of the citric acid cycle is depressed (3, 4). The mitochondria then use the hydrogen equivalents originating from ethanol rather than from oxidation through the citric acid cycle of two carbon fragments derived from fatty acids. Thus, fatty acids that normally serve as the main energy source of the liver are supplanted by ethanol. This decrease in fatty acid oxidation results in the deposition of dietary fat in the liver when available, or fatty acids derived from endogeneous synthesis in the absence of dietary fat (5–7).

The increased NADH to NAD ratio is also associated with a rise in the concentration of sn-glycerol-3-phosphate (8) which favors hepatic triacylglycerol accumulation by trapping fatty acids. Theoretically, lipogenesis at all levels (enhanced synthesis of fatty acid, triacylglycerols and cholesterol and/or its esters) can be considered as metabolic systems whereby excess reducing equivalents formed during the oxidation of ethanol can be utilized.

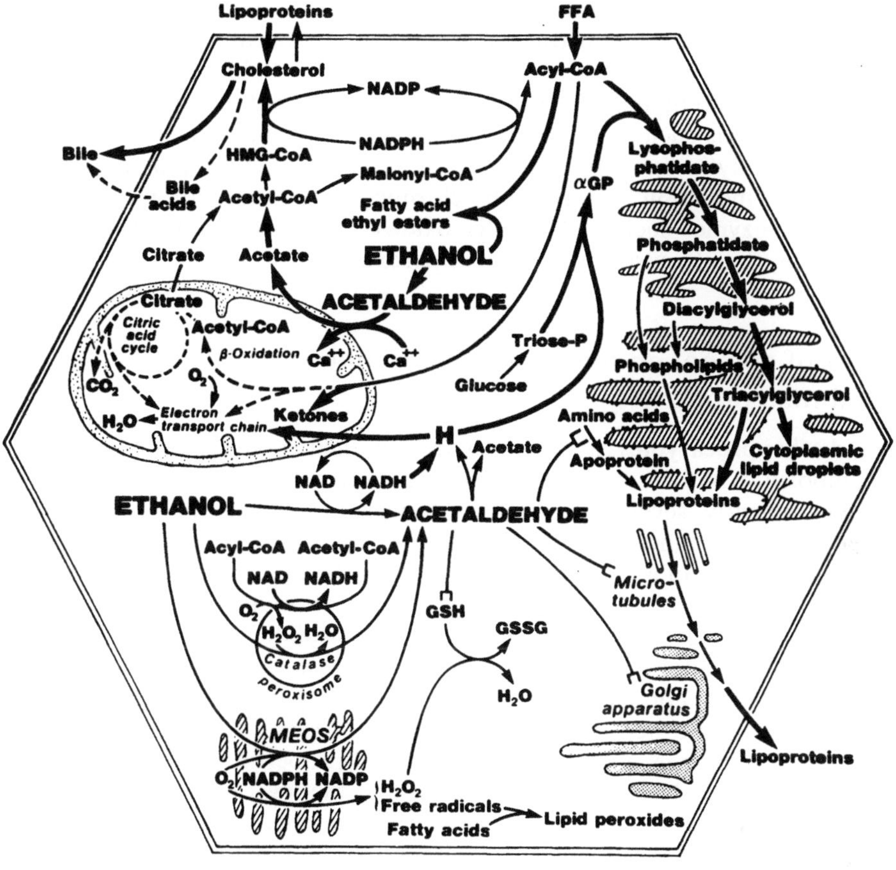

Fig. 1. Oxidation of ethanol in the hepatocyte and link of the redox change and the metabolite acetaldehyde to disturbances in lipid metabolism. Abbreviations: FFA, unesterified ("free") fatty acid; *a* GP, *sn*-glycerol-3-phosphate; triose-P, dihydroxyacetone phosphate or glyceraldehyde 3-phosphate; GSH, reduced glutathione; GSSG, oxidized glutathione; H, reducing equivalent; MEOS, microsomal ethanol-oxidizing system (344).

2.2 Interaction of ethanol with hepatic microsomal lipid metabolism

Most of the enzymes involved in the synthesis of complex lipids (such as triacylglycerols and phospholipids) and lipoproteins are bound to the membranes of the endoplasmic reticulum. Since ethanol is also metabolized by the microsomal enzyme system, it may interfere with lipid metabolism by mechanisms not related to changes in the cytoplasmic or mitochondrial redox shift. The exact mechanism of the ethanol-induced alterations of microsomal function has not been clarified. It is known, however, that chronic ethanol feeding results in proliferation of the membranes of the smooth endoplasmic reticulum, the morphological counterpart of the microsomal fraction obtained by ultracentrifugation. Therefore, the lipid synthesizing capacity of the liver may be enhanced partly by the increase in the mass of the smooth endoplasmic reticulum, although it has been demonstrated that both acute and chronic ethanol administration produce some specific alterations in the activities of enzymes involved in the synthesis of complex lipids and lipoproteins. Moreover, microsomal ethanol oxidation interferes with lipid metabolism by generating oxygen radicals such as O^{2-} and $HO^{\cdot}$ (9, 10) which in turn can initiate the cascade of autocatalytic lipoperoxidation reactions and damage cell membranes (11). Microsomal "induction" produced by chronic ethanol ingestion may increase the activities of enzymes such as NADPH oxidase (9, 12) resulting in enhanced H_2O_2 production, thereby also favoring lipid peroxidation.

2.3 Effect of ethanol on membrane lipids and related membrane properties

Bacteria and other procaryotes can adapt to drugs in their environment by changing the lipid composition of their membranes. This has been extrapolated to mammalian cells, and consequently, several studies have been made of ethanol effects on the lipid composition and functional properties of cellular membranes. It has been shown that the presence of a physiological concentration of ethanol (20–80 mmol/l) increases the fluidity of the plasma membranes of mouse erythrocytes and synaptosomes *in vitro* (13–15). Moreover, membranes isolated from animals chronically treated with ethanol were more resistant to the *in vitro* fluidizing effect of ethanol than membranes prepared from their respective pair-fed controls (14, 15). These results were confirmed in men by a study of erythrocyte membranes in alcoholics (16). The changes in the physico-chemical properties of the membranes have been linked to the development of tolerance at the level of the intact animal and to the disturbed metabolic function of cell organelles at the cellular level. It has been proposed that the changes in mitochondrial function produced as a consequence of prolonged ethanol consumption may be directly related to alteration in the properties of cellular membranes (17, 18). More recent data indicate that no such simple correlation explains the functional changes observed in the mitochondria of ethanol-fed animals (19). It is possible, of course, that other functional changes may be related to some alteration in membrane fluidity. Contrasting with other membranes, chronic ethanol feeding resulted in an increase in hepatic plasma membrane fluidity as assessed by fluorescence anisotropy (20). Other studies have confirmed these results (21–23). Such changes may relate to the specialized role of the liver in bile excretion, lipoprotein metabolism, and ethanol oxidation.

Chronic ethanol feeding alters the lipid composition of cellular membranes. The more constant changes in lipid composition are the increase in cholesterol, the decrease in phospholipids or the increase in the total cholesterol/phospholipid ratio, and the decrease of the arachidonate/linoleate ratio in the phospholipids of extraneural membranes (14, 24–29). Differences in experimental conditions, especially dietary factors, could explain the discrepancies between some studies. In liver plasma membrane, the total or the free cholesterol were found to be decreased (20–21) or normal, with increased cholesteryl esters (22). Arai et al. have shown also a decrease in membrane mitochondrial phospholipids from baboon liver associated with increased phospholipase A_2 activity (27). Some inhibitory effects by alcohol on Δ^5, Δ^6, Δ^9 acyl CoA desaturase activities have been reported in rat liver (30–32); these could explain the relative decrease in unsaturated fatty acyl chains. It is not clear whether these rather small changes are associated with a biologically significant change in the physical state of the membranes (33). The role of phospholipid composition (29) and cholesteryl esters (22, 34) in membrane properties and their modifications by ethanol have been pointed out. Other changes may be produced by the incorporation of ethanol *per se* into membrane lipids. Both *in vitro* and *in vivo*, the presence of ethanol leads to formation of abnormal lipids by incorporation of ethanol into fatty acids or phospholipids via an ester or ether bond (35–38). The same mechanism may be involved in both processes: it has been demonstrated that ethanol can replace water molecules in the reactions catalyzed by cholesterol esterase (39) and phospholipase D (40). In the former reaction ethyl esters of fatty acids are formed, whereas in the latter the product is an ethylated derivative of phosphatidic acid. These compounds do not exist under normal physiological conditions in the absence of ethanol. Especially, the demonstration of the abnormal phospholipid in the various organs of the rat after ethanol administration, if confirmed by other laboratories, may help elucidate the mechanism of ethanol intoxication, because this abnormal lipid occurs in the membranes of the brain only when ethanol is present in the blood (35).

2.4 Effect of acetaldehyde and acetate on lipid metabolism

Acetaldehyde can depress the capacity of liver mitochondria to oxidize fatty acids (41), particularly in organelles derived from alcohol-fed animals (42). The concentrations of acetaldehyde, although relatively high, do probably occur in some circumstances in the liver *in vivo* (43), at least in subcompartments. The impairment of mitochondrial function together with NADH produced by the oxidation of acetaldehyde results in an additional load of reducing equivalents. Furthermore, acetaldehyde reduces the capacity of the liver to export proteins (44, 45), possibly through interference with microtubules; this effect may be associated with a diminution of lipoprotein secretion.

A minor pathway for acetaldehyde oxidation are the reactions catalyzed by aldehyde oxidase and xanthine oxidase (10, 46). Although not quantitatively important in the catabolism of acetaldehyde, these reactions generate oxygen radicals (O^{2-}) which in turn may promote lipid peroxidation (47, 48). Indeed, acetaldehyde has been shown to promote lipid peroxidation in isolated hepatocytes (49). However, the affinity of acetaldehyde for xanthine oxidase is low, and the role of peroxidation in the development of ethanol-induced liver injury is still controversial (50–52).

Acetaldehyde also could promote lipid peroxidation by forming a hemiacetal complex with L-cysteine. Such a complex, L-2-methyl-thiazolidine-4-carboxylic acid, may be a non-toxic detoxification product whose formation protects against acetaldehyde toxicity (53). However, binding of acetaldehyde with cysteine or glutathione (a tripeptide glu-cys-gly) decreases liver glutathione (GSH) content (54), thereby reducing the capacity of scavenging free radicals. It has been shown that a severe reduction in GSH favors peroxidation (55), and the damage may possibly be compounded by the increased generation of active radicals by the "induced"

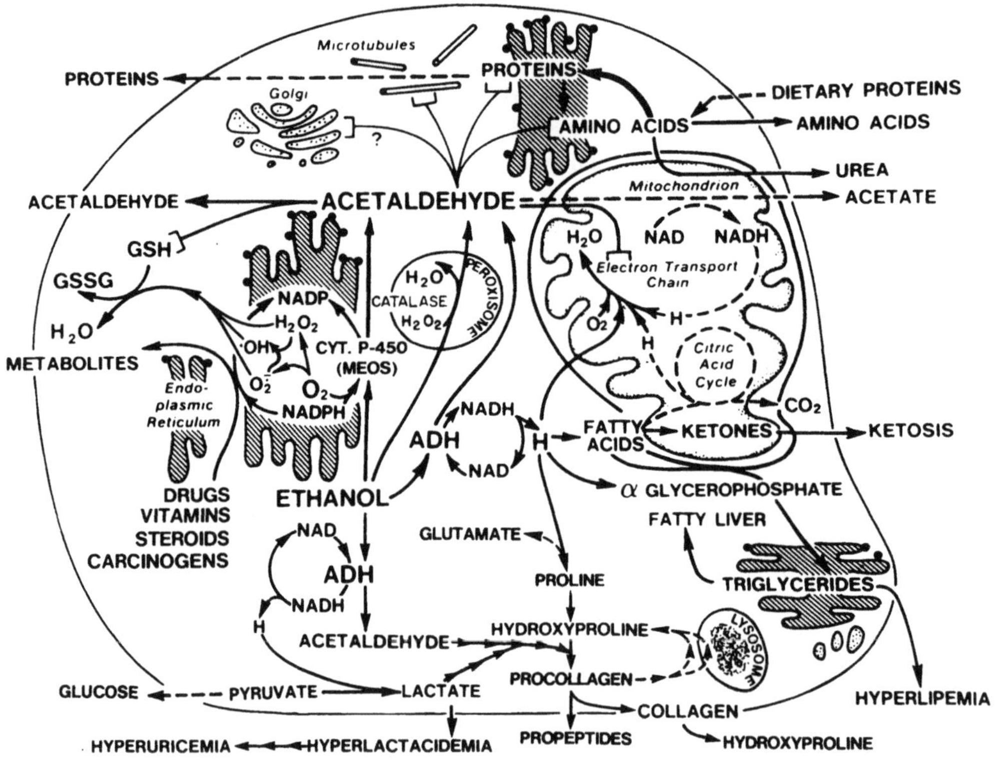

Fig. 2. Oxidation of ethanol in the hepatocyte and link of the two metabolites (acetaldehyde and H) to disturbances in intermediary metabolism, including abnormalities of amino acid and protein metabolism. Abbreviations: NAD, nicotinamide adenine dinucleotide; NADH, reduced NAD; NADP, nicotinamide adenine dinucleotide phosphate; NADPH, reduced NADP; MEOS, microsomal ethanol oxidizing system; ADH, alcohol dehydrogenase. The broken lines indicate pathways that are depressed by ethanol; repeated arrows indicate stimulation.

microsomes following chronic ethanol consumption (as illustrated in Fig. 2). In naive rats, a very large acute dose of ethanol (5 – 6 g/kg) is required to produce lipid peroxidation (48), whereas a smaller dose (3 g/kg) had no effect (56). By contrast, after chronic ethanol administration to the rat, even the smaller dose of ethanol, administered acutely, induced liver peroxidation, and this effect could be prevented, at least in part, by the administration of methionine, a precursor of GSH (56). Moreover, ethanol ingestion enhanced hepatic lipoperoxidation as assessed by the increased spontaneous chemiluminescence of rat liver (57). The involvement of lipid peroxidation in induced cell damage was supported by the absence of evidence in favor of decreased GSH synthesis since ^{35}S-L-methionine incorporation into hepatic and biliary GSH was unchanged or even increased after chronic ethanol feeding (58). The ethanol-induced lipid peroxidation was even more striking in the baboon: Administration of relatively small doses of ethanol (1 – 2 g/kg) after five to six hours produced lipid peroxidation and depletion of GSH. Baboons with alcoholic liver disease (including cirrhosis in some), when chronically fed alcohol, exhibit evidence of enhanced hepatic lipid peroxidation and GSH depletion. These changes were observed following an overnight withdrawal from ethanol and were exacerbated by the readministration of ethanol. Furthermore, supportive evidence for lipid peroxidation has now been produced by two studies in man (54, 59).

It is tempting to speculate that the propensity of primates to develop more severe lesions than the rat after chronic ethanol consumption may in some ways be related, at least in part, to the greater susceptibility to GSH depletion and the initiation of lipid peroxidation. It is apparent, however, that GSH depletion *per se* does not suffice to produce liver damage (60).

2.5 Main sites of the ethanol-induced disturbances in lipid metabolism

The most common disturbance of lipid metabolism produced by ethanol ingestion is fatty liver, characterized by accumulation in the hepatocytes of triacylglycerols, as cytoplasmic lipid droplets. Fatty liver develops within a few hours after the ingestion of ethanol and is dose-dependent (61–63). The deposition of triacylglycerols develops progressively over the first month of alcohol administration and persists if the ingestion of ethanol continues, at least for one year in the rat (64) and for many years in the baboon (65–67). Fatty liver may be also prominent in patients with alcoholic hepatitis or cirrhosis (1). Eventually other structural damage, such as alterations of mitochondria may develop (68, 69). In rats, cholesteryl esters (70, 71) and fatty acyl ethyl esters (38) also accumulate in the liver.

Lipids accumulate also in other tissues such as the heart (36, 72–74), lungs (75), pancreas (76) and in the blood (71, 73), partly because of the ethanol-induced derangement of hepatic lipid metabolism and also because of a direct action of ethanol and its oxidation on lipid metabolism of the peripheral tissues.

3 Mechanisms of Hepatic Fat Accumulation Induced by Ethanol

3.1 Etiologic role of ethanol

It is now well established that ethanol *per se* is responsible for the development of the alcoholic fatty liver. With nutritionally adequate diets, isocaloric replacement of sucrose or other carbohydrate by ethanol consistently produces a 5 to 10 fold increase in hepatic triglycerides whereas isocaloric replacement of carbohydrate by fat instead of ethanol does not produce steatosis (71, 77, 78). Moreover ethanol as well as insulin and dexamethasone were shown to promote accumulation of triacylglycerol in cultured hepatocytes (79). The low dose of ethanol administered is probably the reason for conflicting results obtained in earlier studies (80).

Moreover, individuals with a morphologically normal liver (with or without a history of alcoholism) developed a fatty liver when given ethanol in a variety of non-deficient diets as an isocaloric substitute for carbohydrates (71, 77). Even with a high protein, vitamin-supplemented diet, there was a significant increase in hepatic triglycerides, as measured in percutaneous biopsies. These studies therefore established the fact that even in the absence of dietary deficiency ethanol can produce a fatty liver.

3.2 The influence of dietary factors

3.2.1 **Role of dietary fat**

Reduction in dietary fat to a level of 25 percent (or less) of total calories is accompanied by a significant decrease in the steatosis induced by ethanol (64). The importance of dietary fat was confirmed in volunteers. For a given alcohol intake, much more steatosis developed with diets of normal fat content than with low-fat diets (6). In addition to the amount, the chain length of the dietary fatty acid is also important for the degree of fat deposition in the liver. Replacement of dietary triglycerides containing long-chain fatty acids (LCT) by fat containing medium-chain fatty acids (MCT) reduces the capacity of alcohol to produce a fatty liver in rats (81). The propensity of medium-chain fatty acids to undergo oxidation rather than esterification probably explains this phenomenon (4). A partial preventive effect of arachidonate (82) and prostaglandin PGE_1 (83) on hepatic fat accumulation have been also reported but could not be verified (84). Dietary supplementation with cholesterol increased the alcohol-induced lipid changes (not only cholesterol changes) in liver and blood (70, 85), whereas decreased absorption of cholesterol due to supplementation with dietary fiber produced the opposite effect (86).

3.2.2 **Role of protein and lipotropic factors (choline and methionine)**

In growing rats, deficiencies in dietary protein and lipotropic factors (choline and methionine) can produce a fatty liver (80), whereas primates are far less susceptible to protein and lipotropic deficiency than rodents (87). Clinically, treatment with choline of patients suffering from alcoholic liver injury has been found to be ineffective in the face of continued alcohol abuse (88–90) and, experimentally, massive supplementation with choline failed to prevent the fatty liver produced by alcohol in volunteer subjects (91). This is not surprising, since unlike rat liver, human liver contains very little choline oxidase activity, which may explain the species difference with regard to choline deficiency. There are some key differences between the fatty liver produced by ethanol and that caused by choline deficiency suggesting that choline deficiency is not directly involved in the pathogenesis of the ethanol-induced fatty liver. The two types of fatty liver differ both ultrastructurally and biochemically as described in detail in previous reviews (1).

More recent data indicate that in baboons given dietary choline at twice the recommended amount, chronic alcohol consumption resulted in the production not only of fatty liver but also of fibrosis and cirrhosis, despite additional supplementation with methionine in the diet (92). In other subhuman primates (93), moderate fatty liver but no fibrosis developed with ethanol feeding in the presence of choline supplementation. The authors concluded that choline may have been protective, but it must be pointed out that they did not carry out the crucial experiment, namely the demonstration that feeding of diets with low amounts of choline and ethanol resulted in significantly more severe liver lesions. Such experiments were done in rhesus monkeys by Rogers et al. (94), who concluded that lipotrope deficiency in the baboon model did not account for the development of cirrhosis. The difference in response between baboons and other primates studied is probably due to greater alcohol intake (relative to the size of the animal) although differences in species susceptibility can, of course, not be excluded. Thus, hepatic injury induced by choline deficiency appears to be primarily an experimental disease of rats with little, if any, relevance to human alcoholic liver injury. Even in rats, massive choline supplementation failed to fully prevent this ethanol-induced lesion, whether alcohol was administered acutely (95) or chronically (81). In the baboon, whereas a 10 fold increase in dietary choline failed to prevent alcohol induced fibrosis, signs of toxicity due to choline appeared (96).

The effect of protein deficiency has not yet been clearly delineated in human adults. In children, protein deficiency leads to hepatic steatosis, one of the manifestations of kwashiorkor. In adolescent baboons, protein restriction to 7 percent of total calories did not result either by biochemical analysis or by light and electron microscopic examination in conspicuous liver injury even after 19 months. Significant steatosis was observed only when the protein intake was reduced to 4 percent of total energy (65). On the other hand, an excess of protein (25 percent of total energy, or 2.5 times the recommended amount) did not prevent ethanol from producing fat accumulation in human volunteers (97). Thus, in man, ethanol is capable of producing striking changes in liver lipids even in the presence of a protein-enriched diet. When protein deficiency is present, it may potentiate the effect of ethanol. In the rat, a combination of ethanol and a diet deficient in both protein and lipotropic factors leads to a more pronounced hepatic steatosis than either alone (98, 99). This potentiation is not unexpected, since as discussed subsequently, increased lipoprotein secretion offsets, in part, the alcohol-induced steatosis. Protein or choline deficiency (or both) result in impaired lipoprotein secretion, which can be expected to potentiate hepatic lipid accumulation secondary to alcohol. In addition to its role as lipotrope, methionine may also act as a cysteine and glutathione precursor, thereby affecting lipid peroxidation.

3.3 Origin of fatty acids deposited in alcoholic fatty liver and mechanism of accumulation

Lipids that accumulate in the liver can originate from three main sources: dietary lipids (which reach the bloodstream as chylomicrons); adipose tissue lipids (which are transported to the liver as free fatty acids: FFA); and lipids synthesized in the liver itself. These fatty acids can accumulate in the liver because of a variety of

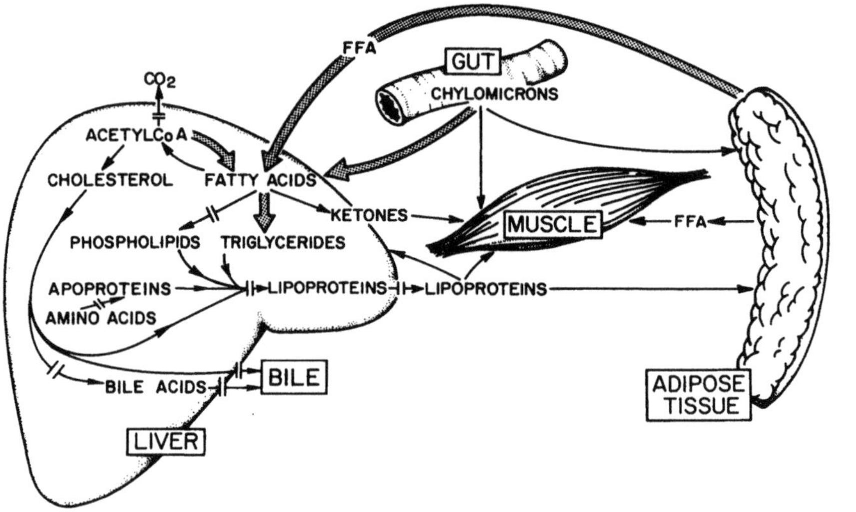

Fig. 3. Possible mechanisms of fatty liver production through either increase (→) or decrease (–||→) of lipid transport and metabolism. See text for explanation (1).

metabolic disturbances. Five major mechanisms may be involved in the accumulation of lipids in the liver, namely (1) increased availability of precursors for hepatic lipid synthesis; (2) increased hepatic lipogenesis because of ethanol-induced stimulation of enzymes; (3) decreased lipid breakdown; (4) decreased hepatic secretion of lipids; and (5) enhanced hepatic uptake of circulating lipids (Fig. 3). Depending on the experimental conditions, any of the three sources and the various mechanisms can be implicated.

3.3.1 **Increased availability of substrates for hepatic lipid biosynthesis**

Ethanol interferes in many ways with the supply of precursors of lipogenesis from both hepatic and extrahepatic sources. First ethanol oxidation provides reducing equivalents and C_2 units for lipid synthesis. Secondly, the reduced redox state inhibits the oxidation of fatty acids and diverts them into esterification which is further enhanced by the increased concentration of *sn*-glycerol-3-phosphate. Thirdly, ethanol affects the amount of fatty acids transported from the adipose tissue and the small intestine into the liver. In the following paragraphs these effects of ethanol are described in detail.

(1) *Increased mobilization of peripheral fat.* In rats given one large, sublethal dose of ethanol, fatty acids resembling those of adipose tissue accumulate in the liver (100, 101). Experimental procedures or agents that reduce the normal rate of peripheral fat mobilization (that is adrenalectomy, spinal cord transection, or ganglioplegic drugs) prevent or decrease this type of hepatic fat accumulation (102, 103). More direct approaches, however, such as studies in rats with prelabeled epididymal fat pads yielded conflicting information, with evidence for increased

(104) or unchanged (105) fatty acid mobilization. Similarly, one single dose of ethanol has been reported to result in increased (99, 106), unchanged (107), or decreased (108) circulating levels of FFA. Since ethanol probably both stimulates fatty acid mobilization (via catecholamine release) and depresses it (via the acetaldehyde and acetate produced), the net effect may depend on the particular experimental conditions, including the dose of ethanol and the nutritional status of the subject. This may account for some of the apparent contradictions of the literature.

(2) *Decreased hepatic fatty acid oxidation.* Decreased fatty acid oxidation caused by ethanol has been demonstrated in liver slices (109, 110), perfused liver (4), isolated hepatocytes (111) and *in vivo* (112). This decrease in fatty acid oxidation results in the deposition of dietary fat in the liver when available, or fatty acids derived from endogenous synthesis in the absence of dietary fat (5 – 7, 101). The inhibition can be attributed to the redox change and the decrease in the activity of the citric acid cycle. After chronic ethanol consumption, the acute inhibition of fatty acid oxidation was less pronounced, in keeping with the attenuation of the redox change (67). An exaggeration of this change takes place in the perivenular zones of the liver, where a greater redox shift is favored by low oxygen tensions prevailing in that zone (113, 114). These oxygen tensions become even lower with progression of the liver injury due to impairment of the sinusoidal circulation. Structural changes of the mitochondria (68, 69) could also be responsible for some alterations in lipid metabolism beyond those produced by the altered redox state. The mitochondria showed decreased oxidation of two-carbon fragments of fatty acids (41, 42, 109, 115, 116) but no changes in β-oxidation (117). In any event, decreased fatty acid oxidation, whether a function of the reduced citric acid cycle activity (secondary to the altered redox potential) or of the permanent changes in mitochondrial structure, offers the most likely explanation for the deposition of fat in the liver, at least during the initial stage of chronic alcohol abuse.

(3) *Increased hepatic fatty acid synthesis.* Earlier *in vitro* studies in rats (109) and in man (118) have demonstrated that, in the presence of ethanol, hepatic fatty acid synthesis is increased. Subsequent studies have revealed a more complex picture. In *in vitro* experiments in naive rats following an acute dose of ethanol no evidence of enhanced fatty acid synthesis was observed (119). In animals consuming ethanol for prolonged periods of time, conflicting results have been obtained too. An increase (73, 120) in hepatic fatty acid synthesis, or no change (121) was observed. A recent study has shown a decrease of fatty acid synthetase activity after chronic consumption (32).

It should be noted, however, that the assessment of the effect of ethanol on hepatic fatty acid synthesis is extremely difficult, mainly for two reasons. First, ethanol by its metabolism causes significant isotope dilution not only when ^{14}C-acetate incorporation into fatty acids is measured, but also in the case of tritium incorporation from ^3H_2O into fatty acids, due to a decrease in the specific activity of the NADPH (122). Secondly, since the fatty acid synthesis rate *in vivo* is very sensitive to the amount of both carbohydrate and fat in the diet, it is very difficult to add a large amount of ethanol into the diet without drastically changing the carbohydrate/fat ratio (121). Although ethanol may not affect directly the rate of fatty acid synthesis significantly, it interferes with intermediary metabolism by providing a large part of the C_2 units for lipogenesis (123).

(4) *Increased supply of sn-glycerol-3-phosphate.* It has been proposed that the partition of fatty acids between oxidation and esterification is determined on the outer membrane of mitochondria by the relative activities of carnitine acyltransferase and *sn*-glycerol-3-phosphate acyltransferase. Therefore, an increase in the concentration of *sn*-glycerol-3-phosphate because of the ethanol-induced redox shift would favor the esterification of fatty acids. Indeed, it has been shown that a correlation exists between the hepatic *sn*-glycerol-3-P concentration and triacylglycerol accumulation both *in vivo* (8) and in perfused livers (124). Both of these phenomena are prevented by thyroxine (125) and clofibrate treatment (106) or by pyrazole or its derivatives (126) suggesting that the concentration of *sn*-glycerol-3-phosphate is an important factor in the pathogenesis of ethanol-induced fatty liver. However, opposite results were also obtained (127, 128). The conflicting results are probably due to differences in the *sn*-glycerol-3-phosphate levels in the various experimental conditions used. Because the *sn*-glycerol-3-phosphate concentration limits the rate of esterification only within a narrow range (129) and because there are probably several pools of *sn*-glycerol-3-phosphate in the hepatocyte, the measurement of the total concentration does not reveal the regulatory role of the pool of *sn*-glycerol-3-phosphate involved in glycerolipid synthesis. Recently decreased steatosis after dietary supplementation with glycerol or pyruvate (130) and carnitine (130) have been reported.

A clinical study, using some markers of lipolysis, showed an increase of hepatic free fatty acids with progression of alcoholic liver injury (131): cytosolic free fatty acids are potentially toxic (132); this accumulation could play a primary role, but it also could be secondary to the liver injury. Usually the free fatty acids are bound to the fatty acyl binding protein (133) and an increase by alcohol of the amount of this protein has been described recently (134). Increased formation of fatty acyl esters tends to counteract the accumulation of free fatty acids.

3.3.2 **Increased activity of the enzymes of hepatic lipid synthesis**

In addition to increasing the supply of precursors, ethanol may enhance the rate of triglyceride synthesis by stimulating the activities of the enzymes catalyzing this event. These enzymes are bound to the membranes of the endoplasmic reticulum, with the exception of glycerophosphate acyltransferase and phosphatidate phosphohydrolase which are in part found also in the outer membrane of mitochondria and in the cytoplasm, respectively. Phosphatidate phosphohydrolase fulfills almost all criteria of a rate-limiting enzyme and its activity changes in most situations parallel the rate of triglyceride synthesis (135). Both acute and chronic ethanol administration stimulate phosphatidate phosphohydrolase in the microsomal and soluble fractions (63, 136—138), and the increase in the enzyme activity precedes the accumulation of hepatic triglycerides (136). The ethanol-induced stimulation of the soluble phosphatidate phosphohydrolase was shown to be partly due to an increase in the amount of the enzyme, but partly also due to activation of the existing phosphohydrolase (139). This dual action may be caused by the mediation of the ethanol effect by two separate mechanisms: the former by glucocorticoids (140, 141) and the latter by low molecular mass molecules, probably by *sn*-glycerol-3-phosphate (142).

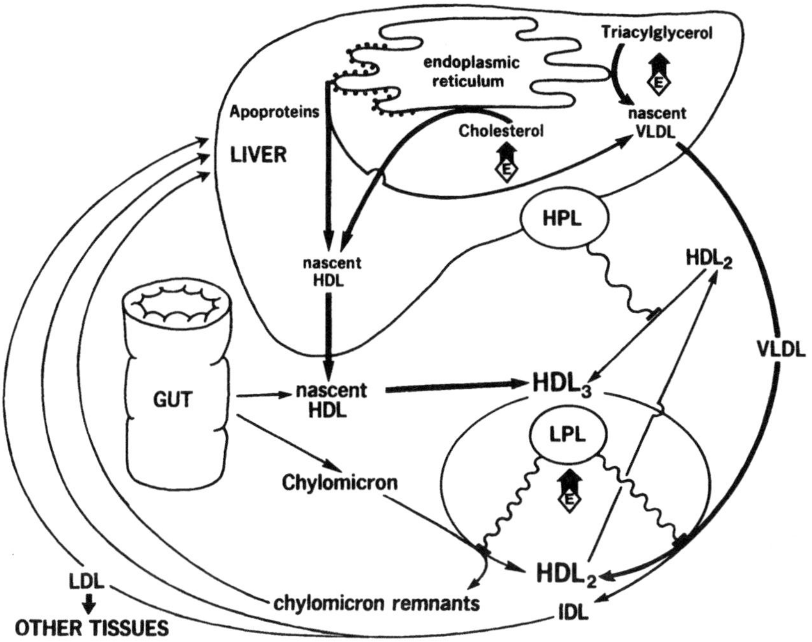

Fig. 4. Mechanism of increased VLDL, HDL_2 and HDL_3 after ethanol. An increased hepatic production of VLDL by the liver will result from enhanced synthetic enzyme activities and supply of precursors. An increased production of HDL by the liver will produce an increased HDL_3 concentration, observed in association with moderate ethanol consumption. The catabolism of VLDL and chylomicrons by extrahepatic lipoprotein lipase is associated with transformation of HDL_3 to HDL_2. An increased production of VLDL by the liver and an increased extrahepatic lipoprotein lipase activity will result in an increased HDL_2 concentration, observed in association with high alcohol intake. LPL: Extrahepatic lipoprotein lipase. HPL, hepatic lipoprotein lipase; ⇑E, stimulation by ethanol; ∿∿∿, enzymatic mediated transformation (346).

The first enzyme of hepatic glycerolipid synthesis, glycerophosphate acyltransferase, is affected only by chronic ethanol ingestion (143). Diacylglycerol transferase, the last enzyme of the triacylglycerol synthesis pathway, is the only one specific for the formation of triacylglycerols; the others share a common pathway with phospholipid synthesis. Therefore, diacylglycerol acyltransferase would be a plausible site for the ethanol effect, since there is a much greater increase in the triacylglycerols than in the phospholipids. Previous studies showed no change after an acute dose of ethanol (137). More recently, a 2-fold increase in the diacylglycerol acyltransferase activity was observed when the cytosolic fraction was included in the assay of the microsomal enzyme (144). In baboons, at the early stages of fatty liver, there were increased microsomal diacylglycerol acyltransferase and cytosolic phosphatidate phosphohydrolase activities, which disappeared with progression of the liver injury (145).

In keeping with the increased phospholipid content of the liver after chronic ethanol administration (77), increased activity of two enzymes involved in the synthesis of phosphatidyl choline, namely choline phosphotransferase and phosphatidyl ethanolamine methyltransferase, was reported (146).

In alcohol fed rats, cholesteryl esters accumulate in the liver (70, 147), and acyl coenzyme A:cholesterol acyltransferase and 3-hydroxy-3-methylglutaryl-coenzyme A (HMG-CoA) reductase activities increase (147). The latter enzyme was found unchanged in activity in rats given alcohol in drinking water (148). Moreover, an incomplete feedback inhibition of HMG-CoA reductase activity by chylomicron remnants from alcohol-fed rats was described (149).

This increased activity of enzymes of hepatic lipid synthesis with the increased supply of free fatty acid could explain the hepatic lipoprotein overproduction associated with alcoholic hyperlipemia (Fig. 4).

3.3.3 Decreased lipid breakdown

The accumulation of esterified lipids in the fatty liver could result from an impaired capacity of lysosomal lipases and esterases to hydrolyze the glycerol esters. However, chronic ethanol administration increases the appearance of lysosomes (150, 151), and the activity of lysosomal enzymes is either increased or unchanged (151, 152). Acutely, ethanol has no uniform effect on these enzymes (153, 154), but it increases the activity of both microsomal and cytosolic phospholipase A (155). Hydrolysis of cholesterol ester was found to be reduced after chronic ethanol consumption (156). Chronic feeding increases the phospholipase A_2 activity in baboons (27). A systematic study of the effects of ethanol on lipases and esterases of other subcellular fractions is presently lacking.

3.3.4 Decreased secretion of lipids from the liver

Lipids may leave the hepatocyte either as lipoproteins secreted into the bloodstream or as constituents of bile. Ethanol increases the output of triacylglycerol-carrying lipoproteins from the liver (see below). But this compensatory mechanism of fat accumulation is relatively inefficient, probably because of impairment of protein secretion secondary to cytoskeletal lesions (44, 45).

For the fate of cholesterol synthesized in the hepatocyte or taken up from the circulation, the secretion into bile (with or without conversion into bile acids) is as important as the secretion into the blood as lipoproteins. Any disturbance in these functions leads to imbalance in the cholesterol content of the liver or other tissues, because the carbon skeleton of cholesterol is not catabolized in the body, but is eliminated almost exclusively by excretion through the bile into the feces. In the rat, ethanol increases cholesterogenesis in the liver (70) and in the small intestine. Furthermore, ethanol feeding decreases degradation of cholesterol to bile acids and excretion of bile acid in feces (70) although after cessation of ethanol administration, the opposite effect was observed in the bile (157). In the presence of alcohol however, biliary secretion of bile acids was depressed (158, 159) depending on the experimental conditions. The reduction of bile acid excretion could be secondary to diminution in cholesterol 7α-hydroxylase activity (148). Contrasting with the rat, the baboon does not accumulate cholesteryl ester in the liver after chronic ethanol consumption (160), possibly due to increased fecal cholesterol elimination (161).

4 Alcoholic Hyperlipemia

Alcohol consumption is associated with changes in all lipid components of the plasma. Hypercholesterolemia was first reported by Ducceschi in 1915 (162) and recently has stimulated great interest because of a possible link between moderate alcohol consumption and the prevention of coronary heart disease. Serum lactescence and hypertriglyceridemia after bouts of excessive drinking were first reported by Feigl in 1918 (163).

4.1 Description of lipid perturbations induced by alcohol

4.1.1 **Plasma triacylglycerols**

The plasma triacylglycerols are the major determinants of the serum lactescence or turbidity found after bouts of excessive drinking. They increase in all lipoprotein fractions, but proportionally more in those lipoproteins which are normally rich in triacylglycerols: VLDL, which migrate as pre-β-lipoproteins on electrophoresis, and chylomicrons or chylomicron-like particles, which stay at the origin of the electrophoretic strip (160). Thus alcoholic hyperlipemia is usually classified as type IV or type V. However, the phenotypic pattern changes rapidly after alcohol withdrawal, from type V to type IV and to type II, because of the rapid clearance of chylomicrons, followed by VLDL and the slower clearance of cholesterol and phospholipids which predominate in LDL or β-lipoproteins. This striking change in the same subject that occurs with time, is illustrated in Figure 5.

The incidence of hypertriglyceridemia varies with the population studied. Among hyperlipemic patients, alcohol constitutes the second major cause, following diabetes (164). In epidemiolologic studies, plasma triacylglycerol levels correlate with alcohol consumption (165, 166). But significant increases in plasma triacylglycerols are rather rarely encountered in patients hospitalized for alcoholism or its complication. Only 27%–28% of hospitalized alcoholics had fasting triacylglycerol levels over normal limits (2 mmol/l) and 17% over 3 mmol/l, with phenotypes IV, II, or V, in that order of frequency (167, 168). The withdrawal state favors hypertriglyceridemia by the reversal of the inhibition of extrahepatic lipoprotein lipase (169). This hypertriglyceridemia is frequently seen in patients with fatty livers but rarely among cirrhotics (170–172). A type III hyperlipoproteinemia can occur in severe liver disease (173). Thus, it is apparent that classification of the hyperlipemia in alcoholics is fraught with difficulties.

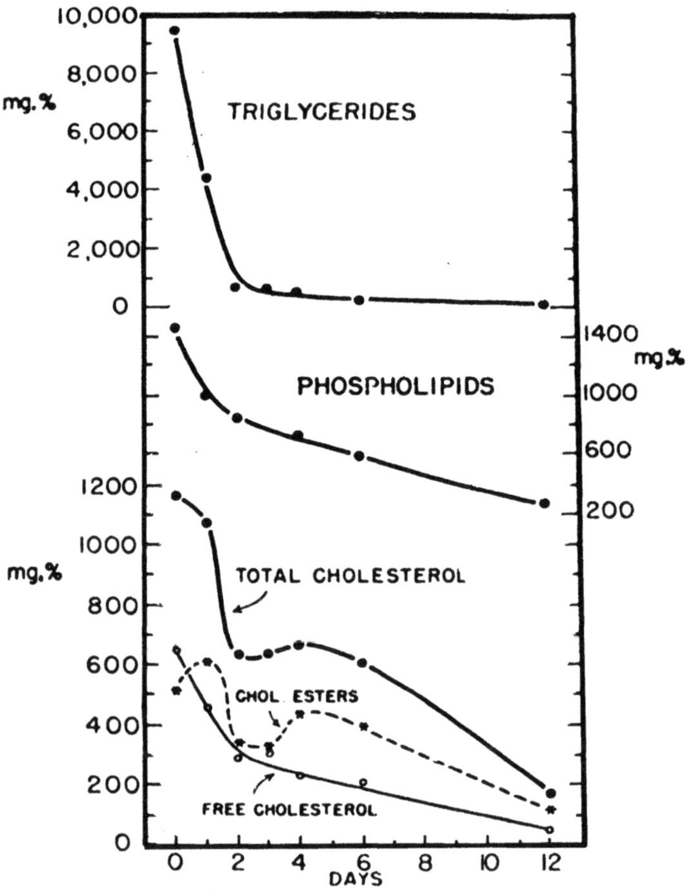

Fig. 5. Changes in plasma lipid fractions during recovery from alcoholic hyperlipemia (203).

4.1.2 Plasma cholesterol

In patients with marked hypertriglyceridemia, plasma cholesterol is also increased. Cholesterol is a component of all lipoprotein fractions, but most of it is transported in LDL and in HDL (or α-lipoproteins). In hospitalized alcoholics, total plasma cholesterol was not significantly increased (168). However, in about 30% of alcoholics seeking medical attention (174) and 86% of patients after a recent drinking bout (175), a prominent α-lipoprotein band was found on electrophoresis. This alteration normalized after approximately 2 weeks of abstinence (175, 176). By contrast, LDL tend to be reduced in alcoholics (177). These changes in cholesterol-carrying lipoproteins are also shown in epidemiologic studies: there is a strong positive correlation between alcohol consumption and plasma HDL cholesterol, and a weaker but significant negative correlation with LDL cholesterol (166, 178).

Since only few other factors (chronic administration of some drugs and vigorous physical exercise) are capable of increasing HDL cholesterol, this has been proposed as one of the biological markers of alcoholism (179–181). This increase occurs in women (who normally have higher HDL cholesterol) as well as men, indicating that this is not merely a consequence of the feminization of the alcoholics.

By contrast, alcoholics with advanced liver disease failed to show the increase in HDL cholesterol shortly after their last drink, nor did they show a significant change after abstinence (176). The main apolipoproteins of HDL, Apo AI and Apo AII, are increased by alcohol (182–184), and Apo AII could serve as a marker of alcohol abuse (185).

HDL is an heterogeneous family of lipoproteins of different sizes, composition, and metabolic characteristics (186). The HDL increase predominates in the light subfractions (HDL_2) or heavy subfractions (HDL_3), depending on the alcohol intake (see below). A minor subfraction that floats as HDL, although it contains Apo B (with a unique antigenic determinant), is named Lp (a) and has been found to be markedly decreased in heavy alcohol consumers (187).

4.1.3 **Other plasma lipids**

Plasma FFA concentration is increased or decreased depending on the alcohol intake and the nutritional conditions (see below). Alcohol consumption increases both total and HDL plasma phospholipid levels (71, 168, 185, 188).

In alcoholics, some modifications of the plasma fatty acyl patterns were described, including a relative increase in oleic acid and a relative decrease in linoleic acid in triacylglycerols and phospholipids (189, 190); others have shown a low level of arachidonic acid (171, 191). In a recent study, Johnson et al. (192) assessed the phospholipid fatty acyl pattern in a large group of patients. They concluded that excessive alcohol consumption alone does not induce essential fatty acid deficiency and that cirrhosis, with or without alcoholism, is accompanied by a deficiency in arachidonic acid.

4.2 Etiologic role of ethanol

The administration of an ethanol dose which results in blood concentrations over 1 g/l and mild intoxication increases triacylglycerols in the plasma of normal volunteers (193–198), whereas the administration of smaller doses does not (199). In about 25% of subjects drinking a "cocktail dose" of alcohol (1 g/kg) in the evening, the fasting serum samples obtained next morning reflect a type IV hyperlipemia (198). Thus prior alcohol ingestion has to be investigated in patients with hyperlipemia. Acute ethanol administration does not, however, produce significant changes in plasma cholesterol (200).

The effects of ethanol on plasma triacylglycerols and plasma cholesterol are exaggerated by chronic ethanol administration. Either as a supplement to a normal diet (Fig. 6) (201, 202) or as substitution for other foods (203), ethanol produces a four fold increase in plasma triacylglycerols and lesser increases in cholesterol and phospholipids. This requires several days of administration and blood ethanol concentrations greater than 1 g/l. Administration of smaller doses to normal subjects

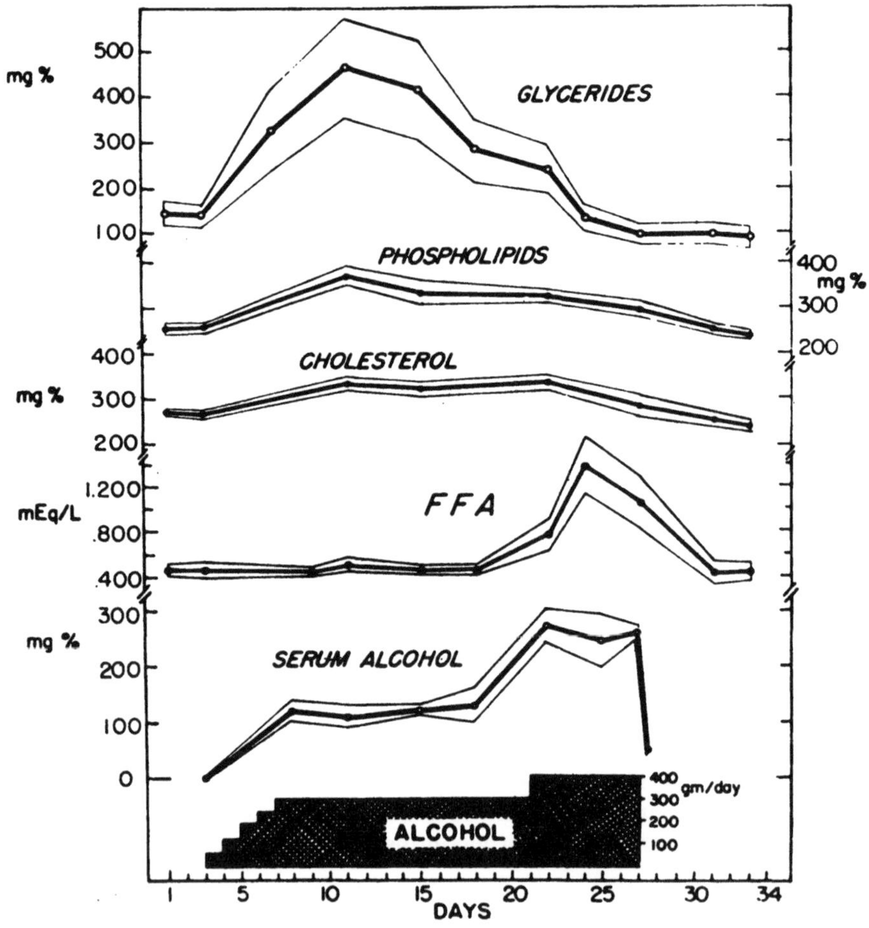

Fig. 6. Effect of prolonged alcohol intake on serum lipids in seven chronic alcoholic individuals (average results ±SEM). The hypertriglyceridemia lasts for several weeks, whereas the rise in FFA only occurs with very large amounts of ethanol (71).

does not produce significant hyperlipemia (204), but HDL cholesterol is elevated (182, 196, 202, 205).

The lipemic response to chronic ethanol consumption in human volunteers is usually transient: after reaching maximal levels in several days or weeks, serum lipids decrease and even normalize despite continuation of the ethanol intake (71, 201, 202, 206). Most of these effects of ethanol can also be produced in experimental animals although there seem to be differences among species. A consistent lipemic response has been produced by acute administration of ethanol to rabbits (207, 208), but only after chronic administration of ethanol-containing diets (with or without an acute ethanol dose) is a significant postprandial hyperlipemia observed in rats (209, 210). Lipids increase in all the fractions, including HDL. In nonhuman primates (which have a lipoprotein composition and metabolism more akin to those of man), chronic ethanol administration produces a sustained increase in fasting triacylglycerol, cholesterol, and phospholipid levels in the plasma, at least during the early stages of alcoholic liver injury (85, 160, 211).

4.3 Clinical manifestations of alcoholic hyperlipemia

Usually, alcoholic hyperlipemia is first suspected because of the incidental observation of serum lactescence or of hypercholesterolemia. It is commonly associated with hepatic, gastrointestinal, or pancreatic complications of alcoholism, and on this basis a combination of varying symptoms is seen, including most frequently anorexia, nausea, vomiting, abdominal pain, fever, jaundice, and transient hemolysis. Indeed, transient hemolytic anemia can accompany episodes of hyperlipemia in alcoholics (the so-called Zieve's Syndrome) (212). In rats chronically fed with ethanol, reticulocytosis and increased osmotic fragility of the erythocytes have been reported. These abnormalities correlate with the degree of hyperlipemia (213), and lipid abnormalities of the red cell membrane have been incriminated (214).

As discussed before, the degree of ethanol-induced hyperlipemia varies but is usually moderate. In a few individuals, plasma lactescence can be severe because of potentiating factors such as an underlying abnormality of lipid metabolism (that is, "forme fruste" of essential hyperlipemia), pancreatitis (215), or diabetes. Massive hypertriacylglyceridemia may result in some clinical manifestations ("chylomicronemia syndrome"), such as abdominal pain (216) with or without pancreatitis (217, 218, 219), lipemia retinalis (220), and eruptive xanthoma (164, 219). The diagnosis of pancreatitis may be difficult because serum lactescence interferes with the enzyme determinations (218, 221). Indeed, alcohol can produce pancreatitis and hyperlipemia but pancreatitis has also been reported in non alcoholic patients with primary dyslipoproteinemia (222, 223).

4.4 Potentiating factors of alcoholic hyperlipemia

4.4.1 Role of nutritional factors

The lipemic response to acute ethanol administration in volunteers is greatly enhanced by dietary fat (224–226). The increase includes both VLDL and chylomicrons (227). In patients with alcoholic fatty liver, administration of a high fat meal, even in the absence of ethanol, produced a striking increase in serum triacylglycerols (228, 229). This effect tends to disappear with the progression of the liver damage (Fig. 7). Thus dietary fat represents a striking potentiating factor for alcoholic hyperlipemia.

4.4.2 Role of underlying alteration of lipid metabolism

The subclinical defects in lipid metabolism most commonly aggravated by alcohol consumption are reduced ability to remove serum lipids and overproduction of VLDL. The first type of alteration is found in subjects with primary (familial) or secondary (usually diabetic) decreased activity of lipoprotein lipase (164, 203). In

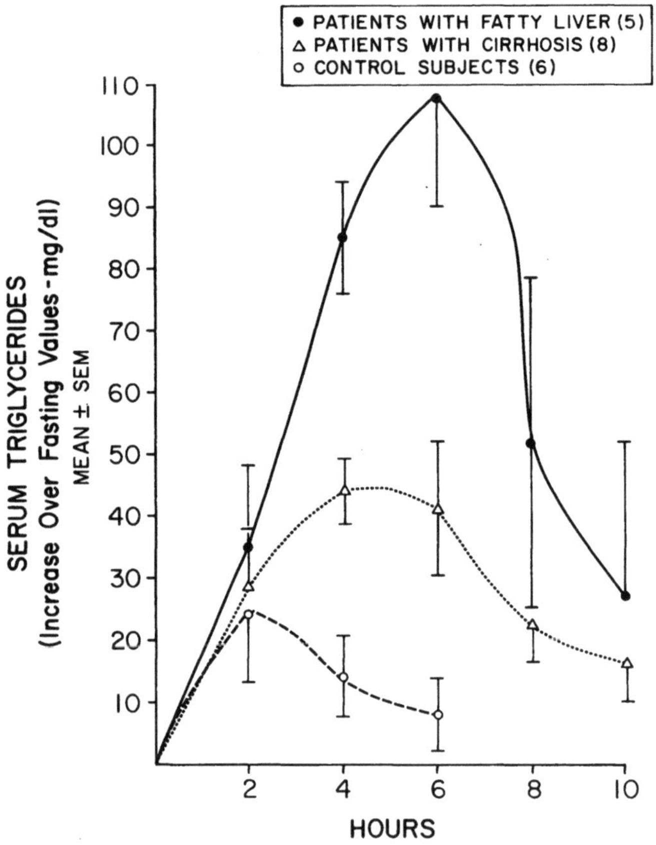

Fig. 7. Serum triglyceride response of control subjects and alcoholic patients after a high-fat meal. Note the marked response of patients with fatty liver and the less striking but still significant response in patients with cirrhosis (228).

another group of individuals, alcohol aggravated either a primary or a secondary (obesity-induced) overproduction of VLDL (type IV) or both VLDL and LDL (type IIb) 198, 230–232). Another possible mechanism could be an increased capacity to secrete serum lipoproteins upon challenge with alcohol. This may account for the observation that some alcoholic patients appear to have an unusual sensitivity to the hyperlipemic effect of ethanol (233). A large percentage of patients with marked and sustained alcoholic hyperlipemia have relatives with hyperlipemia (234). The participation of an underlying defect in lipid metabolism should be suspected in any alcoholic with severe hyperlipemia. Conversely, the lack of response to the usual treatment for hyperlipemia should lead to a search for alcoholism.

4.4.3 **Role of associated liver injury**

An inverse relationship between the magnitude of alcoholic hyperlipemia and the severity of the liver injury has been observed (170, 171). In baboons fed alcohol-containing diets for several years, the hypertriglyceridemia and the hypercholesterolemia decreased or disappeared during the progression from alcoholic fatty liver to the fibrotic stage (145) (Fig. 8). In man, no simple association was evident between a well-defined plasma lipid, lipoprotein and/or apolipoprotein profile and chronic alcoholism in the presence or absence of liver injury (235, 236). The findings however suggested that measurement of the absolute concentrations and ratio of HDL_2 and HDL_3, as well as of the levels and ratio of Apo AI and Apo AII, may permit differentiation of certain of the subpopulation of chronic alcoholics (237).

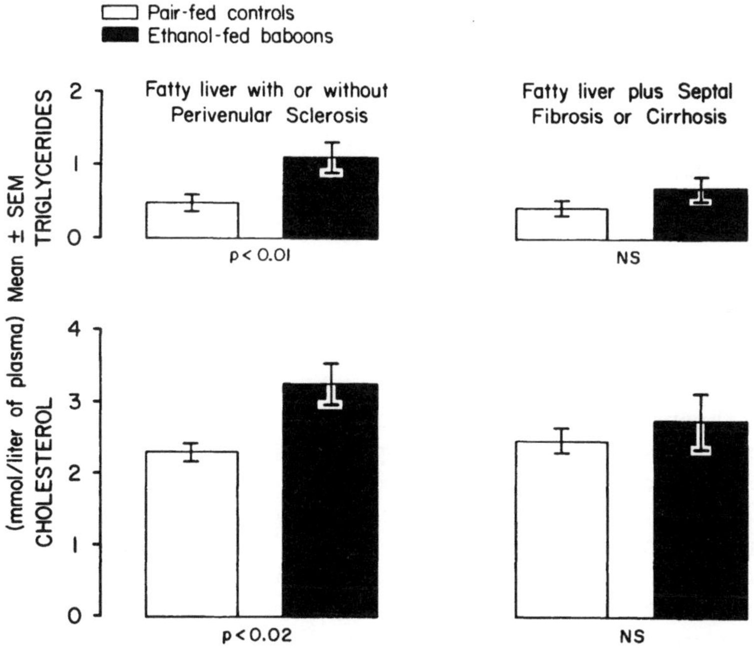

Fig. 8. Effects of the severity of alcoholic liver damage on the magnitude of hypertriglyceridemia and hypercholesterolemia. As liver damage progressed, alcoholic hyperlipemia decreased. NS, not significant (160).

At advanced stages, striking alteration of serum lipoproteins occurs, regardless of the etiology of the liver injury. Electrophoretically, these alterations are characterized by a marked decrease or disappearance of both pre-β and α-lipoproteins with appearance of a broad β-band (238) and, chemically, by a decreased cholesteryl ester/free cholesterol ratio. These changes are associated with decreased lecithin: cholesterol acyltransferase (LCAT) activity (239, 240). Apo AI, the major apoprotein of HDL and activator of LCAT, is markedly decreased (183, 239). Lipoprotein abnormalities associated with cholestasis (241), portacaval anastomosis (242), malnutrition, pancreatitis, or endocrine abnormalities may also complicate the clinical picture (1).

4.5 Pathogenesis

Hyperlipemia results when the rate of entry of lipoprotein lipids into the plasma exceeds the rate of exit from plasma. Thus we will consider the effect of alcohol on hepatic and intestinal lipid production and on plasma removal.

4.5.1 **Hepatic lipoprotein production**

The initial phase of hepatic lipid deposition after ethanol consumption is accompanied by an increased release of lipoproteins into the blood: this tends to counteract lipid accumulation in the liver. The effect is dose-dependent. *In vitro*, high ethanol concentrations may decrease hepatic lipoprotein release (243). When livers were perfused with ethanol in concentrations more in keeping with *in vivo* conditions, no inhibition of lipoprotein secretion was found (244). The stimulatory effect of an acute oral dose of ethanol on hepatic VLDL synthesis in isolated hepatocytes was abolished in thyroidectomized rats but not in adrenalectomized rats (245). Triton WR1339 inhibits lipoprotein lipase activity by complexing with the lipoprotein, resulting in some lipid accumulation. After several weeks of feeding an alcohol containing diet, rats developed marked postprandial hyperlipemia. Complete block of the blood lipid removal produced a greater accumulation of serum lipids in the ethanol fed rat than in pair fed controls (209). Incorporation into lipoprotein of intragastrically administered ^{3}H-palmitate and intravenously injected ^{14}C-lysine was significantly increased by alcohol administration (209). These as well as other studies suggested that the ethanol-induced hyperlipemia results, at least in part, from enhanced lipoprotein production. This contrasts with choline (246) and protein (247, 248) deficiencies, which produce the opposite effect.

More recently direct evidence was provided for this increase in hepatic lipoprotein production (249, 250). Savolainen et al. (250), have shown by catheterization studies that the livers of alcohol-fed baboons secrete more VLDL-TG than the livers of the pair-fed controls and the main part of the increased hepatic output of VLDL is due to an enhanced production of abnormally large VLDL particles which have flotation characteristics similar to chylomicrons ($S_f > 400$). This may account for the observation of chylomicron-like particles in the fasting blood of some alcoholics (164). Moreover, Sane et al. (251) have described an increased fractional catabolic rate and total turnover (production) rate of VLDL triacylglycerol in ten alcoholic men. After withdrawal both the synthetic and catabolic rate return to normal.

In addition to VLDL, the liver normally secretes an Apo-E rich HDL. In alcoholics, as well as after chronic administration of phenobarbital, a correlation between the plasma levels of HDL and hepatic microsomal activities has been pointed out (252, 253). Malmendier et al. (254) have performed kinetic studies of HDL Apo AI in healthy volunteers before and after a 4 weeks period of alcohol intake. The synthetic rate of this apolipoprotein is increased, providing direct evidence in favor of increased HDL production (Fig. 4).

4.5.2 **Lipid absorption and production by the intestine**

The suggestion has been made that after alcohol feeding the intestine releases more lipid into the lymph, either by decreasing oxidation of fatty acids or by increasing the synthesis of lipids from sources other than dietary fat (255, 256). It was shown that intraduodenal infusion of 10% ethanol to the fasted rat in a dose of 5 g/kg produces a mild increase in the VLDL output in the lymph (257). It was postulated that this increase in non-dietary lymph lipid could contribute to the hyperlipemia, although the peak serum rise actually preceded the maximum increase in intestinal lymph lipids. Furthermore, lymph lipoproteins can derive in part from plasma lipoproteins (258). Actually, the acute load of an ethanol-containing diet did not increase lymph lipid output in rats fed alcohol for several weeks compared to their pair-fed controls; however, marked hyperlipemia developed in these alcohol-fed rats. In addition, when a similar lymph lipid load was infused intravenously into alcohol-pretreated and control animals with diversion of intestinal lymph, the alcohol-fed rats developed hyperlipemia, but this did not occur when there was lymph depletion (210). This indicates that, although an adequate supply of dietary lipids represents a permissive factor, changes in lymph lipid output do not seem to play a major role in the lipemic effect of ethanol, and that the site of origin of increased production of serum lipoprotein is not intestinal. Similarly, the contribution of lymph lipids to the hepatic steatosis appears to be a minor one (259).

4.5.3 **Removal of intestinal lipids by the liver**

More recently, the possibility that intestinal lipids could be poorly removed by the liver has been considered. Indeed, after the injection of chylomicrons doubly labeled in the triacylglycerol and cholesteryl ester moieties to alcohol-fed rats, the clearance of chylomicron-cholesteryl esters was impaired to a greater extent than the clearance of chylomicron-triacylglycerol (149, 260). These observations are consistent with the possibility that after hydrolysis of the chylomicron-triacylglycerol by extrahepatic lipoprotein lipases, the cholesteryl ester-enriched remnants are poorly removed by the liver. This alteration could be due to a perturbation of the Apo E specific liver uptake or saturation of the system by catabolic products of endogenous VLDL which share a similar uptake mechanism. In alcohol-fed baboons, the hepatic extraction of labeled chylomicrons and VLDL-triacylglycerols during passage through the splanchnic vascular bed was increased rather than decreased (160). This favors the possibility that an excessive supply of products of both endogenous and dietary origin saturate the hepatic uptake, leading to lipid accumulation in the plasma during the postprandial state.

4.5.4 **Removal of plasma lipoproteins**

The mechanisms of removal are similar for hepatic and intestinal lipoproteins. Most of the early studies focused on the total post heparin lipoprotein lipase activity and found no change after acute (194, 227, 261) or chronic (233) ethanol administration. Recent studies have focused either on tissue measurements of lipoprotein lipase activity or on post heparin measurements after separation of the hepatic and extrahepatic components. A large dose of ethanol can produce a slight decrease in extrahepatic lipoprotein lipase activity and some delay in the clearance of serum triacylglycerols (206, 262–264). Most of the delayed clearance, however, appears to be secondary to the excessive supply of endogenous and exogenous lipids competing for the lipases (206). Even in those alcoholics with severe type V hyperlipemia (in whom lipoprotein lipase activities could be impaired), the turnover rates of VLDL-Apo B and VLDL-triacylglycerol are increased rather than decreased (265). This reflects the predominant role of increased lipoprotein production.

By contrast, chronic alcohol consumption has been found to increase the activity of extrahepatic lipoprotein lipase in the majority of subjects (169, 177, 202), according to a recent kinetic study (251). This change may represent a compensatory adaptation to the increased supply of plasma lipids, which would promote their removal and may account for the transient nature of hypertriglyceridemia during alcohol administration (71, 202). Mordasini et al. (266) documented that the rise in plasma triacylglycerol (mainly VLDL) is associated with an initial decrease in lipoprotein lipase activity, whereas the return of triacylglycerol to normal levels (despite continuous ethanol intake) is associated with increased lipase activity. This adaptation may also account for the modest degree of hypertriglyceridemia found in the majority of alcoholics, and the paradoxical increase in VLDL and LDL during the initial period of abstinence (169). In heavy drinkers with type V hyperlipemia, both post heparin lipoprotein lipase activities were decreased (267), confirming some observations made two decades ago (203) that in the patient without hyperlipemia, there was a "compensatory" increase in hepatic lipase activity (267).

The enhanced production and metabolism of VLDL after chronic alcohol consumption may have important consequences for plasma cholesterol, since VLDL (as well as the chylomicron) supplies free cholesterol to the HDL fraction of the plasma during extrahepatic lipolysis. This process is associated with the conversion of HDL_3 into HDL_2 (Fig. 4). Thus an increase of this lipase activity induces an increased HDL_2, as seen in heavy drinkers. The hepatic lipase is thought to regenerate HDL_3 from HDL_2 (268). An increase in HDL_3 could result from raised production of nascent hepatic HDL, and/or stimulation of the hepatic lipase activity (Fig. 4). In alcoholics with mild liver injury, this hepatic activity was found to be unchanged (169) or increased (177), but it may decrease in more severe forms of liver impairment, such as in alcoholic hepatitis (173, 269). In normal subjects (200, 206, 263, 270), an acute inhibition was described. Thus, the response of this enzyme to alcohol is time dependent and probably dose dependent, as for the extrahepatic lipase.

A recent study suggests that alcoholic hypercholesterolemia could have a predominantly extrahepatic origin. In alcohol-fed baboons, the increased output of VLDL-cholesterol from the liver was markedly exceeded by enhanced hepatic removal of cholesterol from the other lipoprotein fractions (161). Moreover, a decreased removal of cholesteryl ester from HDL, associated with decreased transfers to LDL, was described.

4.6 Treatment

The first step in the treatment is alcohol abstinence or at least a decrease in consumption. Associated factors should be treated: obesity by decreased calorie intake, diabetes by diet and drugs, if needed.

If the hyperlipemia persists after alcohol withdrawal, restriction in dietary fat and cholesterol (type V hyperlipemia) or restriction in simple carbohydrates (type IV hyperlipoproteinemia) is necessary. Exogenous estrogen should be avoided.

If after dietary change and alcohol withdrawal the triacylglycerol remains elevated, drug therapy with niacin, gemfibrozil or clofibrate can be instituted (271, 272).

5 Ethanol, Atherosclerosis, and High Density Lipoproteins

For more than a century, the effect of alcoholism on atherosclerosis has been the subject of a lively debate. At the end of the 19th century, alcohol was generally considered as one of the causes of atherosclerosis. This opinion was challenged by the observation of Cabot (273) who reported that atherosclerosis was uncommon in alcoholics. This "protective" effect of alcohol against atherosclerosis was confirmed by some (274) but denied by others (275). In alcoholics who died of cirrhosis of the liver, some researchers found a low incidence of coronary heart disease (276–278), whereas others failed to observe a significant correlation (279) or found a negative correlation with cirrhosis but not with alcoholism (280). Since atherosclerosis and hypertensive heart disease contributed heavily to all these autopsy series, the paucity of myocardial infarction in cirrhotics could be a result of the biased sample (281, 282). Indeed, in other series composed of victims of automobile accidents or other sudden unexpected deaths, there was either no relationship between alcoholism and atherosclerosis (283) or even a positive correlation (284). Such series of sudden deaths, however, are heavily weighted with alcoholics. Experimentally produced atherosclerosis was found to be either aggravated (285, 286), unaffected (287, 288) or prevented (289, 290) by ethanol. Recently, a protective effect of ethanol (as 36% of energy) was reported in a study of *Macaca nemestrina* (291) mainly on the atherosclerotic lesions induced by a high cholesterol diet. However, the relevance of all these experimental models to human atherosclerosis has been questioned. Prospective clinical studies showed either a positive correlation between alcoholism and atherosclerosis (292–294) or no correlation (295–298). Only when recognized risk factors for coronary heart disease, such as smoking and hypertension (which are positively associated with alcoholism) (299), were excluded, did a negative correlation between moderate alcohol consumption and coronary heart disease emerge in most of the recent epidemiologic studies (299–306). The protective effect of alcohol appeared to be abolished by heavy alcohol consumption (306–308), but not in all the prospective studies (309).

One common problem in many of the reported studies is the difficulty of establishing a proper control group. For example, some studies (299, 306) included in the group of teetotalers, both lifetime teetotalers and ex-drinkers, who have a higher incidence of coronary heart disease death (307, 308). However, removal of the past heavy drinker did not eliminate a statistically significant difference between teetotalers and light drinkers (307). In a recent review, Eichner (310) has pointed out the importance of taking into account the various factors associated with coronary heart disease and choosing an appropriate end-point. Thus, a study of a random sample of all coronary events in Auckland, New Zealand, has shown that heavy drinkers are more likely to die immediately from a heart attack (311). Obviously, a higher incidence of coronary heart disease in teetotalers than in moderate drinkers may be caused not only by a difference in alcohol intake but also by a variety of other factors that are present in the teetotalers (the light use of alcohol could be a marker of behavior only) (310). Popham et al. have shown a negative correlation between consumption of alcohol and of dairy products (312); the association between alcohol intake and cardiovascular mortality disappeared when the consumption of milk proteins was taken into account, thus this association could be only indirect (312, 313). While the relationship between alcohol and coronary heart disease is controversial (310, 314), the combined evidence from all studies still suggests that moderate alcohol consumption may be associated with reduced risk of coronary heart disease. However, further studies are needed before we can accept a causal relationship, especially since it was pointed out by Eichner (310), that in the non-smoking subpopulation of the Kaiser Permanente experience (307), the protective effect of light alcohol consumption disappeared and the non-smoking teetotalers had the lowest coronary heart disease mortality.

One mechanism that has been evoked in the protective effect of alcohol is its action on the blood lipids other than triacylglycerol. The rise in HDL associated with alcohol has been shown both experimentally (160, 209) and in a number of clinical studies (166, 175, 315); the increase includes not only cholesterol but also apolipoproteins AI and AII (182, 184). It occurs in teenagers (316), in both older and younger men (317), but only the inactive men, not in runners whose HDL are already increased (205). This alcohol-induced increase of HDL may be particularly significant with regard to the problem of atherosclerosis, in view of the negative correlation between HDL blood levels and the development of coronary heart disease (318, 319), either because HDL promotes the transport of cholesterol out of the cell (320) or because HDL competes with LDL for cellular uptake, thereby suppressing the increment in cell sterol content induced by LDL. A recent study in baboons supports the former hypothesis: the turnover of HDL free cholesterol was higher in alcohol-fed baboons than in pair-fed controls, with increased uptake by the liver and excretion into feces, suggesting a more active transport of the cholesterol from peripheral tissues to the liver and subsequently into the feces (161). In another study in monkeys, the area of coronary arteries occupied by atherosclerotic lesions was inversely correlated with HDL cholesterol (291). Thus the concept that moderate drinking may exert protective cardiovascular effects was strengthened when elevated levels of circulating HDL emerged as a possible mechanism. When the role of HDL in cholesterol transport and its protective effect against atherosclerosis became apparent, it made sense to postulate that the decreased incidence of coronary heart disease in moderate drinkers might be due to the ethanol-induced elevation of HDL. However, HDL is an heterogeneous group of lipoproteins with two major subclasses: the less dense HDL_2, epidemiologically associated with reduction in coronary heart disease, and the more dense HDL_3, with a weaker link to coronary heart disease (321–323). Moreover, agents or conditions that are thought to affect coronary heart disease through HDL (such as exercise and female sex) have been shown to increase HDL_2, not HDL_3 (186). A more recent study (324) has shown a negative correlation between HDL_3 mass (but not HDL_2 mass) and coronary lesions. It was reported that the increase in HDL after alcohol consumption involved primarily HDL_2 (169, 177, 200). However, these observations were made in individuals with a relatively high intake of alcohol. Similarly, after an acute dose of ethanol (40 g, equivalent to 4 ounces of 86 proof beverage) there was a transient increase in both HDL_2 and HDL_3 (270). It is now well recognized that large amounts of alcohol have adverse effects not only on the liver (71, 92) but also on virtually all the tissues of the body, including the cardiovascular system (1, 325, 326), and it is generally agreed that such high intakes are not associated with protection against coronary heart disease (176, 326). Furthermore, Haskell et al. reported that moderate doses of ethanol (about 0.5–2.2 oz. or 12–51 g of absolute ethanol per day) raised levels of HDL_3 but not levels of HDL_2, and that upon abstention from moderate consumption, levels of HDL_3, not of HDL_2 decreased (327). The relation between alcohol and HDL_3 at moderate alcohol dose was confirmed by two recent studies (328, 329). William et al. pointed out the correlation between nutritional components, such as alcohol and starch, and HDL_3 (but not HDL_2). Thus, we now must revise some of the previously derived implications (330). For instance, the conclusion of Hartung et al. (205) that "nonexercisers can maintain levels of HDL similar to those of individuals who jog regularly by ingesting three beers a day" must be reassessed in light of the fact that the HDL subfractions involved appear to be different in the two groups and that the relation between some of these HDL subfractions (such as HDL_3) and coronary heart disease is not completely established. With the progress of alcoholic liver injury, the HDL fractions tend to decrease and abnormal lipoproteins appear in the bloodstream (176, 183, 228, 239).

The relationship between alcohol, coronary heart disease and HDL could also be explained by the protein moiety (184). The major apoproteins of HDL are Apo AI and Apo AII. Apo AI is thought by some to be a better marker for risk of coronary heart disease than HDL cholesterol (331–333). The association between Apo AII concentration and coronary heart disease is less well established (184). Apo AI is increased, and Apo AII increased or normal in alcoholics (184, 237) as in healthy volunteers after moderate consumption (182, 184). Lipoproteins other than HDL are, of course, also involved in the process of atherosclerosis. It is of interest that a strong positive correlation was observed between the particle size of LDL and the area of atherosclerotic plaques in the aorta and coronary arteries. Therefore, the decreased size of LDL particles observed in alcohol-fed animals may attenuate the atherosclerotic process (291).

The main route by which cholesterol is eliminated from the body is via excretion into feces. Chronic alcohol feeding increased fecal excretion of cholesterol and its derivatives (161, 334, 335). The secretion of neutral sterols was increased more than the secretion of acidic steroids (bile acids).

Mechanisms other than lipoproteins might explain the protective effect of alcohol on cardiovascular disease. Inhibition of platelet aggregation and increased fibrinolytic activity have been suggested as alternative explanations (313, 336). One other mechanism of ethanol could be a block of the vasoconstrictive response to stress associated with a reduced effect of stress on blood pressure (337).

Whatever the mechanism whereby ethanol affects atherosclerosis, it must be pointed out that the protective action was observed in a population comprised of moderate rather than heavy drinkers (302, 338–340). The overall decreased mortality associated with alcohol intake was shown with consumption less than 34 g (340), or 31 ml of pure ethanol (341), or two drinks (307) per day. Large amounts of alcohol exert cardiotoxic effects, may promote hypertension (325, 342) and have been shown to increase total mortality (307, 308, 340, 341).

In conclusion, we can distinguish three partially overlapping stages in the relationship between alcohol, coronary heart disease, and HDL: (1) moderate alcohol intake is associated with a decreased incidence of coronary heart disease and with an increased level of HDL_3 rather than HDL_2. This could be due to a protective effect of HDL_3 or could be secondary to some mechanism unrelated to HDL (such as prostaglandin-mediated vascular effects or enhanced fibrinolysis). Moreover, there are some epidemiologic arguments to support the view that the effect of alcohol on coronary heart disease is unrelated or related only partially to the effect of alcohol on HDL (305, 343). Furthermore, an association does not prove a cause-and-effect relationship; it is possible that teetotalers may differ from moderate drinkers in ways other than alcohol intake; (2) in the absence of severe liver injury, high alcohol intake results in increased levels of HDL, but at that level of alcohol intake there is no evidence for "protection" against coronary heart disease; (3) when alcohol abuse is associated with severe liver disease, the HDL fraction decreases (330).

Acknowledgements

Original studies were supported, in part, by grants from NIAAA, NIH and the Veterans Administration. One of the authors (J.-P.) is a recipient of grants from the Ministère des Relations Extérieures and Fondation pour la Recherche Médicale.

References

1. Lieber, C. S. (1982) Medical Disorders of Alcoholism, Pathogenesis and Treatment. Saunders, Philadelphia.
2. Lieber, C. S. & Davidson, C. S. (1962) Am. J. Med. *33*, 319–327.
3. Forsander, D. A., Mäenpää, P. H. & Salaspuro, M. P. (1965) Acta Chem. Scand. *19*, 1770–1771.
4. Lieber, C. S., Lefevre, A., Spritz, N., Feinman, L. & DeCarli, L. M. (1967) J. Clin. Invest. *46*, 1451–1460.
5. Lieber, C. S., Spritz, N. & DeCarli, L. M. (1966) J. Clin. Invest. *45*, 51–62.
6. Lieber, C. S. & Spritz, N. (1966) J. Clin. Invest. *45*, 1400–1411.
7. Mendenhall, C. L. (1972) J. Lipid Res. *13*, 177–183.
8. Nikkilä, E. A. & Ojala, K. (1963) Proc. Soc. Exp. Biol. Med. *113*, 814–817.
9. Reitz, R. C. (1975) Biochim. Biophys. Acta *380*, 145–154.
10. Chance, B., Sies, H. & Boveris, A. (1979) Physiol. Rev. *59*, 527–605.
11. Tappel, A. L. (1973) Fed. Proc. *33*, 1870–1874.
12. Lieber, C. S. & DeCarli, L. M. (1970) Science *170*, 78–80.

13. Chin, J. H. & Goldstein, D. B. (1977) Mol. Pharmacol. *13*, 435–441.
14. Chin, J. H., Parson, L. M. & Goldstein, D. B. (1978) Biochem. Biophys. Acta *513*, 358–363.
15. Johnson, D. A., Lee, N. M., Cooke, R. & Loh, H. H. (1979) Mol. Pharmacol. *15*, 739–746.
16. Beaugé, F., Stibler, H. & Borg, S. (1985) Alcoholism: Clin. Exp. Res. *9*, 322–326.
17. Rottenberg, H., Waring, A. & Rubin, E. (1981) Science *213*, 583–585.
18. Waring, A. J., Rottenberg, H., Ohnishi, T. & Rubin, E. (1981) Proc. Natl. Acad. Sci. USA *78*, 2582–2586.
19. Gordon, E. R., Rochman, J., Arai, M. & Lieber, C. S. (1982) Science *216*, 1319–1321.
20. Yamada, S. & Lieber, C. S. (1984) J. Clin. Invest. *74*, 2285–2289.
21. Polokoff, M. A., Simon, T. J., Harris, R. A., Simon, F. R. & Iwahashi, M. (1985) Biochem. *24*, 3114–3120.
22. Zysset, T., Polokoff, M. A. & Simon, F. R. (1985) Hepatology *5*, 531–537.
23. Schüller, A., Moscat, J., Diez, E., Fernandez-Checa, J. C., Gavilanes, F. C. & Municio, A. M. (1985) Hepatology *5*, 677–682.
24. Cunningham, C. C., Bottenus, R. E., Spach, P. I. & Rudel, L. L. (1983) Alcoholism: Clin. Exp. Res. *7*, 424–430.
25. John, G. R., Littleton, J. M. & Jones, P. A. (1980) Life Sci. *27*, 545–555.
26. Cunningham, C. C., Filus, S., Bottenus, R. E. & Spach, P. I. (1982) Biochim. Biophys. Acta *712*, 225–233.
27. Arai, M., Gordon, E. R. & Lieber, C. S. (1984) Biochim. Biophys. Acta *797*, 320–327.
28. Rouach, H., Clement, M., Orfanelli, M. T., Janvier, B. & Nordmann, R. (1984) Biochim. Biophys. Acta *795*, 125–129.
29. Sun, G. Y. & Sun, A. Y. (1985) Alcoholism: Clin. Exp. Res. *9*, 164–180.
30. Nervi, A. M., Peluffo, R. O., Brenner, R. R. & Leikin, A. I. (1980) Lipids *15*, 263–268.
31. Wang, D. L. & Reitz, R. C. (1983) Alcoholism: Clin. Exp. Res. *7*, 220–226.
32. Rao, G. A., Lew, G. & Larkin, E. C. (1984) Lipids *19*, 151–153.
33. Taraschi, T. F. & Rubin, E. (1985) Lab. Invest. *52*, 120–131.
34. Dawidowicz, E. A. (1985) Hepatology *5*, 697–699.
35. Alling, C., Gustavsson, L. & Änggärd, E. (1983) FEBS Lett. *152*, 24–28.
36. Lange, L. G., Bergmann, S. R. & Sobel, B. E. (1981) J. Biol. Chem. *256*, 12968–12973.
37. Polokoff, M. A. & Bell, R. M. (1978) J. Biol. Chem. *253*, 7173–7178.
38. Goodman, D. S. & Deykin, D. (1963) Proc. Soc. Exp. Biol. Med. *113*, 65–67.
39. Lange, L. G. (1982) Proc. Natl. Acad. Sci. USA *79*, 3954–3957.
40. Kovatchev, S. & Eibl, H. (1978) Adv. Exp. Med. Biol. *101*, 221–226.
41. Hasumura, Y., Teschke, R. & Lieber, C. S. (1976) J. Biol. Chem. *251*, 4908–4913.
42. Matsuzaki, S. & Lieber, C. S. (1977) Biochem. Biophys. Res. Commun. *75*, 1059–1065.
43. Baraona, E. & Lieber, C. S. (1982) Ann. Rev. Med. *33*, 281–292.
44. Baraona, E., Leo, M. A., Borowsky, S. A. & Lieber, C. S. (1977) J. Clin. Invest. *60*, 546–554.
45. Matsuda, Y., Baraona, E., Salaspuro, M. & Lieber, C. S. (1979) Lab. Invest. *41*, 455–463.
46. Massey, U., Strickland, S., Mayhew, S. G., Howell, L. G., Engle, P. C., Matthews, R. G., Schuman, M. & Sullivan, P. A. (1969) Biochem. Biophys. Res. Commun. *36*, 891–897.
47. DiLuzio, N. R. & Stege, T. E. (1977) in: Alcohol and the Liver, vol. 3 (Fisher, M. M., Rankin, J. G., eds.) pp. 45–62, Plenum Press, New York.
48. DiLuzio, N. R. & Hartman, A. D. (1967) Fed. Proc. *26*, 1436–1442.
49. Stege, T. E. (1982) Res. Commun. Chem. Pathol. Pharmacol. *36*, 287–297.
50. Hashimoto, S. & Recknagel, R. O. (1968) Exp. Mol. Pathol. *8*, 225–242.
51. Scheig, R. & Klatskin, G. (1969) Life Sci. *8*, 855–865.
52. Comporti, M., Burdino, E. & Raja, F. (1971) Life Sci. *10* (Part II), 855–866.
53. Sprince, H., Parker, C. M., Smith, G. & Gonzales, L. J. (1974) Agents Actions *4*, 125–130.
54. Shaw, S., Rubin, K. P. & Lieber, C. S. (1983) Dig. Dis. Sci. *28*, 585–589.
55. Wendel, A., Fenerstein, S. & Konz, K. H. (1979) Biochem. Pharmacol. *28*, 2051–2055.
56. Shaw, S., Jayatilleke, E., Ross, W. A., Gordon, E. R. & Lieber, C. S. (1981) J. Lab. Clin. Med. *98*, 417–425.

57. Videla, L. A., Fraga, C. G., Koch, O. R. & Boveris, A. (1983) Biochem. Pharmacol. *32*, 2822–2825.
58. Vendemiale, C., Jayatilleke, E., Shaw, S. & Lieber, C. S. (1984) Life Sci. *34*, 1065–1073.
59. Suematsu, T., Matsumura, T., Sato, N., Miyamoto, T., Ooka, T., Kamada, T. & Abe, H. (1981) Alcoholism: Clin. Exp. Res. *5*, 427–430.
60. Siegers, C.-P., Schütt, A. & Strubelt, O. (1977) Proc. Eur. Soc. Toxicol. *18*, 160–162.
61. Mallov, S. & Bloch, J. L. (1956) Am. J. Physiol. *184*, 29–34.
62. Horning, M. B., Williams, E. A., Maling, H. M. & Brodie, B. B. (1960) Biochem. Biophys. Res. Commun. *3*, 635–640.
63. Savolainen, M. J. & Hassinen, T. E. (1980) Arch. Biochem. Biophys. *201*, 640–645.
64. Lieber, C. S. & DeCarli, L. M. (1970) Am. J. Clin. Nutr. *23*, 474–478.
65. Lieber, C. S., DeCarli, L. M., Gang, H., Walker, G. & Rubin, E. (1972) in: Medical Primatology, Part III (Goldsmith, E. J. & Moor-Jankowski, J., eds.) pp. 270–278, Karger, Basel.
66. Popper, H. & Lieber, C. S. (1980) Am. J. Pathol. *98*, 695–716.
67. Salaspuro, M. P., Ross, W. A., Jayatilleke, E., Shaw, S. & Lieber, C. S. (1981) Hepatology *1*, 33–38.
68. Iseri, O. A., Lieber, C. S. & Gottlieb, L. S. (1966) Am. J. Pathol. *48*, 535–555.
69. Arai, M., Leo, M. A., Nakano, M., Gordon, E. R. & Lieber, C. S. (1984) Hepatology *4*, 165–174.
70. Lefevre, A. F., DeCarli, L. M. & Lieber, C. S. (1972) J. Lipid Res. *13*, 48–55.
71. Lieber, C. S., Jones, D. P., Mendelson, J. & DeCarli, L. M. (1963) Trans. Assoc, Am. Phys. *76*, 289–300.
72. Lieber, C. S., Spritz, N. & DeCarli, L. M. (1966) J. Clin. Invest. *45*, 51–62.
73. Vasdev, S. C., Subrahmanyam, D., Chakravarti, R. N. & Wahi, P. L. (1974) Biochim. Biophys. Acta *369*, 323–330.
74. Kikuchi, T. & Kako, K. J. (1970) Circulation Res. *26*, 625–634.
75. Liau, D. F., Hashim, S. A., Pierson, R. N. & Ryan, S. F. (1981) J. Lipid Res. *22*, 680–686.
76. Wilson, J. S., Colley, P. W., Sasula, L., Pirola, R. C., Chapman, B. A. & Somer, J. B. (1982) Alcoholism: Clin. Exp. Res. *6*, 117–121.
77. Lieber, C. S., Jones, D. P. & DeCarli, L. M. (1965) J. Clin. Invest. *44*, 1009–1021.
78. DeCarli, L. M. & Lieber, C. S. (1967) J. Nutr. *91*, 331–336.
79. Dich, J., Bro, B., Grunnet, N., Jensen, F. & Kondrup, J. (1983) Biochem. J. *212*, 617–623.
80. Best, C. H., Hartroft, W. S., Lucas, C. S. & Ridout, J. H. (1949) Br. Med. J. *2*, 1001–1006.
81. Lieber, C. S. & DeCarli, L. M. (1966) Gastroenterology *50*, 316–322.
82. Karpe, F., Wejde, J. & Änggärd, E. (1984) Acta Pharmacol. Toxicol. *55*, 95–99.
83. Wilson, D., Engel, J. & Wong, R. (1973) Clin. Res. *21*, 829.
84. Goheen, S. C. Pearson, E. E., Larkin, E. C. & Rao, G. A. (1981) Lipids *16*, 43–51.
85. Leathers, C. W., Bond, M. G., Bullock, B. C. & Rudel, L. L. (1981) Exp. Mol. Pathol. *35*, 285–299.
86. Indira, M. & Kurup, P. A. (1982) Atherosclerosis *41*, 241–246.
87. Hoffbauer, F. W. & Zaki, F. G. (1965) Arch. Pathol. *79*, 364–369.
88. Post, J., Benton, J. G., Breakstone, R. & Hoffman, F. (1952) Gastroenterology *20*, 403–410.
89. Phillips, G. B. & Davidson, C. S. (1954) Ann. NY Acad. Sci. *57*, 812–830.
90. Olson, R. F. (1964) in: Modern Nutrition in Health and Disease (Wohl, M. G. & Goodhart, R. S., eds.) pp. 1037–1050, Lea and Febiger, Philadelphia.
91. Rubin, E. & Lieber, C. S. (1968) N. Eng. J. Med. *278*, 869–876.
92. Lieber, C. S. & DeCarli, L. M. (1974) J. Med. Primat. *3*, 153–163.
93. Mezey, E., Potter, J. J., French, S. W., Tamura, T. & Halsted, C. H. (1983) Hepatology *3*, 41–44.
94. Rogers, A. E., Fox, J. G. & Gottlieb, L. S. (1981) in: Frontiers in Liver Disease, chapter *15* (Berk, P. D. & Chalmers, T. C., eds.) pp. 167–175, Thieme, Stuttgart.
95. DiLuzio, N. R. (1958) Am. J. Physiol. *194*, 453–456.
96. Lieber, C. S., Leo, M. A., Mak, K. M., DeCarli, L. M. & Sato, S. (1985) Hepatology *5*, 561–572.
97. Lieber, C. S. & Rubin, E. (1968) Am. J. Med. *44*, 200–206.
98. Klatskin, G., Krehl, W. A. & Conn, H. D. (1954) J. Exp. Med. *100*, 605–614.
99. Lieber, C. S., Spritz, N. & DeCarli, L. M. (1969) J. Lipid Res. *10*, 283–287.

100. Brodie, B. B., Butler, W. M., Horning, M. G., Maickel, R. P. & Mailing, H. M. (1961) Am. J. Clin. Nutr. *9*, 432–435.
101. Lieber, C. S., Spritz, N. & DeCarli, L. M. (1966) J. Clin. Invest. *45*, 51–62.
102. Mallov, S. (1957) Am. J. Physiol. *189*, 428–438.
103. Reboucas, G. & Isselbacher, K. J. (1961) J. Clin. Invest. *40*, 1355–1362.
104. Kessler, J. I. & Yalovsky-Mishkin, S. (1966) J. Lipid Res. *7*, 772–778.
105. Poggi, M. & DiLuzio, N. R. (1964) J. Lipid Res. *5*, 432–441.
106. Savolainen, M. J., Jauhonen, V. P. & Hassinen, I. E. (1977) Biochem. Pharmacol. *26*, 425–431.
107. Elko, E. E., Wooles, W. R. & DiLuzio, N. R. (1961) Am. J. Physiol. *201*, 923–926.
108. Jauhonen, V. P., Savolainen, M. J. & Hassinen, I. E. (1975) Biochem. Pharmacol. *24*, 1879–1883.
109. Lieber, C. S. & Schmid, R. (1961) J. Clin. Invest. *40*, 394–399.
110. Blomstrand, R., Kager, L. & Lantto, O. (1973) Life Sci. *13*, 1131–1141.
111. Ontko, J. A. (1973) J. Lipid Res. *14*, 78–86.
112. Blomstrand, R. & Kager, L. (1973) Life Sci. *13*, 113–123.
113. Jauhonen, P., Baraona, E., Miyakawa, H. & Lieber, C. S. (1982) Alcoholism: Clin. Exp. Res. *6*, 350–357.
114. Baraona, E., Jauhonen, P., Miyakawa, H. & Lieber, C. S. (1983) Pharmacol. Biochem. Behav. (Suppl. 1) *18*, 449–454.
115. Rubin, E., Beattie, D. S., Toth, A. & Lieber, C. S. (1972) Fed. Proc. *31*, 131–140.
116. Gordon, E. R. (1973) J. Biol. Chem. *248*, 8271–8280.
117. Cederbaum, A. I., Lieber, C. S., Beattie, D. S. & Rubin, E. (1975) J. Biol. Chem. *250*, 5122–5129.
118. Holmström, B. (1969) Arkiv. Kemi. *30*, 333–345.
119. Guynn, R. W., Veloso, D., Harris, R. L. & Veech, R. L. (1973) Biochem. J. *136*, 639–647.
120. Arakawa, M., Taketomi, S., Furuno, K., Matsuo, T., Iwatsuka, H. & Suzuoki, S. (1975) J. Nutr. *105*, 1500–1508.
121. Savolainen, M. J., Hiltunen, J. K. & Hassinen, I. E. (1977) Biochem. J. *164*, 169–177.
122. Selmer, J. & Grunnet, N. (1976) Biochim. Biophys. Acta *428*, 123–137.
123. Brunengraber, H., Boutry, M., Lowenstein, L. & Lowenstein, J. M. (1974) in: Alcohol and Aldehyde Metabolizing Systems (Thurman, R. G., Yonetani, T., Williamson, J. R. & Chance, B., eds.) pp. 329–337, Academic Press, New York.
124. Fellenius, E., Bengtsson, G. & Kiessling, K.-H. (1973) Acta Chem. Scand. *27*, 2893–2901.
125. Ylikahri, R. (1970) Metabolism *19*, 1036–1045.
126. Morgan, J. C. & DiLuzio, N. R. (1970) Proc. Soc. Exp. Biol. Med. *134*, 462–468.
127. Bustos, G. O., Kalant, H., Khanna, J. M. & Loth, J. (1970) Science *168*, 1598–1599.
128. Johnson, O., Hernell, O., Fex, G. & Olivecrona, T. (1971) Life Sci. *10*, 553–559.
129. Declercq, P. E., Debeer, L. J. & Mannaerts, G. P. (1982) Biochem. J. *202*, 803–806.
130. Sachan, D. S., Rhew, T. H. & Ruark, R. A. (1984) Am. J. Clin. Nutr. *39*, 738–744.
131. Mavrelis, P. G., Ammon, H. V., Gleysteen, J. J., Komorowski, R. A. & Charaf, U. K. (1983) Hepatology *3*, 226–231.
132. Shaw, W. (1985) Clin. Chem. *31*, 1109–1115.
133. Ockner, R. K., Manning, J. A. & Kane, J. P. (1982) J. Biol. Chem. *257*, 7872–7878.
134. Pignon, J. P., Bailey, N. A., Baraona, E. & Lieber, C. S. (1987) Hepatology *7*, 865–871.
135. Fallon, H. J., Lamb, R. G. & Jamdar, S. C. (1977) Biochem. Soc. Trans. *5*, 37–40.
136. Savolainen, M. J. (1977) Biochem. Biophys. Res. Commun. *75*, 511–518.
137. Pritchard, P. H., Bowley, M., Burditt, S. D., Cooling, J., Glenny, H. D., Lawson, N., Sturton, R. G. & Brindley, D. N. (1977) Biochem. J. *166*, 639–642.
138. Lamb, R. G., Wood, C. K. & Fallon, H. J. (1979) J. Clin. Invest. *63*, 14–20.
139. Sturton, R. G., Butterwith, S. C., Burditt, S. L. & Brindley, D. N. (1981) FEBS Lett. *126*, 297–300.
140. Glenny, H. P. & Brindley, D. N. (1978) Biochem. J. *176*, 777–784.
141. Lehtonen, M. A., Savolainen, M. J. & Hassinen, I. E. (1979) FEBS Lett. *99*, 162–166.
142. Savolainen, M. J. & Hassinen, I. E. (1978) Biochem. J. *176*, 885–892.
143. Joly, J.-G., Feinman, L., Ishii, H. & Lieber, C. S. (1973) J. Lipid Res. *14*, 337–343.

144. Väänänen, R. A., Pikkukangas, A. H., Savolainen, M. J. & Hassinen, I. E. (1981) Acta Pharmacol. Toxicol. *49*, (Suppl. 4) A42.
145. Savolainen, M. J., Baraona, E., Pikkarainen, P. & Lieber, C. S. (1984) J. Lipid Res. *25*, 813–820.
146. Uthus, E. O., Skurdal, D. N. & Cornatzer, W. E. (1976) Lipids *11*, 641–644.
147. Field, F. J., Boydstun, J. S. & LaBrecque, D. R. (1985) Hepatology *5*, 133–138.
148. Lakshmanan, M. R. & Veech, R. L. (1977) J. Lipid Res. *18*, 325–330.
149. Lakshmanan, M. R. & Ezekiel, M. (1982) Alcoholism: Clin. Exp. Res. *6*, 482–486.
150. Rubin, E. & Lieber, C. S. (1967) Gastroenterology *52*, 1–13.
151. Mezey, E., Potter, J. J., Slusser, R. J., Brandes, D., Romero, J., Tamura, T. & Halsted, C. H. (1980) Lab. Invest. *43*, 88–93.
152. Mezey, E., Potter, J. J. & Ammon, R. A. (1976) Biochem. Pharmacol. *25*, 2663–2667.
153. von Platt, U., Stein, U. & Heissmeyer, H. (1971) Z. Klin. Chem. Klin. Biochem. *9*, 126–129.
154. Berg, T. & Mørland, J. (1973) Acta Pharmacol. Toxicol. *33*, 409–416.
155. Sturton, R. G., Pritchard, P. H., Han, L.-Y. & Brindley, D. N. (1978) Biochem. J. *174*, 667–670.
156. Takeuchi, N., Ito, M. & Yamamura, Y. (1974) Lipids *9*, 353–357.
157. Boyer, J. L. (1972) Gastroenterology *62*, 294–301.
158. Marin, G. A., Karjoo, M., Ward, N. & Rosato, E. (1975) Surg. Gynecol. Obstet. *141*, 353–356.
159. Ideo, G., Dellobuono, A. & Bellati, G. (1978) Biomedicine *29*, 225–227.
160. Baraona, E., Savolainen, M., Karsenty, C. & Lieber, C. S. (1983) Trans. Assoc. Am. Phys. *96*, 306–315.
161. Karsenty, C., Baraona, E., Savolainen, M. & Lieber, C. S. (1985) J. Clin. Invest. *75*, 976–986.
162. Ducceschi, V. (1915) Arch. Fisiol. *13*, 147–153.
163. Feigl, J. (1918) Biochem. Z. *92*, 282–317.
164. Chait, A., Mancini, M., February, A. W. & Lewis, B. (1972) Lancet *2*, 62–64.
165. Ostrander, L. D. Jr., Johnson, B. C. & Block, W. D. (1971) Circulation *44*, 11–59.
166. Castelli, W. P., Gordon, T., Hjortland, M. G., Kagan, A., Doyle, J. T., Hames, C. G., Hulley, S. B. & Zukel, W. J. (1977) Lancet *2*, 153–155.
167. Sirtori, C. R., Agradi, E. & Mariani, C. (1973) Pharmacol. Res. Commun. *5*, 81–85.
168. Böttiger, L. E., Carlson, L. A., Hultman, E. M. & Romanus, V. (1976) Acta Med. Scand. *199*, 357–361.
169. Ekman, R., Fex, G., Johansson, B. G., Nilsson-Ehle, P. & Wadstein, J. (1981) Scand. J. Clin. Lab. Invest. *41*, 709–715.
170. Cachera, R., Lamotte, M. & Lamotte-Barrillon, S. (1950) Sem. Hop. Paris *26*, 3497–3514.
171. Marzo, S., Ghirardi, P., Sardini, D., Prandini, B. D. & Albertini, A. (1970) Klin. Wochen. (Berlin) *48*, 949–950.
172. Patek, A. J. & Earampamoorthy, S. (1976) Am. J. Clin. Nutr. *29*, 1122–1126.
173. Muller, P., Fellin, R. & Lamprecht, J. (1974) Eur. J. Clin. Invest. *4*, 419–428.
174. Johansson, B. G. & Laurell, C. B. (1969) Scand. J. Clin. Lab. Invest. *23*, 231–233.
175. Johansson, B. G. & Medhus, A. (1974) Acta Med. Scand. *195*, 273–277.
176. Devenyi, P., Robinson, G. M. & Roncari, D. A. K. (1980) Can. Med. Assoc. J. *123*, 981–984.
177. Taskinen, M.-R., Välimäki, M., Nikkilä, E. A., Kuusi, T., Ehnholm, C. & Ylikahri, R. (1982) Metabolism *31*, 1168–1174.
178. Hulley, S. B. & Gordon, S. (1981) Circulation (Suppl. 3) *64*, 57–63.
179. Danielsson, B., Ekman, R., Fex, G., Johansson, B. G., Kristensson, H., Nilsson-Ehle, P. & Wadstein, J. (1978) Scand. J. Clin. Lab. Invest. *38*, 113–119.
180. Barboriak, J. J., Jacobson, G. R., Cushman, P., Herrington, R. E., Lipo, R. F., Daley, M. E. & Anderson, A. J. (1980) Alcoholism: Clin. Exp. Res. *4*, 346–349.
181. Sanchez-Graig, M. & Annis, H. M. (1981) Alcoholism: Clin. Exp. Res. *5*, 540–544.
182. Fraser, G. E., Anderson, J. T., Foster, W., Goldberg, R., Jacobs, D. & Blackburn, H. (1983) Atherosclerosis *46*, 275–286.
183. Poynard, T., Abella, A., Pignon, J. P., Naveau, S., Poitrine, A., Agostini, H., Zourabichvili, O., Leluc, R. & Chaput, J. C. (1985) Gastroenterology *88*, 1685.
184. Camargo, C. A., Williams, P. T., Vranizan, K. M., Albers, J. J. & Wood, P. D. (1985) J.A.M.A. *253*, 2854–2857.
185. Puchois, P., Fontan, M., Gentilini, J.-L., Gelez, P. & Fruchard, J.-C. (1984) Clin. Chim. Acta *185*, 185–189.

186. Kraus, R. M. (1982) Med. Clin. N. Amer. *66*, 403–429.
187. Marth, E., Cazzolato, G., Bon, B., Avogaro, P. & Kostner, G. M. (1982) Ann. Nutr. Metab. *26*, 56–62.
188. Cuvelier, I., Steinmetz, J., Mikstacki, T. & Siest, G. (1985) Clin. Chem. *31*, 763–766.
189. Alling, C., Aspenstom, G., Dencker, S. J. & Svennerholm, L. (1979) Acta Med. Scand. (Suppl.) *631*, 1–38.
190. Norbeck, H.-E., Walldius, G., Carlson, L. A. & Ideström, C.-M. (1979) in: Metabolic Effects of Alcohol (Avogaro, P., Sirtori, C. R. & Tremoli, E., eds.) pp. 215–225, Elsevier/North-Holland Biomedical Press, Amsterdam.
191. Holman, R. T. & Johnson, S. (1981) Prog. Lipid Res. *20*, 67–73.
192. Johnson, S. B., Gordon, E., McClain, C., Low, G. & Holman, R. T. (1985) Proc. Natl. Acad. Sci. USA *82*, 1815–1818.
193. Jones, D. P., Losowsky, M. S., Davidson, C. S. & Lieber, C. S. (1963) J. Lab. Clin. Med. *62*, 675–682.
194. Verdy, M. & Gattereau, A. (1967) Am. J. Clin. Nutr. *20*, 997–1004.
195. Kaffarnik, H. & Schneider, J. (1970) Z. Gesamte Exp. Med. *152*, 187–200.
196. Belfrage, P., Berg, B., Cronholm, T., Elmqvist, D., Hägerstrand, I., Johansson, B., Nilsson-Ehle, P., Nordén, G., Sjövall, J. & Wiebe, T. (1973) Acta Med. Scand. (Suppl.) *552*, 5–44.
197. Avogaro, P. & Cazzolato, G. (1975) Metabolism *24*, 1231–1242.
198. Taskinen, M.-R. & Nikkilä, E. A. (1977) Acta Med. Scand. *202*, 173–177.
199. Friedman, M., Rosenman, R. H. & Byers, S. O. (1965) Proc. Soc. Exp. Biol. Med. *120*, 696–698.
200. Taskinen, M.-R., Válimäki, M., Nikkilä, E. A., Kuusi, T. & Ylikahri, R. (1985) Metabolism *34*, 112–119.
201. Schapiro, R. H., Scheig, R. L., Drummey, G. D., Mendelson, J. H. & Isselbacher, K. J. (1965) N. Engl. J. Med. *272*, 610–615.
202. Belfrage, P., Berg, B., Hägerstrand, I., Nilsson-Ehle, P., Tornquist, H. & Wiebe, T. (1977) Eur. J. Clin. Invest. *7*, 127–131.
203. Losowsky, M. S., Jones, D. P., Davidson, C. S. & Lieber, C. S. (1963) Am. J. Med. *35*, 794–803.
204. Glueck, C. J., Hogg, E., Allen, C. & Gartside, P. S. (1980) Am. J. Clin. Nutr. *33*, 2287–2293.
205. Hartung, G. H., Foreyt, J. P., Mitchell, R. E., Mitchell, J. G., Reeves, R. S. & Gotto, A. M. Jr. (1983) J.A.M.A. *249*, 747–750.
206. Schneider, J., Panne, E., Braun, H., Mordasi, R., Kaffarnik, H. (1983) J. Lab. Clin. Med. *101*, 114–122.
207. Bezman-Tarcher, A., Nestel, P. J., Felts, J. M. & Havel, R. J. (1966) J. Lipid Res. *2*, 248–257.
208. Warembourg, H., Biserte, G., Jaillard, J., Sezille, G. & Scherpereel, P. (1970) Rev. Eur. Études Clin. Biol. *15*, 649–653.
209. Baraona, E. & Lieber, C. S. (1970) J. Clin. Invest. *49*, 769–778.
210. Baraona, E., Pirola, R. C. & Lieber, C. S. (1973) J. Clin. Invest. *52*, 296–303.
211. Vasdcv, S. C., Subrahmanyam, D., Chakravarti, R. N. & Wahi, P. L. (1974) Biochim. Biophys. Acta *369*, 323–330.
212. Zieve, L. (1958) Ann. Intern. Med. *48*, 471–496.
213. Baraona, E. & Lieber, C. S. (1969) Am. J. Clin. Nutr. *22*, 356–357.
214. Goebel, L. M., Goebel, F. D., Schubotz, R. & Schneider, J. (1979) J. Lab. Clin. Med. *94*, 124–132.
215. Cameron, J. L., Capuzzi, D. M., Zuidema, G. D. & Margolis, S. (1974) Am. J. Med. *56*, 482–487.
216. Bloomfield, A. L. & Shenson, B. (1947) Stanford Med. Bull. *5*, 185–191.
217. Albrink, M. J. & Klatskin, G. (1957) Am. J. Med. *23*, 26–33.
218. Greenberger, N. J., Hatch, F. T., Drummey, G. D. & Isselbacher, K. J. (1966) Medicine *45*, 161–174.
219. Nestel, P. J. (1967) Aust. Ann. Med. *16*, 139–143.
220. Grossberg, S. J. (1966) NY State J. Med. *66*, 2951–2953.
221. Fallat, R. W., Wester, J. W. & Glueck, C. J. (1973) J.A.M.A. *225*, 1331–1334.
222. Farmer, R. G., Winkelman, E. I., Brown, H. B. & Lewis, L. A. (1973) Am. J. Med. *54*, 161–165.
223. Brunzell, J. D. & Bierman, E. L. (1982) Med. Clin. N. Amer. *66*, 455–468.
224. Talbot, G. D. & Keating, B. M. (1962) Geriatrics *17*, 802–808.
225. Brewster, A. C., Lankford, H. G., Schwartz, M. G. & Sullivan, J. F. (1966) Am. J. Clin. Nutr. *19*, 255–259.
226. Barboriak, J. J. & Meade, R. C. (1968) Am. J. Med. Sci. *255*, 245–251.
227. Wilson, D. E., Schreibman, P. H., Brewster, A. C. & Arky, R. A. (1970) J. Lab. Clin. Med. *75*, 264–274.

228. Borowsky, S. A., Perlow, W., Baraona, E. & Lieber, C. S. (1980) Dig. Dis. Sci. *25*, 22–27.
229. Avgerinos, A., Chu, P., Greenfield, C., Harry, D. S. & McIntyre, N. (1983) Hepatology *3*, 349–355.
230. Mendelson, J. H. & Mello, N. K. (1973) Science *180*, 1372–1374.
231. Ginsberg, H., Olefsky, J., Farquhar, J. W. & Reaven, G. M. (1974) Ann. Intern. Med. *80*, 143–149.
232. Debry, G., Mejean, L., Max, J. P., Lambert, D., Pointel, J.-P. & Drouin, P. (1979) in: Metabolic Effects of Alcohol (Avogaro P., Sirtori, C. R. & Tremoli, F., eds.) pp. 227–234, Elsevier/North Holland Biomedical Press, Amsterdam.
233. Kudzma, D. J. & Schonfeld, G. (1971) J. Lab. Clin. Med. *77*, 384–395.
234. DeGennes, J. L., Thomopoulus, P., Truffert, J. & Labrousse de Tregomain, B. (1972) Nutr. Metabol. *14*, 141–158.
235. Duhamel, G., Nalpas, B., Goldstein, S., Laplaud, P. M., Berthelot, P. & Chapman, M. J. (1984) Hepatology *4*, 577–585.
236. Sabesin, S. M. & Weidman, S. W. (1984) Hepatology *4*, 737–738.
237. Malmendier, C. L., Mailier, E. L., Amerijckx, J. P. & Fischer, M. L. (1983) Hepato-gastroenterol. *30*, 236–239.
238. Papadopoulus, N. M. & Charles, M. A. (1970) Proc. Soc. Exp. Biol. Med. *134*, 797–799.
239. Sabesin, S. M., Hawkins, H. L., Kuiken, L. & Ragland, J. B. (1977) Gastroenterology *72*, 510–518.
240. Day, R. C., Harry, D. S., Owen, J. S., Foo, A. Y. & McIntyre, N. (1979) Clin. Sci. *56*, 575–583.
241. McIntyre, N. (1978) Gut *19*, 526–530.
242. Balint, J. A. (1981) in: Liver Annual-1 (Arias, I. M., Frenkel, M., Wilson, J. H. P., eds.) pp. 31–48 Excerpta Medica, Amsterdam, Oxford, Princeton.
243. Middleton, W. R. J., Carter, E. A., Drummey, G. D. & Isselbacher, K. J. (1971) Gastroenterology *60*, 880–887.
244. Gordon, E. R. (1972) Can. J. Biochem. *50*, 949–957.
245. Lakshmanan, M. R., Felver, M. E. & Ezekiel, M. (1984) Alcoholism: Clin. Exp. Res. *8*, 359–361.
246. Lombardi, B. P., Pani, P. & Schlunk, F. F. (1968) J. Lipid Res. *9*, 437–446.
247. Flores, H., Pak, N., Maccioni, A. & Monckeberg, F. (1970) Br. J. Nutr. *24*, 1005–1011.
248. Seakins, A. & Watelow, J. C. (1972) Biochem. J. *129*, 793–795.
249. Wolfe, B. M., Havel, J. R., Marliss, E. B., Kane, J. P., Seymour, J. & Ahuja, S. P. (1976) J. Clin. Invest. *57*, 329–340.
250. Savolainen, M., Baraona, E., Karsenty, C. & Lieber, C. S. (1982) Hepatology *2*, 681.
251. Sane, T., Nikkilä, E. A., Taskinen, M.-R., Välimäki, M. & Ylikahri, R. (1984) Atherosclerosis *53*, 185–193.
252. Luoma, P. V., Sotaniemi, E. A., Pelkonen, R. O. & Ehnholm, C. (1982) Res. Commun. Chem. Pathol. Pharmacol. *37*, 91–96.
253. Cushman, P., Barboriak, J. J., Liao, A. & Hoffman, N. E. (1982) Life Sci. *30*, 1721–1724.
254. Malmendier, C. L. (1984) in: Diet, Diabetes, and Atherosclerosis (Pozza, G., Micossi, P., Catapano, A. L. & Paoletti, R., eds.) pp. 157–166, Raven Press, New York.
255. Windmueller, H. G. & Levy, R. I. (1968) J. Biol. Chem. *243*, 4878–4884.
256. Ockner, R. K., Hughes, F. B. & Isselbacher, K. J. (1969) J. Clin. Invest. *48*, 2079–2088.
257. Mistilis, S. P. & Ockner, R. K. (1972) J. Lab. Clin. Med. *80*, 34–46.
258. Windmueller, H. G., Herbert, P. N. & Levy, R. I. (1973) J. Lipid Res. *14*, 215–223.
259. Ockner, R. K., Mistilis, S. P., Poppenhausen, R. B. & Stiehl, A. F. (1973) Gastroenterology *64*, 603–609.
260. Redgrave, T. G. & Martin, G. (1977) Atherosclerosis *28*, 69–80.
261. Barboriak, J. J. (1966) Life Sci. *5*, 237–241.
262. Jaillard, J., Dewailly, P., Sezille, G., Fruchart, J. C., Rault, C. & Scherpe-Reel, P. (1974) Atherosclerosis *19*, 493–500.
263. Nikkilä, E. A., Taskinen, M.-R. & Huttunen, J. K. (1978) Horm. Metab. Res. *10*, 220–223.
264. Nilsson-Ehle, P., Carlstrom, S. & Belfrage, P. (1978) Lipids *13*, 433–437.
265. Sigurdsson, G., Nicoll, A. & Lewis, B. (1976) Eur. J. Clin. Invest. *6*, 167–177.
266. Mordasini, R., Kaffarnik, H., Schneider, J. & Riesen, W. (1982) Schweiz. Med. Wochenschr. *112*, 1928–1931.
267. Breier, C., Lisch, H.-J., Drexel, H. & Braunsteiner, H. (1984) Atherosclerosis *52*, 317–328.
268. Nikkilä, E. A., Kuusi, T. & Taskinen, M. R. (1982) in: Metabolic risk factors in

ischemic cardiovascular disease (Carlson, L. A. & Pernow, B., eds.) pp. 205–215, Raven Press, New York.
269. Freeman, M., Kuiken, L., Ragland, J. B. & Sabesin, S. M. (1977) Lipids *12*, 443–445.
270. Goldberg, C. S., Tall, A. R. & Krumholz, S. (1984) J. Lipid Res. *25*, 714–720.
271. Janus, E. D. & Lewis, B. (1978) Clin. Endocrinol. Metab. *7*, 321–332.
272. Schaefer, E. J. & Levy, R. I. (1985) N. Engl. J. Med. *312*, 1300–1310.
273. Cabot, R. C. (1904) J.A.M.A. *43*, 774–775.
274. Martland, H. S. (1932) in: Alcohol and Man (Emerson, ed.) p. 201, MacMillan, New York.
275. Wilens, S. L. (1947) J.A.M.A. *135*, 1136–1139.
276. Hall, E. M., Olsen, A. Y. & Davis, F. E. (1953) Am. J. Pathol. *29*, 993–1022.
277. Grant, W. C., Wasserman, F., Rodensky, P. L. & Thomson, R. V. (1959) Ann. Intern. Med. *51*, 774–779.
278. Howell, W. L. & Manion, W. C. (1960) Am. Heart J. *60*, 321–343.
279. Spain, D. M. & Bradess, V. A. (1957) Arch. Intern. Med. *100*, 228.
280. Hirst, A., Hadley, G. C. & Gore, L. (1965) Am. J. Med. *143*, 45–49.
281. Ruebner, B. H., Miyai, K. & Abbey, H. (1961) Lancet *2*, 1435–1436.
282. Parrish, H. M. & Eberly, A. L. (1961) J. Indiana State Med. Assoc. *54*, 341–347.
283. Viel, B., Donoso, S., Salcedo, D. & Alessandri, R. (1966) Arch. Intern. Med. *117*, 84–91.
284. Lifsic, A. M. (1976) Bull. WHO *53*, 623–630.
285. Gottlieb, L. S., Broitman, S. A., Vitale, J. J. & Zamcheck, N. (1959) J. Lab. Clin. Med. *53*, 433–441.
286. Ramakrishnan, S., Balasubramanian, A. & Subrahmanyam, K. (1973) Indian J. Exp. Biol. *11*, 225–226.
287. Nikkilä, E. A. & Ollila, O. (1959) Circulation Res. *7*, 588–594.
288. Brown, D. F. (1966) Metabolism *15*, 868–873.
289. Eberhard, T. P. (1936) Arch. Pathol. *21*, 616–622.
290. Goto, Y., Kikuchi, H., Abe, K., Naghashi, Y., Ohira, S. & Kudo, H. (1974) Tohoku J. Exp. Med. *114*, 35–43.
291. Rudel, L. L., Leathers, C. W., Bond, M. G. & Bullock, B. C. (1981) Atherosclerosis *1*, 144–155.
292. Tibblin, G., Wilhelmsen, L. & Werkö, L. (1975) Am. J. Cardiol. *35*, 514–522.
293. Hrubec, Z., Cederlof, R. & Friberg, L. (1976) Am. J. Epidemiol. *103*, 16–29.
294. Dyer, A., Stamler, J. & Paul, O. (1976) CVD Epidemiol. Newsletter *20*, 34.
295. Paul, O., Lepper, M. H., Phelan, W. H., Dupertuis, G. W., MacMillan, M., McKean, H. & Park, H. (1963) Circulation *28*, 20–36.
296. Morris, J. N., Kagan, A., Pattison, D. C. & Gardner, M. J. (1966) Lancet *2*, 553–559.
297. Sackett, D. L., Epid, M. S., Gibson, R. W., Bross, I. D. J. & Pickren, J. W. (1968) N. Engl. J. Med. *279*, 1413–1420.
298. Myrhed, M. (1974) Acta Med. Scand. (Suppl.) *567*, 8–93.
299. Klatsky, A. L., Friedman, G. D. & Siegelaub, A. B. (1974) Ann. Intern. Med. *81*, 294–301.
300. Stason, W. B., Neff, R. K., Miettinen, O. S. & Jick, H. (1976) Am. J. Epidemiol. *104*, 603–608.
301. Barboriak, J. J., Rimm, A. A., Anderson, A. J., Schmidhoffer, M. & Tristani, F. E. (1977) Br. Heart J. *39*, 289–293.
302. Yano, K., Rhoads, G. G. & Kagan, A. (1977) N. Engl. J. Med. *297*, 405–409.
303. St. Leger, A. S., Cochrane, A. L. & Moore, F. (1979) Lancet *1*, 1017–1020.
304. Gordon, T., Kagan, A., Garcia-Palmieri, M., Kannel, W. B., Zukel, W. J., Tillotson, J., Sorlie, P. & Hjortland, M. (1981) Circulation *63*, 500–515.
305. Criqui, M. H., Cowan, L. D., Tyroler, H. A., Bangdiwala, S., Heiss, G., Wallace, R. B. & Davis, C. E. (1982) Circulation *66* (part. II), 235.
306. Kittner, S. J., Garcia-Palmieri, M. R., Costas, R. Jr., Cruz-Vidal, M., Abbott, R. D. & Havlik, R. J. (1983) Am. J. Epidemiol. *117*, 538–550.
307. Klatsky, A. L., Friedman, G. D. & Siegelaud, A. B. (1981) Ann. Intern. Med. *95*, 139–145.
308. Dyer, A. R., Stamler, J., Paul, O., Berkson, D. M., Shekelle, R. B., Lepper, M. H., McKean, H., Lindberg, H. A., Garside, D. & Tokich, T. (1981) Circulation *64* (Suppl. III), 20–27.
309. Kozararevic, D., McGee, D., Vojvodic, N., Racic, Z., Dawber, T., Gordon, T. & Zukel, W. (1980) Lancet *1*, 613–616.
310. Eichner, E. R. (1985) Am. J. Med. *79*, 231–240.

311. Fraser, G. E. & Upsdell, M. (1981) Am. J. Epidemiol. *114*, 462–476.
312. Popham, R. E., Schmidt, W. & Israel, Y. (1983) Med. Hypotheses *12*, 321–329.
313. Barboriak, J. J. (1984) Alcohol *1*, 341–345.
314. Gordon, T. & Doyle, J. T. (1985) Am. Heart J. *110*, 331–334.
315. Ernst, N., Fisher, M., Smith, W., Gordon, T., Rifkind, B. M., Little, J. A., Mishkel, M. A. & Williams, O. D. (1980) Circulation *62* (Suppl. IV), 41–52.
316. Glueck, C. J., Heiss, G., Morrison, J. A., Khoury, P. & Moore, M. (1981) Circulation *64* (Suppl. III), 48–56.
317. Barrett-Connor, E. & Sharez, L. (1982) Am. J. Epidemiol. *115*, 888–893.
318. Miller, G. J. & Miller, N. E. (1975) Lancet *1*, 16–19.
319. Miller, N. E., Førde, O. H., Thelle, D. S. & Mjøs, O. D. (1977) Lancet *1*, 965–967.
320. Miller, N. E., La Ville, A. & Crook, D. (1985) Nature *314*, 109–111.
321. Gofman, J. W., Young, W. & Tandy, R. (1966) Circulation *34*, 679–697.
322. Miller, N. E., Hammett, F., Saltissi, S., Rao, S., Van Zeller, H., Coltart, J. & Lewis, B. (1981) Br. Med. J. *282*, 1741–1744.
323. Ballantyne, F. C., Clark, R. S., Simpson, H. S. & Ballantyne, D. (1982) Metabolism *31*, 433–437.
324. Levy, R. I., Brensike, J. F., Epstein, S. E., Kelsey, S. F., Passamani, E. R., Richardson, J. M., Loh, I. K., Stone, N. J., Aldrich, R. F., Battaglini, J. W., Moriarty, D. J., Fisher, M. L., Friedman, L., Friedewald, W. & Detre, K. M. (1984) Circulation *69*, 325–337.
325. Knochel, J. P. (1983) Ann. Intern. Med. 98 (Part 2), 849–854.
326. Friedman, H. S. (1984) Alcohol *1*, 333–339.
327. Haskell, W. L., Camargo, C. Jr., Williams, P. T., Vranizan, K. M., Krauss, R. M., Lindgren, F. T. & Wood, P. D. (1984) N. Engl. J. Med. *310*, 805–810.
328. Haffner, S. M., Applebaum-Bowden, D., Wahl, P. W., Hoover, J. J., Warnick, G. R., Albers, J. J. & Hazzard, W. R. (1985) Arteriosclerosis *5*, 169–177.
329. Williams, P. T., Kraus, R. M., Wood, P. D., Albers, J. J., Dreon, D. & Ellsworth, N. (1985) Metabolism *34*, 524–530.
330. Lieber, C. S. (1984) N. Engl. J. Med. *310*, 846–848.
331. Avogaro, P., Bittolo Bon, G., Cazzolato, G. & Quinci, G. B. (1979) Lancet *1*, 901–903.
332. Riesen, W. F., Mordasini, R., Salzmann, C., Theler, A. & Gurtner, H. P. (1980) Atherosclerosis *37*, 157–162.
333. Maciejko, J. J., Holmes, D. R., Kottke, B. A., Zinsmeister, A. R., Dinh, D. M. & Mao, S. J. T. (1983) N. Engl. J. Med. *309*, 385–389.
334. Cohen, B. I. & Raicht, R. F. (1981) Alcoholism: Clin. Exp. Res. *5*, 225–229.
335. Topping, D. L., Weller, R. A., Nader, C. J., Calvert, G. D. & Illman, R. J. (1982) Am. J. Clin. Nutr. *36*, 245–250.
336. Ashley, M. J. (1982) J. Stud. Alcohol *43*, 869–887.
337. Zeichner, A., Feuerstein, M., Swartzman, L. & Reznick, E. (1983) in: Stress and Alcohol use (Pohorecky, L. A. & Brick, J., eds.) pp. 353–368, Elsevier, New York.
338. Hennekens, C. H., Rosner, B. & Vole, D. S. (1978) Am. J. Epidemiol. *107*, 196–200.
339. Barboriak, J. J., Anderson, A. J., Rimm, A. A. & Tristani, F. E. (1979) Alcoholism: Clin. Exp. Res. *3*, 29–32.
340. Marmot, M. G., Rose, G., Shipley, M. J. & Thomas, B. J. (1981) Lancet *1*, 580–583.
341. Blackwelder, W. C., Yano, K., Rhoads, G. G., Kagan, A., Gordon, T. & Palesch, Y. (1980) Am. J. Med. *68*, 164–169.
342. Klatsky, A. L., Friedman, G. D. & Siegelaub, A. B. (1979) Alcoholism: Clin. Exp. Res. *3*, 33–39.
343. Barboriak, J. J., Anderson, A. J. & Hoffmann, R. G. (1979) J. Lab. Clin. Med. *94*, 348–353.
344. Lieber, C. S. & Savolainen, M. (1984) Alcoholism: Clin. Exp. Res. *3*, 409–423.
345. Lieber, C. S. (1985) Hepatology *5*, 1243–1260.
346. Lieber, C. S. (1987) Alcohol and the liver in: Liver Annual-VI (Arias, I. M., Frenkel, M. S. & Wilson, J. H. P., eds.) pp. 163–240 Excerpta Medica, Amsterdam, Oxford, Princeton.

Chapter 13
Dyslipoproteinemias and Autoimmunity

J. L. Beaumont and V. Beaumont

1 Introduction

Under normal circumstances, the circulating lipoproteins of one individual are "tolerated" by his own immune competent cells. This tolerance applies in the same way, on the one hand to the cell receptors and enzymes which function in lipoprotein metabolism, and on the other to the lipoproteins of cell membranes which may be similar, in their molecular structure, to the circulating lipoproteins.

Because of this state of tolerance, lipoproteins are able to assume their function in lipid transport and exchanges between the gut, the liver and other organs and tissues without triggering immunological reactions; and at the same time the enzymes and receptors are able to assume their regulatory functions. The final state of equilibrium may vary from one individual to the other, under genetic, environmental and especially dietetic conditions.

But as happens for many molecular constituents, immune tolerance for lipoproteins or their regulating factors may be suppressed, and one or several clones may develop and produce immunoglobulins (Igs) which react via their antibody sites with specific structures within lipoproteins, enzymes or cellular receptors. The latter are usually considered to be autoantigens, a correct definition from a molecular point of view, but one which does not infer that the process of suppression of the immune tolerance which leads to the production of autoantibodies is similar to the process of immunization with an heteroantigen.

The pathological consequences of this state of autoimmunity are essentially related to the nature of the antigen recognized by the antibody site which is located on the variable part of the Ig molecule, and to a lesser extent, to the characteristics of the constant part of the Igs, and the place of the producing clone in the repertory of all the clones which might produce autoantibodies.

The location of the reactive antigenic structure in the pathway of lipid and lipoprotein metabolism determines the features and the importance of the metabolic abnormality which arises. If the antigen plays a key role, the metabolic consequences may be catastrophic; if a minor one, the abnormality may not be detectable. For instance, major hyperlipoproteinemia may be induced when antilipoprotein antibodies block sites necessary for enzymatic digestion of a lipoprotein or for its receptor recognition. But no significant changes in the circulating lipoprotein levels may be seen if the antibody blocks other sites. A similar mechanism may also explain why lipids accumulate in tissues. Autoantibodies to mesenchymal structures may be responsible for xanthomatosis and perhaps atherosclerosis in a number of patients. In other cases, the autoantibody may react with a molecule implicated in several functions, and consequently apparently unrelated disorders may be associated with the disturbance of lipid metabolism: for instance, dyslipidemia and thrombosis may result from autoantibodies against heparin.

The characteristics of the stimulated immunoglobulins also play a role in generating the differences observed between patients with autoantibodies to lipoproteins. Immunoglobulins similar with regard to their antibody site and their specificity may differ in the structure of the constant part of their molecule. This was observed in IgA K myelomas: differences in polymerization of the Ig molecule resulted in major differences in blood viscosity and in a variable precipitability of circulating immune complexes from one patient to the other.

Little is known of the importance of the place of the autoantibody producing clone in the immunological repertory of an individual. Nor is it known what kind of stimulation is necessary to abrogate the immune tolerance of this clone. However, it should be noted that autoimmune-induced hyperlipidemia is rarely associated with other immunological disorders, in contrast to systemic lupus erythematosis or rheumatoid arthritis. But it is clear on the other hand, that during courses of immunization with non specific antigens, transient hyperlipidemias are often observed, and have been shown to be due to the induction of antilipoprotein autoantibodies.

As may be seen, the relationship between the immune system and lipoproteins is becoming more and more complex. In order to explore this area we will 1. review the molecules and functions concerned and the nature of the antigen-antibody reaction; 2. remind ourselves of the known clinical and experimental associations which support the hypothesis of a relationship between autoimmunity and lipoproteins; 3. analyse the different clinical and biological models which may be constructed on the basis of the characteristics of the antigen-antibody reaction; 4. consider their possible clinical consequences and their treatment; 5. try to evaluate the incidence of autoimmune hyper- and dyslipidemia in the larger context of hyperlipidemia in general.

2 Cells, Molecules and Functions

An autoimmune disorder of lipid metabolism may result from an abnormality in the interaction of 2 major molecular systems: the lipoprotein system which comprises the circulating lipoproteins and the molecular and cellular factors responsible for their production, function and degradation on one hand; and the immune competent system of circulating immunoglobulins and B and T lymphocytes on the other. The abnormality originates at the molecular level, by the binding of the antibody to structures of the lipoprotein system. Antibodies which react with constituents of their own organism are autoantibodies, and the reacting lipoproteins behave as autoantigens. The reactive part of the lipoprotein structure is specific for the antibody site.

2.1 Immunoglobulins, antibodies, paratopes, and idiotopes

Circulating immunglobulins belong to 3 main classes (isotypes): IgA, IgG, and IgM. Basically an immunoglobulin in the monomer form is made up of a four chain structure comprising 2 light chains (L) and 2 heavy chains (H). There are two kinds of L chains, λ and κ; and 3 kinds of H chains, γ for IgG, α for IgA, μ for IgM. The four chains are linked together by disulfide bonds. IgA and IgG usually circulate in the monomer form. Their molecular mass is about 150 000 daltons. IgM are polymers of IgM monomers held together by a special piece J. They have 5 units and their molecular mass is therefore around 900 000 daltons. Circulating IgA's are sometimes found in the polymer form.

The heavy and light chains of immunoglobulins have both a constant and a variable end fragment. The association of the variable part of an H and L chain governs the shape of the *antibody site* of the molecule. So, each monomer has 2 antibody sites which may be separated after papain digestion of the immunoglobulin molecule into 2 antigen binding fragments (Fab), or following pepsin digestion which generates 1 fragment containing the 2 antibody sites (Fab '2).

Each antibody site is constituted by the foldings of the flexible polypeptidic heavy and light chains. Its configuration is determined by the amino acid sequence of these chains, which controls the structure of the *paratope*, i.e. the part of the antibody which reacts with the antigen molecule.

The antibody site of an IgG comprises approximately 15 amino acids, about 1 per cent of the whole molecule. It occupies about 1 nm^2 at the surface of the IgG. The antibody site of an IgA is usually larger. For instance, the murine IgA Mo Pc 603 reacting with phosphorylcholine is $2 \times 1.5 \times 1.2$ nm (1) (Fig. 1).

The antibody sites of an individual immunoglobulin are identical since they have the same paratope. They possess another structure, specific for the individual

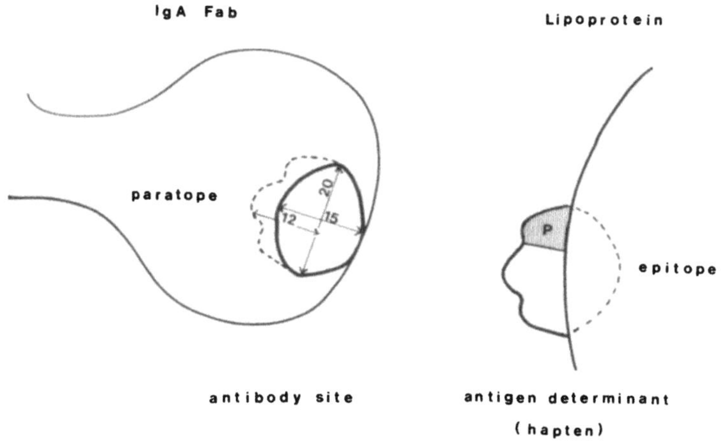

Fig. 1. A schematic antibody site and its corresponding epitope. The measurements given are those of the MOPC 603 Ig A antiphosphorylcholine site. The phosphorylcholine molecule (P) is only part of the "real" epitope. Dimensions in Angströms (0.1 nm). (Fougereau, M., Eléments d'immunologie expérimentale, Masson, Paris 1977, 2nd ed. p. 74.)

molecule, the *idiotope*. The idiotope is always located near the paratope. The specificity of the idiotope, the idiotype, may be recognized by another antibody (antiidiotype antibody).

That component of the immunoglobulin which remains after separation of the Fab fragments, the Fc fragment, comprises the constant parts of the H chains. It has important properties: complement fixation, binding to lymphocyte membranes, binding with the staphylococcal protein A, polymerization. These properties vary with the structure of the Fc fragment. For instance, IgG 1, IgG 2, IgG 3 and IgM are usually complement fixing whereas IgA and IgG 4 are not.

The circulating immunoglobulins are produced by B lymphocytes. They may also remain attached to the surface of the lymphocyte and represent specific membrane receptors. All the lineage of cells (or *clone*) which comes from a B lymphocyte, elicits immunoglobulins with antibody sites of the same structure and specificity. They also have the same idiotype.

Animals are capable of producing a great variety of antibodies by immunization with different antigens indicating that the repertory of the antigenic specificities (epitopes) recognized by the lymphocyte is extraordinarily large. It is at present accepted that for each individual, this repertory gives access to a giant library of independent lymphocyte cells, which may recognize a tremendous number of epitopes, including quite probably structures of the organism to which they belong. That is each individual may well possess B lymphocytes which recognize his or her own antigens (self recognition).

How the lymphocytes of this library are differentiated is still controversial and outside the scope of this review. But the mechanism responsible for the control of their antibody production is central to the problem of autoimmunity, and revolves around the function of the T lymphocytes. They are the first cells which respond to information on the antigen molecules present in the organism. They may or may not activate the antibody producing B lymphocytes. The modalities of this regulation are still obscure. Most likely, the B lymphocytes are suppressed as a result of interaction between their membrane receptors, which are antibody sites, and antiidiotype antibodies which block the antibody sites by reacting with their specific idiotope. T lymphocytes, stimulated by a foreign antigen in case of hetero or isoantibodies, or by a still unknown process in case of autoantibodies, might reverse this suppression and trigger antibody production and release from the B cells.

2.2 Lipoproteins, antigens, and epitopes

The circulating lipoproteins, VLDL, LDL, HDL and their derivatives are macromolecular complexes containing lipids, carbohydrates and polypeptides. The polypeptides or apolipoproteins, are the stabilizing component of these structures. The lipoprotein surface is occupied by the hydrophilic part of the polypeptides and the hydrophilic pole of the lipids: phospholipids, glycolipids, and non esterified cholesterol.

The hydrophilic surface of the particles renders them stable and water soluble. It is in contact with enzymes and cell receptors and also with the cells of the immunocompetent system. From this point of view, the lipoprotein surface appears as a mosaic which reproduces, according to the class of lipoprotein, a variable number of sites, corresponding to numerous antigenic determinants or epitopes.

Each of the lipoprotein classes of an individual has epitopes peculiar to its own lineage and its own species, and also epitopes common to several lineages and several species. Each epitope is repeated to a greater or lesser extent on the molecule surface. In general, the larger the molecule, the greater the number of epitopes on its surface. According to the surface area calculations of Eisenberg (2) for the circulating lipoproteins, and taking into account that an antibody site may cover a surface of 1 to 3 nm^2, the number of possible epitopes at the surface of a lipoprotein molecule might be estimated as hundreds for HDL and thousands for LDL. However, the number of the different epitopes on a molecule is certainly smaller, for it is limited by the number of apolipoproteins contained in the molecule (Table 1).

Table 1. The Surface of Circulating Lipoproteins*

	VLDL	LDL	HDL_2	HDL_3
Diameter (nm)	42.7	21.5	10	8
Surface (nm^2)	5730	1450	310	200
Molecules on surface (mole/mole lipoprotein)				
Apo B	1	1	–	–
Apo E	24	–	–	–
Apo C	180	–	1	±
Apo A_1	–	–	5	3
Apo A_2	6	6	2	1
Phospholipids	4300	743	137	51
Unesterified cholesterol	3700	538	50	13

* Adapted from Eisenberg, S., Plasma lipoprotein conversion, in: Lipoprotein Structure, Ann. N. Y. Acad. Sci., *348*, 30–44 (1980)

Immunization of an animal with the lipoproteins of a donor animal produces antibodies specific for the epitopes the donor species. These antisera provide tools currently in use in laboratories for the analysis and quantitation of lipoproteins and their apoproteins A, B, C and E. They consist of complex mixtures of antibodies against different epitopes. But it is possible, by selection and hybridization of the antibody producing cells, to obtain stable clones, which are able to elicit monoclonal antibodies specific for one epitope. Such tools have a growing place in biological practice.

Within one species, isoimmunization is possible, as may be observed in poly-transfused patients. In 1961, 2 specific epitopes for variants of LDL were reported (the Ag system) by Allison and Blumberg (3), and Berg, in 1971, described the Lp(a) epitope (4). For HDL, an isoantibody reacting with epitope Va was described by Galango (5).

Finally, it should be noted that among the autoantibodies which will be described here, some recognize epitopes specific for a lineage (eg site Lp AS, an allotype present in 5% of humans (6)), some are specific for the species as a whole, while some are common to several species, or even to the whole phylum. Two "phylotypic" sites have already been described, common to all vertebrates from fish to man: Site Pg (7) and site Si e (8). Some of these antibodies react with the polypeptide structures of the apoproteins, while others recognize the lipid structure and more particularly the glycophospolipids of the lipoprotein surface. In both cases, it must be supposed that the lipid epitopes are stabilized by the presence of apolipoproteins.

The number of epitopes recognized by an antibody at the surface of a lipoprotein varies according to the characteristics of the antibodies and lipoprotein class.

In two cases (9) we have been able to identify 60 and 56 epitopes on LDL, and 20 and 22 epitopes on HDL. This study also indicated that the sites were grouped by 3 on HDL and 2 on LDL in one case, and were distant from one another in the other case.

2.3 Structure-function relationships

Circulating lipoproteins share common transport and regulatory functions. Briefly, the transport functions are assumed by chylomicra and their remnant particles for the lipids and lipid soluble substances absorbed by the gastrointestinal tract (vitamin A, K, D and drugs); VLDL and their degradation products IDL and LDL are responsible for transporting the lipids synthesized in the liver; while HDL assimilates and carries tissue lipids.

Lipids, and especially cholesterol are also exchanged between the different lipoproteins via transfer proteins (10).

Lipoproteins also play an important role in regulating the synthesis of lipids like cholesterol and fatty acids within the tissues; and finally some of them have structures common to the lipoproteins of cell membranes.

In all these functions the apoproteins share an important role (Table 2) in conjunction with enzymes and tissue receptors like plasma and liver lipoprotein lipases, lecithin: cholesterol acyl transferase, the LDL receptor, the Apo E receptor and specific receptors present on macrophages (scavenger pathway (11)). Antibodies

Table 2. Functions of Apolipoproteins in Lipoprotein Metabolism*

Function	Apolipoprotein
Lipoprotein biosynthesis/secretion	B
Lipoprotein-lipid metabolism	
Lipoprotein lipase (LpL) activity	CII
Lecithin:cholesterol acyltransferase (LCAT) activity	AI
Cholesteryl ester transfer	D
Recognition of lipoproteins by cell receptors	
Cultured cells, leucocytes, adrenal cortex, gonads	B and E
Liver	B and E
Inhibition of cellular uptake, liver	C's

* From R. J. Havel, Lipoprotein biosynthesis and metabolism, in: Lipoprotein Structure, Ann. N.Y. Acad. Sci. *348*, 25 (1980)

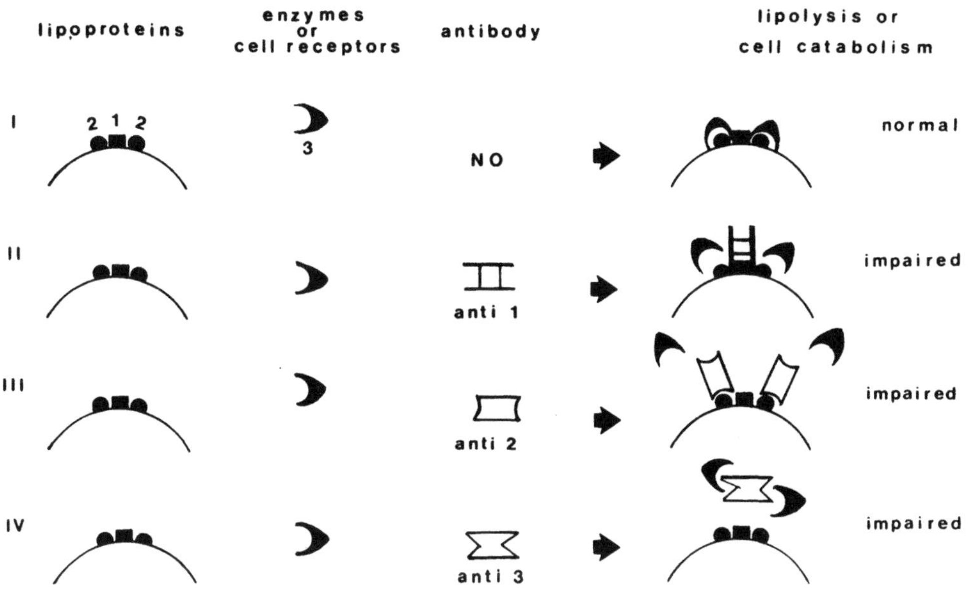

Fig. 2. The binding of an enzyme to a lipoprotein induces normal lipolysis (I). Lipolysis may be impaired either by antibodies to the lipoprotein (II), to the binding site of the enzyme to the lipoprotein (III), or to enzyme (IV). In the same way antibodies may impair the catabolism of lipoproteins when reacting with an epitope identical to or situated near the binding site of a cell receptor (1: lipoprotein epitope 2: lipoprotein epitopes for enzyme or cell receptor binding 3: enzyme or cell receptor).

may interfere with each of the above functions and result in an autoantibody dyslipoproteinemia (Fig. 2).

2.3.1 Antilipoprotein antibodies

Antilipoprotein antibodies can induce hyperlipoproteinemia when the epitopes with which they react are near to or cover the site of attack of the lipoprotein by an enzyme, or the site by which the lipoprotein is recognized by a receptor. In both cases degradation of the circulating lipoprotein is delayed, as has been shown by fat tolerance tests, especially vitamin A tests (12, 13) (Fig. 3) or by isotopic dilutions of ^{14}C-cholesterol (14).

The composition of the accumulated lipid particles differs according to the epitope concerned: they are enriched in triglycerides if the antibody blocks the enzymatic site of a triglyceride lipase (15) or are especially enriched in cholesterol if the epitope is a component of the site by which the LDL binds to the receptor (8, 16). Furthermore if the antibody interferes with lipoprotein sites specific for receptor recognition, the circulating lipoproteins may lose their cell regulating function. In such cases, an accumulation of lipids in the cell may be associated with hyperlipoproteinemia and produce xanthomas, as already reported in several cases of autoimmune hyperlipidemia (14, 17).

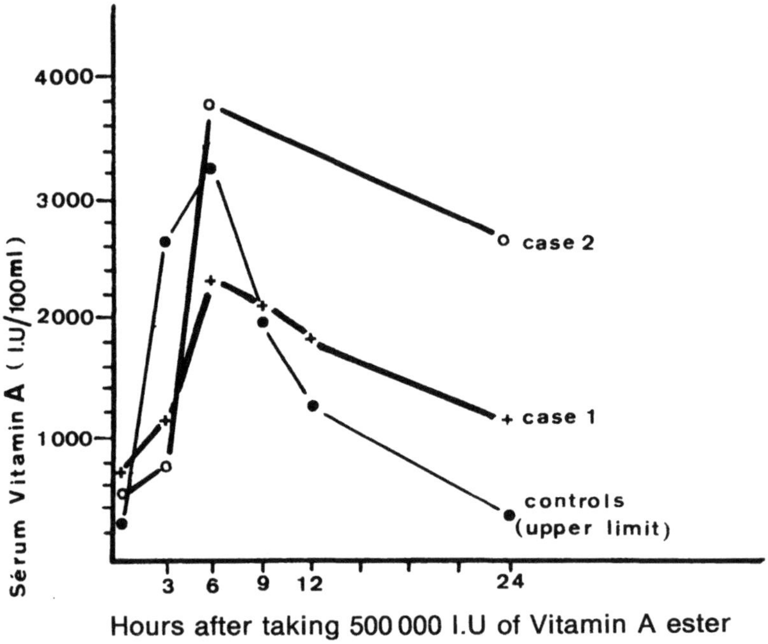

Fig. 3. Vitamin A tolerance test: total vitamin A plasma levels before (time 0) and 3, 6, 9, 12 and 24 hours after ingestion of 500 000 I. U. vitamin A ester in 2 anti Pg Ig A myeloma GER (case 1) and SOR (case 2). Upper limit of normal: M = mean ± 2 SD in 26 controls.

In contrast, when the reacting epitopes are not related to an important function, lipid metabolism may not be disturbed by their reaction with the antibody. Hypolipidemia may also be observed. Finally in all types of autoimmune hyperlipidemia, the possible pathogenicity of the circulating immune complexes formed between antibodies and lipoproteins, and the role of complement must also be considered. However the antibodies studied to date have not been shown to fix complement.

2.3.2 Antienzymes, antireceptors and antibodies to the lipid components of cell membranes

Antibodies may be generated against epitopes located on an enzyme, a receptor necessary for the metabolism of lipoproteins or a lipid transported by lipoproteins. In such cases, an autoimmune hyperlipidemia is induced in the absence of an antilipoprotein antibody. In this regard, antiheparin and antilipoprotein lipase antibodies have already been described (18, 19). Experiments made with hetero immune antibodies have confirmed that antibodies to the LDL receptor interfere with LDL catabolism, in vitro as well as in vivo (83, 84). A well documented case of auto immune hyperlipidemia, induced by an anti LDL receptor antibody was first described in 1986 (85).

If the epitope is a constituent of a cell membrane the viability of the cell may be disturbed. For example, a case of hyperlipidemia associated with anemia has been observed, in which the antibody reacted with circulating HDL and also with a lipoprotein of red blood cell membranes (20).

2.3.3 **Molecular interaction**

The molecular interaction between the lipoprotein epitope and the corresponding antibody has the characteristics of classical immunological reactions, but also displays features specific to the lipoprotein system.

Like all antigen-antibody reactions, lipoproteins and immunoglobulins are not covalently bound. Their binding is reversible, according to the Law of Mass Action, and is determined by complementarity between the antibody site surface and the antigen site. This complementarity is not only steric, but also involves ionic, hydrogen, van der Waals and hydrophobic bonds. These hydrophobic forces are probably important in the lipoprotein system under consideration.

In vivo, when an antibody reacts with circulating lipoproteins, immune complexes form, which are usually soluble. However, when the antibody is polymerized, an hyperviscosity syndrome may follow (8, 9).

In vitro, the immune complexes are usually soluble in plasma and serum. But they often precipitate or trail on gel electrophoresis and lengthened and modified precipitin lines may be seen on immunoelectrophoresis (9). These can be separated from serum by flotation on ultracentrifugation, gel filtration, or precipitation at the increased ionic strength produced by addition of 25% ammonium sulfate.

In rare cases, spontaneous precipitation *in vitro* may be observed, when the serum has been kept at + 4 °C or at − 20 °C, or even simply diluted in saline. Spontaneous agglutination of lipid particles may also be detected under the microscope or by naked eye as turbidity in the serum. This phenomenon may have diagnostic value if the serum is examined using appropriate conditions.

Solubility *in vivo*, and precipitability *in vitro* are dependant on the extent of interaction between the antigen and antibody. In autoimmune diseases, antibodies are usually mono or pauciclonal and recognize a limited number of epitopes on the antigen molecule. So, they are not precipitating antibodies, except when they are polymerized. To detect their activity *in vitro*, more sensitive methods than immunoprecipitation in saline or in gels must be used. In this context, radio immuno assay, ELISA, gel filtration and ultracentrifugation become important.

3 Suggestive Clinical and Experimental Associations

In a number of clinical and experimental conditions, modifications of the immunological status of the subject are associated with disorders of lipid metabolism. These cases have opened the way to our understanding of the autoimmune mechanisms responsible for a number of dyslipoproteinemias.

3.1 Clinical conditions

Human immunological diseases may be associated with dyslipoproteinemia.

Patients with monoclonal gammopathy provide the best examples of this association. Low serum lipid levels are usually found in myeloma (21–27). However, since the first publication of Hill and coworkers in 1948 (28), a growing number of myeloma cases with major hyperlipidemias have been reported (7, 8, 15, 17, 18, 19, 29), with or without xanthomatosis (6, 16, 30–42). The best documented cases are summarized in Table 3.

Disturbances of lipid metabolism may also appear only at the tissue level, and many cases of monoclonal gammopathy with xanthomatosis have been reported in normolipidemic patients (43, 44).

Other types of immunological disturbance may also link dysglobulinemia with modifications of lipid metabolism.

Hyperlipidemia was reported in a patient with cryoglobulinemia (45), in lupus erythematosus (46, 47) and in a case of lymphoma (46).

Immunization of volunteers with influenza vaccine was shown to be followed by a significant increase in serum cholesterol (48).

Conversely hypolipidemia was described by Noseda et al. (49) in patients with rheumatoid arthritis.

In major Ig A deficiencies, reports have appeared of an obstructive biliary syndrome associated with hyperlipidemia and palmar xanthomatosis (50). The hyperlipidemia was usually attributed to biliary obstruction, but a modification in the transport function of the Ig A should also be considered. Bile is known to contain large quantities of Ig A; and serum Ig A, complexed with radiolabeled antigens, was been recovered from bile (51).

Furthermore, 16 of 56 patients with apparently idiopathic hyperlipoproteinemia were shown to have a decrease in serum Ig M, while 11 had a reduction in Ig G and 8 in Ig A. Their lipid disorders were characterized by hypertriglyceridemia in 86% of cases (52).

3.2 Animal experimentation

Arguments supporting an association between immunoglobulins and lipids may also be found in animal experimentation.

Vasectomy in rabbits is followed by the development of antisperm antibodies associated with significant increases in plasma cholesterol and triglycerides (53).

Immunization of rabbits with a variety of antigens also results in an increase in cholesterol and triglyceride which is maximal in the third week of immunization (54, 55).

3.3 History of autoimmune hyperlipidemia

The above examples suggest that there is a relationship between the production of specific immunoglobulins and lipid metabolism. Lennard-Jones (30) first suggested that there might be a link between the production of monoclonal gamma globulin

Table 3. Monoclonal Gammopathy Associated with Major Hyperlipidemia

Reference			abnormal immuno-globulin	Aspect of serum	Hyperlipidemia Serum cholesterol	Serum triglycerides	tuberous	Xanthomatosis plane	tendinous
1960	Lennard-Jones	(30)	IgA	milky	↑	↑	+	+	–
	Heremans	(31)	IgA	milky	↑	↑	–	+	–
1962	Kayden	(32)	IgG	milky	↑	↑	–	+	–
1964	Levin	(33)	IgG κ	?	↑	↑	+	+	–
	Cohen	(34)	IgG λ	milky	↑	↑	+	+	–
1965	Beaumont	(16)	IgA κ	milky	↑	↑	+	+	+
	Savin	(35)	IgA	milky	↑	↑	+	–	–
	Lewis	(36)	IgA λ	milky	↑	↑	+	+	–
1967	Beaumont	(9)	IgA κ	milky	↑	↑	+	+	–
	Beaumont	(6)	IgG κ	milky	↑	↑	+	+	–
1968	Spikes	(38)	IgA	milky	↑	↑	–	+	–
	Watson	(42)	IgA	milky	↑	↑	–	+	–
1970	Ozer	(29)	IgA	?	↑	↑	–	–	–
	Mullinax	(41)	IgA κ	milky	↑	↑	?	?	–
	Beaumont	(18)	IgA λ	milky	↑	↑	–	–	–
1972	Cloarec	(39)	IgG κ	milky	↑	↑	+	+	–
	Glueck	(19)	IgG κ	milky	–	↑	–	–	–
1974	Beaumont	(15)	IgG λ	milky	↑	↑	–	–	–
1976	Beaumont	(20)	IgM κ	clear	↑	–	–	–	–
1979	Descamps	(40)	IgA λ	milky	↑	↑	+	–	–
1980	Baudet	(17)	IgA κ	clear	↑	↑	–	–	–
1985	Kilgore	(8)	IgA κ	milky	↑	↑	–	–	–

and the occurrence of hyperlipidemia and xanthomatosis in patients with myeloma. The molecular interaction between the M protein and serum lipoproteins in some of these cases provided the first physicochemical support for this hypothesis. Heremans (31) proposed that the association of an Ig A myeloma, hyperlipidemia and xanthomatosis could be the result of a special attribute of the M protein. Cohen et al. (34), in turn, demonstrated that the serum of an Ig G myeloma patient with hyperlipidemia contained a factor which prevented the α and β lipoproteins of serum from a normal subject from moving into a starch gel by electrophoresis. The name of "serum lipoprotein altering factor" was proposed by Spikes et al. (38). The first demonstration of an antigen-antibody like interaction was published in 1964 by Beaumont et al. (16) and confirmed later by others.

Different types of immunoglobulins and a variety of lipid disturbances were implicated in this interaction. In 22 patients with hyperlipidemia and myeloma (Table 3) or benign monoclonal gammapathy, the reactive Ig was an Ig G in 7 cases, an Ig A in 14 cases, and an Ig M in 1 case. The light chains, when determined, were either κ (10 cases) or λ (5 cases).

Different specificities were also found. Some Igs behaved as antilipoprotein antibodies, and some as antienzyme antibodies. Consequently, a variety of clinical and physicochemical pictures appeared, leading to the concept of autoimmune hyperlipidemia (AIH).

In 1970, AIH was defined as "a metabolic disease in which hyperlipidemia is induced by the antibody site binding activity of immunoglobulins which interferes with the molecular interactions involved in the normal lipolytic processes regulating the transport and removal of plasma lipids". (56).

A similar mechanism also seems to occur in normolipidic autoimmune xanthomatosis. Finally cases of autoimmune hypolipidemia were reported in which a decrease in circulating lipids and lipoproteins was induced by the antibody activity of Ig's. In this last disease, the lipolytic process may be abnormal but the catabolism of plasma lipids is not impaired.

Different clinical and biological models of this autoimmune desease will be described.

4 Models

4.1 Autoimmune hyperlipidemia

Autoimmune hyperlipidemia (AIH) was shown to be induced by 3 kinds of autoantibodies: antilipoprotein antibodies, antienzyme antibodies and anti receptor antibodies.

4.1.1 Antilipoprotein autoantibodies

Ig A class antibodies

Antilipoprotein autoantibodies have been linked to 3 classes of monoclonal immunoglobulins, Ig A, Ig G and Ig M (Tables 4 and 5)

Table 4. Antilipoprotein Antibodies in Monoclonal Gammopathy: Serum lipoprotein levels

Lp epitope	Pg GER	PgSOR	P[1] Bar	P Sie	Le	P Ra	A[2] Sa	H[3] Lac	SHRBC[4]
Ig Type	IgA κ	IgA κ	IgA κ	IgA κ	IgA λ	IgA λ	IgG κ	IgG κ	IgM κ
Reference	9, 16	9	17	8	13, 36	40	6, 9	15	20
Sex	male	male	female	male	male	female	female	male	female
Age (years)	57	62	58	74	48	57	52	38	63
Serum									
Aspect	milky	milky	turbid	turbid	milky	milky	milky	milky	clear
Cholesterol (mmol/l)	10.4–23.3	20–24	14.5	35.5	16.4–28.6	16.3	9.3–10.4	8.3–13.5	9.0
Triglycerides (g/l)	1.8–14.4	4.3–13.3	4.5	36.2	7.5–23.2	3.8	4–10	3–30	1.9

[1] P: phylotypic [2] A: allotypic [3] H: isotypic for human species [4] SHRBC: stored human red blood cell Lp associated antigen.

Table 5. Antilipoprotein Antibodies in Monoclonal Gammopathy: Immunological features

Lp epitope	Pg Ger	Pg SOR	P Bar	P Sie	Le	P Ra	A Sa	H Lac	SHRBC
Reference	9, 16	9	17	8	13, 36	40	6, 9	15	20
M protein									
Ig type	IgA κ	IgA κ	IgA κ	IgA κ	IgA λ	IgA λ	IgG κ	IgG λ	IgM κ
Active papain fragment	Fab	Fab	?	Fab	?	Fab	Fab	?	Fab
Polymerization	+++	+	–	+++	?	–	–	–	–
Reactive antigen									
Human Lp[1])	VLH	VLH	VHl	VL	VLH	VL	VLH	VH	LH
Animal Lp	+	+	+	+	?	+	–	±	+
Location of epitope									
Lipid	+	+	+	–	?	–	+	–	?
Peptide	–	–	–	Apo C	?		–	Apo B, E	?

[1] V: VLDL; L: LDL; H: HDL.

Several cases of hyperlipoproteinemia associated with monoclonal IgAs have been studied at the molecular level and provide model systems of AIH. Any differences between cases depend on the epitope involved in the reaction. They will be designated according to the following characteristics: Ig A, for the immunoglobulin class of the autoantibody; Lp, indicating that the epitope is a part of a lipoprotein structure; capital letters to identify whether the epitope is an allotype (A), is specific to the human species (H), or is phylotypic and common to other species (P); and finally the first letters of the patient's name will identify each individual case.

The Ig A k anti Lp Pg (9, 16) was the first to be detected in a 57 year old man, Ger..., who presented with a mixed hyperlipidemia, milky serum, cholesterol levels ranging from 10–24 mmol/l and triglyceride levels from 2–16 mmol/l. He had palmar and plantar tuberous and tendinous xanthomas, ischemic heart disease with angina pectoris and myocardial infarction, and, from time to time, an hyperviscosity syndrome which required repeated plasmapheresis until he died at age 59. His circulating chylomicra, VLDL, LDL and HDL were complexed with a polymerized Ig A k. Vitamin A tolerance tests demonstrated that the complexed lipoproteins were very slowly metabolized, a finding which explained the accumulation of lipoproteins in his serum.

The Ig A k was purified by several methods, including ultracentrifugation in the presence of sodium oleate which provided the best results. Its antilipoprotein activity against purified LDL and HDL was demonstrated by precipitation in saline and agarose gels, by passive hemagglutination and by gel-filtration on Sephadex G 200. In different tests the Ig appeared in a more or less polymerized form, only the polymerized molecules being agglutinating. This reaction was attributed to the Fab fragment. The Pg structure was shown to be present on LDL and HDL, and also on VLDL and chylomicra of all patients studied. The site occured 60 times on each LDL and 20 times on each HDL molecule. The Pg site was found to contain phosphorylcholine and in a study using fluorescence quenching was shown also to contain choline, phosphorus and glucose. The Ig A k reacts with glucophosphorylcholine at a concentration of 1.6×10^5 M^{-1}ka and with its Fab fragment at 2.9×10^5 M^{-1} ka (57). This epitope occured on the lipoproteins of all the mammals and vertebrates studied, including fish, cow, goat, rabbit, mice and rat.

A second case of Ig A k anti Lp Pg, Sor... (9) was studied in a man with increased levels of cholesterol and triglycerides, xanthomatosis and atherosclerosis expressed by intermittent claudication and angina pectoris. He had no hyperviscosity syndrome and the Ig A did not appear in a polymerized form. This Ig A reacted with the same epitope present on human HDL, LDL and VLDL, and also had an affinity for other mammalian and vertebrate lipoproteins.

In both of the above cases, the antibody was shown to suppress the ability of LDL to inhibit cholesterol synthesis in cultured fibroblasts, though it did not interfere with binding of LDL to its membrane receptors (17) (Table 6).

An Ig A k anti Lp P Ba (17) was found in a 58 year old female with myeloma, who had mild hyperlipidemia but a highly abnormal vitamin A tolerance test. Xanthomatosis and atherosclerosis were not evident. The autoantibody reacted with a phylotypic epitope common to the HDL, LDL and VLDL from a broad spectrum of vertebrates. But in distinction to the Lp Pg system, the Ig A anti Lp

Table 6. Fixation, internalization and degradation of LDL by normal human fibroblasts in the presence of normal versus myeloma proteins after a 22 hour incubation

	LDL fixation %	LDL internalization %	LDL degradation %
Normal Igs	n. m.[1]	n. m.	n. m.
IgA SOR	n. m.	↓ − 42.3	↓ − 32
IgA GER	↑ + 100	↑ + 33.7	↓ − 39
IgA Bar	n. m.	n. m.	n. m.

[1] not modified.

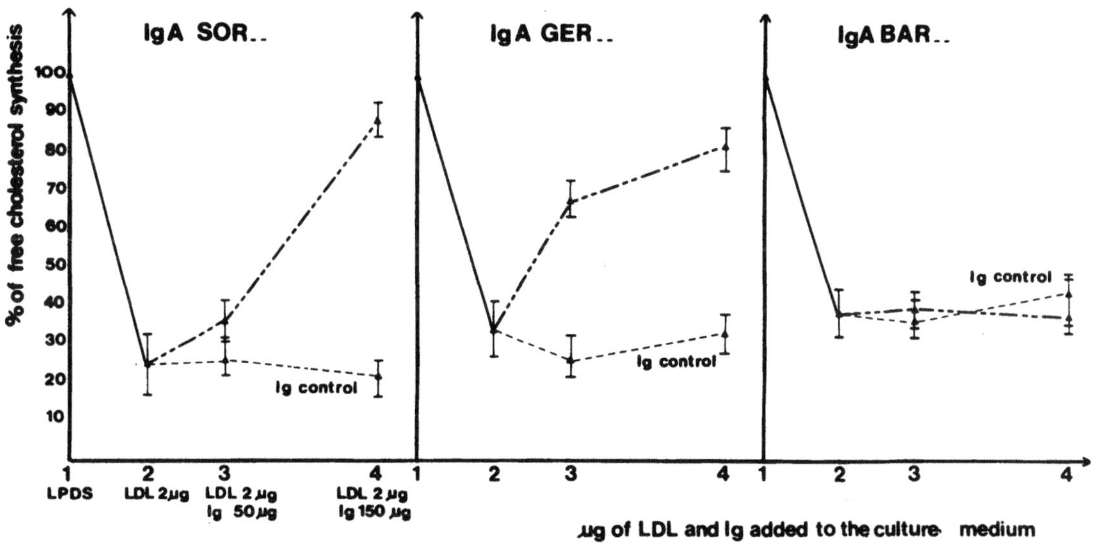

Fig. 4. Cholesterol synthesis in normal human fibroblasts studied by the incorporation of 1−^{14}C acetate to free cholesterol, in the presence of LDL (———), LDL + normal Igs (– – –), LDL + myeloma Ig As (BAR, SOR and GER cases) (— – – – —).

P Ba did not suppress the ability of LDL to inhibit cholesterol synthesis in cultured fibroblasts (Table 6, Fig. 4).

An IgAk anti Lp P Sie (8) was described in a 74 year old man with myeloma, hyperlipidemia and no xanthomas. Circulating agglutinates were visible in the retinal vessels and he had an hyperviscosity syndrome necessitating plasmapheresis. The isolated Ig A k was found to react by its Fab fragment with an epitope common to VLDL, IDL and LDL of all vertebrates tested, humans at the one extreme and fish at the other. But it differed from the anti Lp Pg and anti Lp P Ba described above, in that it did not react with HDL. Furthermore the antibody was found to bind to delipidated Apo B and E, another difference from the previously studied cases.

Ig A λ anti Lp Le (13, 36) was one of the first models studied. Its clinical features were very similar to those of the Pg model, with a major increase in cholesterol and triglyceride levels, a decrease in the catabolism of circulating lipoproteins demonstrated by vitamin A tolerance tests (13) and an increase in cholesterol synthesis demonstrated in vivo by 14 C-cholesterol exchange studies (14). The reacting epitope was common to LDL, VLDL and HDL, like epitope Pg. Its destribution on animal lipoproteins was not studied. This patient, a 48 year old man, had a benign monoclonal gammopathy, xanthomas and atherosclerosis. He was followed for more than 15 years. Remissions in the hyperlipoproteinemia and xanthomatosis were observed when plasmapheresis and lipid lowering drug therapy was instituted.

This epitope is probably different from epitope Pg in that it reacts with an Ig A λ instead of an Ig A k.

Ig A λ anti Lp P Ra (40) is a well studied model, originally described in a 57 year old woman with myeloma, hyperlipidemia and xanthomatosis. The activity of the Ig A λ was shown to be directed against an epitope common to LDL and VLDL of humans and mammals. But in contrast with epitope Pg, it was not present on HDL.

Ig G class antibodies

Detailed studies of two models of Ig G antilipoprotein autoantibody are available. The Ig G k anti Lp A Sa (6, 9) was found in a woman with a mixed hyperlipidemia in which hypertriglyceridemia predominated. She had no xanthomas. When added to purified lipoproteins, the antibody was not precitipating in saline or in agarose gels, and only poorly agglutinating in plasma agglutination tests. However, the activity of the whole Ig G and its Fab fragment was demonstrated by means of Sephadex G 200 gel filtration. The corresponding epitope A Sa was found on the VLDL, LDL and HDL molecules of about 5% of the humans studied, but not in lower animals. This allotypic epitope is different from the above phylotypic epitopes which were found to react with other Ig A antibodies.

The Ig G λ anti Lp H Lac (15) was detected in a case of benign monoclonal Ig G gammapathy associated with nephrotic syndrome and predominant hypertriglyceridemia. There were no xanthomas. The Ig G λ was purified by ultracentrifugation followed by affinity chromatography on apo HDL bound to Sepharose. It could agglutinate HDL-coated red blood cells and inhibit *in vitro* fatty acid release from a plasma activated triolein substrate by post heparin plasma lipoprotein lipase (Fig. 5). The Lac epitope was found to be present on human VLDL and HDL but not on LDL; it was also present on delipidated HDL and probably as a trace contaminant in commercial albumin preparations. It was not found in animal lipoproteins except perhaps as a trace in rabbit HDL. This epitope being restricted to the human species may be called Lp H Lac. It seems necessary for normal lipoprotein triglyceride hydrolysis by lipoprotein lipase. It might be part of an Apo C protein.

Ig M class antibodies

Although Ig M has often been reported to "contain" lipids, hyperlipidemia is rare in Ig M monoclonal gammopathies. Only one model has been studied at present.

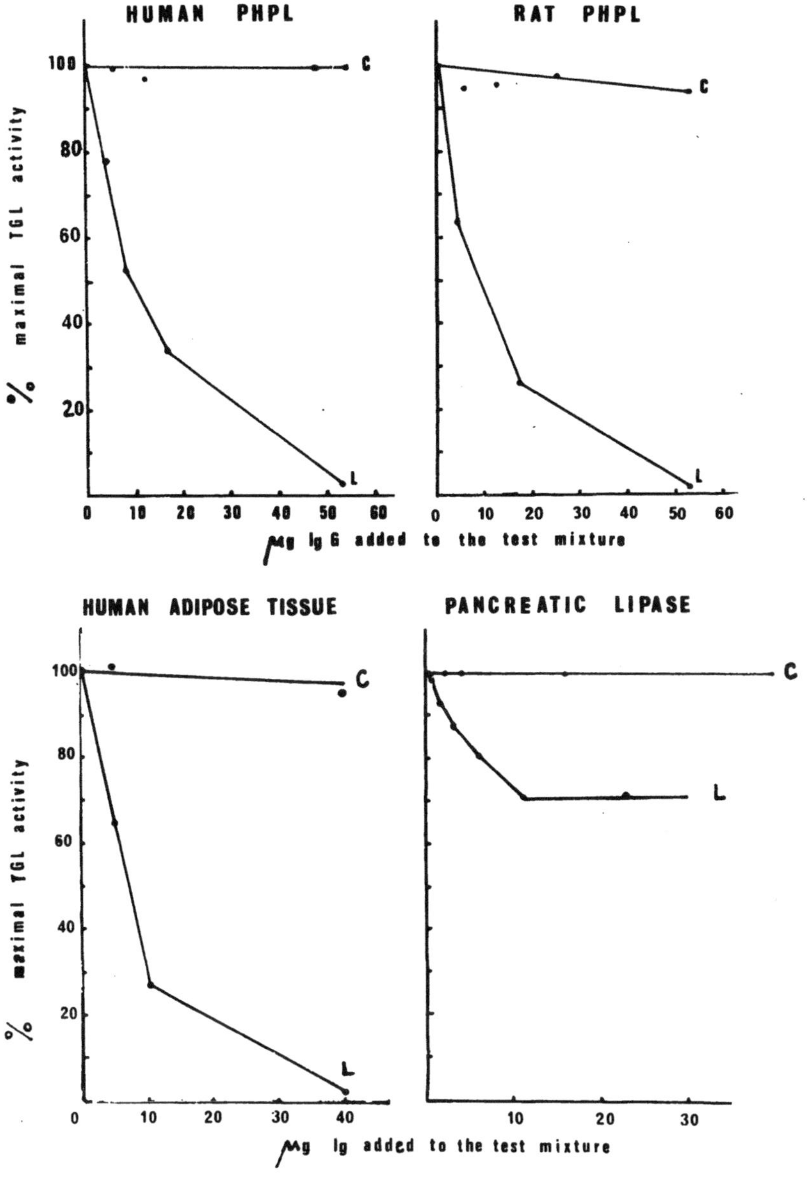

Fig. 5. Effect of lg G λ Lac on the triglyceride lipase activity of several lipases. Fatty acid release after 15 min incubation in mixtures of human serum-activated ^{14}C triolein substrate and lg G λ Lac (L) or pooled human normal lg G (C).

The Ig M anti Lp P Re (20) was described in a 63 years old female with anemia and moderately increased serum cholesterol levels. After purification the Ig M was found to bind to stored human red blood cells (SHRBC) and human LDL. The SHRBC epitope was present, in cryptic form, in all fresh human erythrocytes, but not in fresh or stored animal red blood cells. The SHRBC epitope and LDL were shown to cross react with the Ig M Re and its Fab fragments. But in contrast to the SHRBC epitope, the LDL epitope was not species specific and was also found in rabbit and rat LDL. Furthermore, cross reacting soluble inhibitors were present in human saliva, in a low molecular weight fraction isolated from human and rabbit serum, and in human and animal glomerular membranes. This model is an example of an auto antilipoprotein antibody which recognizes not only an epitope on circulating LDL, but also a similar structure on erythrocyte and other cell membranes. It is noteworthy that this structure is a secreted, non human specific, substance. It is related to other lipoprotein-associated erythrocyte antigens which have already been described, including the Lewis blood group Ag (58), the J substance of cattle (59), the HDL Age antigen which is closely related to blood group substance A (60) and the HB erythrocyte autoantigen of NZB mice (61).

Ig M s reacting with lipoproteins and epitopes present at the surface of synthetic lipid emulsions have also been detected in several cases of Waldenstroëm macroglobulinemia. As the outcome was hypolipidemia rather than hyperlipidemia, they will be discussed later.

4.1.2 **Antienzyme antibodies**

Antienzyme antibodies also induce AIH when the antibodies are directed against the enzymes necessary for normal lipoprotein lipolysis. Several cases are already known. At present, they all are antiheparin antibodies (Table 7).

Table 7. Antiheparin Antibodies in Dysgammaglobulinemia

	Antiheparin Sab	Antiheparin Ed	Antiheparin D. M.	Antiheparin F. M.	Antiheparin RD
Reference	18	19	46	46	46
Sex	female	male	female	female	male
Age (years)	54	60	18	11	24
Lipidemia					
aspect	milky	milky	milky	milky	milky
cholesterol (mmol/l)	7.8 – 15.5	?	9.0	14.3	15.5
triglycerides (mmol/l)	2.8 – 22.6	6.8 – 13.6	43.5	44.3	21.2
Disease	myeloma	myeloma	LE[1]	LE	lymphoma
M protein	IgA λ	IgG κ	IgG	IgG	IgA – IgM
Active papain fragment	FAb	Fab	?	?	?

[1] LE: lupus erythematosus.

An Ig A λ Sab (18) was found in a patient with predominant hypertriglyceridemia, who had experienced several episodes of thrombosis which appeared to be rather resistant to heparin therapy. The Ig A λ Sab was purified by chromatography on DEAE cellulose/Biogel P_2 columns with an intervening front of glycine buffer, pH 10.5; 6 mol/l urea; 0.5 mol/l NaCl. The purified Ig was shown to inhibit both antithrombin and prolipase heparin activity. Heparin fragments, desulfated and hydrolyzed in 0.1 mol/l HCl at 100 °C could still react.

An Ig G k (19) extracted from the plasma of a patient with myeloma, produced moderate hypercholesterolemia and marked hypertriglyceridemia. Post heparin lipoprotein lipase activity was decreased when heparin was injected at the usual dosage, but could be normalized with higher dosages. The activity of this Ig G ED which could be obtained in a crystallized form was able to inhibit both the anticoagulant and prolipase activities of heparin.

Similar cases were found in 2 patients with lupus erythematosus and lymphoma (46). The characteristics of the immunoglobulins were not studied.

4.1.3 **Anti receptor antibodies**

Anti receptor auto antibodies were demonstrated in a 32 years old man with a benign Ig A k paraproteinemia associated with a non familial type II hypercholesterolemia ranging from 15 to 24 mmol/l, xanthomatosis and severe atherosclerosis (85). This Ig A k H did not react with the circulating lipoproteins but was shown to inhibit the binding of LDL to in vitro cultured human fibroblasts. This Ig A interacted directly with the LDL receptor in ligand blotting experiments and the interaction was blocked by treating the receptors with reducing agents.

4.2 Autoimmune hypolipidemia

An autoantibody activity against lipoproteins associated with hypolipidemia was first detected in a patient with multiple myeloma (62). The serum of the patient contained an Ig A k reacting with LDL and VLDL in passive hemagglutination assays. No reaction was obtained with HDL. This Ig A did not react with the lipoproteins of dog, horse, calf, pig and rabbit.

In 3 additional myeloma cases (two of which generated Ig G and one Ig A) with hypolipidemia, it was shown (63) that the reactive part of the immunoglobulins was the papain digested Fab fragment, and that, in contrast with the findings in auto immune hyperlipidemia, the reactive part of the lipoprotein molecule was not a lipid constituent, but the protein moiety.

The binding activity of lipoproteins was subsequently studied in a systematic screening of over 200 sera with a monoclonal gamma globulin. At least 10% gave a positive reaction with human LDL and VLDL. When the gamma globulin had no binding activity, serum lipid levels were normal. In cases with binding activities, low lipid levels were observed. Xanthomas were never reported. It was concluded from this study that the hyperlipidemia which is often observed in patients with paraproteinemia might be due to an auto-immune mechanism. The authors felt that the antigen – antibody complexes might be catabolized faster than unbound lipoproteins.

A similar polyclonal type of autoantibody against lipoproteins has also been detected in rheumatoid arthritis and cancer. The autoantibodies, mainly Ig G's, were active via their Fab fragments and recognized the protein moiety of the lipoprotein (63).

Hypolipidemia was also observed in Waldenstroem macroglobulinemia. In a study of 5 cases (64), it was demonstrated that the Ig M's had binding activities against lipoproteins and lipid emulsion. Low cholesterol levels ranging from 1.9 – 2.9 mmol/l were found in 4 out of 5 cases. The Ig M was shown to form immune complexes with LDL and VLDL which were cryoprecipitating (1 case) or which floated on ultracentrifugation (4 cases). When purified, the Ig M agglutinated lipid emulsions activated with LDL and human albumin. Furthermore, 4 out of the 5 Ig M's gave precipitin lines with human LDL in double diffusion agarose gels. The Ig M Fab fragments were reactive and were able to inhibit the agglutination of activated emulsions and the precipitation of LDL by whole Ig M. An inhibition study showed that the reactive epitopes consisted of a series of related structures present at the surface of synthetic activated lipid emulsion, on normal chylomicra and VLDL, and on a number of albumin molecules which may react via the small molecules that they transport. The epitopes contained phospholipids, especially phosphoinositol derivatives, which have emulsifying properties. The designation 'antiemulsifying" autoantibodies was proposed for these Ig M's (64).

4.3 Experimental models

2 types of autoimmune hyperlipidemia have been discussed in experimental models.

4.3.1 An antilipoprotein antibody detected in rabbits

Increased serum lipoprotein levels are a well known consequence of antigenic stimulation of rabbits (54, 55). The hyperlipidemia is obtained in the absence of dietary cholesterol supplementation. It is independent of the nature of the injected antigen, and may be seen with several sorts of heterologous proteins, tissue extracts and/or Freund's adjuvant. Usually transient and contemporary with the vaccination course, it may persist and become permanent in susceptible rabbits. The hyperlipidemia is of a mixed type and lipoprotein abnormalities are observed, similar to those of autoimmune hyperlipidemia.

In one rabbit which had received injections of Freund's adjuvant over 4 weeks, a severe hyperlipidemia developed after 6 weeks. Serum cholesterol had risen to 8.3 mmol/l and triglyceride to 5.4 mmol/l when the animal was sacrificed at 12 weeks. The presence of a gamma globulin with antilipoprotein activity was demonstrated after adequate separation of the Ig from the circulating lipoproteins of the rabbit (65). The Ig was shown to react with a site present on the lipoprotein of a number of rabbits, but not with human lipoproteins.

4.3.2 An antienzyme autoantibody in hamsters

An antienzyme autoantibody was also detected in the serum of lymphoma bearing golden hamsters. The Greene lymphoma, a subcutaneous tumor homotransplantable in the golden hamster, was shown to be associated with marked hyperlipidemia, serum milkiness, increased serum cholesterol and especially triglyceride levels, abnormal fat tolerance test, and a type IV or V electrophoretic pattern (66). Post heparin lipase activity was strongly depressed (67).

Extracts prepared from the tumor were able to inhibit in vivo and in vitro both the antithrombin effect of heparin and the post heparin lipase activity of a normal hamster plasma. The active fraction of the tumor extract was identified after purification as an Ig G (68). The hyperlipidemia of these lymphoma bearing animals may therefore be considered to represent a model of autoimmune hyperlipidemia of the antienzyme variety.

4.4 Normolipidemic auto immune xanthomatosis

A model of this association was recently described in a 53 year old woman having an Ig Gλ (Ig G Mol.) benign gammopathy associated with normolipidemic diffuse plane cutaneous xanthomas. In this case, the IgG was bound to HDL (apo A) and the circulating complexes were found to be cytotoxic for human fibroblasts in vitro (86).

5 Tissue Consequences of Autoimmune Hyperlipidemia

5.1 Xanthomatosis and dysglobulinemia

A relation between dysglobulinemia and xanthomatosis was reported long ago. But this is a rare association, found almost exclusively in myeloma cases. Two clinical aspects may be seen: xanthomatosis with hyperlipidemia and xanthomatosis without hyperlipidemia (69).

In xanthomatosis with hyperlipidemia, all types of xanthomas may be seen. Plane xanthomas are the most frequent, but tuberous, nodular, eruptive and sometimes tendinous xanthomas have been recorded as well. This xanthomatosis may be associated with all types of hyperlipoproteinemia, except type II. The myeloma may be of various types like for example Ig G, Ig A with either λ or κ chains. In the 23 cases referenced in Table 1, xanthomas were present in 15.

Xanthomatosis may also be present, though less frequently, in myeloma patients with normal serum lipids (43, 44, 45, 70, 86). The xanthomas are always of the plane type. 35 cases were listed (44) of both Ig G κ and Ig G λ type. Only one was Ig A and another Ig M. So, immunoglobulin abnormalities may be associated with accelerated lipoprotein tissue deposition, independently of increased serum lipoprotein levels.

The mechanism of this deposition is controversial.

- Overproduction of lipids by the plasma cells was suggested (71). Inclusion bodies in the cytoplasm of myeloma cells, reacting with phospholipid stains were also detected (72). But the presence of lipoproteins in the plasma cells was not confirmed in other reports.
- Accumulation of lipids in tissues may also result from a decrease in the elimination of the lipid rich Ig complexes from the blood. Such a decrease in the rate of lipoprotein removal was demonstrated in several myeloma cases by fat and vitamin A tolerance tests (12, 13). This apparently led to intracellular infiltration of Ig A/LDL complexes (73).

Table 8. Xanthomatosis, Atherosclerosis and Thrombosis in Patients with Known Antibody Specificity

author	reference	antibody Ig type	Antigen Lp[1]	epitope	Serum lipids	Xanthomatosis tuberous	palmar	plane	tendinous	Athero-sclerosis[2]	Thrombosis
Beaumont	(9,16)	IgA κ	VLH	Lp PG	↑ +++	+	+	−	+	+	−
Aubert	(72)	IgA κ	VLH	Lg PG	↑ +++	+	+	−	−	+	−
Beaumont	(17)	IgA κ	VLH	LpPBa	↑ +++	−	−	−	−	+	−
Kilgore	(8)	IgA κ	VL	LpPSie	↑ +++	−	−	−	−	−	−
Lewis	(13, 37)	IgA λ	VLH	LpPLe	↑ +++	+	+	−	−	−	−
Descamps	(40)	IgA λ	VL	LpPRa	↑ ++	−	−	−	−	−	−
Beaumont	(6, 9)	IgG κ	VLH	Lp As	↑ ++	−	−	−	−	−	−
Beaumont	(15)	IgG λ	VH	Lp Lac	↑ ++	−	−	−	−	−	−
Kodama	(45)	IgG κ + λ	VLH	?	↑ ±	−	−	+	−	−	−
Groszek	(43)	IgG κ	VIC	?	nl[3]	−	−	+	−	−	−
Doutre	(44)	IgG κ	VLH	?	nl	−	−	+	−	−	−
Beaumont	(86)	IgG κ	H	?	nl	−	−	+	−	−	−
Glueck	(19)	IgG κ	heparin	Ed	↑ ++	−	−	−	−	+	−
Beaumont	(18)	IgA λ	heparin	Sab	↑ ++	−	−	−	−	+	+

[1] Lipoproteins V: VLDL; L: LDL; H: HDL; I: IDL; C: chylomicra.
[2] Atherosclerosis: ischemic heart disease or/and peripheral atherosclerosis.
[3] nl: normolipemic.

– Another mechanism can be proposed. It is known (11) that LDL are metabolized by human cultured fibroblasts. The process is integral to regulation of intracellular cholesterol synthesis. An abnormality in the circulating LDL might result in altered binding to its receptors and/or degradation in the cells, which may disturb normal cellular cholesterol synthesis. This was demonstrated in 3 cases of myeloma with hyperlipidemia: cases Ger and Sor, and case Bar (17). Normal Igs introduced into the medium of human cultured fibroblasts did not interfere with cellular cholesterol homeostasis. When Igs from the 3 myeloma patients were added, a decrease in the *in vitro* degradation of LDL was observed for case Ger and Sor who had xanthomas, and in these 2 cases, the inhibitor activity of LDL on cell cholesterol synthesis was suppressed. No decrease was observed for case Bar who had no xanthomas (Table 6, Fig. 4). So it is felt that the specificity of the myeloma protein in AIH may explain the presence or absence of tissue lipid deposition.

Table 8 illustrates the possible role of antibody specificity in relation to xanthomatosis. In 14 patients with a variety of antibody specificities, tuberous and palmoplantar xanthomas were seen in the 3 patients with antilipoprotein antibodies of the Pg type, and plane xanthomas in the 4 normolipidemic IgG κ cases. No xanthomas were found for antilipoprotein antibodies of other types, nor for antiheparin antibodies.

It is felt that xanthomas may appear when the circulating complexed lipoproteins have lost their capacity to regulate cell cholesterol synthesis, its esterification, or its rate of excretion from the cells. It is possible that this mechanism might happen without visible hyperlipidemia. So, a category of autoimmune xanthomatosis without AIH might be expected (74).

5.2 Atherosclerosis and thrombosis

A similar mechanism might be implicated to explain the atherosclerotic lesions reported in the patients with AIH (12, 13), with or without myeloma. Atherosclerosis appears to be frequent and often severe in this situation, in contrast to the known rarity of ischemic disease in most patients with myeloma.

Also, cases of venous thrombosis have been reported particularly in patients with antiheparin antibodies (18, 19).

6 Treatment

In patients with AIH but no myeloma, standard lipid lowering drug and diet therapy should be prescribed. The results are often disappointing. In one case of AIH with major hyperlipidemia and severe atherosclerosis, D-penicillamine showed some efficacy (75).

In patients with myeloma, treatment of the primary disease is often, but not always, followed by a decrease in both the monoclonal immunoglobulin and the serum lipid levels. Plasmapheresis usually has transient effects on lipids (12, 19, 40) although melphalan may be strongly active (19, 76).

In one case for instance, the decrease in the monoclonal immunoglobulin induced by melphalan was associated with a normalization of the hyperlipoproteinemia (76), confirming that there was a direct relation between the immunological and lipid disorders.

7 The Significance of AIH in Disturbances of Lipid Metabolism

In the above reports, cases of AIH were associated with myeloma, SLE, rheumatoid arthritis or nephrotic syndrome. But AIH may also be detected among apparently common hyperlipoproteinemias with no evidence of dysglobulinemia.

Thus antilipoprotein activity has been found in 15 hyperlipoproteinemic patients with type I, IIb, III, IV or V electrophoretic pattern, predominant hypertriglyceridemia and delayed post prandial lipid clearance. The active Igs extracted from the patient's sera were shown to be complexed with lipoproteins. The antibodies were usually not precipitating but gave positive passive haemagglutination tests with VLDL and HDL. In most cases, the antibody was an Ig G or an Ig A, but an Ig M has also been found associated with an Ig G. The type of the reacting antibody could not however be exactly determined because there was much less circulating antibody than in myeloma.

An evaluation of the incidence of AIH in hyperlipoproteinemia is not easy. An accurate diagnosis of AIH may only be obtained by the demonstration of the antilipoprotein or antiheparin activity of the serum immunoglobulins. But if there is no free activity in serum, preliminary steps are necessary to separate the complexed antibodies and antigens.

However, AIH may be suspected from examination of a number of physico-chemical characteristics of the lipoproteins: the complexed lipoproteins are unstable; they may cryoprecipitate; and they also precipitate easily in ammonium sulfate at 25% saturation. Furthermore spontaneous agglutination of lipid containing particles may be detected in serum (77).The agglutinated particles may be observed by the naked eye or under the microscope in fresh serum or in plasma and especially in the creamy supernatant collected after 1 night at +4 °C. This test of spontaneous agglutination of lipid rich particles was applied to 270 patients with hyperlipoproteinemia and more or less milky serum. Two per cent exhibited large agglutinates, and twenty per cent small ones.

Another test has been developed recently (78) to detect Igs bound to circulating lipoproteins. In this test, the Ig/Lp complexes are floated in a ultracentrifuge at $\rho = 1.10$ kg/l in a sucrose density gradient. They are partially delipidated, and their immunoglobulin content estimated by radial immunodiffusion or laser nephelometry with specific antisera. The Ig/Lp test was negative in 40 normolipidic controls, positive in 5 myelomatous patients with patent AIH, and positive in 10 out of 39 patients selected for hyperlipidemia of types II a, II b, IV or V (79).

We still await the development of a more convenient test suitable for epidemiological studies of AIH, but from the above preliminary data, it is felt that circulating Ig/Lp particles might be much more frequent than supposed.

8 Concluding Remarks

Three main partners appear to be implicated in the metabolism of lipids and lipoproteins: lipoproteins themselves and especially their surface structures (epitopes), specialized enzymes, and specific cell receptors.

The information that we have accumulated since 1965 has shown that, in a number of cases, a fourth partner had to be taken in account: immunoglobulins acting via their antibody site. This process of development in our understanding of the mechanism of dyslipoproteinemia may be compared to the evolution of our knowledge of other metabolic diseases like diabetes and a number of endocrine dysfunctions. In these diseases, the first partners to be identified were the hormones, the receptors and the substrates. Today, the roles of specific autoantibodies are being more frequently stressed.

In the last twenty years, our knowledge of the importance of antilipoprotein autoantibodies and their consequences has developed through the study of human monoclonal gammopathies. In the coming years, it may be hoped that the use of monoclonal antibodies, produced in vitro by cell hybridization and the culture of immune competent clones, will allow faster progress.

With these new tools, information previously obtained by the study of human myelomas has already been confirmed. The lipids associated with apoproteins play a role in the structure of the lipoprotein epitopes (80); human and animal lipoproteins have not only species specific epitopes, but also common, phylotypic epitopes; these phylotypic epitopes, which have been preserved throughout evolution and differentiation of the species, have an important function in lipoprotein metabolism, especially in their recognition by cell receptors (81), a fact that we have previously observed for the phylotypic epitopes described in human myeloma cases.

Monoclonal antibodies might also prove useful for the detection of epitopes associated with vascular risk in different forms of dyslipoproteinemia (82).

References

1. Fougereau, M. (1977) Eléments d'immunologie expérimentale, p. 74, Masson, Paris.
2. Eisenberg, S. (1980) in: Ann. New York Acad. Sciences. Lipoprotein Structure (A. Scanu and F. Landsberger, eds.) p. 30–34.
3. Allison, A. C. & Blumberg, B. S. (1961) Lancet *1*, 634–637.
4. Berg, K. (1971) Genetic polymorphism of lipoproteins. 19th annual Coll. Prot. Biol. Fluids (Peeters, ed.) p. 169, Brügge Pergamon Press.
5. Galango, M. L. (1971) Nature, New Biol. *234*, 111–113.
6. Beaumont, J. L. (1969) C. R. Acad. Sci. Paris, série D, *269*, 107–110.

7. Beaumont, J. L. (1969) Ann. Biol. Clin. *27*, 611–635.
8. Kilgore, L. L., Patterson, B. W., Parenti, D. M. & Fisher, W. R. (1985) J. Clin. Invest. *76*, 225–232.
9. Beaumont, J. L., Beaumont, V., Antonucci, M. & Lemort, N. (1970) Ann. Biol. Clin. *28*, 387–399.
10. Eisenberg, S. (1985) J. Lipid Res. *26*, 487–494.
11. Goldstein, J. L. & Brown, M. S. (1977) Ann. Rev. Biochem. *46*, 897–930.
12. Beaumont, J. L., Jacotot, B. &Beaumont, V. (1967) Presse Med. *75*, 2315–2320.
13. Lewis, L. A., DeWolfe, V. G., Butkus, A. & Page, I. H. (1975) Am. J. Med. *59*, 208–218.
14. Ho, K. J., DeWolfe, G., Siler, W. & Lewis, L. A. (1976) J. Lab. Clin. Med. *88*, 769–779.
15. Beaumont, J. L., Berard, M., Antonucci, M., Delplanque, B. & Vranckx, R. (1977) Atherosclerosis *26*, 67–77.
16. Beaumont, J. L., Jacotot, B., Vilain, C. & Beaumont, V. (1965) C. R. Acad. Sci. Paris, Série D, *260*, 5960–5962.
17. Baudet, M. F., Dachet, C. & Beaumont, J. L. (1980) Clin. Exp. Immunol. *39*, 455–460.
18. Beaumont, J. L. & Lemort, N. (1970) C. R. Acad. Sci. Paris, Série D, *271*, 2452–2454,
19. Glueck, H. I., MacKenzie, M. R. & Glueck, C. J. (1972) J. Lab. Clin. Med. *79*, 731–743.
20. Beaumont, J. L., Lorenzelli, L. & Delplanque, B. (1976) Vox Sanguinis *30*, 36–49.
21. Waldenstroem, J. (1970) Diagnosis and treatment of multiple myeloma. Grune and Stratton, New York, p. 73.
22. Lewis, L. A. & Page, I. H. (1954) Amer. J. Med. *17*, 670–678.
23. Gross, P. & Weicker, H. (1954) Klin. Wschr. *32*, 509–512.
24. Liebetsaeder, F., Hugentobler, F., Wunderly, C. & Wuhrmann, F. (1951) Wien Z. Inn. Med. *32*, 1–10.
25. Kanzow, U. (1957) in: European Society of Haematology, Transactions of the 6th Congress, Copenhagen, Part. 2, p. 103, Karger, Basel.
26. Waldenstroem, J. (1952) Advanc. Int. Med. *5*, 398–406.
27. Polonovski, J., Douste-Blazy, L. & Valdiguie, P. (1963) Lipides plasmatiques dans les dysprotéinémies, in: Les Dyslipoprotéinémies, p. 207, Expansion Scientifique, Paris.
28. Hill, R. M., Mulligan, R. M. & Dunlop, S. G. (1948) Amer. J. Path. *24*, 688–690.
29. Ozer, F. L., Telatar, H., Telatar, F. & Muftuoglou, E. (1970) Am. J. Med. *49*, 841–844.
30. Lennard-Jones, J. E. (1960) British Med. J. *1*, 781–783.
31. Heremans, J. (1960) Les globulines sériques du système gamma. Leur nature et leur pathologie, p. 227–232, Arscia Bruxelles.
32. Kayden, H. J., Franklin, E. C. & Rozenberg, B. (1962) Circulation *26*, 659–667.
33. Levin, W. C., Aboumrad, M. H., Ritzmann, S. E. & Brantly, C. (1964) Arch. Int. Med. *144*, 688–693.
34. Cohen, L., Blaisdell, R. K., Djordevitch, J., Ormiste, V. & Dobrilovic, L. (1966) Am. J. Med. *40*, 299–317.
35. Savin, R. C. (1965) Arch. Dermat. Syph. *92*, 679–686.
36. Lewis, L. A. & Page, I. H. (1965) Am. J. Med. *38*, 286–297.
37. Beaumont, J. L., Poullin, M. F., Jacotot, B. & Beaumont, V. (1967) Nlle Rev. Fr. Hematol. *7*, 481–498.
38. Spikes, Jr., J. R., Cohen, L. & Djordevitch, J. (1968) Clin. Chim. Acta *20*, 413–421.
39. Cloarec, M., Etienne, J. & Debray, J. (1972) Sem. Hôp. Paris *48*, 3459–3466.
40. Descamps, J., Miled, A. & Goudemand, J. (1979) Nlle Rev. Fr. Hématol. *21*, 171–183.
41. Mullinax, F., Himrod, R. & Berry, E. R. (1970) Clin. Res. *18*, 429.
42. Watson, J. & Harland, W. R. (1968) Plasma cell disease and xanthomatosis, Transaction of the American Clinical and Climatological Association, 18th annual meeting, Waverly Press New York.
43. Groszek, E., Abrams, J. J. & Grundy, S. M. (1981) Metabolism *30*, 927–935.
44. Doutre, M. S., Conri, C., Beylot, C., Fleury, B., Chapoulard, H. & Bioulac, P. (1985) Dermatologica *170*, 157–164.
45. Kodama, H. (1977) Clin. Exp. Immunol. *28*, 437–444.
46. Glueck, C. J., Kaplan, A. P., Levy, R. I., Greten, H., Gralwick, H. & Fredrickson, D. S. (1969) Ann. Int. Med. *71*, 1051–1062.
47. Stern, H. P., Kolterman, O. G., Devitt, H. & Reaven, G. M. (1972) Arch. Int. Med. *130*, 817–821.
48. Matthews, J. D. & Feery, B. J. (1978) Lancet *2*, 1212–1213.

49. Noseda, G., Riesen, W., Schlumpf, E. & Morell, A. (1972) Europ. J. Clin. Invest. *2*, 342–347.
50. Isenberg, J. N., Hanson, R. F., Williams, G. C., Zavoral, J., Page, A. R. & Sharp, H. L. (1976) Am. J. Med. *61*, 393–400.
51. Russel, M. W., Brown, T. A. & Mestecky, J. (1981) J. Exp. Med. *153*, 968–976.
52. Storch, H., Konig, C. & Reuter, W. (1982) Z. ges. Inn. Med. *37*, 567–571.
53. Magundar, S., Bhatia, N. D. & Chakrararti, R. N. (1981) Indian J. Med. Res. *73*, 910–915.
54. Beaumont, V. & Beaumont, J. L. (1968) Path. Biol. *16*, 869–876.
55. Di Perri, T., Forconi, S. & Auteri, A. (1980) Experimental evidence for immunological mechanism in athero-arteriosclerosis, in: Immunity and atherosclerosis (Constantinidès, P., ed.) p. 7, London.
56. Beaumont, J. L. & Beaumont, V. (1977) Atherosclerosis *26*, 405–418.
57. Baudet, M. F., Delplanque, B. & Beaumont, J. L. (1977) Path. Biol., *25*, 330–340.
58. Marcus, D. M. & Cass, L. E. (1962) Science N.Y. *164*, 553–555.
59. Blakeslee, D. & Stone, W. H. (1971) Vox Sanguinis *21*, 269–283.
60. Brendemoen, O. J. (1952) Acta Microbiol. Scand. *31*, 574–579.
61. Linder, E. & Eddington, T. S. (1973) J. Immunol. *110*, 53–62.
62. Noseda, G., Butler, R., Schlumpf, E., Brunner, E. & Riesen, W. (1971) Schweiz. Med. Wsch. *24*, 893–899.
63. Riesen, W., Noseda, G. & Butler, R. (1972) Vox Sanguinis *22*, 420–431.
64. Beaumont, J. L., Lorenzelli, L. & Antonucci, M. (1973) Path. Biol. *21*, 241–254.
65. Beaumont, J. L., Beaumont, V. & Antonucci, M. (1969) C. R. Acad. Sci. Paris *268*, 1830–1832.
66. Beaumont, V. & Beaumont, J. L. (1974) Biomédecine *20*, 68–73.
67. Beaumont, V., Berard, M. & Boisnier, C. (1975) C. R. Acad. Sci. Paris *280*, 665–668.
68. Beaumont, J. L., Beaumont, V., Boisnier, C., Szondy, E. & Vranckz, R. (1976) C. R. Acad. Sci. Paris, Série D, *283*, 1369–1374.
69. Bazek, A., Dupre, A. & Christol-Jalby, B. (1970) Bull. Derm. Syphiligraphie *77*, 607–620.
70. Tannhauser, S. J. (1958) Lipidoses: diseases of the intracellular lipid metabolism. Grune and Stratton, New York.
71. Neufeld, A. H., Morton, H. S. & Halpenny, G. N. (1964) Canad. Med. Ass. *91*, 374–380.
72. Aubert, L., Arroyo, H., Detolle, P., Cavalier, C., Picard, D. & Cotte, G. (1967) Sem. Hôp. Paris *43*, 3014–3021.
73. Lewis, L. A. & Lazzarini-Robertson, A. Jr. (1974) Hyperimmunoglobulinemia, lipoproteinemia and atherosclerosis, in: Atherosclerosis III, (Schettler, G., Weigel, A., eds.) p. 595, Springer, Berlin, New York.
74. Beaumont, J. L., Baudet, M. F., Dachet, C. & Beaumont, V. (1980) in: Atherosclerosis V, vol. 1 (A. M. Gotto, L. C. Smith, B. Allen, eds.) p. 334, Springer, Berlin.
75. Beaumont, J. L., Beaumont, V. & Jacotot, B. (1967) Bull. Mémoires Soc. Med. Hôp. Paris *118*, 709–716.
76. Sobel, A. T., Antonucci, M., Intrator, L., Bernard, D., Beaumont, J. L. & Lagrue, G. (1976) Nlle Presse Med. *5*, 2375–2378.
77. Beaumont, J. L. & Lorenzelli, L. (1972) Path. Biol. *20*, 357–367.
78. Lorenzelli-Edouard, L., Marie, F. & Beaumont, J. L. (1980) Biomedecine 33, 160–163.
79. Beaumont, J. L., Buxtorf, J. C. & Lorenzelli-Edouard, L. (1980) Atherosclerosis *36*, 315–321.
80. Tikkanen, M. J., Cole, T. G. & Schonfeld, G. (1983) J. Lipid Res. *24*, 1494–1499.
81. Nelson, C. A., Tasch, M. A., Tikkanen, M., Dargar, R. & Schonfeld, G. (1984) J. Lipid Res. *25*, 821–830.
82. Fievet, C., Demarquilly, C., Luyeye, I., Fievet, P., Moschetto, Y. & Fruchart, J. C. (1985) Ann. Biol. Clin. *43*, 493–517.
83. Beiseigel, U., Schneider, W. J., Goldstein, J. L., Anderson, R. G. and Brown, M. S. (1981) J. Biol. Chem., *256*, 11923–11931.
84. Kita, T., Beiseigel, U., Goldstein, J. L., Schneider, W. J. and Brown, M. S. (1981) J. Biol. Chem., *256*, 4701–4703.
85. Corsini, A., Roma, P., Sommariva, D., Fumagalli, R. and Catapano, A. L. (1986) J. Clin. Invest., *78*, 940–946.
86. Berard, M., Antonucci, M. and Beaumont, J. L. (1986) Atherosclerosis, *62*, 111–115.

Chapter 14
Diet and Hyperlipidemia – Current Questions and Controversies

P. J. Nestel

1 Introduction

That dietary management is the initial step in the treatment of hyperlipidemia is accepted generally. This applies even when later drug therapy will almost certainly be required. The responsiveness to dietary change is unpredictable and sometimes unexpectedly gratifying. Tests that may be developed in the future, such as a simple assessment of an individual's low density lipoprotein (LDL) receptor status, may allow such predictions to be made in hypercholesterolemic patients.

Of greater concern is the question of advice to be given to the wider community. It is now clear that hypercholesterolemia carries a quantifiable risk for coronary heart disease (CHD), and that in populations such as that of the United States, the risk for CHD is continuously related to the plasma cholesterol concentration at least from 4.7 mmol/l (180 mg/dl) upward (1). Since this includes well over half of all adults in Western countries, the propriety of advising whole communities to modify food habits and the nature of such advice, take on a new significance. Such a strategy would not displace the need to identify individuals at unduly high risk and treat those vigorously. However it has become recognized that numerically, far more subjects with CHD arise from among subjects with moderate hypercholesterolemia. It is also being accepted that relatively small reductions in the average plasma cholesterol concentration of an entire population, would result in a dramatic fall in the prevalence of CHD. A recent 20-year longitudinal study of some 1000 men in the "Irish-Boston Heart Study" showed that quite small differences in food habits separated those who subsequently died of CHD or survived (2). For dietary cholesterol the difference was a mere 50 mg or so per day, yet this difference was statistically significant in differentiating the two groups.

The increasing emphasis on community based intervention (3) requires a much more knowledgeable approach to the consequences of any dietary change that might be advised. For individuals at high risk of CHD because of their degree of hyperlipidemia the need for lowering plasma lipids is clear. Major dietary modifications are justifiable in such subjects and the minor possibility of adverse effects can be ignored.

The reverse is true when advising a whole population. Each aspect of the diet which is singled out for change must be scrupulously examined for its likely effects. There are for instance a number of options for lowering plasma cholesterol concentration. Some are however associated with a distortion in the lipoprotein pattern which are in the main beneficial but potentially undesirable in some aspects. Just as a minor downward move in the average plasma cholesterol concentration of a population may have major health benefits, so conversely, a minor adverse change may lead to unexpected problems.

This review has therefore placed particular emphasis on the more uncertain and controversial questions in this area. Only from a careful consideration of the quantitative effects of modifying individual nutrients and of the mechanisms through which this occurs, can an optimal diet be developed.

2 Nutritional Epidemiology

This is an appropriate point at which to review relevant cross-sectional and prospective epidemiological data. Much of this information is quite recent. The observations relate to nutrients and combinations of nutrients in a national diet and their association with the prevalence of coronary heart disease in that community. Liu et al. (4) reviewed the food disappearance data in 20 industrialized countries for the years 1954–1969, during which time coronary heart disease mortality reached its peak. On univariate analysis, the significant positive correlations with CHD were the "dietary lipid score" (based on per capita consumption of saturated fat, cholesterol and PUFA derived from the Keys' equations), and sugar. Fiber intake, calculated from available fruits, vegetables, grains and legumes, was negatively correlated. However, since the consumption of lipid is associated with high sugar and low fiber intakes, a multivariate analysis was carried out, which left only the lipid score as influencing CHD significantly.

More informative data have been provided by several major prospective studies which involved large numbers of subjects observed for up to 20 years. The "Western Electric Study" in Chicago (5) followed the outcome of 1900 men whose dietary habits had been assessed 19 and 20 years previously. The risk of dying from CHD was significantly related to the lipid score (sum of dietary cholesterol plus saturated fatty acid, minus PUFA). One-third more men died whose scores were in the upper than in the lower tertile. Consumption of plant foods, reflecting PUFA and fiber was inversely correlated with CHD mortality. Similar conclusions were drawn from the Boston-Irish study (2), in which over 1000 brothers of Irish descent were kept under observation over two decades in two countries. CHD mortality was significantly higher when dietary lipid scores were high, and plant foods and fiber were low. These are very important observations because of earlier beliefs that CHD prevalence could not be related to those dietary factors which give rise to raised plasma cholesterol levels (6).

The Oslo Heart Study (7) and the Lipid Research Clinics' Coronary Primary Prevention Trial (CPPT) (8) were intervention trials in which dietary change was the major intervention in Oslo but played a secondary role in the CPPT. The Norwegian Study showed that a composite diet designed to lower the plasma cholesterol concentration, maintained a reduced cholesterol level for the six years of the trial and led to significant lowering of heart attack rate.

In the CPPT (8), an initial cross-sectional survey of a random sample of the 4374 participants showed a significant positive correlation between the LDL cholesterol level and the consumption of fat. The corollary finding was that carbohydrate intake was negatively associated with the LDL concentration. Further important observations (9) were made after the entire population (both the placebo and the cholestyramine treated groups) was instructed to change to a diet with a reduced cholesterol and saturated fat content. The changes in the plasma cholesterol concentration were significantly correlated with weight loss, decreased intakes of saturated fatty acids and cholesterol, and increased intake of PUFA.

The importance of the last observation was that nutritional changes known to be effective in the metabolic ward could be successfully transferred to a community of several thousand individuals. Further, the quantitative relationships previously established for individual nutrients and groups of nutrients with respect to their influence on the plasma cholesterol level, were reproducible in a much wider setting. Lewis and coworkers (10) have also demonstrated the magnitude of the additive effects which can be achieved by changing several nutrients. In Lewis's study total fat intake was reduced to 27% of energy with a proportional decrease in cholesterol intake, an increased P/S ratio of 1.0 and an increase in starch and fibrous foods. This merely emphasises the obvious fact that changes in plasma cholesterol are exerted by changing groups of foods and thereby altering the entire diet.

3 Effects of Dietary Fatty Acids

The effects of long-chain saturated fatty acids and of polyunsaturated fatty acids (PUFA) of the n-6 series (mainly linoleic acid) on the plasma cholesterol concentration have been known for some time. The plasma cholesterol lowering effect of a high polyunsaturated/saturated (P/S) ratio diet is one of the most consistent findings in nutrition. What remains controversial is (a) the relative advantages of raising the P/S ratio alone versus a substantial reduction in fat intake; (b) the disadvantages of increasing the consumption of PUFA; (c) the merits of monounsaturated fatty acids; (d) the role of PUFA of the n-3 variety, found in some plants (linolenic acid) but mostly in marine oils (eicosapentaenoate and docosahexaenoate).

The least controversial aspect of the dietary management of hypercholesterolemia is the effectiveness of lowering the saturated fatty acid and increasing the PUFA content. The pioneering work of the Minnesota and Rockefeller University groups established the quantitative relationships more than 20 years ago. The Keys (11) equation Δ-cholesterol $= 1.3\ (2\Delta S - \Delta P)$, expresses the change in the plasma cholesterol level (mg/dl) in terms of changes in saturated and polyunsaturated fatty acids (% total energy). This is quantitatively much more effective than lowering cholesterol consumption, at least in the context of current Western diets which are characterized by high intake of saturated fatty acids. The above equation predicts that halving the present consumption of 18% energy as saturated fatty acid, and substituting this with starchy carbohydrate would reduce the plasma cholesterol by over 20 mg/dl. A partial substitution with PUFA (e. g. from cereal products and cooking oils), together with the benefits from halving the cholesterol intake would lower the plasma cholesterol by a total of 35 – 40 mg/dl. Since this represents at least 15% of the mean adult plasma cholesterol concentration, the corresponding reduction in heart attack rate extrapolated from the CPPT data, could be around 30%.

One uncertainty about the relative merits of a low-fat diet versus a diet with a higher P/S ratio stems from the need to increase carbohydrate intake to compensate for the removal of fat. Although this is the pattern in much of the less affluent part of the world, it is feared that in the context of the relatively high energy intake of Western populations, undesirable changes in HDL cholesterol and in VLDL triglyceride may follow. The effects on these lipoproteins may even negate the benefits from reducing LDL cholesterol.

The amount of dietary fat consumed is important for a number of reasons. First, fatty foods are high energy foods and predispose to overweight which is undesirable in itself, but also stimulate cholesterol and triglyceride synthesis. Overweight also lowers the HDL cholesterol concentration. Second, a low fat diet very effectively lowers serum cholesterol even if much of the remaining fat is saturated because it is the proportion of saturated fat as a source of total energy that counts (11). Third, important benefits may accrue from reducing fat intake which have little to do with plasma cholesterol or CHD. It has been proposed that other disorders such as certain cancers, as well as hypertension, gallstones and diabetes are influenced by dietary fat.

Studies carried out under strict metabolic ward conditions consistently show that the LDL cholesterol concentration is low when little fat is eaten. This reduction in LDL is generally greater than can be achieved by raising the P/S ratio alone. For instance in a short-term study involving normal subjects, Schaefer et al. (12) observed a 23% fall in total cholesterol at a very low fat intake, compared with a fall of 17% when the P/S ratio was substantially raised to 2.0, but dietary fat intake was high. Importantly, with both diets, the falls in plasma cholesterol could be attributed to reductions in both HDL cholesterol and LDL cholesterol.

The fall in HDL cholesterol, which was of similar magnitude to that in LDL cholesterol has raised questions about the desirability of these approaches to lowering the plasma cholesterol. It has been closely scrutinized in several recent publications. Kuusi et al. (13) have deliberately compared these alternative dietary measures in a well supervised study of 40 families in Finland. Reducing fat (and cholesterol) consumption from 38% to 23% of energy achieved a 16% lowering in the plasma cholesterol concentration, which could not be significantly improved by raising the P/S ratio modestly from 0.4 to 0.9. Once again both LDL and HDL contributed to the overall fall in cholesterol; in fact HDL cholesterol fell 25%.

Similar reductions in HDL and in HDL_2 cholesterol levels, accompanying falls in LDL cholesterol have been reported by others (14 – 18), with both kinds of dietary intervention (lowering fat alone or raising the P/S ratio alone). This finding is however not invariable. Interestingly, the HDL cholesterol concentration did not fall in the Oslo Heart Study (7), despite a substantial drop in LDL, over a six year period of dietary intervention, predominated by a high P/S ratio. Weisweiler et al. (19) have reported a fall in cholesterol confined to LDL alone, through raising the P/S ratio, with both a moderately low fat diet (32% energy, P/S = 1.0) and a higher fat diet (42% energy, P/S = 1.0). Brussard et al. (15) have also reported substantial LDL cholesterol reduction with little effect on HDL cholesterol when dietary fat reduction was modest (about 30% energy) and the P/S ratio was raised. However, the HDL cholesterol levels did fall with further restriction of fat to 20% energy (15).

The following general conclusions can be drawn: 1. LDL cholesterol levels are substantially lowered by marked reduction in fat intake, which is largely independent of P/S ratio. 2. Quantitatively similar lowering of LDL is achieved by a more modest restriction of fat intake, to about 30% energy, provided the P/S ratio is raised to at least 1. 3. HDL cholesterol may also fall by almost as much as the LDL cholesterol by increasing PUFA intake, although this is by no means an invariable finding. 4. A low fat diet (20% of energy or less) usually lowers HDL cholesterol by as much as LDL cholesterol.

Findings such as these on HDL cholesterol, have prompted a re-examination of the value of increasing monounsaturated fatty acids as an alternative to raising PUFA, i. e. mainly linoleic acid. The equations derived by Keys et al. (11) and Hegsted et al. (20) had shown that substituting monounsaturated fatty acids for saturated fatty acids reduces serum cholesterol levels. This was however attributed to the lower saturates since monounsaturates were regarded as "neutral" being less effective in lowering cholesterol levels than were PUFA. However, since these early studies did not differentiate the relative contributions from LDL and HDL, it is possible that the greater cholesterol lowering potential of PUFA than of monounsaturates was due to the additional depression of the HDL cholesterol concentration. Since this maybe undesirable, Mattson and Grundy (21) undertook direct comparisons of two sources of safflower oil which contained differing proportions of monounsaturates and PUFA. This was extended to a comparison of a monounsaturate rich, high fat (40% energy) diet, with a low fat (20% energy) diet (22). Of the various comparisons, the change from saturates (palm oil or coconut oil) to monounsaturates led to a lowering of LDL cholesterol which, importantly, was at least as great as occurred with the PUFA-rich or the low-fat diets. Furthermore, HDL cholesterol was not affected unfavourably by the monounsaturate diet. An earlier study by Becker et al. (23) did not however demonstrate a superiority of monounsaturates over PUFA with respect to either total or HDL cholesterol. The use of monounsaturates does however deserve reappraisal, especially in view of its wide usage in Mediterranean countries where CHD prevalence has been low.

One further potential disadvantage to the widespread introduction of unduly low-fat diets (below 25% energy), is possible hypertriglyceridemia induced by a high carbohydrate intake (24). Although this does not appear to be a problem in Asian countries where starch is the dietary staple, hypertriglyceridemia may be prevented there by the relatively low total energy intake. Furthermore, carbohydrate-inducible hypertriglyceridemia may be temporary, lasting only weeks, in otherwise normolipidemic individuals (24).

Nevertheless, overproduction of triglyceride fatty acids is a feature of short-term carbohydrate rich diets (24, 25). Although the extent of the triglyceride rise is influenced by the nature and the amount of the carbohydrate diet, it occurs with most sources of simple and complex saccharides. The likelihood that this effect may persist, at least in Western populations, derives from dietary surveys of cross-sections of the population. The LRC CPPT Survey (26) showed that the plasma triglyceride concentration was indeed inversely related to total fat consumption, and hence directly to the proportion of dietary carbohydrate. (Overweight and alcohol were other nutritional factors contributing directly to the triglyceride concentration.)

The relevance of this concern is that the plasma triglyceride concentration may also be a risk factor for CHD. Although frequently associated with CHD prevalence on univariate analysis, the triglyceride level has been regarded as merely reflecting a low HDL cholesterol, with which hypertriglyceridemia is also often associated. Recently however, a reassessment of the evidence has changed sufficiently for a diet-induced rise in triglyceride to be viewed with disquiet (27–29).

It is therefore important to recognize that most studies of low-fat diets have shown a rise in VLDL triglyceride, which may compound the potentially undesired effect of lowering HDL cholesterol. Such rises in the triglyceride concentration have tended to occur when fat intake has been restricted to less than 25% energy (25, 30). At fat intakes of 30–40% energy, VLDL levels are in fact generally lowered by diets rich in PUFA (10, 17, 19, 31–33), although sometimes such changes are not great.

To summarize, there is no certainty about the optimal nutritional approach to the lowering of LDL cholesterol. The controversy relates to the nature and to the magnitude of coincidental changes in VLDL and HDL concentrations, which together may detract from the benefits of lowering LDL. These potentially adverse effects can be minimized by moderating the nutritional changes, whether these be to lower fat intake or increase the PUFA content of the diet.

It is also important to review the possible mechanisms through which the above changes in LDL, VLDL and HDL are brought about. This information has been largely derived from measurements of lipoprotein protein kinetics in which lipoproteins, radiolabelled in their protein moiety have been reinjected into subjects so that the removal of the lipoprotein could be traced. An alternative approach has been to measure cholesterol and bile acid kinetics which estimates the metabolism of the body's total exchangeable cholesterol rather than the fraction within circulating lipoproteins.

The overall effect on VLDL Apo B metabolism depends on the extent of the substitution of carbohydrate for fat. If fat reduction is modest (> 25% versus 40% en), changes in VLDL metabolism are slight (32). However, greater changes (reducing fat to < 20%) can lead to VLDL overproduction and expansion in the VLDL Apo B and triglyceride pools (30). Clearly removal mechanisms are also affected adversely (34), at least during the relatively short periods of these studies. It is not known whether this represents changes in VLDL composition or in the nature of VLDL removal. Lipid to protein ratios are higher with carbohydrate induced VLDL overproduction which may reduce the susceptibility of the particles to lipolysis. The nature of VLDL removal is also probably altered since increased production of large VLDL leads to the direct removal of VLDL remnants by the liver, in preference to further catabolism to LDL (35).

The synthesis of LDL Apo B has been reported to be diminished by low-fat diets, whether or not dietary cholesterol is also reduced (32). This may reflect the reduced catabolism of VLDL to LDL. A further factor in the lowering of LDL cholesterol by a low-fat diet, is the corresponding increase in carbohydrate intake. A diet high in sucrose and starch but low in fat has been shown to increase the rate of LDL Apo B removal (34), which has also been demonstrated under other circumstances of increased VLDL production such as in obesity (36). The higher insulin levels induced by dietary carbohydrate may be a factor since receptor mediated uptake of LDL is enhanced by insulin at least *in vitro* (37). The nature of the LDL particle is also changed by carbohydrate-rich diets and this may affect removal.

Concentrations of HDL protein and of HDL cholesterol, especially of HDL_2 cholesterol, are also reduced with low-fat, high-carbohydrate diets. As with LDL Apo B, the removal rate of HDL Apo A is increased (38). This has also been reported in vegetarians who tend to have low HDL cholesterol levels (39). The reason for this is not known, but it is likely that the nature of the HDL particles is altered in an environment of reduced fat transport. It is therefore possible that such particles are removed more quickly than the cholesterol-enriched larger HDL which are formed during the catabolism of other lipid-transporting lipoproteins.

Considerable interests has focused on the mechanisms determining the general reduction in lipoprotein levels by PUFA. The formation of VLDL Apo B is less with linoleate-rich than with more saturated fat diets (32). That this may reflect the diversion of linoleate from pathways of esterification in the liver will be considered in the section dealing with marine oils. Plasma triglyceride levels can therefore be reduced with PUFA, although this is not as consistent as for plasma cholesterol. The variable response may be due to the heterogeneity of VLDL particles in patients with hypertriglyceridemia; possibly PUFA are effective only when the hypertriglyceridemia is due to overproduction of VLDL triglyceride.

It is unlikely that the composition and structure of VLDL are altered by linoleate-rich diets, although this may be a matter of the degree of PUFA enrichment. Both Shepherd et al. (40) and Baudet et al. (41) observed a relative depletion of cholesterol in the LDL particle, whereas Vega et al. (17) reported no compositional change (other than of the fatty acids). Kinetic studies of LDL Apo B have failed to show a consistent effect of PUFA on LDL metabolism which would account for the substantial reduction in LDL concentration. Whereas LDL Apo B synthesis has been reported as diminished in some individuals, the fractional removal rate is accelerated in others (40). In several studies which have not specifically investigated the mode of action of PUFA, the FCR of LDL Apo B has been higher than average against a dietary background of a high P/S ratio. However in these studies of young men the LDL Apo B levels were very low and might alone have accounted for the high FCR (42).

The possibility that LDL receptor function may become altered as the result of changed dietary fatty acids has received support from a study in hamsters in which receptor mediated LDL removal in vivo was suppressed by saturated fatty acid feeding (43). Attempts to show enhanced uptake by cells *in vitro*, of LDL obtained from subjects eating PUFA have been inconclusive (41, 44). However such studies had not used liver cells which would appear to be the relevant cell type. The consistent reduction in HDL dietary PUFA has been attributed in one study to diminished Apo A synthesis (45).

Preceding the era of lipoprotein kinetics, interest in this area had concentrated on sterol metabolism. Since changes in cholesterol consumption had generally resulted in clear-cut compensatory metabolic events, similar effects were sought for changes in dietary fatty acids. The results had however failed to give a consistent pattern which might satisfactorily explain the PUFA induced reduction in cholesterol. It is nevertheless likely that at least in normal subjects, there is a transient increase in sterol excretion which coincides with the fall in plasma cholesterol (46, 47). This effect does not appear to persist beyond a few weeks by which time at least the plasma cholesterol has reached a new steady state (48). However, patients with familial hypercholesterolemia do not exhibit these metabolic changes even when they respond to dietary PUFA. Under less controlled, outpatient studies, the effects of PUFA on neutral steroid or bile acid excretion have been minor.

Finally any assessment of linoleic acid enriched diets must take into account possible adverse effects in areas unrelated to lipoproteins. It must also consider beneficial effects in biological systems other than lipid metabolism. It is for instance possible that the significant reduction in heart attack rate reported in the Oslo Heart Study was not entirely due to the substantial cholesterol lowering effect of the diet which was rich in linoleic acid. Substituting linoleic acid for more saturated fatty acids has led to healthier coagulation profiles in studies carried out in humans and in animals (49, 50). The likelihood of a fatal cardiac arrhythmia developing in an animal following coronary artery ligation is also much less with safflower oil than with saturated sheep fat as the predominant fat (51). Blood pressure may also be advantageously reduced with linoleic acid in humans and in animals (52). Benefits such as these which lower cardiovascular risk, require confirmation and further study.

By contrast, serious adverse effects have been attributed to PUFA. At least in hypertriglyceridemic patients, who typically excrete excessive cholesterol in the bile, PUFA-rich diets may exaggerate cholesterol secretion to the point of producing gallstones (53). A more serious allegation relates to cancer formation. High fat diets promote certain cancers in susceptible animals and PUFA may be marginally more potent than more saturated fatty acids in this regard. The possibility of increased carcinogenesis from a PUFA-rich diet was also raised in a trial of elderly men. More detailed analysis of other major trials comparing linoleate enriched with saturated fatty acid rich diets failed however to demonstrate any increased predisposition to cancers (54). The linoleate concentration in adipose tissue (and presumably in other tissues) has been relatively high in the population of the USA, yet there is no epidemiological evidence for a change in cancer incidence which may be related to this.

Finally, it must be noted that there is no evidence to support the belief that the reduction in HDL cholesterol which accompanies a proportionate reduction in LDL cholesterol is harmful. Populations characterized by very low CHD rates frequently have HDL levels which are sustantially lower than in Western countries with high rates of CHD (55). Furthermore, urbanization of primitive societies is associated with rising concentrations of HDL cholesterol (56), and also with a higher prevalence of CHD. If the same argument were extended to dietary cholesterol, a reduction in cholesterol consumption could not be recommended on the grounds that this lowers HDL cholesterol.

3.1 The n-3 fatty acids

The most exciting recent development in the area of dietary fatty acids concerns the n-3 group of fatty acids. Although the biological effects of α-linolenic acid (18:3 n-3) had been studied for many years, the great potential of n-3 fatty acids in health and disease has been recognized only recently. This has come about through the rediscovery of the abundance of n-3 fatty acids in fish and their possible role in preventing heart disease.

While α-linolenic acid is widely distributed in terrestrial plants within the chloroplast membrane rather than in the seed oils, the longer, polyenoic fatty acids of marine oils are present in much higher concentration in fish. From a nutritional point of view, larger amounts of n-3 fatty acid can be obtained by eating fatty fish or fish oil concentrates than by eating plants or edible seed oils. The nature of these fatty acids varies widely among species of fish. Fish from the cold northern oceans contain more eicosapentaenoic acid (EPA or 20:5), than do fish from temperate southern seas, which contain relatively more docosahexaenoic acid (DHA or 22:6) (57). This may be important because these two major fatty acids do not have identical metabolic effects.

Adequate comparisons have not been made between the potential of α-linolenic acid and the longer, more unsaturated marine fatty acids in human biology. Although it is clear that the n-3 fatty acids of marine oils are more rapidly incorporated into plasma and membrane lipids and exert more profound and rapid biological effects than does dietary α-linolenic acid, it is by no means certain that this would obtain during life-long consumption of these two sources of n-3 fatty acids.

It has been suggested that there may be a desirable ratio of dietary n-6 to n-3 fatty acids of around 5 (58). There is a preponderance of n-6 fatty acids in the recent diet of western countries due to the high consumption of seed oils. By contrast the lipids of wild plants and non-domesticated animals eaten by non-agrarian peoples contain a much lower n-6/n-3 ratio (from 2:1 to 4:1, in contrast to a ratio of above 10 in our present diet). The biological importance of this is uncertain although it is clear that n-6 and n-3 fatty acids compete for the enzymes that lead to their elongation and desaturation, which in the case of 18:3 is to the type of fatty acid characteristic of fish oils. Furthermore the n-6 and n-3 fatty acids gave rise to different families of eicosanoids such as prostanoids and leukotrienes, which are regulators and modulators of many biological systems in man.

The n-6 fatty acids are elongated and desaturated to arachidonic acid (20:4 n-6). Arachidonic acid then gives rise to prostanoids and leukotrienes through the cyclooxygenase and lipoxygenase pathways respectively. The corresponding major compounds so formed, prostacyclin (PGI_2), thromboxane (TXA_2) and the leukotriene (LTB_4), differ from those derived from EPA. These are respectively PGI_3, TXA_3 and LTB_5. Each of these is biologically less active than the corresponding compounds derived from arachidonic acid (59). DHA (22:6 n-3) and EPA (20:5) do differ significantly as substrates for leukotriene formation and the consequent capacity to interfere with arachidonate-derived leukotriene activity.

There are major differences among animal species in the incorporation of marine fatty acids into individual phospholipids in membranes. This may be critical because the rate at which arachidonic acid is released from phospholipids and becomes available for prostanoid formation depends on the type of phospholipid in which it is stored. There are also important differences between the *in vivo* and *in vitro* findings in man. Whereas studies with platelets *ex vivo* suggest that EPA substantially suppresses the formation of thromboxane A_2, the *in vivo* reduction is very much less (60). Prostacyclin, or PGI_2, production is also not inhibited *in vivo* even at high intakes of

EPA, despite evidence for this obtained *in vitro* (60). Dietary EPA does however give rise to the formation of the "three" series of prostanoids (PGI_3 and TXA_3) in man (60). This leads to a marked change in the "two" or "three" series of prostanoids, with major consequences for at least thrombogenic mechanisms (61). Many other effects with potential for changing the development of atherosclerosis and other disorders, have been reported to result from consuming n-3 fatty acids.

The possibility for this had been suggested by Dyerberg et al. (62) from their observations of Greenland Eskimos who appear to be relatively immune from atherosclerosis. The daily consumption of fish averages 400 g which contains several grams of EPA. The Japanese who also suffer little from coronary heart disease, eat an average of 100 g fish daily. That as little as 30 g fish per day may be protective is suggested by a recently published survey from Holland in which the 20 year mortality from heart attacks was some 2.5 times less than amongst men who ate little fish (63). Yet the quantity of EPA eaten in such small amounts of fish (a mixture of lean white and fatty fishes) would have been well under 1 g daily which is very much less than is required to alter the coagulation profile of blood.

An increase in bleeding time of whole blood and a decrease in the aggregability of platelets has been reported only at higher intakes (2–20 g EPA daily) (61). An anti-inflammatory effect, mediated through the inhibition of leukotriene B_4 production in leucocytes, also required at least 3 g of EPA daily (64). This may also be relevant to atherosclerosis since the influx of cholesterol-laden monocytes into the arterial endothelium may also be influenced by marine fatty acids. Studies in experimental animals indicate that feeding fish oil also opposes the arrhythmogenicity of heart muscle subjected to ischemia (65). A blood pressure lowering effect has also been reported in man (66).

While these observations are not relevant to the control of hyperlipidemia by n-3 fatty acids, they form an important background to the discussion of their lipid-lowering potential. The aim in controlling hyperlipidemia is mostly to prevent atherosclerosis. The likelihood that this can be augmented by the use of fish oils, makes this group of fatty acids so promising.

In summary

1. Dietary n-3 fatty acids profoundly influence biological functions in the human body. These include blood coagulation, the production of both thrombogenic and antithrombogenic compounds by blood platelets and the endothelium of the artery.
2. These factors and others may be responsible for the low rates of coronary heart disease in populations eating large amounts of n-3 fatty acids.
3. The relative benefits of the three commoner n-3 fatty acids, α-linolenic acid, eicosapentaenoic acid and docosahexanoic acid are not known. The significance of the differences in their biological properties remains to be determined.
4. The possible hazards associated with increased consumption of n-3 fatty acids need investigation.
5. The optimal doses of fatty acids have not been established. Until answers to these questions are known, advice about the widespread usage of n-3 fatty acid concentrates should be discouraged. The preferred source at the present time is through consumption of fish which may provide protection against coronary heart disease through means other than n-3 fatty acids.

Effects on lipids and lipoproteins

The triglyceride lowering effect of this group of fatty acids is dramatic and probably exceeds the capacity of most current drugs. Substantial reductions in plasma triglyceride levels have been reported almost without exception. The extent of this effect is dose related although the optimal daily intake has not been established. It would appear to be no less than 2–3 g n-3 fatty acids (EPA + DHA) daily but is probably closer to 5 g (67–72). The response is likely to be influenced by the nature of the hyperlipemia, although fish oils have been very effective whenever VLDL are elevated, in lipoprotein phenotypes II b, IV and V (68–71). It is probably ineffective when chylomicronemia is responsible for the hypertriglyceridemia. This in itself suggests that these fatty acids do not stimulate removal mechanisms and indeed diabetic rats which become lipemic through lack of lipoprotein lipase activity, become more lipemic when fed fish oil (73).

The triglyceride lowering effect is mediated mainly through inhibition of hepatic formation of VLDL. In normal and hypertriglyceridemic subjects, Nestel et al. (69) have shown that the rate of VLDL triglyceride and of VLDL Apo B production can be greatly reduced by large doses of fish oil (15–20 g of n-3 fatty acids). Sanders et al. (71) have reported reduced triglyceride production with just under 5 g n-3 fatty acid daily. Livers of rats fed fish oil show a number of effects which collectively account for the substantial reduction in VLDL secretion. Using perfused livers and isolated hepatocytes, Wong et al. (74) have reported diminished fatty acid synthesis but increased fatty acid oxidation, which together greatly reduced the availability of fatty acids for esterification to glycerides. Observations by Wong & Nestel (74 a) indicate that human HepG2 liver cells are similarly affected when preincubated with either EPA or DHA. Both n-3 fatty acids inhibited triglyceride and Apo B secretion within VLDL. It is likely that these fatty acids affect other metabolic steps in the liver.

These observations support the conclusion that fish oils are highly effective whenever hypertriglyceridemia is due to overproduction of VLDL. Since this is a component of the hyperlipidemia in most subjects with types 2B, 4 and 5 hyperlipoproteinemia, fish oils have lowered triglycerides in such patients by between 20% and 70% depending on the dose. A striking example of the effectiveness of these oils in reversing triglyceride overproduction is the prevention of hypertriglyceridemia induced by a high carbohydrate diet (75).

The usefulness of fish oils in treating hypercholesterolemia is doubtful, except when the excess cholesterol is in VLDL as in Type V hyperlipoproteinemia. Under those circumstances, Phillipson et al. (70) reported a 40% reduction in plasma total cholesterol, despite an actual increase in the LDL cholesterol concentration. In subjects with other lipoprotein phenotypes and in normal subjects, the picture is less clear. At lower doses of fish oils (2–8 g n-3 fatty acid), Simons et al. (68), Sanders et al. (71) and Bronsgeest-Schoute et al. (67) reported minimal effects on total and on LDL cholesterol. At much higher doses (15–30 g n-3 fatty acid) Illingworth et al. (72), Harris et al. (76) and Phillipson et al. (70), observed falls in LDL cholesterol in normal subjects and in patients with Type II b hyperlipoproteinemia. Others have reported that LDL cholesterol may even rise (76 a). This may reflect the nature of the initial hyperlipidemia. As noted before a higher proportion of VLDL is converted to LDL at low production rates of VLDL (35); a diet induced reduction in VLDL secretion may lead to a greater fractional conversion of the VLDL flux to LDL. Fish oils appear to oppose the rise in LDL cholesterol normally induced by dietary cholesterol (77). Illingworth et al. (72) have reported findings on LDL metabolism with high doses of fish oils: rates of formation were significantly reduced probably reflecting suppression of VLDL secretion. The rise in LDL levels may indicate down-regulation of removal mechanisms (77a).

The effects of fish oils on HDL cholesterol are equally confusing. To some extent this has been due to the failure of some workers to recognize the not negligible cholesterol content of the fish oils. Thus more cholesterol may have been consumed with the fish oil diets giving rise to higher HDL cholesterol levels. When this factor is taken into account, HDL cholesterol and HDL Apo A concentrations may fall quite substantially with large doses of n-3 fatty acids (69, 75). Others have reported either no change or an increase in HDL cholesterol at lower doses (68, 76). Sanders et al. (71) have noted that the increase reflects a rise in HDL_3 cholesterol, with HDL_2 cholesterol actually falling. Such an effect is consistent with reduced transport and catabolism of VLDL.

In summary, the daily consumption of between 10–15 g of fish oil extract, representing 3–5 g n-3 fatty acid is probably adequate to control moderate hypertriglyceridemia. More severe forms of hypertriglyceridemia appear to require higher intakes. However, fish oils are probably not effective if the lipemia is predominantly due to chylomicron accumulation secondary to diminished lipoprotein lipase activity, as in diabetes mellitus. Fish oils do not appear to have an important role in the dietary management of hypercholesterolemia, except when this represents excess VLDL cholesterol. The effects on HDL cholesterol may be relatively unfavorable at high doses of fish oils although this may merely follow the low rate of lipid transport through plasma.

4 The Effects of Dietary Cholesterol

Despite the very large number of studies undertaken during the last 30 years this issue remains controversial. This reflects the complexity of the several factors that determine the response: the variability among individuals, the significant fluctuation in an individual's plasma cholesterol, the interaction between dietary cholesterol and other foods. Added to this are the exaggerated expectations from experiments in animals, variable experimental design, inexact measurements and the confounding effects of different changes in cholesterol levels in the major lipoproteins.

Nevertheless, the sum of observations, derived from studies that have taken these variables into account, clearly points to a consistent average rise in plasma cholesterol in response to dietary cholesterol when a group of individuals is studied.

Whereas much has been written about individual variability, it is likely that the serum cholesterol response to dietary cholesterol is almost normally distributed among metabolically normal subjects (78). This allows identification of both hypo-responders and hyper-responders. Some of this variability can be explained on the basis of cholesterol regulation and on the effectiveness of metabolic compensation in the face of increased cholesterol absorption. To this extent the variability is likely to reflect multiple genetic points of regulation, justifying the term "polygenic hypercholesterolemia" to distinguish this common disorder from the much less frequent LDL receptor deficiency state. It is not clear whether those individuals in a given population who hyper-respond to dietary cholesterol are merely to the right of a normal distribution curve (78) or represent a truly separate population (79, 80). Equations which predict reliably the relationship between changes in dietary cholesterol and changes in plasma cholesterol which have been derived by several investigators, suggest a normal distribution. Yet findings from metabolic studies indicate otherwise, with up to one-third of subjects showing a degree of hyper-responsiveness which is outside the average range (81, 81 a) Although the number of subjects so studied is small, it does reflect the approximate prevalence of polygenic hypercholesterolemia in the wider population.

The original equation published in 1965 by Keys and Parlin (82) was originally derived from metabolic ward studies on only 22 healthy men but was extended later. The equation is: Δ cholesterol (mg/dl) $= 1.5\ (Z_2 - Z_1)$ where the subscripts refer to any two diets and Z is the square root of the dietary cholesterol, expressed as mg/1000 cal. The observed and predicted values correlated in their initial study with an r value of 0.95. Similar studies, reported by others over the next 20 years, have on the whole been remarkably consistent and supportive of Keys' observations. However, the controversy stems from the occasional failures to confirm these findings. In general this has arisen through the failure to recognize that the relationship between dietary and serum cholesterol shows a threshold and is therefore nonlinear. That is, the equation holds only over a range of intakes, being steepest at lower levels of cholesterol consumption. So the difference in the serum cholesterol between daily intakes of 100 mg (low) and 500 mg (average) is greater than that between 1000 and 1400 mg (both being high). This leads to two conclusions. First, that lowering the cholesterol intake from the "average" English, American or Australian values would lower the "average" plasma cholesterol levels. For example, halving the cholesterol intake from a diet that provides 2000 Cals and 600 mg cholesterol, would lead to an average reduction of 8 mg/dl in the plasma cholesterol level ($\simeq$ 4%). Extrapolating from the CPPT Study (83), this might lower coronary disease by 8%. Second, that when cholesterol intake is already high, adding a further egg to the diet (250 mg cholesterol) may be undetectable in terms of plasma cholesterol change.

Keys and Parlin (82) had also stressed the not insignificant variability in an individual's plasma cholesterol, as much as 20 mg/dl (0.5 mmol/l) in free-living subjects. An individual's response cannot therefore be expected to be predictable from single measurements before and after a single egg is added daily to a diet already containing a great deal of cholesterol. This goes some way to understanding those reports which have claimed no relationship between cholesterol intake and plasma cholesterol concentration.

Two reviews (84, 85) of this subject conclude that virtually all properly conducted studies concur that dietary cholesterol exerts a significant effect on the plasma cholesterol in most people in the range of cholesterol intakes habitually consumed in Western populations. However two recent studies, in normocholesterol emic or mildly hypercholesterolemic subjects showed only small (86) or minimal increases (87) with about 2 egg yolks daily.

A threshold effect of around 300–400 mg of dietary cholesterol per day can also be inferred from Connor's studies with the Tarahumara Indians of Mexico, who normally eat very little cholesterol (88). The plasma cholesterol rose most steeply as the dietary cholesterol was experimentally increased to 400 mg daily, beyond which little further increase in the plasma cholesterol occurred. While McGandy and Hegsted (89), and Mattson (90) derived equations not unlike that of Keys et al., they did not allow for the diminishing effect at higher cholesterol intakes. This explains the higher predicted values given by these equations.

A further explanation for some discordant results is the failure to appreciate the interaction between dietary cholesterol and other nutrients, notably the proportions of saturated and polyunsaturated fatty acids. This has become evident in more recent studies. Changes in dietary fatty acid saturation influence the plasma cholesterol more powerfully than does dietary cholesterol.

There are some good examples of a lesser serum cholesterol-raising effect by dietary cholesterol when polyunsaturated fatty acid intake is increased. Schonfeld et al. (91) have reported significantly lower effects on the serum cholesterol than might have been expected from the Keys equation, as the dietary P/S ratio was increased. They noted a significant attenuation of the LDL cholesterol raising effect of dietary cholesterol at a P/S ratio of 0.25–0.4. When the cholesterol intake was raised from 300 mg to 1050 mg, the rise in LDL cholesterol on the most highly saturated fatty acid diet was 16 mg/dl (0.4 mmol/l) compared with 9 mg/dl when the P/S ratio was 0.8, and 6 mg/dl when it was 2.5. This has however not been confirmed in a more recent study (87), although a similar interaction was observed by Nestel with fish oil supplementation (76). Neither the LDL cholesterol nor the plasma Apo B concentration rose despite an increase in cholesterol consumption from 190 to 940 mg daily.

This interaction between dietary cholesterol and other nutrients may extend beyond dietary fatty acids. Crustaceans and shellfish may not raise the plasma cholesterol in proportion to their cholesterol content, even when eaten in an otherwise high fat diet (92). The sterols in shellfish comprise only 40% as cholesterol; the remainder (mainly brassicasterol, 22-dehydrocholesterol and 24-methylene cholesterol) resemble plant sterols which are poorly absorbed when eaten. However, crustacean sterols are predominantly cholesterol; the apparently less than predicted cholesterol raising effect is unlikely to be explicable by the small amounts of n-3 fatty acids.

These interactions raise the important notion that emphasis should be placed on particular *foods* rather than on individual *nutrients*. Masarei et al. (93) have pointed out that nutrients which either raise or lower plasma lipids are grouped within foods with very high degrees of correlation. Total fat, saturated fat, cholesterol and energy comprise one group, whereas another group is characterized by more fiber and polyunsaturates but by less protein. Interestingly, they observed, in a free-living population, a greater effect (decrease in plasma cholesterol) with the polyunsaturate, fiber-rich foods than an increase with saturate, cholesterol-rich foods. Others have also stressed this association of high energy, high saturates and high cholesterol in meat and full-cream dairy products. The obvious lesson from this is that this group of foods exerts an adverse effect through three separate factors. By contrast, plant foods have a beneficial effect also by virtue of several factors (low energy, mostly high polysaturates, absence of cholesterol, high fiber). This underscores the need to view dietary cholesterol in terms of the overall diet in practical dietetics.

Further reference must be made to the questions of the variability in the plasma cholesterol in response to dietary cholesterol (and to other nutrients). This is true for an individual from day-to-day and also from experiment to experiment. Jacobs et al. (78) suggest that the average standard deviation for this variability was around 15 mg/dl (0.4 mmol/l). This has important implications in dietary counselling in that it explains unexpected results, such as apparent deterioration in a patient who has complied. The problem can be overcome by multiple samplings, initially to establish a baseline, and subsequently to follow progress. By reducing the confounding effects of intra-individual variability, through multiple samplings on which duplicate analyses were carried out, Jacobs et al. (78) found that only 3% of 58 subjects who were studied, were "non-responders" to dietary changes in cholesterol and fats. Two-thirds of subjects responded to within 30% of the predicted change. Nevertheless, Beynen and Katan (79) have shown that individuals may truly hyper-respond or hyporespond and that this characteristic persists upon retesting. They observed a low but significant correlation between two responses to changing cholesterol intake tested six years apart. Interestingly, hyper-responders were leaner and had higher total, as well as HDL cholesterol levels, than average or hypo-responders. Oh and Miller (80), in an outpatient study of 21 healthy men found that one-third (7) remained persistently hyporesponsive to dietary cholesterol. The phenomenon of hyper- and hyporesponsiveness is well recognized in animals in which the genetic basis is readily demonstrable.

The rise in the plasma cholesterol due to increased cholesterol consumption reflects changes in cholesterol in several lipoproteins. Although this occurs mostly in LDL, both IDL and HDL levels may also rise (91, 94–97). Results for VLDL have been variable; Nestel et al. (97) have isolated VLDL that were small, enriched in cholesterol and in Apo E, and which were metabolically the immediate precursors of IDL. The most consistent change with dietary cholesterol is the rise in LDL cholesterol Apo B (91, 94, 95). A significant increase in LDL cholesterol may occur even when the total cholesterol fails to rise significantly, due to concomitant small falls in other lipoproteins, notably VLDL (95, 97). Thus measurement of plasma total cholesterol alone may be misleading.

The nature and composition of the LDL remain unchanged, indicating an increase in the number of LDL particles (94). IDL cholesterol also increases (98). The effect on HDL is variable, but there have been several reports of an increase in HDL, especially HDL_2, cholesterol (91, 94, 97). This has been associated with a rise also in the Apo A_I level. Total Apo E generally does not increase, despite enrichment of at least VLDL with Apo E (97). This is in contrast to findings in some species of animals in which much greater increases in the total cholesterol concentration lead to the appearance of unusual species of Apo E-enriched lipoproteins. These include β-VLDL and HDL_C as well as LDL of abnormal molecular weight, size and composition (99).

Neither β-VLDL nor HDL_C have been unequivocally isolated from the plasma of normal subjects eating large amounts of cholesterol. However, the presence of at least HDL_C can be inferred from a study by Mahley et al. (100) in which HDL from cholesterol fed subjects interacted with the LDL receptor on fibroblasts. Since this does not occur with normal HDL, it presumably reflected the presence of Apo E-rich HDL, or HDL_C. This lipoprotein may be formed in interstitial fluid through the interaction of HDL_3 or even smaller HDL with cholesterol enriched membranes of cells. HDL_C is seen as an important mechanism for redistributing excess cellular cholesterol among lipoproteins and the liver.

The mechanisms responsible for the increase in plasma cholesterol with cholesterol consumption are multiple and complex. Put most simply there is evidence for both increased secretion and diminished removal of lipoproteins. Nestel and Billington (101) showed that IDL production and secretion was raised in 7 of 8 subjects eating 1500 mg additional cholesterol and that the rise in the plasma cholesterol concentration could be correlated with the higher rate of IDL formation. They speculated that these particles may have been preferentially secreted from the liver (which has been demonstrated in cholesterol fed animals), since the transport rate of VLDL, a common precursor of IDL, was in fact reduced. Packard et al. (94) have also reported increased production of LDL (which are largely derived from IDL). Furthermore, they showed that the removal of LDL by the LDL-receptor route was impaired, whereas non-receptor mediated removal remained unaffected.

However, there is evidence from experimental animal work that increased delivery of cholesterol via chylomicron remnants suppresses hepatic LDL receptors (102). This in turn reduces the removal of circulating LDL. However, the magnitude of several metabolic responses to experimental cholesterol feeding is much greater in some species of animals than in man. Diminished hepatic LDL receptor function has not been directly observed in man, although the increased formation of IDL and of LDL could be a consequence of the failure of their immediate precursors (e. g. small VLDL) to be cleared by the liver. However, it is unlikely that diminished receptor mediated removal is the predominant event, since two reports have clearly shown that patients with reduced LDL receptor numbers (familial hypercholesterolemia) do not show an exaggerated response to eating cholesterol (103, 104).

Compensatory responses involving cholesterol metabolism, in contradistinction to lipoprotein metabolism have been well described. Reduced synthesis and increased excretion of cholesterol and increased oxidation of cholesterol to bile acids operate with varying degrees of efficiency (81). The rise in serum cholesterol varies in animals in inverse proportion to the efficiency of those compensatory controls. This may also be true in man. Studies of New Guinea Highlanders revealed changes in synthesis and excretion which quantitatively matched the increase in absorbed cholesterol (105). This was thought to explain the failure of the plasma cholesterol to rise in the face of additional dietary cholesterol. However, such a response was observed in the Tarahumara Indians despite appropriate adjustments in sterol metabolism (106). Nevertheless, Nestel and Poyser (81) and McNamara et al. (81 a)did find that the adequacy of such metabolic regulation determined the likelihood of a rise in the plasma cholesterol. Mistry et al. (96) observed a decrease in both LDL uptake and HMG CoA reductase activity in blood mononuclear cells of subjects eating an additional 1500 mg cholesterol daily; furthermore the rise in the LDL cholesterol level was inversely related to the capacity of the cells to degrade LDL. At present compensatory responses appear to provide a part but not the complete explanation for the variability in the LDL cholesterol change with dietary cholesterol. Regulation of sterol metabolism through dietary cholesterol has been observed as early as infancy (107).

In summary, the effects of dietary cholesterol have been described in some detail because of the heated controversy surrounding this topic. However there is no doubt that under well controlled circumstances, there is an increase in serum cholesterol, and especially in LDL cholesterol, in response to dietary cholesterol in about one-third of all subjects. Although there is inter-individual variability, much of this is normally distributed, so that both hypo-responders and hyper-responders might be expected. The largest intervention study (9), in over 4000 men, shows a clear relationship between the change in the cholesterol intake and the change in the plasma cholesterol concentration. The slope of this change is less than for weight loss, or the changes in saturated/polyunsaturated fatty acid intakes but is nevertheless significant both for the individual and for a population. Since this study (CPPT) was carried out in hypercholesterolemic men, this group may be more sensitive to dietary cholesterol. Subjects with the Apo E_4 phenotype are also hyperresponders (107 a). The risk for heart attack would be reduced by halving our present intake of cholesterol.

5 Other Dietary Factors Influencing Plasma Lipids

5.1 Body weight

The importance of maintaining normal weight to optimize plasma lipoprotein concentrations is well established. Increasing body weight can become quickly reflected in higher levels of plasma triglyceride and VLDL. The degree of hypertriglyceridemia depends on the quantities of fatty acids becoming available to the liver. The large flux of free fatty acids from the larger adipose mass (108), coupled with increased lipogenesis stimulate VLDL triglyceride and Apo B production (109, 110). However, an increased flux of VLDL need not in itself lead to hypertriglyceridemia. This occurs when catabolism also becomes affected: uptake and esterification of triglyceride fatty acids by muscle and adipose tissue may become impaired possibly due to diminishing sensitivity of these tissues to the action of insulin (111). VLDL triglyceride production and concentration fall quickly with weight reduction (109).

The influence of overweight and of weight loss on plasma cholesterol levels are less consistent, partly because the effects on the different lipoproteins may move cholesterol concentrations in opposing directions. Overweight may raise cholesterol in VLDL and lower cholesterol in HDL (112).

Population studies have consistently shown body weight and plasma cholesterol concentration to be related. The quantitative aspects have been recently examined in the Dutch (Zutphen) study (113) and in the CPPT Study (9). In the former, spontaneous variations in body weight were found to correlate significantly with changes in plasma cholesterol. Multivariate analysis showed that a change of 1 kg weight raised or lowered the plasma cholesterol by 2 mg/dl (0.05 mmol/l). The design of the CPPT included an initial dietary intervention in over 4000 participants. Of the several dietary variables which influenced the plasma cholesterol, reducing body weight led to the greatest fall in cholesterol.

Blacket (114) had earlier stressed the value of weight loss as a major means for treating hypercholesterolemia. Although this is clearly desirable, the evidence for an independent effect of weight loss is equivocal. Wolf and Grundy (112) found that the initial reductions in plasma total and in LDL cholesterol induced by weight loss, were not sustained when weights reached a new steady state. On the other hand HDL cholesterol which was initially low, returned to normal values at the new, lower levels of weight.

5.2 Alcohol

Excessive alcohol intake typically gives rise to hypertriglyceridemia. HDL cholesterol levels also rise provided liver function is not greatly disturbed. Production of VLDL triglyceride is stimulated by drinking alcohol and this is accentuated by overweight (115). VLDL triglyceride levels rise quickly when alcohol is drunk. Initially the triglyceride and phospholipid, but not cholesterol concentrations rise also in LDL and HDL_2 (116). The lipemia produced after a fatty meal is greater and more prolonged if alcohol is consumed at the same time.

Although there is general agreement that HDL cholesterol levels are on average higher than normal in habitual drinkers, there are conflicting reports about the relative contributions from HDL_2 and HDL_3 (117–119). Chronic alcoholics usually show elevated HDL_2 cholesterol although population surveys have shown a better correlation with HDL_3 cholesterol. In one study, only HDL_3 cholesterol levels fell when habitual drinkers abstained for 6 weeks (118).

Plasma total cholesterol levels are not directly affected by drinking alcohol, although disturbances in bile acid metabolism have been reported (120). LDL cholesterol does not rise and in fact, more commonly falls with alcohol consumption in studies in which other dietary variables are well controlled.

With the development of alcoholic hepatitis, profound disturbances in lipoprotein metabolism occur. HDL and VLDL levels fall and lipoprotein particles throughout the lipoprotein spectrum become markedly altered in their lipid and protein composition (121).

5.3 Fiber

The amount and nature of dietary fiber can influence plasma lipid especially cholesterol concentrations. There have been consistent reports of lower plasma cholesterol levels with some but not other forms of fiber. The term edible fiber includes a wide range of compounds with differing physical and chemical properties. They are mostly derived from the cell walls of plants. Some, such as pectins, gums and mucilages are apparently water soluble to the extent that they form gels. Other fibers such as lignins, celluloses and most hemicelluloses do not form gels and are insoluble.

In general, the soluble or gel-forming fibers have been more consistently capable of lowering plasma cholesterol levels in man (122–125). This may be partly related to their property of binding bile acids and so increasing their excretion. Oat bran contains relatively more soluble fiber than other brans and this may explain the superior cholesterol lowering effect of oats. Wheat bran which contains mainly insoluble fiber is relatively ineffective in lowering cholesterol (126), although even celluloses and hemicelluloses have been reported to reduce plasma cholesterol levels in some subjects. Responsiveness appears to be associated with a rapid transit time through the gut (123).

Thus diets rich in fruits, in vegetables, especially legumes, and in oats effectively reduce the plasma cholesterol concentration (124). The influence on triglyceride levels is less but has been reported with diets rich in leguminous seeds in particular (125).

The additional benefits on glucose and insulin metabolism augment the value of fiber rich foods in reducing cardiovascular risk (122).

5.4 Protein

In general, there is little evidence to support earlier suggestions that the source of dietary protein influences plasma lipid levels in man (127 – 129). At most, very small differences in favor of plant based protein usually derived from soy bean, can be claimed (130 – 132). Plasma cholesterol concentrations have been reported as being slightly lower with diets based on plant protein compared with those containing more animal than plant protein. However, more commonly the differences have not been statistically significant. The possibility has however not been excluded that the effect on total cholesterol masks a fall in LDL cholesterol which is counter-balanced by a similar rise in HDL cholesterol, an observation made by van Raaij et al. (127). Furthermore, Huff et al. (133) have demonstrated that soy bean protein consumption may increase VLDL transport without changing its concentration.

The small, albeit inconsistent, reduction in LDL cholesterol, which may occur when soy bean protein substitutes for meat and dairy protein, cannot explain the low plasma cholesterol concentration of vegetarians (134). It may however be an adjunct to the management of hypercholesterolemia. This is in any case likely with most diets that have a substantially reduced animal fat content.

5.5 Other dietary factors

Many claims have been made and rejected for the lipid-lowering value of nutrients and foods, other than those discussed above. The claims for vitamin supplements, such as C (135) and E (136), have no foundation. However, coffee consumption does appear to influence the plasma cholesterol concentration. The evidence is derived from cross-sectional surveys of populations and from a recent intervention study. The latter, a 10-week study of 33 habitual coffee drinkers, showed that abstention led to significant falls in the plasma cholesterol. Unfortunately the diet was unsupervised (137). The effect is greater with unfiltered coffee of the kind consumed in Scandinavian countries.

6 Conclusion: A Single Diet for Hyperlipidemia?

The concept of a specific diet for each phenotypic form of hyperlipoproteinemia evolved with the initial categorization of these disorders, two decades ago. Some disorders appeared fat-inducible and others carbohydrate-inducible. It appeared logical to withhold either fat or carbohydrate. The original classification was based on the spectrum of hyperlipoproteinemic patients referred to a specialized unit because of the severity of the disorder and failure to respond to treatment.

It is now clear that the vast majority of hyperlipidemic subjects manifest milder degrees of hypertriglyceridemia and/or hypercholesterolemia. The causes are multiple, a combination of minor dysfunctions at several regulatory points and environmental stresses, mostly nutritional in nature. The precise metabolic causes for these common groups of disorders is known only in part. Then the phenotype in any given individual may change rapidly, frequently and sometimes for no apparent reason. Multiple phenotypic manifestations may cluster in a single family. The penotypic expression in an individual with polygenic hypercholesterolemia may resemble that of a heterozygote with a single gene defect, involving the LDL receptor. For these reasons dietary changes based solely on phenotypic expression have been largely discarded.

That is not to say that a great deal is not known about several particular forms of hyperlipoproteinemia. These include the hypercholesterolemia due to absence or dysfunction of the LDL receptor, the dyslipoproteinemias due to Apo E abnormalities and the chylomicronemias due to a variety of lipase and lipase cofactor deficiencies (discussed in chapters 1 and 6). However, from the point of view of practical nutrition, the vast majority of hyperlipidemic patients do not belong to these categories. The emphasis should be placed on dietary changes which are likely to achieve satisfactory reductions in the concentrations of atherogenic lipoproteins and are readily instituted and accepted (138).

From this has arisen the concept of a single diet for most subjects with mild to moderate hypercholesterolemia, mild to moderate hypertriglyceridemia, or both (139). As discussed in detail in preceding sections, the measures to be taken initially which are applicable to each of the above hyperlipidemias, are as follows: 1. Achieving desirable weight or at least shedding as much surplus weight as is possible. 2. Reducing total fat intake to around 30% but not below 25% of energy, i. e. cutting back by between a quarter and a third of dietary fat. 3. Emphasising reduction of saturated fatty acids, which are present mostly, but not exclusively in animal derived foods. 4. Changing the nature of the dietary fats to a mixture of monounsaturated and polyunsaturated fatty acids but avoiding any undue increase in the latter. 5. Reducing cholesterol intake to less than 300 mg daily. This can be recommended even to those whose abnormality is confined to raised triglyceride levels, since the risk of developing coronary artery disease is related continuously with the plasma cholesterol concentration from 4.7 mmol/l upward. 6. Substituting the eliminated fat with complex carbohydrate. This will increase the consumption of plant protein, fibre and polyunsaturated fatty acid. This is an appropriate diet even for those who are hypertriglyceridemic, provided the reduction in fat is not excessive ($<$ 30% energy) and not requiring an excessive increase in carbohydrate ($>$ 55%).

Clearly other adjustments must be made to this basic diet, as appropriate. If alcohol contributes to the hypertriglyceridemia, this must be reduced. Whenever chylomicronemia is due to lipase or cofactor deficiency (rare disorders, collectively known as Type I hyperlipoproteinemia) fat intake must be reduced to a minimum, although some can be replaced by medium chain fatty acid preparations, which are metabolised differently. However when chylomicronemia occurs as part of the much commoner disorder, Type V hyperlipoproteinemia, energy, alcohol and fat reductions can be allied with fish oil supplementation. In fact, fish oils may become part of the basic dietary treatment in future, especially for hypertriglyceridemia. It is interesting that the basic diet as outlined above is frequently effective even for patients with the uncommon disorder Type III dyslipoproteinemia. Patients with genetic hypercholesterolemia require the same basic diet, although fat and cholesterol restriction may need to be even stricter.

The general framework of dietary intervention has been simplified in recent years and found to be effective in commonly occurring forms of hyperlipidemia. It also represents a readily understood message for the large sections of Western communities for whom dietary change is recommended. However important questions remain and new developments arise. These are the aspects which have been discussed in detail in this review.

References

1. Martin, M. J., Hulley, S. B., Browner, W. S., Muller, L. H. & Wentworth, D. (1986) Lancet *2*, 933–936..
2. Kushi, L. H., Lew, R. A., Stare, F. J., Ellison, C. R., Lozy, M., Bourke, G., Daly, L., Graham, I., Hickey, N., Mulcahy, R. & Kevaney, J. (1985) New Engl. Med. *312*, 811–818.
3. Grundy, S. M., Arky, R., Bray, G. A., Brown, W. V., Ernst, N. D., Kwiterovich, P. O., Mattson, F., Weidman, W. H., Schonfeld, G., Strong, J. P. & Weinberger, M. (1985) Arteriosclerosis *5*, 678A–682A.
4. Liu, K., Stamler, J., Trevisan, M. & Moss, D. (1982) Arteriosclerosis *2*, 221–227.
5. Shekelle, R. B., Shyrock, A. M., Paul, O., Lepper, M., Stamler, J., Liu, S. & Raynor, W. J. (1981) New Engl. J. Med. *304*, 65–70.
6. McGill, H. C. Jr. (1979) Am. J. Clin. Nutr. *32*, 2664–2702.
7. Hjermann, I., Velve Byre, K., Holme, I. & Leren, P. (1981) Lancet *2*, 1303–1310.
8. Gordon, T., Fisher, M., Ernst, N. & Rifkind, B. M. (1982) Arteriosclerosis *2*, 502–512.
9. Gordon, D. J., Salz, K. M., Roggenkamp, K. J. & Franklin, F. A. (1982) Arteriosclerosis *2*, 537–548.
10. Lewis, B., Hammett, F., Katan, M., Kay, R. M., Merkx, I., Nobels, A., Miller, N. E. & Swan, A. V. (1981) Lancet *2*, 1310–1313.
11. Keys, A., Anderson, J. T. & Grande, F. (1965) Metabolism *14*, 747–787.
12. Schaefer, E. J., Levy, R. I., Ernst, N. D., Van Sant, F. D. & Brewer, H. B. (1981) Am. J. Clin. Nutr. *34*, 1758–1763.
13. Kuusi, T., Ehnholm, F., Huttunen, J. K., Kostiainen, E., Pietinen, P., Leino, U., Uusitalo, U., Nikkari, T., Iacono, J. M. & Puska, P. (1985) J. Lipid Res. *26*, 360–367.
14. Ehnholm, C., Huttunen, J. K., Pietinen, P., Leino, U., Mutanen, M., Kostiainen, E., Iacono, J. M., Dougherty, R. & Puska, P. (1984) Arteriosclerosis *4*, 265–269.
15. Brussaard, J. H., Katan, M. B., Groot, P. H. E., Havekes, L. M. & Hautvast, J. G. A. J. (1982) Atherosclerosis *42*, 205–219.
16. Shepherd, J., Packard, C. J., Patsch, J. R., Gotto, A. M. Jr. & Taunton, O. D. (1978) J. Clin. Invest. *61*, 1582–1592.
17. Vega, G. L., Groszek, E., Wolf, R. & Grundy, S. M. (1982) J. Lipid Res. *23*, 811–822.
18. Vessby, B., Gustafsson, I. B., Boberg, J., Karlstrom, B., Lithell, H. & Werner, I. (1980) Eur. J. Clin. Invest. *10*, 193–202.
19. Weisweiler, P., Janetschek, P. & Schwandt, P. (1985) Metabolism *34*, 83–86.
20. Hegsted, D. M., McGrandy, R. B., Myers, M. L. & Stare, F. J. (1965) Am. J. Clin. Nutr. *17*, 281–295.
21. Mattson, F. H. & Grundy, S. M. (1985) J. Lipid Res. *26*, 194–202.
22. Grundy, S. M. (1986) New Engl. J. Med. *314*, 745–748.
23. Becker, N., Illingworth, D. R., Alaupovic, P., Connor, W. E. & Sundberg, E. E. (1983) Am. J. Clin. Nutr. *37*, 355–360.

24. Nestel, P. J. (1973) Progr. biochem. Pharmacol. *8*, 125–160.
25. Ginsberg, H. N., Le, N.-A., Melish, J., Steinberg, D. & Brown, W. V. (1981) Metabolism *30*, 347–353.
26. Cowan, L. D., Wilcosky, T., Criqui, M. H., Barrett-Connor, E., Suchindran, C. M., Wallace, R., Laskarzewski, P. & Walden, C. (1985) Arteriosclerosis *5*, 466–480.
27. Carlson, L. A. & Bottiger, L. E. (1985) Acta Med. Scand. *218*, 207–211.
28. Reardon, M. F., Nestel, P. J., Craig, I. H. & Harper, R. W. (1985) Circulation *71*, 881–888.
29. Aberg, H., Lithell, H., Selinius, I. & Hedstrand, H. (1985) Atherosclerosis *54*, 87–97.
30. Huff, M. W. & Nestel, P. J. (1982) Metabolism *31*, 493–498.
31. Nestel, P. J., Carroll, K. F. & Havenstein, N. (1970) Metabolism *19*, 1–18.
32. Cortese, C., Levy, Y., Jaunus, E. D., Turner, P. R., Rao, S. N., Miller, N. E. & Lewis, B. (1983) Eur. J. Clin. Invest. *13*, 79–85.
33. Chait, A., Onitiri, A., Nicoll, A., Rabaya, E., Davies, J. & Lewis, B. (1974) Atherosclerosis *20*, 347–364.
34. Nestel, P. J., Reardon, M. & Fidge, N. H. (1979) Metabolism *28*, 531–535.
35. Howard, B. V., Egusa, G., Beltz, W. F., Kesaniemi, Y. A. & Grundy, S. M. (1986) J. Lipid Res. *27*, 11–20.
36. Kesaniemi, Y. A. & Grundy, S. M. (1983) Arteriosclerosis *3*, 170–177.
37. Chait, A., Bierman, E. L. & Albers, J. J. (1979) J. Clin. Invest. *64*, 1309–1319.
38. Blum, C. B., Levy, R. I., Eisenberg, S., Hall, M., Goebel, R. H. & Berman, M. (1977) J. Clin. Invest. *60*, 795–807.
39. Nestel, P. J., Billington, T. & Smith, B. (1981) Metabolism *30*, 941–945.
40. Shepherd, J., Packard, C. J., Grundy, S. M., Yeshurun, D., Gotto, A. M. Jr. & Taunton, O. D. (1980) J. Lipid Res. *21*, 91–99.
41. Baudet, M. F., Dachet, C., Lasserre, M., Esteva, O. & Jacotot, B. (1984) J. Lipid Res. *25*, 456–468.
42. Grundy, S. M., Vega, G. L. & Bilheimer, D. W. (1985) Arteriosclerosis *5*, 623–630.
43. Spady, D. K. & Dietschy, J. M. (1985) Proc. Natl. Acad. Sci. USA *82*, 4526–4530.
44. Nestel, P. J. & Ma, J. (1979) Artery *6*, 20–27.
45. Shepherd, J., Packard, C. J., Patsch, J. R., Gotto, A. M. & Taunton, O. D. (1978) J. Clin. Invest. *58*, 1582–1592.
46. Connor, W. E., Witiak, D. T., Stone, D. B. & Armstrong, M. L. (1969) J. Clin. Invest. *48*, 1363–1375.
47. Nestel, P. J., Havenstein, N., Whyte, H. M., Scott, T. J. & Cook, L. J. (1973) New Engl. J. Med. *288*, 379–382.
48. Oh, S. Y. & Monaco, P. A. (1985) Am. J. Clin. Nutr. *42*, 399–413.
49. Kay, R. McP., Jacobs, M., Katan, M. B., Miller, N. E. & Lewis, B. (1985) Atherosclerosis *55*, 15–23.
50. Renaud, S., Morazain, R., Godsey, F., Dumont, E., Thevenon, C., Martin, J. L. & Mendy, F. (1986) Atherosclerosis *60*, 37–48.
51. McLennan, P. L., Aberwaydena, M. Y. & Charnock, J. S. (1987) Aust. N. Z. J. Med. *17 (2)*, 597.
52. Margetts, B. M., Beilin, L. J., Armstrong, B. K., Rouse, I. L., Vandongen, R., Croft, K. D. & McMurchie, E. J. (1985) Clin. Sci. *69*, 165–175.
53. Grundy, S. M. (1975) J. Clin. Invest. *55*, 269–282.
54. Ederer, F., Leren, P., Turpeinen, O. & Frantz, I. D. Jr. (1971) Lancet *2*, 203–206.
55. Connor, W. E., Cergueira, M. T., Connor, R. W., Wallace, R. B., Malinow, M. R. & Casdorph, H. R. (1978) Am. J. Clin. Nutr. *31*, 1131–1142.
56. Nestel, P., Ringrose, H., Taylor, R., Zimmet, P. & Sloman, G. (1983) Arteriosclerosis *3*, 132–137.
57. Gibson, R. A. (1983) Lipids *18*, 743–752.
58. Budowski, P. & Crawford, M. A. (1985) Proc. Nutr. Soc. *44*, 221–229.
59. Needleman, P., Raz, A., Minkes, M. S., Ferrendelli, J. A. & Sprecher, H. (1979) Proc. Natl. Acad. Sci. USA *76*, 944–948.
60. Knapp, H. R., Reilly, I. A. G., Alessandri, P. & FitzGerald, G. A. (1986) N. Engl. J. Med. *314*, 937–942.
61. Mortality from Coronary Heart Disease is inversely related to Fish Consumption in The Netherlands. Nutr. Rev. *43*, 271–273.
62. Dyerberg, J., Mortensen, J. Z., Nielsen, A. H. & Schmidt, E. B. (1982) Lancet *2*, 614.
63. Kromhout, D., Bosschieter, E. B. & Coulander, C. (1985) N. Engl. J. Med. *312*, 1205–1209.

64. Lee, T. H., Hoover, R. L., Williams, J. D., Sperling, R. I., Ravalese, J., Spur, B. W., Robinson, D. R., Corey, E. J., Lewis, R. A. & Austen, K. F. (1985) N. Engl. J. Med. *312*, 1217–1224.
65. Charnock, J. S. (1985) Proc. Nutr. Soc. Aust. *10*, 25–33.
66. Singer, P., Wirth, M., Berger, I., Voigt, S., Gerike, U., Godicke, W., Koberle, U. & Heine, H. (1985) Atherosclerosis *56*, 111–118.
67. Bronsgeest-Schoute, H. C., Van Gent, C. M., Luten, J. B. & Ruiter, A. (1981) Am. J. Clin. Nutr. *34*, 1752–1757.
68. Simons, L. A., Hickie, J. B. & Balasubramaniam, S. (1985) Atherosclerosis *54*, 75–88.
69. Nestel, P. J., Connor, W. E., Reardon, M. F., Connor, S., Wong, S. & Boston, R. (1984) J. Clin. Invest. *74*, 82–89.
70. Phillipson, B. E., Rothrock, D. W., Connor, W. E., Harris, W. S. & Illingworth, D. R. (1985) N. Engl. J. Med. *312*, 1210–1216.
71. Sanders, T. A. B., Sullivan, D. R., Reeve, J. & Thompson, G. R. (1985) Arteriosclerosis *5*, 459–465.
72. Illingworth, D. R., Harris, W. S. & Connor, W. E. (1984) Arteriosclerosis *4*, 270–275.
73. Illman, R. J., Trimble, R. P., Storer, G. B., Topping, D. L. & Oliver, J. R. (1986) Atherosclerosis *59*, 313–321.
74. Wong, S. H., Nestel, P. J., Trimble, R. P., Storer, G. B., Illman, R. J. & Topping, D. L. (1984) Biochim. Biophys. Acta *792*, 103–109.
74a. Wong, S. H. & Nestel, P. J. (1987) Atherosclerosis *64*, 139–146.
75. Harris, W. S., Connor, W. E., Inkeles, S. B. & Illingworth, D. R. (1984) Metabolism *33*, 1016–1019.
76. Harris, W. S., Connor, W. E. & McMurry, M. P. (1983) Metabolism *32*, 179–184.
76a. Sanders, T. A. B., Sullivan, D. R., Reeve, J. & Thompson, G. R. (1985) Arteriosclerosis *5*, 459–465.
77. Nestel, P. J. (1986) Am. J. Clin. Nutr. *43*, 752–757.
77a. Roach, P. D., Kambouris, A. M., Trimble, R. P., Topping, D. L. & Nestel, P. J. (1987) FEBS Lett. *222*, 159–162.
78. Jacobs, D. R., Anderson, J. T., Hannan, P., Keys, A. & Blackburn, H. (1983) Arteriosclerosis *3*, 349–356.
79. Beynen, A. C. & Katan, M. B. (1985) Atherosclerosis *57*, 19–31.
80. Oh, S. Y. & Miller, L. T. (1985) Am. J. Clin. Nutr. *42*, 421–431.
81. Nestel, P. J. & Poyser, R. (1976) Metabolism *25*, 1591–1599.
81a. McNamara, D. J., Kolb, R. Parker, T. S. Batwin, H., Samuel, P., Brown, C. D. & Ahrens, E. H. (1981) J. Clin. Invest. *79*, 1729–1739.
82. Keys, A. & Parlin, R. W. (1965) Metabolism *14*, 759–765.
83. The Lipid Research Clinics Coronary Primary Prevention Trial Results. 1. Reduction in Incidence of Coronary Heart Disease (1984) JAMA *251*, 351–364; 365–374.
84. Roberts, S. L., McMurry, M. P. & Connor, W. E. (1981) Am. J. Clin. Nutr. *34*, 2092–2099.
85. Keys, A. (1984) Am. J. Clin. Nutr. *40*, 351–359.
86. Edington, J., Geekie, M., Carter, R., Benfield, L., Fisher, K., Ball, M. & Mann, J. (1981) Br. Med. J. *294*, 333–336.
87. Kestin, M., Clifton, P. M., Rouse, I. L. & Nestel, P. J. (1989) Am. J. Clin. Nutr. (in press).
88. McMurry, M. P., Connor, W. E. & Cerqueira, M. T. (1982) Am. J. Clin. Nutr. *35*, 741–744.
89. Hegsted, D. M., McGandy, R. B., Myers, M. L. & Stare, F. J. (1965) Am. J. Clin. Nutr. *17*, 281–295.
90. Mattson, F. H., Erickson, B. A. & Kligman, A. M. (1972) Am. J. Clin. Nutr. *25*, 589–594.
91. Schonfeld, G., Patsch, W., Rudel, L. L., Nelson, C., Epstein, M. & Olson, R. E. (1982) J. Clin. Invest. *69*, 1072–1080.
92. Connor, W. E. & Lin, D. S. (1982) Metabolism *31*, 1046–1051.
93. Masarei, J. R. L., Rouse, I. L., Lynch, W. J., Robertson, K., Vandongen, R. & Beilin, L. J. (1984) Am. J. Clin. Nutr. *40*, 468–479.
94. Packard, C. J., McKinney, L., Carr, K. & Shepherd, J. (1983) J. Clin. Invest. *72*, 45–51.
95. Sacks, F. M., Salazar, J., Miller, L., Foster, J. M., Sutherland, M., Samonds, K. W., Albers, J. J. & Kass, E. H. (1984) Lancet *1*, 647–649.
96. Mistry, P., Miller, N. E., Laker, M., Hazzard, W. R. & Lewis, B. (1981) J. Clin. Invest. *67*, 493–502.
97. Nestel, P., Tada, N., Billington, T., Huff, M. & Fidge, N. (1982) Metabolism *31*, 398–405.

98. Beynen, A. C. & Katan, M. B. (1985) Atherosclerosis *54*, 157–166.
99. Mahley, R. W. (1978) in: Disturbances in Lipid and Lipoprotein Metabolism (Dietschy, J. M., Gotto, A. M. & Ontko, J. A., eds.) pp. 181–197. Williams & Wilkins, Baltimore.
100. Mahley, R. W., Innerarity, T. L, Bersot, T. P., Lipson, A. & Margolis, S. (1978) Lancet *2*, 807–809.
101. Nestel, P. J. & Billington, T. (1983) Metabolism *32*, 320–322.
102. Spady, D. K. & Dietschy, J. M. (1985) Proc. Natl. Acad. Sci. USA *82*, 4526–4530.
103. Martin, G. M. & Nestel, P. J. (1979) Clin. Sci. *56*, 377–380.
104. Cole, T. G., Pfleger, B., Hitchins, O. & Schonfeld, G. (1985) Metabolism *34*, 486–896.
105. Whyte, M., Nestel, P. & MacGregor, A. (1977) Eur. J. Clin. Invest. *7*, 53–60.
106. McMurry, M. P., Connor, W. E., Lin, D. S., Cerqueria, M. T. & Connor, S. L. (1985) Am. J. Clin. Nutr. *41*, 1289–1298.
107. Nestel, P. J., Poyser, A. & Boulton, T. J. C. (1979) Am. J. Clin. Nutr. *32*, 2177–2182.
107a. Kesaniemi, Y. A., Ehholm, C. & Miettinen, T. A. (1987) J. Clin. invest. *80*, 578–581.
108. Nestel, P. J. & Whyte, H. M. (1968) Metabolism *17*, 1122–1128.
109. Kesaniemi, Y. A., Beltz, W. F. & Grundy, S. M. (1985) Arteriosclerosis *5*, 153–161.
110. Grundy, S. M. (1984) J. Lipid Res. *25*, 1611–1618.
111. Nestel, P. J., Ishikawa, T. & Goldrick, R. B. (1978) Metabolism *27*, 589–597.
112. Wolf, R. N. & Grundy, S. M. (1983) Arteriosclerosis *3*, 160–169.
113. Kromhout, D. (1983) Am. J. Clin. Nutr. *38*, 591–598.
114. Blacket, R. B., Leelarthaepin, B., McGilchrist, C. A., Palmer, A. J. & Woodhill, J. M. (1979) Aust. NZ J. Med. *9*, 521–529.
115. Crouse, J. R. & Grundy, S. M. (1984) J. Lipid Res. *25*, 486–496.
116. Taskinen, M.-R., Valimaki, M., Nikkila, E. A., Kuusi, T. & Ylikahri, R. (1985) Metabolism *34*, 112–119.
117. Valimaki, M., Nikkila, E. A., Taskinen, M.-R. & Ylikahri, R. (1986) Atherosclerosis *59*, 147–153.
118. Haskell, W. L., Camargo, C., Williams, P. T., Vranizan, K. M., Krauss, R. M., Lindgren, F. T. & Wood, P. D. (1984) N. Engl. J. Med. *310*, 805–810.
119. Masarei, J. R. L., Puddey, I. B., Rouse, I. L., Lynch, W. J., Vandongen, R. & Beilin, L. J. (1986) Atherosclerosis *60*, 79–87.
120. Nestel, P. J., Simons, L. A. & Homma, Y. (1976) Am. J. Clin. Nutr. *29*, 1007–1015.
121. Tada, N., Nestel, P. J., Fidge, N. & Campbell, G. (1981) Biochim. Biophys. Acta *664*, 207–220.
122. Jenkins, D. J. A. (1981) in: Nutritional Pharmacology (Spiller, G. A., ed.) pp. 117–145. Liss, New York.
123. Kies, C. (1985) Lipids *20*, 802–807.
124. Mistry, P., Miller, N. E., Laker, M., Hazzard, W. R. & Lewis, B. (1981) J. Clin. Invest. *67*, 493–502.
125. Jenkins, D. J. A., Wong, G. S., Patten, R., Bird, J., Hall, M., Buckley, G. C., McGuire, V., Reichert, R. & Little, J. A. (1983) Am. J. Clin. Nutr. *38*, 567–573.
126. Eastwood, M. A. & Passmore, R. (1983) Lancet *2*, 202–205.
127. van Raaij, J. M. A., Katan, M. B., Hautvast, J. G. A. J. & Hermus, R. J. J. (1981) Am. J. Clin. Nutr. *34*, 1261–1271.
128. Sacks, F. M., Breslow, J. L., Wood, P. G. & Kass, E. H. (1983) J. Lipid Res. *24*, 1012–1020.
129. Wiebe, S. L., Bruce, V. M. & McDonald, B. E. (1984) Am. J. Clin. Nutr. *40*, 982–989.
130. Carroll, K. K., Giovannetti, P. M., Huff, M. W., Moase, O., Roberts, D. C. K. & Wolfe, B. M. (1978) Am. J. Clin. Nutr. *31*, 1312–1321.
131. Goldberg, A. P., Lim, A., Kolar, J. B., Grundhauser, J. J., Steinke, F. H. & Schonfeld, G. (1982) Atherosclerosis *43*, 355–368.
132. Fumagalli, R., Soleri, L., Farina, R., Musanti, R., Mantero, O., Noseda, G., Gatti, E. & Sirtori, C. R. (1982) Atherosclerosis *43*, 341–353.
133. Huff, M. W., Giovannetti, P. M. & Wolfe, B. M. (1984) Am. J. Clin. Nutr. *39*, 888–897.
134. Masarei, J. R. L., Robertson, K., Rouse, I. L., Vandongen, R., Lynch, W. J. & Beilin, L. J. (1984) Aust. NZ J. Med. *14*, 400–404.
135. Wahlberg, G. & Walldius, G. (1982) Atherosclerosis *43*, 283–288.
136. Kesaniemi, Y. A. & Grundy, S. M. (1982) Am. J. Clin. Nutr. *36*, 224–228.

137. Forde, O. H., Knutsen, S. F., Arnesen, E. & Thelle, D. S. (1985) Br. Med. J. *290*, 893–895.
138. Connor, W. E., Connor, S. L. (1982) Med. Clin. North Am. *66*, 485–518. Illingworth, D. R. (1985) N. Engl. J. Med. *312*, 1210–1216.
139. Choudhury, S., Jackson, P., Katan, M. B., Marenah, C. B., Cortese, C., Miller, N. E. & Lewis, B. (1984) Atherosclerosis *50*, 93–103.

Chapter 15
Drug Therapy of Dyslipoproteinemias

J. R. Guyton and A. M. Gotto Jr.

1 Clinical Efficacy of Hypolipidemic Drugs

Evidence is now available to show that pharmacologic agents affecting plasma lipoproteins can prevent the development of symptomatic disease. In cases of extreme hypertriglyceridemia, the prevention of pancreatitis and abdominal pain by a regimen that reduces and maintains triglyceride levels below 17 to 23 mmol/l (1500 – 2000mg/dl) is easy to demonstrate in individual patients. For the vastly more prevalent problem of moderate to severe hypercholesterolemia, the efficacy of pharmacologic intervention has been established in controlled clinical trials, as discussed below. Low levels of high density-lipoprotein (HDL) cholesterol, moderately high levels of plasma triglycerides, and certain alterations of apolipoproteins are predictive of the development of atherosclerotic cardiovascular disease, but the benefits of pharmacologic intervention remain to be shown.

The Coronary Primary Prevention Trial (CPPT) of the U.S. Lipid Research Clinics demonstrated that cholestyramine treatment of hypercholesterolemia reduces the risk of coronary heart disease in middle-aged men (1). The trial recruited men with plasma cholesterol levels of 6.85 mmol/l (265 mg/dl) or greater (95th percentile for the US population) and with no history of coronary heart disease. After an initial period of dietary instruction and management, subjects with low density-lipoprotein (LDL) cholesterol levels remaining at 4.53 mmol/l (175 mg/dl) or greater were entered into a double-blind, placebo-controlled trial. The mean plasma cholesterol of all trial participants after initial dietary therapy was 7.24 mmol/l (280 mg/dl). During the experimental phase lasting 7.4 years, total plasma cholesterol and LDL cholesterol were reduced 8.5% and 12.6% more, respectively, in subjects assigned to cholestyramine than in subjects assigned to placebo. Risk for definite myocardial infarction and coronary deaths (the primary endpoint) was reduced by 19% in the cholestyramine group ($p < 0.05$, Fig. 1). Included in the risk reduction was a 24% reduction in definite coronary death, which in itself was not statistically significant, due to the small numbers involved. Other endpoints indicative of coronary heart disease – i. e., the requirement for bypass surgery and the development of angina or positive exercise tests – were affected similarly. Cholestyramine did not appear to increase noncardiovascular disease rates.

The CPPT has been criticized on the basis that the "t" test used for statistical analysis was one-sided rather than two-sided. This means that the hypothesis that lowering cholesterol would reduce coronary heart disease was tested; the hypothesis

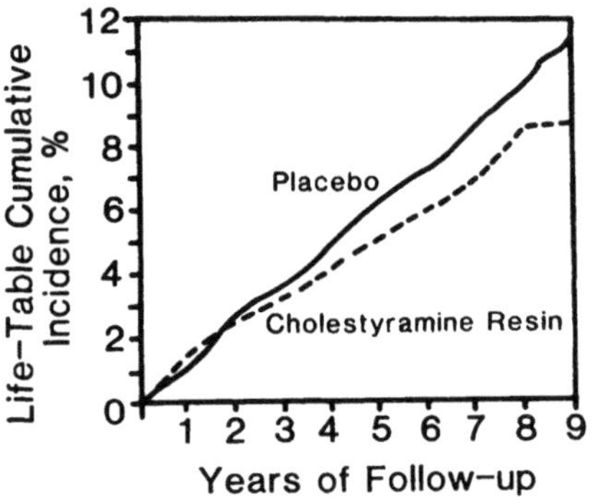

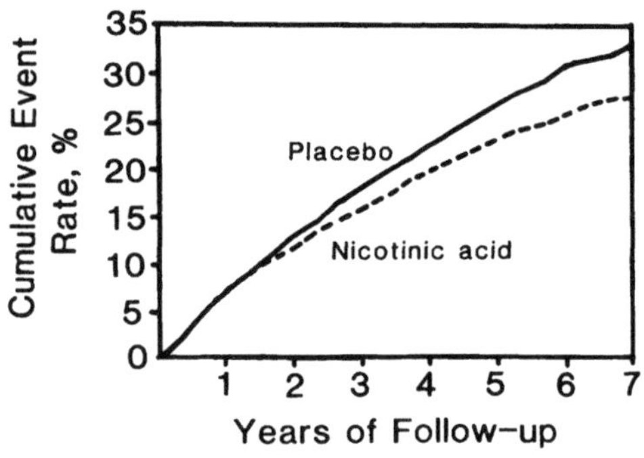

Fig. 1. Effect of cholestyramine on cumulative incidence of definite coronary heart disease death and/or nonfatal myocardial infarction in the Lipid Research Clinics Coronary Primary Prevention Trial. Redrawn with permission from J. Am. Med. Assoc. 251, *356* (1984).

Fig. 2. Effect of nicotinic acid on cumulative event rate for coronary death and/or definite nonfatal myocardial infarction in the Coronary Drug Project. Redrawn with permission from J. Am. Med. Assoc. 231, *365* (1975).

that lowering cholesterol would increase coronary heart disease was not tested. A one-sided approach is acceptable statistically, if the proposal is rational and is agreed upon by the investigators beforehand. The CPPT fulfilled these criteria, and thus its conclusion of a statistically significant reduction in the primary endpoint for coronary heart disease ($p < 0.05$) seems valid. All of the manifestations of coronary heart disease were changed in a favorable direction. Furthermore, the magnitude of the changes was related specifically to the degree of cholesterol lowering – that is, coronary risk did not change if there was no cholesterol reduction, and there was about a 2% decrease in risk for each 1% lowering of cholesterol, up to a 19% lowering of plasma cholesterol and 40% reduction of the primary endpoint for coronary disease.

In a recently completed secondary intervention study, the efficacy of cholestyramine in preventing the progression of angiographically defined coronary atherosclerosis was demonstrated in Type II hyperlipoproteinemic subjects who underwent repeated coronary angiography at an interval of five years (2).

Nicotinic acid may also be credited with successfully preventing coronary heart disease, according to two similar studies performed in survivors of myocardial infarction (3, 4). In each study about one-half of the patients had elevated plasma cholesterol and/or triglyceride levels. Both hyperlipidemic and normolipidemic patients were randomized and treated with lipid-lowering medication or placebo. In the U.S. Coronary Drug Project nicotinic acid was administered alone and in the Swedish study in combination with clofibrate. Total and cardiovascular mortality were essentially unaltered, but nonfatal myocardial infarction was significantly reduced in each study. Figure 2 shows data from the Coronary Drug Project plotted in a manner similar to Figure 1, with comparable relative reduction in risk. A follow-up study on participants in the Coronary Drug Project, ascertaining mortality rates during a five- to nine-year period after the end of the trial, suggested that nicotinic acid exerted a significant beneficial effect on long-term mortality (5).

The results of the above cited studies are consistent with less definitive trends observed in numerous other investigations. There is remarkable agreement that reduction of coronary risk of approximately 2% may be achieved for each 1% lowering of total plasma cholesterol or LDL cholesterol. The ability of the physician to tailor a hypocholesterolemic regimen to an individual patient's response, therefore, could theoretically reduce the risk of coronary heart disease by one-half in the typical hypercholesterolemic man (1).

In addition to cardiovascular endpoints, the shrinkage of tendinous xanthomas serves as a marker of tissue response to hypocholesterolemic medication. Such shrinkage, generally observed over one to three years of treatment, has been achieved with cholestyramine (6–8), colestipol (8), nicotinic acid (9–10), and probucol (11–14). Clofibrate causes regression of tuberoeruptive xanthomas in patients with the rare genetic disorder, dysbetalipoproteinemia (Type III hyperlipoproteinemia), but xanthomas in Type II hyperlipoproteinemia generally do not respond to clofibrate (15).

Selection of hypolipidemic medication must be judiciously done, because negative and even adverse results have been reported with certain agents. Dextrothyroxine, though effective in lowering plasma cholesterol, was withdrawn prior to the completion of the Coronary Drug Project because of excess mortality in patients taking the drug, possibly due to hypermetabolic side effects or to cardiac arrhythmias (16). Clofibrate administration in the same study led to only a small and nonsignificant reduction in the rate of nonfatal myocardial infarction (3). An adverse experience with clofibrate was reported from the World Health Organization study (17), a placebo-controlled trial of primary prevention of coronary heart disease in men whose plasma cholesterol levels were in the upper third of the population distribution. Clofibrate administration was associated with reduced incidence of nonfatal myocardial infarction, but with increased overall mortality, which resulted from noncardiovascular diseases.

The remainder of this chapter will be concerned chiefly with individual drugs and their pharmacokinetics, mechanisms of action, effects on lipids and lipoproteins, and side effects. Combination therapy then will be considered briefly, and finally the approaches to management of various lipoprotein disorders will be summarized.

2 Pharmacologic Agents

2.1 Bile acid sequestering resins

Cholestyramine and colestipol are anion exchange resins that lower plasma cholesterol by binding bile acids in the gut and preventing their reabsorption. The structures of these high molecular weight polymers differ (Fig. 3), but their modes of action, efficacy, and side effects are virtually indistinguishable. After oral administration, the resins remain in the gastrointestinal tract and are not absorbed.

CHOLESTYRAMINE

a

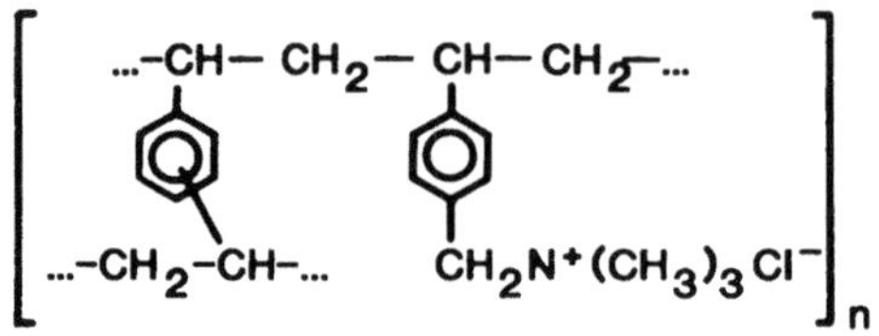

COLESTIPOL

b

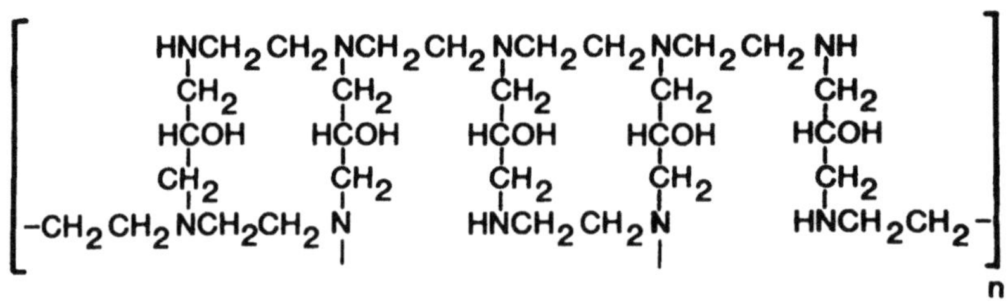

Fig. 3. Structures of cholestyramine and colestipol.

2.1.1 Changes in bile acid metabolism

Cholestyramine binds conjugated and unconjugated bile salt anions with apparent dissociation constants of $20-100 \times 10^{-6}$ M; these binding affinities are rather modest by comparison with biologic ligand binding. The capacity of these resins for binding bile salt anions is high, such that a typical 24 g daily dose of cholestyramine could bind as much as 10 g of cholate under simulated physiologic conditions in vitro. The nonspecificity of binding is indicated by the fact that fatty acid salt anions bind with equal or greater affinity (18, 19). Binding of bile salt anions by colestipol is similar to that of cholestyramine (20).

Bile acids are normally absorbed through the distal ileal mucosa and are recirculated to the liver via the portal vein. When the normal enterohepatic circulation of bile acids is interrupted, hepatic cholesterol is diverted increasingly toward conversion into bile acids to replace fecal losses. Normal production of bile acids from cholesterol in the liver, as reviewed by Grundy (21), is 100 to 500 mg/d; with resin administration this may increase to 1000 to 3000 mg/d. In addition, since cholesterol absorption from the gut depends upon the presence of bile acids (22), the intestinal uptake of both dietary and endogenous cholesterol may be modestly inhibited. The combination of these effects, especially the increased synthesis of bile acids, establishes a substantial drain on total body cholesterol metabolism.

Cholestyramine and colestipol are highly effective, but perhaps not ideal agents for lowering plasma cholesterol by interference with bile acid reabsorption. In a group of severely hypercholesterolemic patients, Miettinen and Lempinen (23) determined the effect of a ten-day course of high-dose cholestyramine on fecal steroid excretion, in comparison with the effect of a subsequent surgical procedure to bypass the distal one-third of small intestine. Fecal steroids increased by an average 1261 mg/d during cholestyramine administration, compared to 2176 mg/day after partial ileal bypass. Cholestyramine decreased serum cholesterol by 17%, significantly less than the 33% decrease found after the operation. These data suggest that the strategy of lowering plasma cholesterol by inhibiting bile acid reabsorption may not be fully exploited with the current resins. Very high doses of resin, even if tolerated symptomatically, can cause steatorrhea. Two theoretical differences exist between the pharmacologic and surgical approaches to inhibition of bile acid reabsorption. First, the resins make bile acids unavailable for fat micelle formation throughout the intestinal tract, while ileal resection only inhibits bile acid reabsorption. Second, the resins in very high doses may cause fat malabsorption directly by binding fatty acids.

2.1.2 **Changes in cholesterol metabolism**

The efficient links between hepatic bile acid and cholesterol metabolism are incompletely understood. The liver responds to bile acid depletion after resin administration by increasing the activity of 3-hydroxy-3-methylglutaryl-CoA reductase, thus increasing cholesterol synthesis, and by increasing the numbers of low density-lipoprotein (LDL) receptors on hepatocyte cell surfaces (24). Both responses make more cholesterol available for bile acid synthesis. It is uncertain whether the increased synthesis of cholesterol is directly related to low hepatic levels of bile acids, or is caused primarily by depletion of liver cholesterol (21). Increased amounts of cholesterol synthesized in the liver during resin treatment are routed toward at least two major competing pathways – the synthesis of very low density-lipoproteins (VLDL) and conversion to bile acids. Most of the increment in cholesterol synthesis appears destined for conversion to bile acids. VLDL triglyceride production is variably increased (25), and as a result, a modest increase in the plasma triglyceride concentration is often found.

Despite the fact that VLDL production increases in some cases, production of LDL appears to be unchanged by resin treatment (26, 27). The desired result of decreased LDL cholesterol levels in the circulation is achieved entirely through an increased fractional catabolic rate for LDL. The specific receptor-mediated pathway for LDL catabolism is enhanced (26), an effect presumably due to increased numbers of hepatic LDL receptors (24).

2.1.3 **Plasma lipid and lipoprotein changes**

The daily administration of 24 g of cholestyramine (active drug) or 30 g of colestipol, given in divided doses, has caused an average decrease of total plasma cholesterol of 18–25% in most studies, whether patients were normocholesterolemic or hypercholesterolemic (1, 21, 28, 29). The responses of individual patients vary more widely. Twice-daily doses appear at least as effective as more frequent regimes in most patients and should enhance compliance (30).

The bile acid sequestering resins specifically lower the LDL fraction of plasma cholesterol, decreasing average LDL cholesterol levels by 25–35% at the doses cited above (1, 28, 29). Plasma apolipoprotein B (Apo B) levels do not fall proportionately in hypercholesterolemic patients. Since Apo B is found predominantly within LDL in these patients, an increased ratio of Apo B to cholesterol in LDL may be inferred (31). Most studies have reported a small increase in plasma triglycerides following resin administration (1, 28, 29). This change is generally clinically insignificant, but a rare patient with elevated baseline triglycerides may experience a significant adverse rise. In one study the increased triglyceride levels were transient, returning to baseline in almost all patients by two months (31). Some studies have shown slightly increased HDL cholesterol levels after resin administration (1, 29). Favorable, though small, changes in HDL subfractions are reflected by increased HDL_2/HDL_3 ratios (29, 32) and increased plasma Apo AI/Apo AII ratios (31, 33).

2.1.4 **Side effects and compliance**

Patients may find the bile acid sequestering resins initially difficult to accept, because the bulky powder does not dissolve in water, fruit juices, or other liquids or semi-solid foods with which it must be mixed. Adequate explanation beforehand or administration of a dose under supervision can improve patient acceptance. Belching, bloating, or intestinal gas may be experienced shortly after drug administration; in some cases these symptoms are caused or exacerbated by air swallowing as the resin-liquid mixture is ingested too rapidly. Increased stool bulk and dryness occurs regularly during use of the resins. Constipation occurs in 10–30% of patients and may require dosage reduction or drug cessation, but is generally controllable with the aid of stool softeners or gentle laxative agents such as dietary fiber, psyllium, or milk of magnesia. As expected for drugs that remain within the gastrointestional tract, bile acid sequestering resins have minimal effects on laboratory chemistry values. Modest increases in serum alkaline phosphatase levels, generally remaining within the normal range, have been reported (34).

Cholestyramine and colestipol can significantly impair the absorption of various drugs that bind to anion exchange resins, including thyroxine, coumarin compounds, digitoxin, phenobarbital, tetracycline, and thiazide diuretics (35). Timing the administration of other medication one hour before or four hours after resin administration may avert potential absorption problems. The resins can have unpredictable effects in patients taking coumarin, since either the anticoagulant or vitamin K may be relatively malabsorbed. The absorption of various vitamins during resin administration has been carefully studied, and cases of resin-induced vitamin deficiency are extremely rare (36–38). Routine monitoring of prothrombin time is not necessary. Folate supplementation is indicated only in pediatric use (36, 37).

2.2 Neomycin

Neomycin is an almost nonabsorbable aminoglycoside antibiotic, which has been known since 1959 (39) to have a moderate hypocholesterolemic effect when given orally. It has attracted increased interest recently because of its efficacy alone or in combination therapy, and its low profile of side effects (40, 41). Theoretical concerns about long-term auditory and renal toxicity have not been borne out at the relatively low doses used for lowering cholesterol.

2.2.1 Mechanism of action

Less than 3% of an oral dose of neomycin is absorbed, and the major site of action is presumed to be the gut (42–44). Although neomycin can precipitate dihydroxy bile acids *in vitro*, its major effect when given in small doses of 1.5 g/d appears to be an inhibition of cholesterol absorption (45). Aspiration of intestinal contents after the administration of neomycin with a test meal to human subjects revealed that the drug interacted with cholesterol and fatty acid anions to form a mixed precipitate. Bile acid anions were relatively spared from precipitation (46). There is a substantial enterohepatic circulation for cholesterol, just as there is for bile acids. Therefore, the inhibition of cholesterol absorption can establish a drain on total body cholesterol metabolism and effectively lower plasma cholesterol levels, even in patients already on a low cholesterol diet. Whether the sole effect of neomycin on cholesterol metabolism is the direct inhibition of intestinal absorption is still uncertain. Because neomycin is an antibiotic, an alteration of bacterial flora has been considered as a possible mechanism of the hypocholesterolemic effect, whether through inhibition of cholesterol absorption or other means. Some studies have suggested that the lowering of bacterial 7-α-dehydroxylase activity by neomycin in the gut may relate to its hypocholesterolemic effect (42). However, the determination of bacterial sterol conversion products in feces suggested eventual bacterial resistance to neomycin in long-term therapy in some patients, while efficacy for cholesterol lowering remained (45).

2.2.2 Plasma lipid and lipoprotein changes

Neomycin at 1.5–2 g/d consistently lowered plasma cholesterol by an average 21% in patients already on a cholesterol-lowering diet (40, 45). LDL-cholesterol levels were reduced by 24%. Triglycerides and HDL-cholesterol were not affected.

2.2.3 Dosage and side effects

Neomycin is given for hypercholesterolemia in a daily dose of 1.5 to 2 g in two or three divided doses. Over the first 10 to 14 days of therapy, up to one-half of patients experience transient abdominal cramping or diarrhea, generally controllable with antimotility agents such as diphenoxylate combined with atropine. In long-term use the drug is well tolerated. Inhibition of digoxin absorption has been described after large doses of neomycin; digoxin is best given one hour before neomycin. In patients with severe liver disease treated for hepatic encephalopathy with maximum neomycin doses of 4 to 8 g/d, serious ototoxicity leading to deafness has been described (47). Hepatic dysfunction is a relative contraindication to neomycin use for hypercholesterolemia. There is only one report of a single patient treated for hypercholesterolemia, who experienced a decrease in speech reception threshold while taking 2 g of neomycin daily for six months (42). Because neomycin is potentially nephrotoxic and because any neomycin absorbed from the gut must be excreted by the kidney, the presence of renal insufficiency is a contraindication to use of the drug.

2.3 Nicotinic acid

Nicotinic acid (Fig. 4), also known as niacin, is a B vitamin which is highly effective in lowering both triglyceride and cholesterol levels in plasma, when it is given in doses far exceeding the daily vitamin requirement. However, the usefulness of nicotinic acid is often limited by side effects.

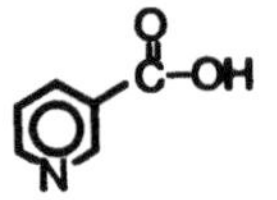

Fig. 4. Structure of nicotinic acid.

2.3.1 **Pharmacokinetics**

Even when large doses are given, the intestinal absorption of nicotinic acid appears to be almost complete. Levels of free nicotinic acid in plasma reach a peak value of approximately 30 mg/l by 30 – 60 minutes after the administration of a single oral dose of 1 g (48). The urinary excretion of unchanged nicotinic acid appears to account for the rapid diminution of plasma levels within six hours (49, 50).

2.3.2 **Mechanism of action**

The explanation of plasma lipid lowering by nicotinic acid remains enigmatic. Its early use was prompted by the hypothesis that an increased supply of nicotine adenine nucleotide might accelerate the oxidative conversion of cholesterol to bile acids (51). However, bile acid excretion does not increase. Nicotinamide, an equally effective precursor for the nucleotide (and thus equally effective as a vitamin), has very little hypolipidemic action. Thus the early hypothesis proved false (52).

The clearest information on nicotinic acid effects comes from studies of lipoprotein kinetics. With the use of radioiodinated LDL, decreased input of LDL to the plasma compartment was found in patients taking nicotinic acid, while fractional clearance of LDL was unchanged (27). The decreased production of LDL may be due to a decreased production of very low density-lipoprotein (VLDL), although other possibilities have also been cited. Using multicompartmental kinetic modeling to analyze the behavior of injected ^{3}H-glycerol, Grundy and colleagues (53) estimated a 21% decrease in synthesis of VLDL-triglyceride following the administration of 3 g of nicotinic acid per day. Another factor in plasma triglyceride lowering by nicotinic acid appeared to be an increased fractional catabolic rate for VLDL-triglyceride. This was consistent with earlier work demonstrating that nicotinic acid administration to human subjects increased substantially the clearance rate of an intravenous fat load. Nicotinic acid also increased lipoprotein lipase activity of adipose tissue in rats. However, postheparin plasma lipoprotein lipase activity in man did not change after nicotinic acid administration (54, 55).

Since the discovery by Carlson and Oro in 1962 (56) that the release of free fatty acids from peripheral tissue in man is markedly inhibited by nicotinic acid, there has been continuing speculation that the drug may exert its hypolipidemic effects partly through a decreased availability of fatty acids for triglyceride synthesis in the liver. The inhibition of peripheral fatty acid release is caused by a lowering of cyclic AMP levels (57). However, the inhibition of fatty acid release is only transient, lasting about two hours after the administration of an oral dose of nicotinic acid, and there is a rebound and overshoot of free fatty acid levels later. In extended studies with normal diet and activity, the average daily free fatty acid level has actually been found somewhat increased in patients taking nicotinic acid (58 – 60). Furthermore, in Grundy's patients on nicotinic acid the ratio of triglyceride to cholesterol within VLDL rose substantially, a finding that would be unexpected if triglyceride synthesis were a limiting factor (53). A complex mechanism of action is further suggested by the differing responses of plasma triglycerides and cholesterol to changes in the time course of nicotinic acid administration. In patients complying fully with a dosage of 1 g of nicotinic acid given three times a day, a timed-release preparation lowered cholesterol to a degree similar to that achieved with unmodified drug, but only the unmodified drug achieved substantial triglyceride lowering (61). These results were quite similar to earlier experience with a timed-release form of nicotinic alcohol (β-pyridylcarbinol), a drug that is converted rapidly in the body to nicotinic acid (62).

Radioisotopic studies have indicated that cholesterol synthesis is decreased during nicotinic acid treatment in man (63, 64). Most, but not all, *in vitro* studies of liver tissue from animals have also shown an inhibitory effect of nicotinic acid on cholesterol synthesis (52). The interpretation of apparently decreased cholesterol synthesis must be reconciled with the fact that sterol balance studies have failed to show a decrease in net sterol output from the body (53, 63). The mobilization of tissue stores of cholesterol could cause, for a considerable period of time, an unchanged or increased sterol output, even if the cholesterol synthetic rate were decreased by nicotinic acid. The large size of the total body cholesterol pool, at least 80 – 90 g (65), suggests that such a nonequilibrium state might persist for months after patients start taking the drug.

2.3.3 **Plasma lipid and lipoprotein changes**

Patients with primary hypercholesterolemia given 3 g/d of nicotinic acid generally experience a reduction of total plasma cholesterol of 10 – 20%. Reductions in LDL cholesterol levels have averaged near 20% (61, 66). Highly favorable increases in HDL-cholesterol have been found in normolipidemic volunteers and most hyperlipidemic patients, but not in Type IIB hyperlipoproteinemic patients. When an HDL response occurs, the HDL_2 subfraction is consistently elevated. It is controversial whether the HDL_3 subfraction may also be elevated to a similar degree (61, 67). The degree of triglyceride lowering by nicotinic acid is dependent upon the initial level of triglycerides in plasma. Normotriglyceridemic volunteers and patients experience only minor reductions in plasma triglycerides. In patients with severe hypertriglyceridemia resulting from Types III, IV or V hyperlipoproteinemia, plasma triglyceride reductions averaging between 55% and 90% were found after 3 g of nicotinic acid per day (68).

2.3.4 **Administration and side effects**

The usual dose of nicotinic acid for control of plasma lipids is 3–4.5 g/d, although higher doses have been used. The drug is given in three or four divided doses daily and is best administered near the end of meals or with a bedtime snack or milk to minimize side effects. The flushing reaction, often accompanied by itching or "prickly heat" paresthesia, occurs transiently about thirty minutes after taking the drug and is experienced by the great majority of patients at some time. The reaction diminishes with continued use as physiologic tolerance develops. This fact has led to drug initiation with as little as 100–300 mg daily, and small dosage increments every three days until full dosage is reached in four to six weeks. Avoidance of hot beverages is recommended. On this regimen patients may not experience the flush until they miss taking the drug temporarily and resume at full dosage. Since the flush is mediated by prostaglandins, one aspirin tablet one-half hour prior to dosing can have a prophylactic effect. Timed-release forms of nicotinic acid may reduce the flushing reaction, but sometimes are associated with increased gastric irritation, the other common side effect of the drug (61). Several nicotinic acid derivatives, including nicotinyl alcohol, *D*-glucitol hexanicotinate (sorbinicate) and pentaerythritol tetranicotinate (niceritrol), available in some countries, are said to be effective at lower doses and with milder side effects (69). Other possible side effects of nicotinic acid are elevation of plasma uric acid levels, activation of peptic ulcer disease (70) and worsening of glucose tolerance or a diabetic state in some patients (71). Mild elevations of plasma activities of hepatic aminotransferases are usually transient. They occur relatively often and are not necessarily an indication for discontinuation of the drug. If due to nicotinic acid, the elevated aminotransferase levels reverse rapidly within two weeks of discontinuation of the drug. Rarely jaundice, usually cholestatic, and hepatic fibrosis have been recorded (72). Mild dermatologic problems that may develop with chronic effective use of nicotinic acid include dry skin, hyperpigmentation, telangiectasia, and acanthosis nigricans.

While patient adherence to nicotinic acid therapy may be a problem, effective adherence can nevertheless be achieved in the majority of patients. In the Coronary Drug Project, a multicenter trial of men who had experienced myocardial infarction, 66% of patients in the nicotinic acid group remained with the study and adhered to their medication, compared to 78% for the placebo group (3).

2.4 Probucol

The hypocholesterolemic effect of probucol (Fig. 5) was discovered through animal screening studies (73). Although it lowers total and LDL-cholesterol, the degree of lowering in a given individual is not highly predictable. Also, the efficacy of probucol

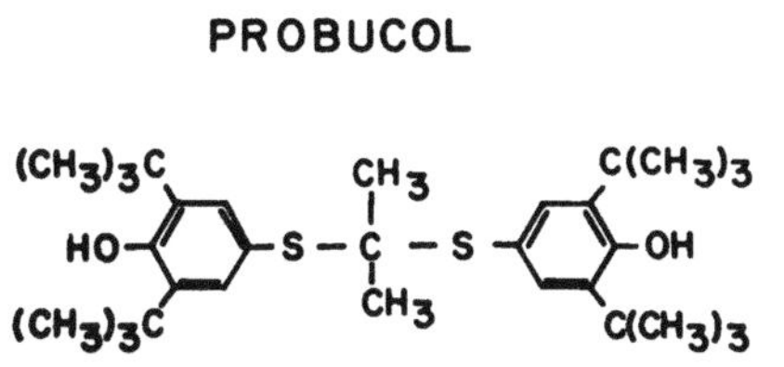

Fig. 5. Structure of probucol.

in preventing atherogenesis has been questioned on theoretical grounds, because it produces concomitant decreases in HDL-cholesterol and apoprotein levels.

2.4.1 **Pharmacokinetics**

The absorption of probucol from the intestine in man is probably only a small fraction of the administered dose, as it is in animals. With chronic administration of the drug, plasma levels do not reach a steady state until three or four months. This is due to the slow accumulation of the drug in adipose tissue, where levels 100-fold greater than in plasma can be found. The major route of excretion of the drug is in the bile (74).

2.4.2 **Mechanism of action**

Early work showed a tendency for the fractional catabolic rate of LDL to increase with probucol administration (75), but conflicting results have been obtained (76). Increases in fecal excretion of bile acids have been seen (75). Effects on hepatic cholesterol synthesis have been variable, but an effect on lipoprotein formation or secretion is suggested by some evidence (74). Lowering of HDL-levels is associated with decreased synthesis of Apo AI (75). Recent studies utilizing the Watanabe LDL receptor-deficient rabbit have demonstrated that probucol may cause a change in the LDL particle itself, leading to a more rapid disappearance of LDL from plasma in vivo and more rapid uptake by fibroblasts in tissue culture. The increased uptake rates were apparently not dependent on the LDL receptor pathway (77).

2.4.3 **Plasma lipid and lipoprotein changes**

Probucol lowers plasma cholesterol in the average patient by 10–15% and LDL-cholesterol by a similar amount. Triglyceride levels are not affected. HDL-cholesterol is reduced by about 20%, and most studies have reported a small increase in the LDL/HDL-cholesterol ratio (28, 78, 79). Reductions in plasma Apo AI and Apo AII have been demonstrated (75, 79). Recently, probucol has been shown to decrease cholesterol in the HDL_2 subfraction (80–82). Based on the inverse relationship between HDL_2 and atherogenic risk, a theoretical disadvantage might be ascribed to probucol therapy. However, in metabolic terms the impact of HDL on atherogenesis is not yet clear; it is uncertain whether maintenance of HDL levels during drug therapy is a necessary goal. Long-term probucol therapy has been successful in reducing the size of cholesterol-laden xanthomata in patients with both heterozygous and homozygous familial hypercholesterolemia, with particular success in the latter (11 – 14). This removal of tissue cholesterol occurred despite supposedly "unfavorable" changes in HDL-cholesterol and LDL/HDL ratios.

2.4.4 Compliance and side effects

Probucol is generally given in the form of two 250 mg tablets, twice daily. The drug is well tolerated. Diarrhea may occur, but is usually transient. Animal toxicity studies have shown heightened cardiac adrenergic sensitivity and increased tendency to ventricular fibrillation. Probucol has thus far not been linked to arrhythmia generation in humans, although a modest lengthening of the QT interval can be noted on electrocardiograms (74).

2.5 Fibric acid derivatives

Drugs classified as fibric acid derivatives show a structural similarity with clofibrate (ethyl *p*-chlorophenoxyisobutyrate), the original drug in this class. Clofibrate was discovered through deliberate pharmaceutical screening of analogues of branched chain fatty acids; the theoretical basis for expecting hypolipidemic effects was through interference with fatty acid metabolism (83). In humans clofibrate lowered plasma triglycerides effectively, but had relatively small effects on cholesterol (84). The lack of immediate side effects made it well received and widely used. In recent years, the occurrence of serious long-term side effects of clofibrate has led to recommendations that its use be sharply restricted. A newer drug, gemfibrozil, shares the triglyceride-lowering action of clofibrate, but has the additional effect of raising HDL-cholesterol levels substantially. Bezafibrate, fenofibrate (procetofen), and ciprofibrate are newer, more potent medications which lower triglycerides, raise HDL-cholesterol, and often effectively lower LDL-cholesterol. Figure 6 shows the structures of some of these compounds.

2.5.1 Pharmacokinetics

Clofibrate is completely absorbed from the intestine and rapidly de-esterified to form the active compound *p*-chlorophenoxyisobutyric acid. There is extensive binding to plasma albumin, which can displace and raise plasma levels of other acidic drugs. The active compound has a biphasic disappearance from the plasma with a terminal half-life of 15 hours. Essentially all of the compound is excreted in the urine, 60% as the glucuronide (85). Gemfibrozil is well absorbed. Its clearance from plasma with a half-life of 1.3 hours is due to extensive conjugation and other routes of drug metabolism, with urinary excretion of the metabolites (86). Bezafibrate is well absorbed and has a plasma half-life of 2.1 hours. It is cleared approximately one-half by metabolism and one-half by urinary excretion of un-

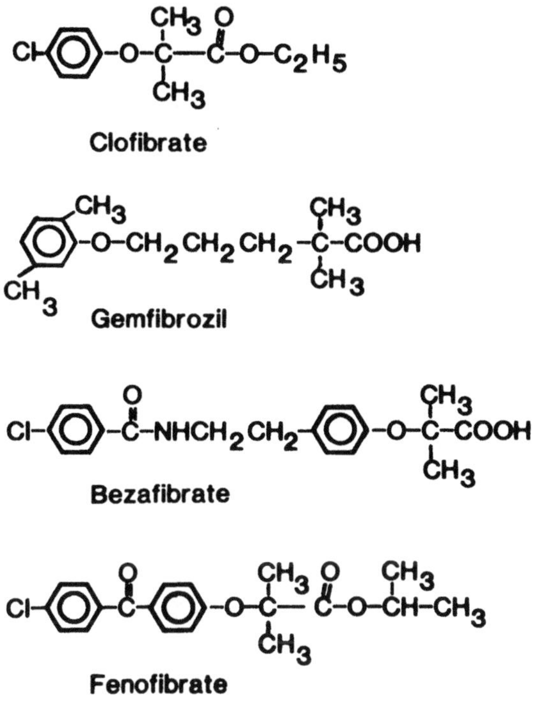

Fig. 6. Structures of clofibrate, gemfibrozil, bezafibrate, and fenofibrate.

changed drug (87). Intestinal absorption of fenofibrate is variable, averaging 30% of the ingested dose. *In vivo* it is rapidly activated by hydrolysis of an ester bond. The resulting active compound is tightly bound to albumin in the plasma. There is a prolonged biphasic disappearance from plasma with half-lives of 5 and 27 hours. Clearance is mostly metabolic, via conjugation of the active compound and urinary excretion (88, 89).

2.5.2 **Mechanism of action**

Strong evidence now exists that the lowering of plasma triglycerides by fibric acid derivatives is due mainly to enhanced peripheral catabolism of triglyceride-rich lipoproteins (90–93). Lipoprotein lipase activity in postheparin plasma or in tissue extracts rises considerably after drug treatment, while hepatic lipase activity is unchanged (94–98). Other potential effects of fibric acid derivatives on triglyceride metabolism, such as decreased delivery of free fatty acids to the liver and decreased production of VLDL triglyceride are less well established and may depend upon the particular drug studied (91, 92, 99).

Although clofibrate lowers plasma cholesterol levels only modestly, distinct changes in body cholesterol metabolism have been noted. Fecal excretion of neutral steroids is enhanced. A net increase in excretion of total steroids has been ascribed to mobilization of cholesterol from tissue pools rather than to increased cholesterol synthesis by the liver (100). However, clofibrate does not appear to induce resolution of cholesterol xanthomas in Type IIa patients as effectively as other drugs. The effect of newer agents on cholesterol xanthomas is as yet unclear. Administration of gemfibrozil in one study was associated with increased fecal neutral steroids and decreased fecal acidic steroids (bile acids), with no net change in total fecal steroid excretion (92).

Bezafibrate, fenofibrate, and ciprofibrate exert a substantial hypocholesterolemic effect in most patients. Available data suggest a similar mechanism of action for at least the first two of these drugs. A reduction in VLDL-cholesterol accounts for a substantial portion of the total plasma cholesterol reduction. The effect on LDL is more complex. A consistent feature is an increased production of LDL particles, which may be due to an increased proportion of VLDL particles escaping direct catabolism and entering the LDL particle population (101). In normotriglyceridemic patients, LDL cholesterol levels fall due to a substantial increase in catabolism mediated by the specific LDL receptor pathway (102, 103). Bezafibrate shares with clofibric acid the ability to inhibit hydroxymethylglutaryl-CoA reductase activity in rats (104), and fenofibrate appears to have similar effects in rat and human experiments (105). On the other hand, it has been suggested that LDL isolated from patients treated with bezafibrate are more effective in inhibiting cholesterol synthesis in cultured fibroblasts because of compositional differences. Thus, bezafibrate may inhibit cholesterol synthesis indirectly in some patients (106). In patients with marked hypertriglyceridemia, bezafibrate and fenofibrate sometimes have the paradoxical effect of raising LDL-cholesterol levels. In these patients, before treatment, LDL catabolism via routes other than the specific LDL receptor appears to be greatly increased, and drug therapy normalizes this "nonreceptor" catabolism. An increase in LDL cholesterol concentrations in plasma is thereby effected (107).

High density lipoprotein metabolism is also altered substantially by the newer fibric acid derivatives. In hypertriglyceridemic patients with a Type V phenotype, gemfibrozil significantly increased plasma HDL-cholesterol and apolipoproteins AI and AII. The synthetic rates of Apo AI and Apo AII were increased, without a change in their fractional catabolic rates (98). In Type IIA hypercholesterolemic patients fenofibrate was shown to increase Apo AI levels. Both the synthetic rate and the fractional catabolic rate of Apo AI were increased, but a much greater increase occurred in the synthetic rate (103).

2.5.3 **Plasma lipid and lipoprotein changes**

All of the fibric acid derivatives reduce plasma triglyceride levels by about 30–50%. Greater fractional reductions can be seen in some individual patients who begin with extreme hypertriglyceridemia. The effect on plasma cholesterol levels is highly dependent upon the type of hyperlipidemia being treated. In Type IIA hypercholesterolemia, gemfibrozil has caused an average reduction of 9% in plasma cholesterol and LDL-cholesterol, while fenofibrate, bezafibrate, and ciprofibrate lowered these levels by 18–25% (108–114). In patients with hypertriglyceridemia, whether or not plasma cholesterol is initially elevated, the average reductions in LDL cholesterol are considerably less and may be negligible. In these patients, total plasma cholesterol levels may fall as a result of VLDL-cholesterol lowering, or occasionally may rise as a result of an increased conversion of VLDL to LDL. All of the newer fibric acid derivatives have a tendency to elevate HDL-cholesterol levels by 10–20%. Apolipoprotein AI and, when measured, apolipoprotein AII have increased to a similar degree.

2.5.4 **Side effects**

The fibric acid derivatives are among the best tolerated lipid-lowering drugs currently available. Nausea or diarrhea may occur. Uncommon side effects of clofibrate identified in two large studies included decreased libido, impotence, skin rash, increased appetite, and weight gain (3, 17). Clofibrate may induce elevations of serum creatine phosphokinase and transaminase, sometimes associated with malaise and muscle aching, stiffness, and weakness (115). A serious, long-term problem with clofibrate is a two-fold increase in the development of gallstones or gallbladder disease during the first five to eight years of treatment; this effect is thought to result from increased biliary secretion of cholesterol (3, 17). Biliary cholesterol is known to rise in patients treated with the newer fibric acid derivatives as well (116–119), though probably not as much as during treatment with clofibrate. Specific screening revealed a substantial incidence (1.0–1.5% per year) of new gallstones in subjects beginning gemfibrozil, but the significance of this data is uncertain because placebo controls were not included (120). Fenofibrate has been suspected to cause drug-induced hepatitis exhibiting anti-smooth muscle antibodies in plasma (121).

In the World Health Organization trial for primary prevention of heart disease (17), clofibrate use in a large group of male subjects with mildly to severely elevated cholesterol levels (upper third of population distribution) was compared to placebo use in patients with similar cholesterol levels and to a third group of subjects with relatively low cholesterol levels. Although clofibrate reduced nonfatal myocardial infarction, the total death rate among clofibrate-treated patients was 28% higher than that in the high cholesterol control group. The significance level of the effect on mortality was $p < 0.05$. The increment in mortality was confined to the category of noncardiovascular and nontraumatic causes. Neoplastic and other conditions related to the liver, biliary and intestinal systems showed the greatest proportionate excess of deaths. These findings have led to the recommendation that clofibrate use be sharply restricted. While the recommendation seems valid, it must be accepted with some reservation, since the high cholesterol control group in this study may have had a fortuitously low incidence of noncardiovascular fatalities. While the newer fibric acid derivatives lead to lipid profiles predicted to be more protective against atherosclerotic disease than those attained with clofibrate, the possibility must be entertained that these agents share some unknown adverse potential with clofibrate. However, this possibility has not been established for the newer drugs. Current and future trials will address this important question.

2.6 Para-aminosalicylic acid

The hypocholesterolemic effect of this drug was first observed in patients treated for tuberculosis. A subsequent reformulation of the drug (122) reduced, but did not eliminate, side effects of epigastric discomfort and diarrhea. At doses of 6–8 g/d, given in four divided doses, various groups of hypercholesterolemic patients have shown lowering of plasma cholesterol levels averaging 6%, 15% and 19% (33, 122). Effects on triglycerides were variable. No changes in HDL-cholesterol have been demonstrated (33). The mechanism of action is unknown; higher drug doses have caused steatorrhea.

2.7 Mevastatin and lovastatin

Mevastatin (compactin, 123) and lovastatin (mevinolin, 124) are fungal fermentation products (Fig. 7) which proved to be extremely potent inhibitors of 3-hydroxy-3-methylglutaryl-coenzyme A (HMG-CoA) reductase, the rate-limiting enzyme in cholesterol synthesis. The two compounds differ only by the presence of one additional methyl group in lovastatin. Both have been used in premarketing clinical trials which appeared very promising. Simvastatin, a structurally related drug, is now being tested in clinical experiments.

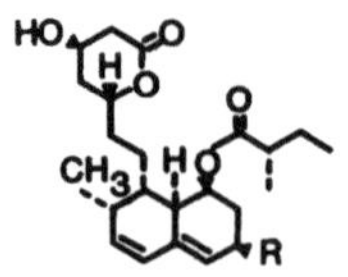

Fig. 7. Structure of mevastatin (compactin) (-R represents -H) and lovastatin (mevinolin) (-R represents -CH_3).

2.7.1 **Mechanism of action**

Mevastatin and lovastatin reversibly inhibit HMG-CoA reductase with K_i's of 1.4 $\times 10^{-9}$ M and 6 $\times 10^{-10}$ M, respectively. In human subjects the lack of cholesterol available to liver cells apparently stimulates the production of hepatic LDL receptors, thus increasing the activity of the receptor-specific pathway for LDL catabolism. LDL production rates are variable, but decreased in most patients (125). Grundy and Bilheimer (126) found that effective lowering of plasma cholesterol by lovastatin was associated with only small decreases in apparent total body cholesterol synthesis, as deduced from measurement of dietary and fecal steroids. Thus the concept emerges that small net reductions in cholesterol synthesis can trigger powerful homeostatic mechanisms – i. e. induction of hepatic LDL receptors – that result in lowering of plasma cholesterol.

2.7.2 Plasma lipid and lipoprotein changes

Daily doses of 30 – 60 mg mevastatin or 40 – 80 mg lovastatin, given in 2 – 3 divided doses through the day, have been given to patients with familial hypercholesterolemia. Mevastatin lowered total cholesterol by 22% and LDL cholesterol by 29% (127). Lovastatin at the lower dose lowered total cholesterol by 30% and LDL cholesterol by 35% and at the higher dose achieved only slightly greater reductions (128). Triglyceride levels, initially in the normal range in these patients, were reduced by about one-third by both drugs. There was no effect on HDL-cholesterol levels. Thus the early data indicate that mevastatin and lovastatin may be the most effective single agents yet found for reducing plasma cholesterol.

2.7.3 Compliance and side effects

Data on potential side effects of treatment with mevastatin and lovastatin are sketchy and incomplete. Apparently, testing of mevastatin has been suspended or stopped due to toxicity in animals. Nevertheless, both drugs have been generally well tolerated according to published reports. A theoretical concern is inhibition of steroid hormone synthesis, which depends upon the availability of cholesterol. Initial studies, including a normal response to prolonged infusion of adrenocorticotrophic hormone, have largely alleviated this concern (129, 130). A further possibility exists that metabolites of HMG-CoA reductase other than cholesterol might be adversely affected. Serum levels of one of these metabolites, ubiquinone-10, did not change appreciably during treatment with mevastatin (127). Reversible changes in liver function tests have been found in clinical use of lovastatin in man.

2.8 Synthetic steroids

The 17-α-alkylated derivatives of nortestosterone, oxandrolone and norethindrone acetate, are useful agents for the treatment of severe hypertriglyceridemic states. They do not have official approval for this purpose in the United States. Oxandrolone, an anabolic steroid, is given to men at a typical dose of 2.5 mg three times daily. Norethindrone acetate, a progrestational agent, has been administered chronically to women at 5 mg daily for 25 days of each menstrual cycle, beginning on the second day of menses (131). Triglyceride reductions in Type V hyperlipidemic patients have ranged from 26% to 89%; the reductions are often sufficient to alleviate symptomatic fasting chylomicronemia. Type I patients have not responded (132). Plasma clearance of triglyceride-rich lipoproteins is increased by these agents, an effect possibly related to an increase in heparin-releasable hepatic triglyceride lipase (133). HDL cholesterol levels may be substantially lowered by oxandrolone (33, 134); the effect of norethindrone acetate appears to be similar (135). The drugs are well tolerated. Both drugs have potential side effects common to 17-α-alkylated steroids, including hepatic abnormalities and fluid retention. Norethindrone acetate may cause a thrombotic tendency.

3 Summary of Sites of Drug Action

The diverse sites of action of hypolipidemic drugs are summarized in Figure 8.

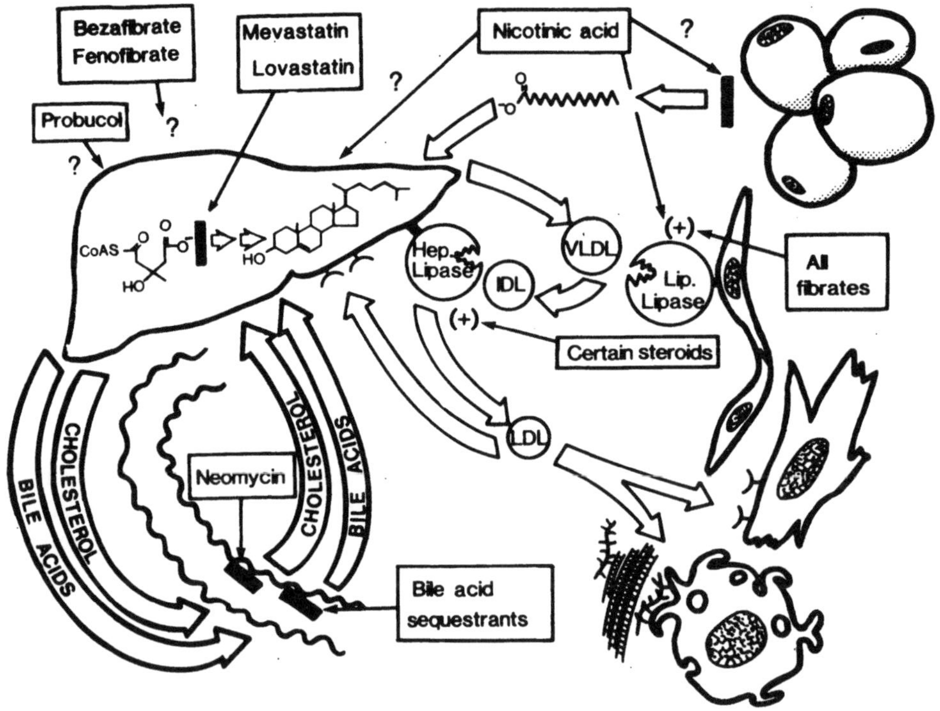

Fig. 8. Major sites of action of drugs affecting lipid metabolism. The bile acid sequestering resins, cholestyramine and colestipol, act entirely in the gut to interrupt the enterohepatic circulation of bile acids. Neomycin interrupts the enterohepatic circulation of cholesterol. The action of probucol is poorly understood, but an effect on liver cholesterol metabolism is postulated. Bezafibrate and fenofibrate probably reduce cholesterol synthesis, and in addition, these agents share with other fibric acid derivatives the ability to increase the activity of peripheral lipoprotein lipase. This enzyme is located on the surface of peripheral vascular endothelial cells. Mevastatin and lovastatin inhibit the activity of hydroxymethylglutaryl-CoA reductase, the rate-limiting enzyme in cholesterol synthesis. Nicotinic acid has multiple effects, including enhancement of peripheral lipolysis of triglyceride-rich lipoproteins, inhibition of free fatty acid release from adipose tissue, and hepatic effects not yet fully characterized. The role of inhibition of free fatty acid release in the lowering of plasma cholesterol and triglyceride by nicotinic acid is questionable, since this effect is short-lived after each dose. The 17-α-alkylated derivatives of nortestosterone, including oxandrolone and norethindrone acetate, appear to increase the activity of hepatic lipase. Also shown in the figure are hepatic and peripheral receptors for LDL. Any maneuver that decreases the availability of cholesterol to the liver may cause an increase in hepatic LDL receptors and thus enhance the clearance of LDL from plasma. For simplicity, pathways of chylomicron metabolism (with peripheral lipoprotein lipase as a key enzyme) and HDL metabolism (complex and incompletely understood) have been omitted from the figure.

4 Combination Therapy

For the treatment of severe hypercholesterolemia it is usually necessary to combine two drugs. Normalization of plasma total and LDL-cholesterol rather than mere reduction is a reasonable goal. Representative combined regimens and their effects on average plasma lipoprotein/lipid levels are shown in Table 1. All the studies incorporated a preliminary dietary phase before drawing baseline lipids. Substantial and favorable additive effects were evident for the combinations of colestipol + nicotinic acid, neomycin + nicotinic acid, and colestipol + mevinolin. The combination of cholestyramine and neomycin was judged to be no better than cholestyramine alone, since only a small additional decrement in total cholesterol (6% of baseline) was achieved by adding neomycin, and this decrement was due largely to a reduction of HDL-cholesterol (137). In the study of colestipol + probucol, the reductions in total and LDL-cholesterol achieved by colestipol alone were 18% and 23%, respectively. Comparison with values listed in Table 1 reveals that the addition of probucol caused further lowering of both total and LDL-cholesterol. However, because of the marked effect of probucol in lowering HDL-cholesterol, the theoretically favorable reduction of LDL/HDL ratio induced by colestipol was negated by the addition of probucol (138).

Combination therapy may sometimes reduce side effects of lipid-lowering drugs. Neomycin ameliorated constipation associated with cholestyramine use (137), and there were far fewer gastrointestinal side effects with the combination of probucol + colestipol than with either drug alone (138).

Table 1. Effects of combined drug regimens on plasma lipids and lipoproteins in patients with severe hypercholesterolemia

Drug combination	Daily dose[1])	Total cholesterol	LDL-C	HDL-C	TG	Reference
Nicotinic acid Colestipol[2])	6–7.5 g 30 g	−45%	−55%	+17%	−38%	136
Nicotinic acid[3]) Neomycin	3 g 2 g	−36%	−45%	+14%	−33%	41
Cholestyramine[2]) Neomycin	24 g 2 g	−29%	−33%	−11%	− 4%	137
Probucol Colestipol[2])	1 g 20 g	−28%	−29%	−29%	+11%	138
Lovastatin (mevinolin) Colestipol[2])	30 mg 20 g	−45%	−51%	− 2%	+ 7%	139

[1]) All drugs were given in divided doses through the day.
[2]) Clinical experience indicates that one can substitute cholestyramine for colestipol, or vice versa, in combination with other drugs.
[3]) Eleven out of 25 patients could not adhere to full nicotinic dosage because of side effects. The percentage lipoprotein/lipid changes represent only those derived from the 14 drug adherers.

When both LDL and VLDL are elevated (Type IIB hyperlipoproteinemia), nicotinic acid may be given alone or in combination with a cholesterol-lowering agent. Alternatively, a cholesterol-lowering agent may be combined with an agent that primarily reduces triglycerides, such as gemfibrozil (see below).

The use of three or four hypolipidemic drugs in combination has been described. Patients treated in this manner have had severe hypercholesterolemia usually accompanied by marked hypertriglyceridemia, and either a personal history or a strong family history of atherosclerotic disease (140).

5 Drug Treatment of Specific Lipid Disorders

Before any patient with hyperlipidemia is committed to drug therapy, potential causes of secondary hyperlipidemia should be ascertained and treated. Sometimes when the underlying metabolic abnormality cannot be corrected fully – e. g., certain cases of nephrotic syndrome, diabetes mellitus, obesity – pharmacologic therapy for secondary hyperlipidemia may be initiated. The secondary hyperlipidemias are covered extensively elsewhere in this book.

It must be emphasized that the first step in treatment of any dyslipoproteinemia is dietary management. In most instances a dietary trial of two months or greater duration should be carried out before considering drugs. When reduction of risk for atherosclerotic disease is the goal of therapy in the dyslipoproteinemic patient, considerable attention should be given to other risk factors such as smoking and hypertension. Cardiovascular risk tends to multiply, and not simply to add, when multiple factors are present.

5.1 Hypercholesterolemia

The results of recent efficacy studies suggest that hypercholesterolemic drug therapy should be used more widely than in the past (141). Individuals over the age of 40 with initial plasma cholesterol levels greater than 6.72 mmol/l (260 mg/dl) are at high risk for the development of atherosclerotic disease. Pharmacologic therapy may be recommended for these individuals, if initial dietary management does not reduce cholesterol (as it often does not) to a level of 5.17 – 5.95 mmol/l (200 – 230 mg/dl). Adults with initial plasma cholesterol levels of 6.21 – 6.72 mmol/l (240 – 260 mg/dl), the 75th to 90th percentiles in the U.S. Lipid Research Clinics Study, should infrequently be candidates for drug therapy, since dietary management is often effective at these levels. These levels are only guidelines, and clinical judgment must be used in every individual case. The presence or absence of existing atherosclerotic disease, strong family history, or other risk factors should greatly influence the decision to use or withhold drug therapy. The level of HDL-cholesterol or the use of lipoprotein cholesterol ratios can be helpful in a given case. Average cholesterol levels in the population rise between the ages of 20 and 40, and a somewhat lower threshold for drug use is appropriate in younger adults. On the other hand, drug therapy may be withheld in moderately hypercholesterolemic elderly individuals without signs of cardiovascular disease, since the expected benefit as well as their tolerance for drug side effects may be relatively low.

The bile acid sequestering resins can be recommended for initial therapy of hypercholesterolemia without concomitant triglyceride elevation (Type IIA), although nicotinic acid represents a good low-cost alternative. When there are relative contraindications to the use of resins (e. g., constipation) or nicotinic acid (e. g., diabetes, recent peptic ulcer disease, or hyperuricemia), then initial therapy may be provided with neomycin, probucol, or, where available, bezafibrate or fenofibrate. Para-aminosalicylate is rarely used. Although patient acceptance of these second-line agents may be better, their routine use in initial therapy of hypercholesterolemia should be discouraged until documentation of their efficacy in preventing atherosclerotic disease is obtained. Since side effects are so common with the first-line agents, and since combination therapy is often desirable, the relative advantages of the second-line drugs should be considered. Lovastatin is well tolerated and highly effective for lowering cholesterol levels; current studies are evaluating its efficacy against atherosclerosis. There is considerable clinical experience with both neomycin and probucol. Neomycin has the advantage of remaining almost entirely in the gut and is relatively well tolerated. Repetitive quantitative testing of auditory function as well as serum creatinine levels is recommended for patients taking neomycin. A therapeutic dilemma may be posed if the patient develops mild progressive decline of neurosensory auditory function, common with aging, but potentially a side effect of neomycin use. Probucol has the demonstrated advantage of reducing the size of tendinous xanthomas in patients with familial hypercholesterolemia, despite the fact that HDL-cholesterol levels are reduced by this drug. Clinical experience is relatively short for bezafibrate and fenofibrate, but these agents have the theoretical advantage of concomitantly decreasing LDL-cholesterol and increasing HDL-cholesterol.

Hypercholesterolemia in the pediatric population may occasionally need treatment with bile acid sequestering resins. The appropriate patient for such treatment would have an hereditary hypercholesterolemia with plasma cholesterol levels well above 6.21 mmol/l (240 mg/dl) and unresponsive to strict dietary intervention. Folic acid supplementation is recommended in children.

5.2 Hypertriglyceridemia

Dietary management is the mainstay of treatment of hypertriglyceridemia. Any form of drug therapy may be ineffective in the obese patient who does not lose weight or in the heavy ethanol user. In patients with poorly controlled diabetes mellitus, the use of insulin or an oral hypoglycemic agent to achieve better glycemic control will also lower triglyceride levels.

The decision to use drugs to treat hypertriglyceridemia in an otherwise optimally managed patient depends upon the plasma triglyceride level, the presence of fasting chylomicrons or a history of pancreatitis, and certain other considerations (142). Patients with triglyceride levels of 17 mmol/l (1500 mg/dl) or greater generally have fasting chylomicrons, are at high risk for acute pancreatitis, and deserve immediate attention to lowering triglyceride levels. Accumulation of chylomicrons and development of pancreatitis may occasionally be seen in patients with triglyceride levels between 5.6 and 17 mmol/l (500 and 1500 mg/dl). All patients with triglyceride levels in this range have some risk of pancreatitis, but a prolonged trial of dietary management (several months to a year) is appropriate before resorting to drug therapy. The patient with triglyceride levels between 2.8 and 5.6 mmol/l (250 and 500 mg/dl) has "borderline hypertriglyceridemia". This patient is not at risk from pancreatitis, but triglyceride levels in this range have been associated with the development of atherosclerotic disease. This association may be due wholly or in part to the inverse correlation between triglycerides and HDL-cholesterol. In epidemiologic studies in which both factors were measured, the HDL-cholesterol, but not triglycerides, served as an independent predictor of atherosclerotic disease. When considering drug therapy, the relative importance of hypertriglyceridemia versus low HDL-cholesterol in atherogenesis is less important, because the commonly employed agents, nicotinic acid and the newer fibric acid derivatives, all raise HDL-cholesterol substantially while reducing triglycerides. The important question is whether the altered lipid metabolism manifested by hypertriglyceridemia in an individual patient is an atherogenic dyslipoproteinemia or not. In the case of familial combined hyperlipidemia, sometimes presenting as an elevation of triglycerides alone, the answer seems to be "yes". Most patients with hypertriglyceridemia, however, do not have this disorder. In hypertriglyceridemic patients who are otherwise at low risk for the development of atherosclerotic disease, the pharmacologic treatment of triglyceride levels less than 5.6 mmol/l (500 mg/dl) seems unnecessary.

Nicotinic acid is the most potent agent for lowering plasma triglycerides. It is the drug of choice for patients with severe hypertriglyceridemia, when a readily treatable cause such as ethanol abuse or poorly controlled diabetes is not apparent. Gemfibrozil and other new fibric acid derivatives are more easily administered and tolerated than nicotinic acid. These drugs play an important role in the treatment of mild to moderate hypertriglyceridemia. Their effect in patients with triglyceride levels above 23 mmol/l (2000 mg/dl), though variable, may occasionally be sufficient to reduce the risk of pancreatitis. Oxandrolone and norethindrone acetate may adversely affect the atherosclerotic profile via their effects on HDL-cholesterol and blood clotting. They should be used only when there is a high risk of pancreatitis or chylomicron-related abdominal pain.

5.3 Familial combined hyperlipidemia

This is an autosomal dominant trait which may express itself by the elevation of plasma cholesterol, triglycerides, or both lipids simultaneously. It must be diagnosed by family studies. Regardless of the manner of expression in an individual patient, it appears to be an atherogenic trait. In familial combined hyperlipidemia the reduction of plasma triglycerides by a fibric acid derivative may be accompanied by a rise in LDL-cholesterol. Nicotinic acid is a preferred agent, since it reduces both cholesterol and triglycerides. If it cannot be used, then a combination of bile acid sequestering resin and fibric acid derivative may be employed. Even if evidence for a familial disorder is lacking, certain groups of patients should probably be managed in the same manner as patients with familial combined hyperlipidemia. These groups include patients with hypercholesterolemia and substantial hypertriglyceridemia, patients whose cholesterol rises upon treatment of hypertriglyceridemia, and patients with premature atherosclerotic disease in whom moderate hypertriglyceridemia exists as an isolated risk factor.

5.4 Dysbetalipoproteinemia (Type III hyperlipoproteinemia)

When both the plasma cholesterol and triglyceride levels are substantially elevated, a rare patient will be found to have this disorder, which represents an accumulation of intermediate density lipoproteins in the plasma. The lipid abnormality is highly atherogenic, but is also responsive to treatment. Wide variation in plasma lipid levels may be found to correspond with dietary changes. However, drug therapy is usually also required. Clofibrate is an effective medication for dysbetalipoproteinemia; this is the only situation in which clofibrate is still recommended. Other fibric acid derivatives and nicotinic acid also appear to be effective, although experience with these agents is more limited.

5.5 Low HDL-cholesterol

The inverse statistical association between coronary heart disease and HDL-cholesterol has often been stronger in epidemiologic studies than the direct relationship with total or LDL-cholesterol. However, due to the complex nature of HDL and its metabolism, the rationale for directing specific treatment toward raising HDL-cholesterol levels is not nearly as strong as the rationale aimed at lowering LDL-cholesterol. It does seem appropriate to consider effects on HDL-cholesterol when choosing agents to treat hypercholesterolemia or hypertriglyceridemia. Lifestyle and dietary changes aimed at raising HDL-cholesterol can also be supported on the basis of available data. However, pharmacologic treatment to raise a low HDL-cholesterol, when it is the only lipoprotein abnormality, has no proven efficacy and probably should not be employed outside of the research setting. Clinical trials designed to evaluate the effectiveness of such therapy in preventing coronary heart disease are in progress.

References

1. Lipid Research Clinics Program (1984) J. Amer. Med. Assoc. *251*, 351–374.
2. Brensike, J. F., Levy, R. I., Kelsey, S. F., Passamani, E. R., Richardson, J. M., Loh, I. K., Stone, N. J., Aldrich, R. F., Battaghini, J. W., Moriarty, D. J., Fisher, M. R., Friedman, L., Friedewald, W., Detre, K. M. & Epstein, S. E. (1984) Circulation *69*, 313–324.
3. The Coronary Drug Project Research Group (1975) J. Amer. Med. Assoc. *231*, 360–381.

4. Carlson, L. A., Danielson, M., Ekberg, I., Klintemar, B. & Rosenhamer, G. (1977) Atherosclerosis *28*, 81–86.
5. Canner, P. L. (1985) J. Amer. Coll. Cardiol. *5*, 442.
6. Hashim, S. A. & Van Itallie, T. B. (1965) J. Amer. Med. Assoc. *192*, 289–293.
7. Kachadurian, A. K. (1968) J. Atherosclerosis Res. *8*, 177–188.
8. Glueck, C. J., Ford, S., Scheel, D. & Steiner, P. (1972) J. Amer. Med. Assoc. *222*, 676–681.
9. Parsons, W. B., Jr. (1961) Arch. Intern. Med. *107*, 634–642.
10. Carlson, L. A. (1969) Adv. Exp. Med. Biol. *4*, 327–338.
11. Enjalbert, M., Lussier-Cacan, S., DuBreuil-Quidoz, S., LeLoier, J. & Davignon, J. (1980) Canad. Med. Assoc. J. *123*, 754–757.
12. Baker, S. G., Joffe, B. I., Mendelsohn, D. & Seftel, H. C. (1982) S. African Med. J. *62*, 7–11.
13. Jacotot, B., Beaumont, J. L., Buxtorf, J. C., Beaumont, V. & Edouard, L. (1985) Adv. Exp. Med. Biol. *183*, 413.
14. Watanabe, A., Mabuchi, H., Sakai, Y., Sakai, T., Yoshimura, A., Wakasugi, T., Koizumi, J., Miyamoto, S. & Takeda, R. (1985) Adv. Exp. Med. Biol. *183*, 471.
15. Levy, R. I., Fredrickson, D. S., Shulman, R., Bilheimer, D. W., Breslow, J. L., Stone, N. J., Lux, S. E., Sloan, H. R., Kraus, R. M. & Herbert, P. N. (1972) Ann. Intern. Med. *77*, 267–294.
16. The Coronary Drug Project Research Group (1972) J. Amer. Med. Assoc. *220*, 996–1008.
17. Committee of Principal Investigators (1978) Brit. Ht. J. *40*, 1069–1118.
18. Johns, W. H. & Bates, T. R. (1969) J. Pharmaceut. Sci. *58*, 179–183.
19. Johns, W. H. & Bates, T. R. (1970) J. Pharmaceut. Sci. *59*, 329–333.
20. Kritchevsky, D. & Story, J. A. (1974) J. Nutr. *104*, 458–462.
21. Grundy, S. M. (1972) Arch. Intern. Med. *130*, 638–648.
22. Siperstein, M. D., Chaikoff, I. L. & Reinhart, W. (1952) J. Biol. Chem. *198*, 111–114.
23. Miettinen, A. & Lempinen, M. (1977) Eur. J. Clin. Invest. *7*, 509–514.
24. Kovanen, P. T., Bilheimer, D. W., Goldstein, J. L., Jaramillo, J. J. & Brown, M. S. (1981) Proc. Nat. Acad. Sci. U.S.A. *78*, 1194–1198.
25. Beil, U., Crouse, J. R., Einarsson, K. & Grundy, S. M. (1982) Metabolism *31*, 438–444.
26. Shepherd, J., Parkard, C. J., Bicker, S., Lawrie, T. D. V. & Morgan, H. G. (1980) New Eng. J. Med. *302*, 1219–1222.
27. Levy, R. I. & Langer, T. (1972) Adv. Exp. Med. Biol. *26*, 155–163.
28. Glueck, C. J. (1982) Ann. Intern. Med. *96*, 475–482.
29. Levy, R. I., Brensike, J. F., Epstein, S. F., Kelsey, S. F., Passamani, E. R., Richardson, J. M., Loh, I. K., Stone, N. J., Battaglini, J. W., Moriarty, D. J., Fisher, M. L., Friedman, L., Friedewald, W. & Detre, K. M. (1984) Circulation *69*, 325–337.
30. Blum, C. B., Havlik, R. J. & Morganroth, J. (1976) Ann. Intern. Med. *85*, 287–289.
31. Witztum, J. L., Schonfeld, G., Weidman, S. W., Giese, W. E. & Dillingham, M. A. (1979) Metabolism *28*, 221–229.
32. Shepherd, J., Packard, C. J., Morgan, H. G., Third, J. H. L. C., Stewart, J. M. & Lawrie, T. D. V. (1979) Atherosclerosis *33*, 433–444.
33. Cheung, M. C., Albers, J. J., Wahl, P. W. & Hazzard, W. R. (1980) Atherosclerosis *35*, 215–228.
34. Probstfield, J. L., Statland, B. E., Gorman, L. & Hunninghake, D. B. (1983) Metabolism *32*, 818–821.
35. Levy, R. I., Morganroth, J. & Rifkind, B. M. (1974) New. Eng. J. Med. *290*, 1295–1301.
36. West, R. J., Fosbrooke, A. S. & Lloyd, J. K. (1975) Postgrad. Med. J. *51* (Suppl. 8), 82–86.
37. Schwarz, K. B., Goldstein, P. D., Witztum, J. L. & Schonfeld, G. (1980) Pediatrics *65*, 243–250.
38. Glueck, C. J. (1983) Prev. Med. *12*, 835–847.
39. Samuel, P. & Steiner, A. (1959) Proc. Soc. Exp. Biol. Med. *100*, 193–195.
40. Hoeg, J. M., Schaefer, E. J., Romano, C. A., Bou, E., Pikus, A. M., Zech, L. A., Bailey, K. R., Gregg, R. E., Wilson, P. W. F., Sprecker, D. L., Grimes, A. M., Sebring, N. G., Ayres, E. J., John, C. E. & Brewer, H. B. Jr. (1984) Clin. Pharmacol. Ther. *36*, 555–565.
41. Hoeg, J. M., Maher, M. B., Bou, E., Zech, L. A., Bailey, K. R., Gregg, R. E., Sprecher, D. L., Susser, J. K., Pikus, A. M. & Brewer, H. B. Jr. (1984) Circulation *70*, 1004–1011.

42. Samuel, P. (1979) New. Eng. J. Med. *301*, 595–597.
43. Kunin, C. M., Chalmers, T. C., Leevy, C. M., Sebastyen, S. C., Lieber, C. S. & Finland, M. (1960) New. Eng. J. Med. *262*, 380–385.
44. Breen, K. J., Bryant, R. E., Levinson, J. D. & Schenker, S. (1972) Ann. Intern. Med. *76*, 211–218.
45. Miettinen, T. (1979) J. Clin. Invest. *64*, 1485–1493.
46. Thompson, G. R., Barrowman, J., Gutierrez, L. & Dowling, R. H. (1971) J. Clin. Invest. *50*, 319–323.
47. Berk, D. P. & Chalmers, T. (1970) Ann. Intern. Med. *73*, 393–396.
48. Carlson, L. A., Oro, L. & Ostman, J. (1968) Acta Med. Scand. *183*, 457–465.
49. Greengard, P. (1970) in: The Pharmacologic Basis of Therapeutics (Goodman, L. S. & Gilman, A., eds.) p. 1654, MacMillan, New York.
50. Fumagalli, R. (1971) in: Metabolic Effects of Nicotinic Acid and Its Derivatives (Gey, K. F. & Carlson, L. A., eds.) pp. 33–49, Huber, Bern.
51. Altschul, R., Hoffer, A. & Stephen, J. D. (1955) Arch. Biochem. Biophys. *54*, 558–559.
52. Kritchevsky, D. (1971) in: Metabolic Effects of Nicotinic Acid and Its Derivatives (Gey, K. F. & Carlson, L. A., eds.) pp. 541–555, Huber, Bern.
53. Grundy, S. M., Mok, H. Y. I., Zech, L. & Berman, M. (1981) J. Lipid Res. *22*, 24–35.
54. Nikkila, E. A. (1971) in: Metabolic Effects of Nicotinic Acid and Its Derivatives (Gey, K. F. & Carlson, L. A., eds.) pp. 487–495, Huber, Bern.
55. Boberg, J., Carlson, L. A., Broberg, S., Olsson, A., Oro, L. & Rossner, S. (1971) in: Metabolic Effects of Nicotinic Acid and Its Derivatives (Gey, K. F. & Carlson, L. A., eds.) pp. 465–470, Huber, Bern.
56. Carlson, L. A. & Oro, L. (1962) Acta Med. Scand. *172*, 641–645.
57. Butcher, R. W., Baird, C. E. & Sutherland, E. W. (1968) J. Biol. Chem. *243*, 1705–1712.
58. Froberg, S. O., Boberg, J., Carlson, L. A. & Eriksson, M. (1971) in: Metabolic Effects of Nicotinic Acid and Its Derivatives (Gey, K. F. & Carlson, L. A., eds.) pp. 167–181, Huber, Bern.
59. Gundersen, K. & Demissianos, H. V. (1969) Adv. Exp. Med. Biol. *4*, 213–226.
60. Carlstrom, L. A. & Laurell, S. (1968) Acta Med. Scand. *184*, 121–123.
61. Knopp, R. H., Ginsberg, J., Albers, J. J., Hoff, C., Ogilvie, J. T., Warnick, G. R., Burrows, E., Retzlaff, B. & Poole, M. (1985) Metabolism *34*, 652–650.
62. Marks, V. (1971)) in: Metabolic Effects of Nicotinic Acid and Its Derivatives (Gey, K. F. & Carlson, L. A., eds.) pp. 433–434, Huber, Bern.
63. Miettinen, T. (1968) Clin. Chim. Acta *20*, 43–51.
64. Kudchokar, B. J., Sodhi, H. S., Horlick, L. & Mason, D. T. (1978) Clin. Pharmacol. Ther. *24*, 354–373.
65. Goodman, D. S., Smith, F. R., Seplowitz, A. H., Ramakrishnan, R. & Dell, R. B. (1980) J. Lipid Res. *21*, 699–713.
66. Yovos, J. G., Patel, S. T., Falko, J. M., Newman, H. A. I. & Hill, D. S. (1982) J. Clin. Endocrinol. Metab. *54*, 1210–1215.
67. Shepherd, J., Packard, C. J., Patsch, J. R., Gotto, A. M., Jr. & Taunton, O. D. (1979) J. Clin. Invest. *63*, 858–867.
68. Carlson, L. A. & Oro, L. (1973) Atherosclerosis *18*, 1–9.
69. Paoletti, R., Franceschini, G. & Sirtori, C. R. (1983) Amer. J. Cardiol. *52*, 21B–27B.
70. Parsons, W. B., Jr. (1960) J. Amer. Med. Assoc. *173*, 1466–1470.
71. Molnar, G. O., Berge, K. G., Rosevear, J. W., McGuckin, W. F. & Achor, R. W. P. (1964) Metabolism *13*, 181–190.
72. Kohn, R. M. & Montes, M. (1969) Am. J. Med. Sci. *258*, 94–99.
73. Barnhart, J. W., Sefranka, J. A., & McIntosh, D. D. (1970) J. Clin. Nutr. *23*, 1229–1233.
74. Heel, R. C., Brogden, R. N., Speight, T. M. & Avery, G. S. (1978) Drugs *15*, 409–428.
75. Nestel, P. J. & Billington, T. (1981) Atherosclerosis *38*, 203–209.
76. Atmeh, R. F., Stewart, J. M., Boag, D. E., Packard, C. J., Lorimer, A. R. & Shepherd, J. (1983) J. Lipid Res. *24*, 588–595.
77. Naruszewicz, M., Carew, T. E., Pittman, R. C., Witztum, J. L. & Steinberg, D. (1984) J. Lipid Res. *25*, 1206–1213.
78. Hunninghake, D. B., Bell, C. & Olson, L. (1980) Atherosclerosis *37*, 469–474.
79. Mellies, M. J., Gartside, P. S., Glatfelter, L., Vink, P., Guy, G., Schonfeld, G. & Glueck, C. J. (1980) Metabolism *29*, 956–964.

80. Lock, D. R., Kuisk, I., Gonen, B., Patsch, W. & Schonfeld, G. (1983) Atherosclerosis *47*, 271–278.
81. Jones, D. B., Simpson, H. C., Slaughter, P., Lousley, S., Carter, R. D., Cobbe, S. M. & Mann, J. I. (1984) Atherosclerosis *53*, 1–7.
82. Durrington, P. N. & Miller, J. P. (1985) Atherosclerosis *55*, 187–194.
83. Thorp, J. J. (1963) J. Atheroscler. Res. *3*, 351–360.
84. Hunninghake, D. B., Probstfield, J. L., Crow, L. O. & Isaacson, S.-O. (1981) Metabolism *30*, 605–609.
85. Brown, M. S. & Goldstein, J. L. (1985) in: The Pharmacological Basis of Therapeutics (Gilman, A. G., Goodman, L. S., Rall, T. W. & Murad, F., eds.) pp. 836–837, MacMillan, New York.
86. Okerholm, R. A., Keeley, F. J., Peterson, F. E. & Glazko, A. J. (1976) Proc. Roy. Soc. Med. *69*, 11–14.
87. Abshagen, U., Bablok, W., Koch, K., Lang, P. D., Schmidt, H. A. E., Senn, M. & Stork, H. (1979) Eur. J. Clin. Pharmacol. *16*, 31–38.
88. Brodie, R. R., Chasseaud, L. F., Elsom, F. F., Franklin, E. R. & Taylor, T. (1976) Arzneimitt. Forsch. *26*, 896–901.
89. Desager, J. P. & Harvengt, C. (1978) Int. J. Clin. Pharmacol. *16*, 570–574.
90. Wolfe, B. M., Kane, J. P., Havel, R. J. & Brewster, H. P. (1973) J. Clin. Invest. *52*, 2146–2159.
91. Kissebah, A. H., Adams, P. W., Harrigan, P. & Wynn, V. (1974) Europ. J. Clin. Invest. *4*, 163–174.
92. Kesaniemi, Y. A. & Grundy, S. M. (1984) J. Amer. Med. Assoc. *251*, 2241–2246.
93. Kesaniemi, Y. A., Beltz, W. F. & Grundy, S. M. (1985) Arteriosclerosis *5*, 153–161.
94. Nikkila, E. A., Huttunen, J. K. & Ehnholm, C. (1977) Metabolism *26*, 179–186.
95. Boberg, J., Boberg, M., Gross, R., Grundy, S., Augustin, J. & Brown, V. (1977) Atherosclerosis *27*, 499–503.
96. Vessby, B., Lithell, H. & Ledermann, H. (1982) Atherosclerosis *44*, 113–118.
97. Eisenberg, S., Gavish, D., Oschry, Y., Fainuru, M. & Deckelbaum, R. J. (1984) J. Clin. Invest. *74*, 470–482.
98. Saku, K., Gartside, P. S., Hynd, B. A. & Kashyap, M. L. (1985) J. Clin. Invest. *75*, 1702–1712.
99. Bierman, E. L., Brunzell, J. D., Bagdade, J. D., Lerner, R. L., Hazzard, W. R. & Port, D., Jr. (1970) Trans. Assoc. Amer. Physcians *53*, 211–224.
100. Grundy, S. M., Ahrens, E. H., Jr. & Salen, G. (1972) J. Lipid Res. *13*, 531–551.
101. Packard, C. J., Munro, A., Lorimer, A. R., Gotto, A. M., Jr. & Shepherd, J. (1984) J. Clin. Invest. *74*, 2178–2192.
102. Stewart, J. M., Packard, C. J., Lorimer, A. R., Borg, D. E. & Shepherd, J. (1982) Atherosclerosis *44*, 355–365.
103. Malmendier, C. L. & Delcroix, C. (1985) Atherosclerosis *55*, 161–169.
104. Berndt, J., Gaumert, R. & Still, J. (1978) Atherosclerosis *30*, 147–152.
105. Schneider, A., Stange, E. F., Ditschuneit, H. H. & Ditschuneit, H. (1985) Atherosclerosis *56*, 257–262.
106. Kleinman, Y., Eisenberg, S., Oschry, Y., Gavis, D., Stein, O. & Stein, Y. (1985) J. Clin. Invest. *75*, 1796–1803.
107. Shepherd, J., Caslake, M. J., Lorimer, A. R., Vallance, B. V. & Packard, C. J. (1985) Arteriosclerosis *5*, 162–168.
108. Pickering, J. E. (1983) Amer. J. Cardiol. *52*, 39B–41B.
109. Rouffy, J., Chanu, B., Bakir, R., Djian, F. & Goy-Loeper, J. (1985) Atherosclerosis *54*, 273–281.
110. Weisweiler, P., Merk, W., Janetschek, P. & Schwandt, P. (1984) Atherosclerosis *53*, 321–325.
111. Vessby, B., Lithell, H., Hellsing, K., Ostlund-Lindquist, A.-M., Gustafsson, I.-B., Boberg, J. & Lederman, H. (1980) Atherosclerosis *37*, 257–289.
112. Mordasini, R., Reisen, W., Oster, P., Kellner, M., Middelhoff, G. & Lang, P. O. (1981) Atherosclerosis *40*, 153–158.
113. Olsson, A. G., Land, P. S. & Vollmar, J. (1985) Atherosclerosis *55*, 195–203.
114. Olsson, A. G. & Oro, L. (1982) Atherosclerosis *42*, 229–243.
115. Langer, T. & Levy, R. I. (1968) New Eng. J. Med. *279*, 856–868.
116. Hall, M. J., Nelson, L. M., Russell, R. J. & Howard, A. N. (1981) Atherosclerosis *39*, 511–516.
117. Von Bergman, K. & Leiss, O. (1984) Eur. J. Clin. Invest. *14*, 150–154.
118. Leiss, O., Von Bergman, K., Gnasso, A. & Augustin, J. (1985) Metabolism *34*, 74–82.
119. Podda, M. & Zuin, M. (1985) Atherosclerosis *55*, 135–142.
120. Marcus, E. (1983) Amer. J. Cardiol. *52*, 41B.

121. Homberg, J.-C., Abuaf, N., Helmy-Kahalil, S., Biour, M., Poupon, R., Islam, S., Darnis, F., Levy, V. G., Opolon, P., Beaugrand, M., Toulet, J., Danan, G. & Berhamou, J.-P. (1985) Hepatology *5*, 722–727.
122. Barter, P. J., Connor, W. E., Spector, A. A., Armstrong, M., Conner, C. L. & Newman, M. A. (1974) Ann. Intern. Med. *81*, 619–624.
123. Endo, A., Kuroda, M. & Tanzawa, K. (1976) FEBS Lett. *72*, 323–326.
124. Alberts, A. W., Chen, J., Kuron, G., Hunt, V., Huff, J., Hoffman, C., Rothrock, J., Lopez, J., Joshua, H., Harris, E., Patchett, A., Monaghan, R., Currie, S., Stapley, E., Albets-Schonberg, G., Hensens, O., Hurshfield, J., Hoogster, K., Liesch, J. & Springer, J. (1980) Proc. Natl. Acad. Sci. USA *77*, 3957–3961.
125. Bilheimer, D. W., Grundy, S. M., Brown, M. S. & Goldstein, J. L. (1983) Proc. Natl. Acad. Sci. USA *80*, 4124–4128.
126. Grundy, S. M. & Bilheimer, D. W. (1984) Proc. Natl. Acad. Sci. USA *81*, 2538–2542.
127. Mabuchi, H., Haba, T., Tatami, R., Miyamoto, S., Sakai, Y., Wakasugi, T., Watanabe, A., Koizumi, J. & Takeda, R. (1981) New Eng. J. Med. *305*, 478–482.
128. Illingworth, D. R. & Sexton, G. J. (1984) J. Clin. Invest. *74*, 1972–1978.
129. Tobert, J. A., Bell, G. D., Birtwell, J., James, I., Kukovetz, W. R., Pryor, J. S., Buntinx, A., Holmes, I. B., Chao, Y.-S. & Bolognese, J. A. (1982) J. Clin. Invest. *69*, 913–919.
130. Illingworth, D. R. & Corbin, D. (1985) Proc. Natl. Acad. Sci. USA *82*, 6291–6294.
131. Glueck, C. J., Levy, R. I. & Fredrickson, D. S. (1971) Ann. Intern. Med. *75*, 345–352.
132. Sachs, B. A. & Wolfman, L. (1968) Metab. *17*, 400–410.
133. Ehnholm, C., Huttunen, J. K., Kinnunen, P. J., Miettinen, T. A. & Nikkila, E. A. (1975) New Eng. J. Med. *292*, 1314–1317.
134. Glueck, C. J., Stein, E. A. & Kashyap, M. L. (1979) Artery *5*, 453–472.
135. Wahl, P., Walden, C., Knopp, R., Hoover, J., Wallace, R., Heiss, G. & Rifkind, B. (1983) New Eng. J. Med. *308*, 863–867.
136. Kane, J. P., Malloy, M. J., Tun, P., Phillips, N. R., Freedman, D. D., Williams, M. L., Rowe, J. S. & Havel, R. J. (1981) N. Eng. J. Med. *304*, 251–258.
137. Hoeg, J. M., Maher, M. B., Bailey, K. R., Zech, L. A., Gregg, R. E., Sprecher, D. L. & Brewer, H. B. (1985) Amer. J. Cardiol. *55*, 1282–1286.
138. Dujovne, C. A., Krehbiel, P., Decoursey, S., Jackson, B., Chernoff, S. B., Pitterman, A. & Garty, M. (1984) Ann. Intern. Med. *100*, 477–482.
139. Illingworth, D. R. (1984) Ann. Intern. Med. *101*, 598–604.
140. Nash, D. T. (1983) Postgrad. Med. *73*, 75–82.
141. National Institutes of Health Consensus Development Panel (1985) Arteriosclerosis *5*, 404–412.
142. National Institutes of Health Consensus Development Panel (1984) Arteriosclerosis *4*, 296–301.

Chapter 16
Adrenergic Blocking Drugs

A. Lehtonen

1 Adrenoceptors

The catecholamines synthesized and secreted by the adrenal medulla regulate a wide spectrum of metabolic processes. Gluconeogenesis, glycogenolysis, and lipolysis, for example, are strongly influenced by these hormones.

Endogenous catecholamines have not only α- and β-receptor agonist properties, but also varying β_1- and β_2-receptor affinities, the former being greater for noradrenaline and the latter for adrenaline. A feature shared by β-receptor-mediated response is that activation of β-receptors generally results in an increase in the intracellular cyclic AMP level following activation of adenylate cyclase (1), which, via intermediate steps leads to phosphorylation and activation of the intracellular lipolytic enzyme, hormone-sensitive triglyceride lipase.

Of special interest are those systems displaying responses regulated by both α- and β-receptors. Hepatic glycogenolysis, for example, falls into this category, the α-response being cyclic AMP-independent and the β-response being cyclic AMP-dependent. In human fat cells, by contrast, activation of α-receptors inhibits, whereas activation of β-receptors stimulates cyclic AMP accumulation and lipolysis (2–4). Robinson et al. (1) proposed that α and β receptors might be modulating adenylate cyclase activity in opposing manners. The overall metabolic responsiveness to catecholamines may depend not only on the regulation of the β-receptor mediated effects, but also on the degree of stimulation of the α-receptors. Under most normal conditions the concomitant stimulation by catecholamines of inhibitory α-receptors seems to be of minor importance for overall lipolytic regulation (5). However, in the human fat cell, as in other cells displaying α- and β-receptors mediating opposing effects, the observed biologic response could, in theory, reflect the existing balance between these two opposing actions.

Increased α-receptor effects by catecholamines have been reported after prolonged fasting (6) and in certain human adipose tissue sites (7). A changed α- to β-receptor responsiveness to endogenous catecholamines represents a way to regulate adrenergic lipolysis. However, the physiological role of α-adrenergic receptors in human adipocytes is a matter of speculation. It is possible that the ratio of functional α- and β-adrenergic receptors can be influenced by such factors as age and nutritional and hormonal status. Lafontan (8) presented evidence that the α-adrenergic receptor population in the adipose tissue of rabbits increases with age. It is concluded that α-adrenergic agonists act as antilipolytic agents in human adipocytes and that this action may result from the interaction of these compounds with a population of α_2-adrenergic receptors (9).

Endogenous catecholamines have not only α- and β-receptor agonist properties but also varying β_1- and β_2-receptor affinities, the former being greater for adrenaline. The β-receptor in human fat cells may be characterized predominantly as β_1 (10). However, some lipolytic effect of β_2-stimulation has been shown in human fat cells (11). It has also been shown that non-selective β-blocking drugs are more potent antilipolytic agents than selective β-blocking agents during periods of increased catecholamine release such as exercise or hypoglycaemia (12).

The main function of adipose tissue is to supply fatty acids for utilization by different peripheral tissues, and glycerol, which is a gluconeogenic substrate in the liver. Catecholamine stimulation of hormone sensitive lipase and lipolysis in man appears to be mediated by β_1 and β_2 adrenoceptors. Adrenoceptor blockade might therefore be expected to inhibit lipolysis, reduce free fatty acid concentrations and limit availability of glucose for hepatic lipoprotein production.

In contrast to hormone sensitive lipase, lipoprotein lipase (chapter 6) is an enzyme that is located on the endothelial surface of extrahepatic capillary beds: fasting triglyceride concentrations are inversely correlated with its activity, which in turn, correlates with the fractional removal rate of endogenous triglycerides from the plasma (13). Lipoprotein lipase (LPL) degrades triglyceride-rich lipoproteins and transfers surface material to HDL, converting HDL_3 to HDL_2 (14). Hepatic lipase (chapter 6) , on the other hand, degrades HDL_2 phospholipids and participates in the hepatic uptake of HDL_2 cholesterol (15). It is postulated that alpha-adrenergic stimulation may have an important role in suppressing adipose tissue lipase, leading to a reduction in plasma HDL cholesterol, and an increment in triglycerides (16).

2 Adrenergic Blocking Drugs

Antihypertensive treatment regimes which produce comparable blood pressure control could have varying effects upon the development of coronary artery disease. The purpose of this paper is to review the literature on the effect of adrenergic blocking drugs on plasma lipids.

2.1 Non-selective beta blockade

2.1.1 **Propranolol**

Many studies have been with propranolol, but they are generally of short duration (from 4 weeks to 6 months) and involve small groups of patients. No change in the concentration of serum total cholesterol was found in most of these (Table 1).

Table 1. Percentage changes in serum total cholesterol during beta adrenergic blockade.

Beta blocker	Reference		Duration (weeks)	Total cholesterol
Non-selective				
Propranolol	Tanaka et al. 1976	[53]	8	NS
Propranolol	Shaw et al. 1978	[38]	4	NS
Propranolol	Day et al. 1979	[83]	26	NS
Propranolol	Leren et al. 1980	[45]	8	NS
Propranolol	Berglund & Anderssson 1981	[19]	312	NS
Propranolol	Gemma et al. 1981	[20]	8	NS
Propranolol	Montanari et al. 1982	[21]	8	NS
Propranolol	Greenberg et al. 1984	[17]	156	+1
Propranolol	Veterans Administration 1982	[18]	52	+4
Propranolol	Leon et al. 1984	[24]	12	NS
Sotalol	Lehtonen & Viikari 1979	[25]	52	+16
Nadolol	Veterans Administration 1983	[26]	12	NS
Selective				
Atenolol	Shaw et al. 1978	[38]	4	NS
Atenolol	Day et al. 1982	[16]	12	NS
Atenolol	Day et al. 1979	[83]	26	NS
Atenolol	Lehtonen & Marniemi 1984	[28]	26	+8
Atenolol	Rössner & Weiner 1983	[29]	12	NS
Atenolol	Rouffy & Jaillard 1984	[27]	12	+3
Metoprolol	Day et al. 1982	[16]	12	NS
Metoprolol	Shaw et al. 1978	[38]	4	NS
Metoprolol	Pasotti et al. 1982	[31]	12	NS
Metoprolol	Rössner & Weiner 1983	[29]	12	NS
Metoprolol	Frishman et al. 1983	[32]	12	−7
With ISA				
Pindolol	Shaw et al. 1978	[38]	4	NS
Pindolol	Lehtonen 1984	[36]	52	NS
Pindolol	Leren et al. 1981	[23]	8	NS
Pindolol	Pasotti et al. 1982	[31]	12	NS
Oxprenolol	Sommerset al. 1981	[40]	3	−13
Acebutolol	Lehtonen 1984	[33]	26	NS
Acebutolol	Giuntoli et al. 1984	[34]	52	NS
Alpha-beta blocker				
Labetalol	Frishman et al. 1983	[32]	12	NS
Labetalol	McGonigle et al. 1981	[41]	52	NS
Labetalol	Flamenbaum et al. 1985	[22]	16	NS

NS = not significant; ISA = intrinsic sympathomimetic activity

Total cholesterol increased slightly but significantly in two studies (17, 18) which were of 12 months or longer in duration. In the 6-year study of Berglund and Andersson (19) no change in the level of serum cholesterol concentration was found.

Table 2. Percentage changes in serum LDL cholesterol during beta adrenergic blockade

Beta blocker	Reference		Duration (weeks)	Total cholesterol
Non-selective				
Propranolol	Day et al. 1982	[16]	12	NS
Propranolol	Leren et al. 1980	[45]	8	NS
Propranolol	Gemma et al. 1981	[20]	8	NS
Sotalol	Lehtonen & Viikari 1979	[25]	52	+32
Selective				
Atenolol	Day et al. 1982	[16]	12	NS
Atenolol	Rössner & Weiner 1983	[29]	12	NS
Atenolol	Rouffy & Jaillard 1984	[27]	12	+6
Metoprolol	Day et al. 1982	[16]	12	NS
Metoprolol	Rössner & Weiner 1983	[29]	12	NS
Metoprolol	Pasotti et al. 1982	[31]	12	NS
Metoprolol	Frishman et al. 1983	[32]	12	−7
With ISA				
Pindolol	Lehtonen 1984	[36]	52	NS
Pindolol	Leren et al. 1981	[23]	8	NS
Pindolol	Pasotti et al. 1982	[31]	12	NS
Acebutolol	Lehtonen 1984	[33]	26	NS
Acebutolol	Giuntoli et al. 1984	[34]	52	NS
Alpha-beta blocker				
Labetalol	Frishman et al. 1983	[32]	12	NS

NS = not significant; ISA = intrinsic sympathomimetic activity

The general observation is that treatment with propranolol increases plasma triglyceride concentration from 10 to 65 per cent. No changes in the levels of plasma triglycerides were found in the investigation of Berglund and Andersson (19). Triglycerides were increased in the very low density lipoprotein (VLDL) fraction in 2 studies in which this component was determined (20, 21).

The changes in plasma low density lipoprotein (LDL) cholesterol concentrations are small (Table 2), but plasma HDL cholesterol levels seems to decrease significantly during treatment with propranolol (Table 3). In the study of Flamenbaum et al. (22) the concentration of HDL cholesterol did not change, but systolic blood pressure was not lowered by the drug. The ratio of serum total cholesterol to HDL cholesterol increased in two studies (23, 24).

2.1.2 **Sotalol**

Plasma total cholesterol (Table 1), triglycerides (Table 4) and phospholipids increased significantly during 12 months of sotalol therapy (25). These changes did not reach statistical significance until the drug had been given for 6 months. HDL cholesterol decreased rapidly, a significant reduction being already evident at 1 month (Table 3). Plasma total cholesterol increased by 17% over the course of a

Table 3. Percentage changes in serum HDL cholesterol during beta adrenergic blockade.

Beta blocker	Reference		Duration (weeks)	HDL cholesterol
Non-selective				
Propranolol	Tanaka et al. 1976	[53]	8	−29
Propranolol	Day et al. 1982	[16]	12	−17
Propranolol	Leren et al. 1980	[45]	8	−13
Propranolol	Gemma et al. 1981	[20]	8	NS
Propranolol	Montanari et al. 1982	[21]	8	NS
Propranolol	Greenberg et al. 1984	[17]	156	NS
Propranolol	Leon et al. 1984	[24]	12	−10
Sotalol	Lehtonen &Viikari 1979	[25]	52	−26
Selective				
Atenolol	Day et al. 1982	[16]	12	−7
Atenolol	Rössner & Weiner 1983	[29]	12	NS
Atenolol	Lehtonen & Marniemi 1984	[28]	26	−12
Atenolol	Rouffy & Jaillard 1984	[27]	12	−10
Metoprolol	Day et al. 1982	[16]	12	−13
Metoprolol	Rössner & Weiner 1983	[29]	12	NS
Metoprolol	Pasotti et al. 1982	[31]	12	NS
Metoprolol	Frishman et al. 1983	[32]	12	NS
With ISA				
Pindolol	Lehtonen 1984	[36]	52	NS
Pindolol	Leren et al. 1981	[23]	8	NS
Pindolol	Pasotti et al. 1982	[31]	12	+20
Oxprenolol	Sommers et al. 1981	[40]	3	NS
Acebutolol	Lehtonen 1984	[33]	26	NS
Acebutolol	Giuntoli et al. 1984	[34]	52	NS
Alpha-beta blocker				
Labetalol	Frishman et al. 1983	[32]	12	NS
Labetalol	Flamenbaum et al. 1985	[22]	16	NS

NS = not significant; ISA = intrinsic sympathomimetic activity

year. This was due to an increase of VLDL + LDL cholesterol (Table 2); HDL cholesterol decreased by 26% at the same time. And so, the ratio of HDL cholesterol to total cholesterol decreased during treatment from 0.28 to 0.18. Plasma triglyceride concentration increased by 39% during the first 3 months of treatment and by about 66% at 1 year. A decrease of plasma free fatty acid concentration was evident within one month, and the concentration of free fatty acids remained lower during the whole year of treatment. This study provides an illustration of how the pattern of lipid alteration may change with time.

2.1.3 **Nadolol**

Increased concentrations of serum triglycerides were found after 3 months therapy with nadolol, whereas total cholesterol remained unchanged (26).

Table 4. Percentage changes in serum triglycerides during beta adrenergic blockade.

Beta blocker	Reference		Duration (weeks)	Triglycerides
Non-selective				
Propranolol	Tanaka et al. 1976	[53]	8	NS
Propranolol	Shaw et al. 1978	[38]	4	+37
Propranolol	Day et al. 1979	[83]	26	+65
Propranolol	Leren et al. 1980	[45]	8	+24
Propranolol	Berglund & Andersson 1981	[19]	312	NS
Propranolol	Gemma et al. 1981	[20]	8	+29
Propranolol	Montanari et al. 1982	[21]	8	+30
Propranolol	Leon et al. 1984	[24]	12	+24
Propranolol	Veterans Administration 1982	[18]	52	+25
Sotalol	Lehtonen & Viikari 1979	[25]	52	+66
Nadolol	Veterans Administration 1983	[26]	12	+22
Selective				
Atenolol	Shaw et al. 1982	[38]	4	+62
Atenolol	Day et al. 1982	[16]	12	+24
Atenolol	Day et al. 1979	[83]	26	+24
Atenolol	Lehtonen & Marniemi 1984	[28]	26	+37
Atenolol	Rössner & Weiner 1983	[29]	12	NS
Atenolol	Rouffy & Jaillard 1984	[27]	12	+18
Metoprolol	Day et al. 1982	[16]	12	+14
Metoprolol	Shaw et al. 1987	[38]	4	+33
Metoprolol	Pasotti et al. 1982	[31]	12	NS
Metoprolol	Rössner & Weiner 1983	[29]	12	+21
Metoprolol	Frishman et al. 1983	[32]	12	NS
With ISA				
Pindolol	Shaw et al. 1978	[38]	4	+28
Pindolol	Lehtonen et al. 1984	[36]	52	NS
Pindolol	Leren et al. 1981	[23]	8	NS
Pindolol	Pasotti et al. 1982	[31]	12	NS
Oxprenolol	Sommers et al. 1981	[40]	3	NS
Acebutolol	Lehtonen 1984	[33]	26	NS
Acebutolol	Giuntoli et al. 1984	[34]	52	NS
Alpha-beta blocker				
Labetalol	Frishman et al. 1983	[32]	12	NS
Labetalol	McGonigle et al. 1981	[41]	52	NS
Labetalol	Flamenbaum et al. 1985	[22]	12	NS

NS = not significant; ISA = intrinsic sympathomimetic activity

2.2 Selective beta blockade

2.2.1 **Atenolol**

Serum total cholesterol rose in 3 studies with atenolol from 3 to 8 per cent (23, 27, 28). LDL cholesterol increased by 6% in the report of Rouffy and Jaillard (27). In most studies the level of serum triglycerides tends to rise, and often by as much as 18 to 62% (Table 4). The level of HDL cholesterol during atenolol treatment was significantly lower, although the level increased again after discontinuation of treatment (28). The concentration of HDL_2 cholesterol decreased ($p < 0.01$) during the first month of therapy and remained more or less constant thereafter. The level of HDL_3 cholesterol decreased slightly but not significantly. The adverse change in HDL cholesterol has been found in two other studies (16, 27). Atenolol has not reduced HDL cholesterol in all studies (29, 30), but the dose of atenolol was smaller (50 mg/day) in the study of Rössner and Weiner (29).

2.2.2 **Metoprolol**

Metoprolol has been reported to increase plasma triglyceride concentration from 14 to 33 per cent (Table 4). In the studies of Pasotti et al. (31) and Frishman et al. (32) serum triglyceride was unaffected by metoprolol. One report indicated a fall in HDL cholesterol (16), but no changes were found in some other studies (29, 31, 32). Frishman et al. (32) noted a decrease in total cholesterol and LDL cholesterol in patients treated with metoprolol for 3 months.

2.3 Beta blockade and intrinsic sympathomimetic activity (ISA)

2.3.1 **Acebutolol**

There were no significant changes in the concentrations of plasma total cholesterol, LDL cholesterol, VLDL cholesterol and triglycerides during acebutolol treatment (33). The level of HDL cholesterol decreased slightly but not significantly. Consequently, the ratio of HDL cholesterol to total cholesterol did not change. The concentration of serum free fatty acids decreased significantly ($p < 0.05$) after 1 month's treatment with acebutolol and remained thereafter fairly constant. No changes in serum lipids during acebutolol treatment was found in the study of Giuntoli et al. (34).

2.3.2 **Pindolol**

The concentration of serum free fatty acids and triglycerides remained constant during 12-month treatment of hypertension with pindolol (35, 36). The total serum cholesterol level was slightly lower after 6 months than after 1 month of therapy. The concentration of HDL cholesterol was increased after the first month of therapy, but there were no significant differences between the levels of HDL cholesterol before the treatment and at 3 and 6 months of therapy. The ratio of total cholesterol to HDL cholesterol decreased slightly, but this decrease was significant ($p < 0.05$) only after 6 months of treatment. Discontinuation of the drug, reduced the ratio of total cholesterol to HDL cholesterol by 6 %. The concentration of serum HDL cholesterol and the ratio of HDL cholesterol to total cholesterol are reported to increase slightly during pindolol treatment (31, 35–37). No significant changes in the concentrations of apolipoprotein AI or AII were found during therapy with pindolol (36).

Several studies show that the mean increase in plasma triglycerides was small for pindolol (23, 31, 35, 36, 37). In the short-time study total triglycerides increased by 28% (38).

2.3.3 **Oxprenolol**

Oxprenolol with weak intrinsic sympathomimetic activity (ISA) has been reported to increase serum triglycerides and to decrease HDL cholesterol (16). Ballantyne et al. (39) did not find significant changes in serum lipids after 4 months treatment with oxprenolol, but trends were similar to the results of Day et al. (16). In the study of Sommers et al. (40) a decrease in total cholesterol and in the ratio of total cholesterol to HDL cholesterol was found during oxprenolol treatment.

2.4 Combined alpha-beta blocker

No significant changes in serum lipids were found when patients were treated for 3 or 4 months with labetalol (22, 32). During one year's therapy with labetalol the levels of serum cholesterol and triglycerides remained unchanged (41).

Total serum cholesterol concentrations were constant during therapy with a cardioselective beta blocker which also possesses an α-receptor blocking effect (42). The concentration of serum HDL cholesterol seemed to be slightly lowered during the therapy, and there was a decrease in the ratio of HDL cholesterol to total cholesterol. The levels of serum triglycerides were raised at 3 and 6 months of treatment. The serum free fatty acid concentration was significantly lowered only during the first month of treatment. Thereafter free fatty acid levels increased again almost to their initial level.

2.5 Combination of diuretics with beta blockers

The principal effect of the thiazide diuretics on lipoproteins is to increase both plasma total cholesterol and plasma total triglycerides concentrations. The possible interaction between beta blockers and diuretics is interesting because the combination is commonly used in clinical practice. The characteristic changes which follow beta blocker monotherapy, namely a tendency for increased total triglycerides and mild decreases in HDL cholesterol, has often been apparent during combination treatment with thiazide-type diuretics (39, 43–45).

In the Multiple Risk Factor Intervention Trial (MRFIT), at the end of 6 years, participants receiving the combination of a diuretic and propranolol had similar changes in total cholesterol, but higher triglycerides and lower HDL cholesterol than those treated with diuretic alone (46). The addition of metoprolol increased total triglycerides by 10% in patients receiving a diuretic. HDL cholesterol decreased by 13%, and the ratio of total cholesterol to HDL cholesterol increased. Apolipoprotein A also decreased by 10% (47). However, in two separate studies, the increases seen with short-term diuretic monotherapy in men or post-menopausal women were prevented or reversed by the concomitant administration of different beta blockers (43, 48). Conversely, in some mixed male-female study groups, total cholesterol and LDL cholesterol values on average did not differ distinctly between short-term diuretic monotherapy and combination treatment with beta blockers (49–51).

3 Comments

Many beta blockers have been shown to increase serum triglyceride levels and to decrease HDL cholesterol concentration. The stimulatory effect of catecholamines on lipolysis is mediated by their β-receptor stimulating property. The importance of catecholamines in regulating lipolysis is evident from the fact that administration of beta blocking drugs will reduce the free fatty acid levels during different lipolytic conditions such as fasting, exercise and following hyperglycaemia (12). According to the studies with pindolol (35, 36), pindolol with strong ISA does not reduce the concentrations of serum free fatty acids; these results therefore indicate that pindolol is not antilipolytic. In the adipocyte the observed biologic response could reflect the balance between two opposing actions, β-blockade and ISA. During the treatment with α/β-blockers (42) the unchanged concentrations of free fatty acids could reflect the existing balance between the opposing actions of α- and β-blockade in the adipose cell.

The blood levels of glucose and insulin represent, besides free fatty acids, the most important stimuli for hepatic lipoprotein production. The beta cells of the pancreas have adrenergic receptors that affect insulin secretion. Serum insulin levels have been reported to be unaltered during beta blocker therapy (33, 35, 52) but there are also reports of impaired glucose tolerance (33, 35).

Empirically, propranolol decreases plasma post-heparin lipase activity (53) so that the catabolism of triglyceride-rich lipoproteins in plasma may be slightly disturbed (16, 54). Metoprolol, a relatively selective β_1 antagonist, has no effect on plasma post-heparin lipase activity (55). Inhibition of lipase could be achieved through either a direct inhibitory action of adrenergic blocking agents themselves, or by secondary unopposed alpha-adrenergic stimulation. It is postulated that alpha-adrenergic stimulation may have an important role in suppressing adipose tissue lipase, leading to a reduction in plasma HDL cholesterol and an increment in triglycerides (16). However, no significant changes in the adipose tissue lipoprotein lipase activities were found upon cessation of sotalol (56) even although there was a uniform increase in serum HDL cholesterol in all subjects. Serum triglycerides decreased by 20% when sotalol was stopped. During treatment no correlation was present between HDL cholesterol and lipoprotein lipase activity, whereas with – drawal of therapy resulted in a positive correlation between these parameters. In normal persons and athletes, adipocyte and skeletal muscle lipoprotein lipase activities are correlated with high density lipoprotein$_2$ levels (57). No significant change in adipose tissue LPL activities was found after discontinuing pindolol treatment (35). On the basis of these findings, it may be suggested, that the decrease of HDL caused by beta blockers is not mediated by their effects on LPL.

Catabolism of high density lipoprotein components takes place at the surface of hepatocytes as a result of hepatic lipase activity (58). Both the lipid and protein components of high density lipoprotein may be released by the action of this enzyme. Conditions that decrease hepatic lipase, such as exercise, have been shown to raise plasma high density lipoprotein levels (59). The effect of beta blockers on hepatic lipase is not well known. Propranolol did not directly affect hepatic lipase in vivo or in cultured hepatic cells (60). However, propranolol reduced plasma high density lipoprotein levels. Thus, that study does not support a role for hepatic lipase in mediating the reduction of high density lipoprotein levels with beta blockers.

High density lipoprotein acquires unesterified cholesterol from cells (61, 62). This unesterified high density lipoprotein cholesterol is then esterified by lecithin:cholesterol acyltransferase (LCAT) and the resulting cholesterol esters move from the surface to the core of the high density lipoprotein particle, where they are susceptible to the actions of transfer proteins in plasma which are responsible for transfering them into LDL and making them available for cellular catabolism. The activity of lecithin:cholesterol acyltransferase is decreased by beta antagonists (49) which may contribute to the decrease in high density lipoprotein found in patients who receive propranolol. A marked decrease in serum LCAT activity was found when pindolol treatment was discontinued (35). It is possible that pindolol influences blood lipoprotein levels possibly due to interaction of the drug with the enzymes (especially LCAT) regulating lipid metabolism. The slight increase in HDL cholesterol during pindolol therapy is accompanied by a rise in apoprotein AI concentration (36).

In conclusion, most beta blockers raise serum triglyceride concentrations and lower HDL cholesterol. Notable exceptions are pindolol and acebutolol, which do not seem to alter serum lipoprotein levels. It remains to be established, however, whether this difference is, indeed, related to the degree of intrinsic sympathomimetic activity per se.

4 α_1-Adrenergic Blocking Drugs

Monotherapy with the alpha adrenoceptor blocker prazosin produced no change (21, 63, 64) or even a slight decrease (23, 27, 65) in total triglyceride total cholesterol, LDL cholesterol and VLDL cholesterol. (Table 5). HDL cholesterol concentration is unaltered (21, 23, 64, 65) or marginally raised (27, 63) so that the most common improvement has been a lowering of the ratio of total cholesterol to HDL cholesterol (23, 27, 63, 64).

The other α_1-adrenoceptor antagonist, trimazosin, has no deleterious effect on serum lipid levels (66, 67).

During 6 months treatment with doxazosin total and LDL cholesterol decreased by 9% and 17% respectively whereas HDL cholesterol increased (68). A 13% decrease was also noted in the ratio of total cholesterol to HDL cholesterol in a 3 months study with doxazosin (64).

Table 5. Percentage changes in serum total cholesterol, HDL cholesterol and triglycerides during α-blocker therapy.

α-blocker	Reference		Duration (months)	Total cholesterol	HDL cholesterol	Triglycer-ides
Prazosin	Leren et al. 1980	[45]	2	−9	NS	−16
Prazosin	Montanari et al. 1982	[21]	2	NS	NS	NS
Prazosin	Takabateke et al. 1984	[63]	12	NS	+17	NS
Prazosin	Shirai et al. 1983	[65]	6	−12	NS	−22
Prazosin	Rouffy & Jaillard 1984	[27]	3	−8	+13	−10
Prazosin	Torvik & Madsbu 1986	[64]	3	NS	NS	NS
Trimazosin	Singleton & Taylor 1983	[66]	7	NS	NS	NS
Trimazosin	Lehtonen 1985	[67]	6	NS	NS	NS
Doxazosin	Torvik & Madsbu 1986	[64]	3	NS	−10	−13
Doxazosin	Lehtonen et al. 1986	[68]	5	−9	+12	NS

NS = not significant

α-adrenergic agonists are antilipolytic (9) while α-adrenergic antagonists act as lipolytic agents (69) in human adipocytes, and this action may result from binding of these compounds to α_2-adrenergic receptors (9). The low affinity of prazosin for the cyclase coupled α_2-receptors of human adipose tissue therefore provides an explanation for its lack of effect on lipid mobilization from tissue to blood.

Little is known of how antihypertensive agents influence lipoprotein metabolism (70). It is postulated that alpha-adrenergic stimulation may have an important role in suppressing adipose tissue lipase leading to a secondary reduction in plasma HDL cholesterol and an increment in triglycerides. In most studies, however, no effect on plasma post-heparin lipase activity with beta blocker therapy has been found. Prazosin, on the other hand, increases extrahepatic lipoprotein lipase activity (55). Adrenaline increases synthesis of lipoprotein lipase by adipocytes (71). Therefore adrenergic agents may influence lipoprotein lipase activity by regulating its synthesis by these tissues. However, the exact contribution of alpha or beta receptors to this effect is unknown. A possible mechanism by which alpha-adrenergic or beta-adrenergic agents influence endothelial lipoprotein lipase activity is via the alteration of blood flow to peripheral beds. Precapillary arteriolar sphincters are constricted by alpha agonists and dilated by beta agonists. Thus it is plausible that alpha and beta receptors have contrary effects on the net action of extrahepatic capillary lipoprotein lipase on nascent chylomicrons and very low density lipoproteins. Alpha inhibition and beta stimulation of lipase could explain why prazosin lowers plasma triglyceride levels and raises high density lipoprotein levels whereas beta blockers produce the opposite effects.

Conditions that decrease hepatic lipase activity raise plasma high density lipoprotein levels (59). Prazosin did not affect hepatic lipase in vivo or in cultured hepatic cells (60). However, adrenaline decreased hepatic lipase. Prazosin reversed the effect of adrenaline, indicating involvement of the α_1 receptor. Thus, that study did not support a role for hepatic lipase in mediating the elevation of high density lipoprotein levels with prazosin.

With regard to low density lipoprotein metabolism, the α_1-adrenergic antagonist, doxazosin, has been shown in increase the activity of low density lipoprotein receptors in normal fibroblasts (72).

The results concerning the effect of prazosin on blood glucose and insulin levels are controversial. Prazosin decreased acute insulin secretion and increased plasma glucose during an intravenous glucose test (73). It was suggested that there might have been a prazosin-induced rise in catecholamines that stimulated pancreatic α_2-receptors. However, prazosin improved glucose tolerance in patients with mild diabetes, when the oral glucose tolerance test was used (74). However, no direct effect of prazosin on glucose and insulin levels has been found (75).

5 The Significance of Lipid Alterations

At present, any of the lipoprotein changes induced by antihypertensive drugs should be categorized as biochemical side effects. It is of considerable interest that most antihypertensive drugs alter plasma lipoprotein levels in a direction that is theoretically deleterious in relation to atherogenesis. However, classification of their possible pathogenic and prognostic relevance is needed. Many complications of hypertension, such as stroke and congestive heart failure, are largely pressure dependent. The satisfactory lowering of blood pressure decreases the pressure-mediated complications (76, 77).

Most of the large randomised trials in the treatment of mild hypertension have shown no reduction in the incidence of coronary heart disease (78, 79). Only two of the large clinical trials have noted a decrease in coronary heart disease (77, 80). A common feature of these studies is a beneficial effect in elderly patients. There is a pattern of a continuous, graded, strong relationship between serum cholesterol and age-adjusted coronary heart disease death rate for men aged 35 to 57 years (81). Nonfasting HDL cholesterol levels are inversely, and total cholesterol levels directly related to development of coronary heart disease in both men and women over 49 years and older (82).

The present information regarding long-term effects of antihypertensive drugs on serum lipoproteins is limited. Long-term studies are required in which the effect of an antihypertensive that is known not to produce deleterious changes in lipoprotein levels is compared with one that does not affect lipoproteins. To test the risk factor theory of coronary heart disease, the effects of antihypertensive drugs on serum lipoproteins must be known, so that optimal therapy can be chosen. Drug regimens having maximal benefits on risk factors of coronary heart disease, seem to be preferable in the treatment of hypertension.

References

1. Robinson, G. A., Butcher, R. W. & Sutherland, E. W. (1967) Adenyl cyclase as an adrenergic receptor. Ann. N.Y. Acad. Sci. *139*, 703–723.
2. Burns, W. T. & Langley, P. E. (1970) Lipolysis by human adipose tissue: the role of cyclic 3′-5′-adenosine monophosphate and adrenergic receptor sites. J. Lab. Clin. Med. *75*, 983–997.
3. Östman, J. & Efendic, S. (1970) Catecholamines and metabolisms of human adipose tissue. Acta Med. Scand. *187*, 471–476.
4. Burns, W. T., Langley, P. E. & Robinson, G. A. (1972) Studies of the role of cyclic AMP in human lipolysis. Adv. Cyclic Nucleotide Res. *1*, 63–85.
5. Kather, H., Pries, J., Schrader, V. & Simon, B. (1980) Inhibition of human fat cell adenylate cyclase mediated via alpha-adrenoceptors. Eur. J. Clin. Invest. *10*, 345–348.
6. Arnes, P., Engfeldt, P. & Östman, J. (1979) Relationship between lipolysis, cyclic AMP, and fat-cell size in human adipose tissue during fasting and in diabetes mellitus. Metabolism *28*, 198–209.
7. Östman, J., Arnes, P., Engfeldt, P. & Kager, L. (1979) Regional differences in the control of lipolysis in human adipose tissue. Metabolism *26*, 1198–1205.
8. Lafontan, M. (1979) Inhibition of epinephrine-induced lipolysis in isolated white adipocytes of aging rabbits by increased alpha-adrenergic responsiveness. J. Lipid Res. *20*, 208–216.
9. Wright, E. E. & Simpson, E. R. (1981) Inhibition of the lipolytic action of β-adrenergic agonists in human adipocytes by α-adrenergic agonists. J. Lipid Res. *22*, 1265–1270.
10. Frisk-Holmberg, M. & Östman, J. (1977) Differential inhibition of lipolysis in human adipose tissue by adrenergic β-receptor blocking drugs. J. Pharmacol. Exp. Ther. *200*, 598–605.
11. Levitzki, S. (1978) Catecholamine receptors. Rev. Physiol. Biochem. Pharmacol. *82*, 1–26.
12. Lager, J., Blohme, G. & Smith, V. (1979) Effect of cardioselective and non-selective beta-blockade on the hypoglycemic response in insulin-dependent diabetics. Lancet *1*, 458–462.
13. Huttunen, J. K., Ehnholm, C., Kekki, M. & Nikkilä, E. A. (1976) Postheparin plasma lipoprotein lipase and hepatic lipase in normal subjects and in patients with hypertriglyceridaemia: correlation to sex, age and various parameters of triglyceride metabolism. Clin. Sci. Mol. Med. *50*, 249–260.
14. Patsch, J. R., Gotto, A. M., Olivecrona, T. & Eisenberg, S. (1978) Formation of high density lipoprotein-like particles during lipolysis of very low density lipoproteins in vitro. Proc. Natl. Acad. Sci. *75*, 4519–4523.
15. Kuusi, T., Saarinen, P. & Nikkilä, E. A. (1980) Evidence for the role of hepatic endothelial lipase in the metabolism of plasma high density lipoprotein in man. Atherosclerosis *36*, 589–593.
16. Day, J. L., Metcalfe, J. & Simpson, C. N. (1982) Adrenergic mechanisms in control of plasma lipid concentrations. Br. Med. J. *284*, 1145–1148.
17. Greenberg, G., Brennan, P. J. & Miall, W. E. (1984) Effects of diuretic and beta-blocker therapy in the Medical Research Council Trial. Am. J. Med. *76*, (2A), 45–51.
18. Veterans Administration Cooperative Study Group on Antihypertensive Agents. (1982) Comparison of propranolol and hydrochlorothiazide for the initial treatment of hypertension. II. Results of long-term therapy. JAMA *248*, 2004–2011.
19. Berglund, G. & Andersson, O. (1981) Beta-blockers or diuretics in hypertension? A six year follow-up of blood pressure and metabolic effects. Lancet *1*, 744–747.
20. Gemma, G., Montanari, G., Suppa, G., Paralova, A., Franceschini, G. et al. (1981) Plasma lipid and lipoprotein changes in hypertensive patients treated with propranolol and prazosin. J. Cardiovasc. Pharmacol. *3*, (Suppl. 3), S207–S215.
21. Montanari, G., Gianfranceschi, G., Suppa, G., Paralovo, A., Franceschini, G. et al. (1982) Plasma lipid and lipoprotein changes in hypertensive patients treated with propranolol and prazosin. In: Lipoproteins and coronary atherosclerosis 1982 (Noseda et al., eds.) pp. 389–398, Elsevier Biomedical Press, Amsterdam.
22. Flamenbaum, W., Weber, M. A., McMahon, F. G., Materson, B. J., Carr, A. A. et al. (1985) Monotherapy with labetalol compared with propranolol: differential effects by race. J. Clin. Hypertension *1*, 56–69.

23. Leren, P., Eide, I., Foss, P. O., Helgeland, A., Hjermann, I. et al. (1981) Antihypertensive drugs and blood lipids. The Oslo Study. J. Cardiovasc. Pharmacol. *3* (Suppl. 3), S187–S192.
24. Leon, A. S., Agre, J., McNally, C., Bell, C., Neibling, M. et al. (1984) Blood lipid effects of antihypertensive therapy: a double-blind comparison of the effects of methyldopa and propranolol. J. Clin. Pharmacol. *24*, 209–217.
25. Lehtonen, A. & Viikari, J. (1979) Long-term effect of sotalol on plasma lipids. Clin. Sci. *57* (Suppl.), 405–407.
26. Veterans Administration Cooperative Study Group on Antihypertensive Agents. (1983) Efficacy of nadolol alone and combined with bendroflumethiazide and hydralazine for systemic hypertension. Am. J. Cardiol. *52*, 1230–1237.
27. Rouffy, J. & Jaillard, J. (1984) Comparative effects of prazosin and atenolol on plasma lipids in hypertensive patients. Am. J. Med. *76*, (2A), 105–108.
28. Lehtonen, A. & Marniemi, J. (1984) Effect of atenolol on plasma HDL cholesterol subfractions. Atherosclerosis *51*, 335–338.
29. Rössner, S. & Weiner, L. (1983) Atenolol and metoprolol – Comparison of effects on blood pressure and serum lipoproteins, and side effects. Eur. J. Clin. Pharmacol. *24*, 573–577.
30. Välimäki, M., Maass, L., Harno, K. & Nikkilä, E. A. (1986) Lipoprotein lipids and apoproteins during beta-blocker administration: Comparison of penbutolol and atenolol. Eur. J. Clin. Pharmacol. *30*, 17–20.
31. Pasotti, C., Capra, A., Fiorella, G., Vibelli, C. & Chierichetti, S. M. (1982) Effects of pindol and metoprolol on plasma lipids and lipoproteins. Br. J. Clin. Pharmacol. *13* (Suppl. 2), 435–439.
32. Frishman, W. H., Michelson, E. L., Johnson, B. F. & Poland, M. P. (1983) Multiclinic comparison of labetalol to metoprolol in treatment of mild to moderate systemic hypertension. Am. J. Med. *75* (Suppl.), 54–67.
33. Lehtonen, A. (1984) The effect of acebutolol on plasma lipids, blood glucose and serum insulin levels. Acta Med. Scand. *216*, 57–60.
34. Giuntoli, F., Scalabrino, A., Galeone, F., Birindelli, A., Natali, A. et al. (1984) Antihypertensive and metabolic effects of a long-term treatment with acebutolol. Curr. Ther. Res. *36*, 188–194.
35. Lehtonen, A., Hietanen, E., Marniemi, J., Peltonen, P. & Niskanen, J. (1982) Effect of pindolol on serum lipids and lipid metabolizing enzymes. Br. J. Clin. Pharmacol. *13*, (Suppl. 2), 445–447.
36. Lehtonen, A. (1984) Long-term effect of pindolol on plasma lipids, apoproteins A, blood glucose and serum insulin levels. Int. J. Clin. Pharmacol. Ther. Toxic. *22*, 269–272.
37. Miettinen, T. A., Vanhanen, H., Huttunen, J. K. Naukkarinen, V., Mattila, S., Strandberg, T. & Kumlin, T. (1982) HDL cholesterol and β-adrenoceptor blocking agents in a 5 year multifactorial primary prevention trial. Br. J. Clin. Pharmacol. *13*, (Suppl. 2), 431–434.
38. Shaw, J., England, J. D. F. & Hua, A. S. P. (1978) Beta blockers and plasma triglycerides. Br. Med. J. *1*, 986.
39. Ballantyne, D., Ballantyne, F. C. & McMurdo, M. (1981) Effect of slow oxprenolol and a combination of slow oxprenolol and cyclopenthiazide on plasma lipoproteins. Atherosclerosis *39*, 301–306.
40. Sommers, D. K., DeVilliers, L. S., Van Wyk, M. & Schoeman, H. S. (1981) The effects of labetalol and oxprenolol on blood lipids. South African Med. J. *60*, 379–380.
41. McGonigle, R. J. S., Williams, L., Murphy, M. J. & Parsons, V. (1981) Labetalol and lipids. Lancet *1*, 163.
42. Lehtonen, A. (1981) Effect of α/β-blockers on serum lipids. Int. J. Clin. Pharmacol. Ther. Toxic. *19*, 228–230.
43. Schiffl, H., Weidmann, P., Mordasini, R., Riesen, W. & Bachmann, C. (1982) Reversal of diuretic-induced increases in serum low-density lipoprotein cholesterol by the beta-blocker pindolol. Metabolism *31*, 411–4125.
44. Bauer, J. H., Brooks C. S., Weinstein, J., Wilcox, H. H., Heimberg, M. et al. (1981) Effects of diuretic and propranolol on plasma lipoprotein lipids. Clin. Pharmacol. Ther. *30*, 35–43.
45. Leren, P., Helgeland, A., Holme, I., Foss, P. O., Hjermann, I. et al. (1980) Effect of propranolol and prazosin on blood lipids. The Oslo Study. Lancet *2*, 4–6.
46. Lasser, N. L., Grandits, G., Gaggiula, A. W., Cutler, J. A., Grimm Jr., R. H. et al. (1984) Effects of antihypertensive therapy on plasma lipids and lipoproteins in the

Multiple Risk Factor Intervention Trial. Am. J. Med. *76*, 52–66.
47. England, J. D. E., Simons, L. A., Gibson, J. C. & Carlton, M. (1980) The effects of metoprolol and atenolol on plasma high density lipoprotein levels in man. Clin. Exp. Pharmac. Physiolol. *7*, 329–333.
48. Meier, A., Weidmann, P., Mordasini, R., Riesin, W. & Bachmann, C. (1982) Reversal or prevention of diuretic-induced alterations in serum lipoproteins with beta-blockers. Atherosclerosis *41*, 415–419.
49. Goto, Y, (1984) Effects of alpha- and beta-blocker antihypertensive therapy on blood lipid: a multicenter trial. Am. J. Med. *76*, (2A), 72–78.
50. Johnson, B. F., Romero, L., Johnson, J. & Marwaha, R. (1984) Comparative effects of propranolol and prazosin upon serum lipids in thiazide-treated hypertensive patients. Am. J. Med. *76*, (2A), 109–112.
51. Lowenstein, J. & Neusy, A. J. (1984) Effects of prazosin and propranolol on serum lipids in patients with essential hypertension. Am. J. Med. *76* (2A), 79–84.
52. Holm, G., Johansson, S., Vedin, A., Wilhelmsson, C. & Smith, U. (1980) The effect of beta-blockade on glucose tolerance and insulin release in adult diabetics. Acta Med. Scand. *208*, 187–191.
53. Tanaka, N., Sakaguchi, S., Oshige, K., Niimura, T. & Kanehisa, T. (1976) Effect of chronic administration of propranolol on lipoprotein composition. Metabolism *25*, 1071–1075.
54. Barboriak, J. J. & Friedberg, H. D. (1973) Propranolol and hypertriglyceridaemia. Atherosclerosis *17*, 31–35.
55. Ferrara, L. A., Marotta, T., Rubba, P., De Simone, B., Leccia, G., Soro, S. & Mancini, M. (1986) Effects of alpha-adrenergic receptor blockade on lipid metabolism. Am. J. Med. *80* (Suppl. 2A), 104–108.
56. Lehtonen, A., Hietanen, E., Marniemi, J., Peltonen, P. & Nikkilä, E. A. (1983) Effect of sotalol withdrawl on serum lipids and lipoprotein lipase activity. Int. J. Clin. Pharmacol. Ther. Toxic. *21*, 73–76.
57. Nikkilä, E. A., Taskinen, M.-R., Rehunen, S. & Härkönen, M. (1978) Lipoprotein lipase activity in adipose tissue and skeletal muscle of runners: relation to serum lipoproteins. Metabolism *27*, 1661–1671.
58. Bamberger, M., Glick, J. M. & Rothblat, G. H. (1983) Hepatic lipase stimulates the uptake of high-density lipoprotein cholesterol by hepatoma cells. J. Lipid Res. *24*, 869–876.
59. Herbert, P. N., Bernier, D. N., Cullinane, E. M., Edelstein, L., Kantor, M. A. & Thompson, P. D. (1984) High-density lipoprotein metabolism in runners and sedentary men. JAMA *252*, 1034–1037.
60. Schoonderwoerd, K., Hulsmann, W. C. & Jansen, H. (1984) Regulation of liver lipase: II. Involvement of the $alpha_1$ receptor. Biochem. Biophys. Acta *795*, 481–486.
61. Eisenberg, S. (1984) High-density lipoprotein metabolism. J. Lipid Res. *25*, 1017–1058.
62. Leiss, O., Muranowski, V. & Egge, H. (1978) Lecithin cholesterol acyltransferase activity in relation to lipoprotein concentration and lipid composition. Scand. J. Clin. Lab. Invest. *38* (Suppl. 150), 77–84.
63. Takabatake, T., Ohta, H., Maekawa, M., Yamamoto, Y., Ishida, Y. et al. (1984) Effects of long-term prazosin therapy on lipoprotein metabolism in hypertensive patients. Am. J. Med. *76*, (2A), 113–116.
64. Torvik, D. & Madsbu, H. P. (1986) Multicentre 12-week double-blind comparison of doxazosin, prazosin and placebo in patients with mild to moderate essential hypertension. Br. J. Clin. Pharmacol. *21*, 695–705.
65. Shirai, K. et al. (1983) Changes in serum lipids in hypertensive patients treated with prazosin. Progress in Medicine *3*, 1501–1504.
66. Singleton, W. & Taylor, C. R. (1983) Effect of trimazosin on serum lipid profiles in hypertensive patients. Am. Heart J. *106*, 1265–1268.
67. Lehtonen, A. (1985) The effects of trimazosin and pindolol on serum lipids, blood glucose and serum insulin levels. Acta Med. Scand. *218*, 213–216.
68. Lehtonen, A., Himanen, P., Saraste, M., Niittymäki, K. & Marniemi, J. (1986) Double-blind comparison of the effects of long-term treatment with doxazosin or atenolol on serum lipoproteins. Br. J. Clin. Pharmacol. *21*, 77–81.
69. Kather, H. & Daerr, W. (1982) Effects of prazosin on human fat-cell lipolysis and adenylate cyclase activity in vitro. J. Cardiovasc. Pharmacol. *4* (Suppl. 2), 245–247.
70. Sacks, F. M. & Dzau, V. J. (1986) Adrenergic effects on plasma lipoprotein metabolism. Speculation on mechanisms of action. Am. J. Med. *80* (Suppl. 2A), 71–81.

71. Ashby, P. & Robinson, D. S. (1980) Effects of insulin, glucocorticoids and adrenaline on the activity of rat adipose-tissue lipoprotein lipase. Biochem. J. *188*, 185–192.
72. Leren, T. P. (1985) Doxazosin increases low-density lipoprotein receptor activity. Acta Pharmacol. Toxicol. (Copenh.) *56*, 269–272.
73. Waern, A. U., Berne, C., Wibell, L. & Lithell, H. (1982) Short-term influence of a postsynaptic alpha-adrenoceptor blocking drug (prazosin) on carbohydrate metabolism. Acta Med. Scand. *665* (Suppl.), 75–77.
74. Barbieri, C., Ferrari, C., Borzio, M. et al. (1980) Metabolic effects of chronic prazosin treatment. Horm. Metab. Res. *12*, 331–334.
75. Ferlito, S., Puelo, F., Carra, G. et al. (1983) Effect of alpha$_1$-blockade by prazosin on blood sugar, insulin, and glucagon levels in normals and non-insulin-dependent diabetics. J. Endocrinol. Invest. *6*, 199–202.
76. The Australian therapeutic trial in mild hypertension. (1980) Lancet *1*, 1261–1267.
77. Hypertension Detection and Follow-up Program Cooperative Group. (1979) Five-year findings of the Hypertension Detection and Follow-up program: I. Reduction in mortality of persons with high blood pressure, including mild hypertension. JAMA *242*, 2562–2571.
78. Medical Research Council Working Party. (1985) MRC trial of treatment of mild hypertension: principal results. Br. Med. J. *291*, 97–104.
79. International Primary Prevention Study in Hypertension (IPPSH) Collaborative Group. (1985) Cardiovascular risk and risk factors in a randomised trial of treatment based on the beta-blocker oxprenolol: the International Prospective Primary Prevention Study in Hypertension (IPPPSH). J. Hypertension *3*, 379–392.
80. Amery, A., Birkenhager, W., Brixko, P., Bulpitt, C., Clement, D. et al. (1985) Mortality and morbidity results from the European Working Party on High Blood Pressure in the Elderly. Lancet *2*, 1349–1354.
81. Stamler, J., Wentworth, D., Neaton, J. D., for the MRFIT Research Group. (1986) Is relationship between serum cholesterol and risk of premature death from coronary heart disease continuous and graded? Findings in 356 222 primary screenees of the Multiple Risk Factor Intervention Trial (MRFIT). JAMA *256*, 2823–2828.
82. Castelli, W. P., Garrison, R. J., Wilson, P. W. F., Abbott, R. D., Kalousdian, S. & Kannel, W. B. (1986) Incidence of coronary heart disease and lipoprotein cholesterol levels. The Framingham Study. JAMA *256*, 2835–2838.
83. Day, J. d., Metcalfe, J., Simpson, N. & Page, R. d. (1979) Metabolic consequences of atenolol and propranolol in treatment of hypertension. Br. Med. J. *1*, 77–80.

Chapter 17
Hyperlipoproteinemia and Coronary Heart Disease

Ph. Douste-Blazy and H. U. Kloer

1 Introduction

Numerous epidemologic and genetic studies have shown that hyperlipoproteinemia, in particular hypercholesterolemia, is one of the principal risk factors for coronary heart disease (1 – 10). It is important to state that, on the basis of these studies, there does not seem to be a particular threshold. Rather, the risk seems to increase proportionally with increasing concentration of cholesterol-rich and, to a lesser extent, triglyceride-rich lipoprotein particles (11). This relationship between hypercholesterolemia and coronary risk was established by comparing population groups, and also by studying individuals within the same population (12). As far as population studies are concerned, the Seven Countries Study (13) very clearly delineated the relationship between the average cholesterol value and coronary heart disease mortality over 10 years in 16 different population groups from Europe, Japan, and the United States. High cholesterol values and high coronary mortality in the United States and Northern Europe contrasted with very low mortality and low cholesterol levels in Japan and the Southern European countries.

Among the studies designed to examine the realtionship between cholesterol value and coronary risk within the same population the "Étude Prospective Parisienne", the "Seven Countries Study" and the "Pooling Project" showed a very similar proportional increase of risk, an increase of 22% in coronary risk corresponding to an elevation of cholesterol levels in the order of 25 mg/dl (12, 14).

2 Plasma Lipoproteins and Coronary Heart Disease (CHD)

In the last two decades more and more attention has focused on the role of different plasma lipoprotein classes and subclasses and their effect on coronary risk. The advance of analytical techniques to differentiate lipoprotein classes and subclasses, not only by density flotation, but also by other physical, chemical, and, more recently, by immunological techniques has led to new perspectives of the role of individual lipoproteins in atherogenesis. The Framingham study was one of the first to go beyond measuring plasma lipids as the only risk parameter (11, 15, 16). It became obvious that coronary risk could be more accurately predicted in subjects older than 50 years if at least some differentation of lipoprotein lipid was attempted. As an outcome of these efforts, the inverse correlation between coronary risk and the cholesterol moiety transported in HDL, the so-called "HDL-cholesterol" value was established. This led to the concept of separating, in terms of risk evaluation, the hypercholesterolemia of hyper-alpha-lipoproteinemia from that of hyper-beta-lipoproteinemia (17). More recently, Miller et al. tried to put this relationship into a more precise perspective by demonstrating that the cholesterol contained in the HDL_2 subfraction showed a very strong negative correlation with the degree of coronary atherosclerosis as proven by angiography, while HDL_3 cholesterol showed no such relationship (18). On the other hand, the Framingham study and, subsequently, a number of other epidemiological studies in Europe and North America very clearly established that there was a strong positive correlation between LDL-cholesterol and the incidence of CHD without any apparent threshold value.

All these studies were performed on an epidemological scale with genetically and metabolically mixed populations. Obviously, they were unable to delineate individual pathophysiological mechanisms linking disturbed lipoprotein metabolism to atherosclerosis. Studies of more precisely defined, albeit quite small populations of familial hyperlipoproteinemias led to a better understanding of the pathophysiology of this process.

Familial hypercholesterolemia (FH) is probably the most frequent genetic disease in Western countries. Its prevalence has been estimated to be 1 in 500. Clinically, it is characterized by vast deposits of cholesterol both in the vascular wall, causing atheromas and atherosclerotic disease and, extravascularly, in the form of xanthomas. Tendinous xanthomas are mostly found on the Achilles tendon and the extensor tendons of the hand. Other forms of lesion include tuberous xanthomas, xanthelasmas and arcus corneae (19). LDL and apolipoprotein B is grossly elevated in plasma. In terms of inheritance, it is a monogenic autosomal dominant disease.

Pathophysiologically, it seems that the main defect is the accumulation of plasma LDL because of a reduction of effective LDL receptors (20, 21, 22). Situated on the surface of the majority of cells in the body, the LDL receptor interacts with plasma LDL, which leads to the uptake of these lipoprotein particles. Subsequently, the LDL particle is degraded by lysosomal action and the liberated cholesterol is used for the needs of the cell, such as synthesis of membranes, steroid hormone synthesis (in the case of the adrenal), and bile acids (in the case of the hepatocyte). As a further consequence of receptor-mediated uptake of LDL, cellular cholesterol production is tightly regulated. Furthermore, the number of available receptors is normally regulated by the intracellular cholesterol concentration.

The LDL receptor is synthesized in the endoplasmatic reticulum as a precursor protein with a mass of approximately 96 000 daltons. After addition of a carbohydrate chain in the Golgi apparatus, the mature protein is transported to the plasma membrane. Several disturbances of this process have been described depending on whether the affected subject is heterozygous or homozygous (23). The LDL receptors of these subjects are either defective 5−30% of normal LDL-binding activity) or negative (less than 5% of normal binding capacity). In certain cases, LDL binding is apparently normal but the LDL receptor complex is not properly internalized and further processed (24). Recently, several classes of mutations have been described in homozygous subjects, leading to an apparent receptor negativity (25). Class I is characterized by defective synthesis of the precursor. In the Class II mutation, the precursor is synthesised but is not transformed into a final receptor molecule. In this category, three subclasses have been indentified according to the resulting molecular mass. In Class III mutations, the final receptor molecule is formed in the cell but does not reach its final position in the membrane (see Chapter 5) (26).

The pathophysiological mechanism becomes more complex if one correlates the results of LDL-turnover studies in FH with genetic studies (19, 27–32). Healthy individuals carry about 10 000 LDL-receptors on the surface of a fibroblast and catabolize about 45% of their LDL pool per day. Fibroblasts from FH heterozygotes have about 4500 LDL-receptors per cell, and, accordingly, catabolize only 25–30% of their LDL pool per day. Finally, there are no detectable receptors on the fibroblast surface of homozygote FH subjects, but these patients do still eliminate about 15% of their LDL plasma pool per day. Therefore, one has to assume that there are 2 pathways of LDL catabolism, one depending on the specific LDL receptor and the other one working independently (33). In FH homozygotes, the LDL receptor independent or "non-specific" uptake mechanism is the only one capable of reducing plasma LDL concentration. Most of this non-specific uptake capability is thought to reside within cells of the reticuloendothelial system, such as macrophages or histiocytes. Considering the phagocytic nature of this cellular system the non-specific pathway of LDL catabolism has also been called the "scavenger pathway" (33, 34). As a clinical proof of the existence of this pathway, FH homozygotes accumulate cholesterol in fairly large quantities in spleen macrophages, Kupffer cells, and the histiocytes of the bone marrow.

While it appears that the reduced receptor-mediated uptake of LDL could account for most of the hypercholesterolemia, most LDL turnover studies of LDL performed in FH homozygotes or heterozygotes have shown that LDL synthesis also seems to be increased two to three times (28, 35). Such an overproduction of LDL has also been shown in an animal model of hypercholesterolemia, the Watanabe heritable hyperlipidemic (WHHL) rabbit. In this model it appears that the hepatic uptake of remnant particles is so much impaired that, eventually, almost all remnant particles are transformed into LDL (36).

It had been proposed for a long time that the liver plays the key role in the uptake of LDL particles from plasma. The validity of this concept for familial hypercholesterolemia has recently been shown in a spectacular clinical and therapeutic experiment. When a healthy liver together with a heart was transplanted into a child suffering from extremely severe homozygous FH, it turned out that, within a short period of time, the plasma LDL levels showed a dramatic decrease in the order of 70–80% of the initial value. On the basis of in vitro and animal studies, exactly that percentage had been postulated as the hepatic contribution to the clearance of LDL from the plasma compartment (37).

3 Familial Dysbetalipoproteinemia (Type III)

While the role of FH as a predisposing condition for coronary heart disease has never been seriously questioned, the situation is still somewhat unclear concerning the role of an increased concentration of relatively triglyceride-rich lipoprotein particles, such as those intermediates of chylomicron and VLDL metabolism, which may temporarily increase postprandially in normal subjects. However, the situation is quite different in a genetic disorder called familial dysbetalipoproteinemia or hyperliproproteinemia type III, where these particles are permanently elevated. There is no doubt that this condition leads to premature atherosclerosis in the coronary and peripheral arteries besides the clinical stigmata of planar palmar xanthomas, tuberous xanthomas and tubero-eruptive xanthomas in different areas of the skin (19). In terms of gross abnormalities of plasma lipids and lipoproteins, this condition is characterized by an elevated triglyceride and cholesterol and by an increased concentration of Apo E. The main feature in the lipoprotein spectrum is the accumulation of β-VLDL particles which are cholesterol-rich (38).

Genetically, the disorder is characterized by a defect in Apo E, an apolipoprotein with a polypeptide chain of 299 amino acids (39). This protein is coded for by a gene present on chromosome 19 (40) with several alleles on one locus (41, 42). Utermann and coworkers discovered the genetic polymorphism of Apo E in this disorder in 1977 (43). Apo E can be separated into three variants (Apo E_2, E_3 and E_4) by isoelectric focusing according to the charge of the polymorphic forms. The primary sequence of Apo E and the amino acid substitutions have been described as follows: in Apo E_2 cysteine occupies positions 112 and 158. In Apo E_3 cysteine is located in position 112 and an arginine residue in position 158. Arginine is found both in position 112 and 158 in Apo E_4. This polymorphism leads to three homozygous (E_2/E_2, E_3/E_3, E_4/E_4) and to three heterozygous constellations (E_2/E_3, E_2/E_4 and E_3/E_4) (44). Almost all subjects with familial dysbetalipoproteinemia (95%) are homozygous for E_2/E_2 (45). It is clear that Apo E plays a major role as a receptor ligand for both peripheral cells and hepatocytes. *In vitro* studies using fibroblasts and other peripheral cells in culture, demonstrated that Apo E was bound to the LDL receptor with an affinity which was 10–20 times superior to that of Apo B-100 (46–48). In the liver, there seems to be a specific so called "remnant receptor" which recognizes Apo E-rich particles. The different isoforms of Apo E are bound to a different extent by these receptors (49). Apo E_2 incorporated into liposomes and incubated with liver cells only shows 1% of the affinity of Apo E_3 for the hepatic receptor.

It should also be mentioned that besides the classical Apo E_2 isoform, several other variant Apo E_2 forms have been recently described (40, 41). The isoform coded for by allele $\varepsilon 2^*$ has an arginine for cysteine substitution in position 145, while the second one (allele $\varepsilon 2^{**}$) has a lysine for glutamine substitution in position 146. Interestingly, these Apo E_2 subspecies interact more strongly with the Apo E receptor when compared with the Apo E_2 coded for by allele $\varepsilon 2$. Pursuing that line of evidence it has been recently shown with monoclonal antibodies directed to different regions of the Apo E molecule that the receptor binding domain seems to be located between residues 140 and 150 (51, 52). The type III phenotype has also been described in other mutations (35). In these cases Apo E_3 was either biologically active (53) or inactive (54), or it showed significant modifications of its electrical charge (55). The type III phenotype has also been described in apolipoprotein E deficiency (56).

In terms of the epidemiology of this genetic disorder it is interesting to note that the phenotype Apo E_2/E_2 occurs widely in caucasian populations (1 in 100 Europeans or Americans). On the other hand, familial dysbetalipoproteinemia is much less prevalent (1 case in 1000 subjects) (19). This apparent discrepancy between gene frequency and clinical manifestation necessitates the implication of other factors in the pathogenesis of the disease. These may be other genetic factors such as heterozygosity for familial combined hyperlipidemia (57) or FH (58).

In some cases, glucose intolerance, overweight or endocrine disorders seem to play a role. In females, in particular, the decrease of natural estrogens during the menopause may lead to the manifestation of type III, while therapy with synthetic estrogens ameliorates the hyperlipoproteinemia (59). In animal experiments it has been demonstrated that estradiol increases LDL receptor activity in hepatocytes (60). Thyroid hormones appear to have a similar effect on hepatic receptors, which explains why hypothyroidism facilitates the clinical manifestation of type III hyperlipoproteinemia (61). It has also been shown that ageing seems to reduce the relative number of LDL receptors which could also explain why type III hyperlipoproteinemia usually becomes clinically apparent at a more advanced age than, for example, familial hypercholesterolemia (62).

4 Hypertriglyceridemia

While there is good clinical evidence both in familial hypercholesterolemia and in familial dysbetalipoproteinemia for an increased risk of coronary heart disease, the role of pure hypertriglyceridemia is much more controversial. Among the very triglyceride-rich and cholesterol-poor particles chylomicrons are probably least atherogenic. Premature atherosclerosis does not seem to exist in familial hyperchylomicronemia (type I hyperlipoproteinemia). (19). In type V hyperlipoproteinemia, where fasting chylomicronemia is combined with elevated concentrations of other triglyceride-rich lipoproteins of the VLDL spectrum, the situation is not as clear-cut. In these two conditions, the clinical consequences seem to be less directed at the arterial wall but rather against organs such as the pancreas (63–68). At any rate, premature atherosclerosis seems to be the exception rather than the rule even in the most severe forms of type V hyperlipoproteinemia.

The most commonly encountered disturbance of triglyceride metabolism, i. e. an increase of VLDL alone, (called phenotype IV), is also very heterogeneous in terms of lipoprotein particle population. Some prospective studies have shown a direct correlation between the VLDL concentration as a whole and coronary risk using different methods of analysis (3, 5, 69). On the other hand, the Framingham and Göteborg studies, for example (5, 16), did not show any direct correlation between plasma triglycerides and coronary risk independent of other factors such as plasma cholesterol, HDL cholesterol, obesity, or diabetes. However, it has to be emphasized that these studies did not evaluate VLDL concentrations directly. This is important to note because it has been recently pointed out that there is a significant heterogeneity of VLDL particles in type IV hyperlipoproteinemia with some particles being enriched with cholesterol while others are not (70, 71). In this context, it has also to be mentioned that familial combined hyperlipoproteinemia, i. e. the existence of multiple lipoprotein phenotypes in a family seems to carry a significant risk of coronary heart disease (72–75). This may be, again, associated with the existence of atherogenic particles in the total VLDL population such as the slow pre-β band described by De Gennes et al. (76).

5 Apolipoproteins and Coronary Heart Disease

In patients with coronary heart disease, total plasma Apo B levels are usually significantly elevated in the presence (77, 79, 80, 82, 84) or absence (78, 81, 83, 85) of a reduction of total plasma Apo AI. In normolipidemic patients with coronary heart disease plasma Apo B levels have been found to be elevated (77, 83). The biochemical basis of this constellation has been defined by Sniderman et al., who found that in hyperapobetalipoproteinemia the LDL particles were enriched in Apo B in relationship to their cholesterol content (86). These findings underscore the importance of the consideration of lipoprotein subfractions as defined by their apolipoprotein content and prove the importance of this approach beyond the mere measurement of total plasma and lipoprotein lipids. In other studies an overproduction of Apo B in LDL was shown in normolipidemic subjects with a normal LDL cholesterol level but with premature coronary heart disease (71, 87).

Two different molecular forms of Apo B, i. e. Apo B of hepatic origin (Apo B-100) and Apo B of mainly intestinal origin (Apo B-48) have been identified. The presence of Apo B is essential for coordinated metabolism of both triglyceride-rich and cholesterol-rich particles in the VLDL, IDL and LDL density range. Apo B mediates the removal of these particles by the LDL receptor (88–90).

In FH, an overproduction of Apo B-100 has been shown (91). It is interesting to note that an overproduction of Apo B has also been shown in familial hypertriglyceridemia, where the increased output of Apo B is associated with that of triglyceride-rich particles (71, 92, 93). This constellation seems to be quite atherogenic. Patients with familial combined hyperlipidemia overproduce Apo B, which, in part, might explain their very high coronary risk (71, 94).

On the basis of these studies it appears that a high level of circulating Apo B means a significant coronary risk. It is not as clear, however, whether the significant reduction of Apo AI such as the levels found in Tangier disease are to be considered as a risk factor. Conversely, the total absence of Apo AI or the combined absence of Apo AI and CIII (95–97) is very clearly associated with premature coronary heart disease. Familial Apo AII deficiency, on the other hand, leads to hypertriglyceridemia and hyperchylomicronemia imitating type I hyperlipoproteinemia. Premature atherosclerosis, however, is not found in this metabolic disorder (98). The relative increase of Apo C containing lipoprotein particles in VLDL and LDL over those of HDL density has been shown to represent an increased risk of angiographically proven heart disease (99). Correlation of plasma AII and coronary risk did not yield any clear cut results (82, 100). The importance of Apo E has been outlined when describing familial dyslipoproteinemia. It should be reemphasized that even in normolipidemic populations the isoform Apo E_2 seems to be an indicator of "latent dyslipoproteinemia" (101, 102, 103).

All these data suggest that a profile of plasma apolipoprotein values may be useful in the evaluation of coronary heart disease risk, although the final goal should be to identify by means of their apolipoprotein components atherogenic and non atherogenic lipoprotein particles (104–112). The use of monoclonal and oligoclonal antilipoprotein antibodies may improve differentiation of atherogenic and non atherogenic lipoprotein particles (107, 109, 113–115). If atherogenic and protective particles could indeed be characterized by these immunological approaches future prospective immunological trials could benefit from this new technique (110, 116).

6 Dyslipoproteinemia and Primary Prevention of Coronary Heart Disease

Having identified populations and individuals at risk by their lipid, lipoprotein and apolipoprotein profiles it is obviously necessary to prove that intervention directed at correcting these metabolic abnormalities can, indeed, influence the development of coronary heart disease. This has been attempted by studies trying to intervene on multiple risk factors and, also, by trying to alter individual risk factors.

6.1 Multiple risk intervention trials

The Oslo Heart Study, which utilized a medical caretaker system, tried to intervene by dietary treatment of hypercholesterolemia and abstention from smoking. This study very clearly showed that, by doing this, the incidence of coronary heart disease could be significantly reduced. In particular, a clear correlation between the relative reduction of cholesterol and the improvement in risk could be demonstrated (106). Another multiple risk intervention trial was organized by the European group of the World Health Organization. In this study a community health care system was used to provide services designed to change certain parameters in the intervention group as compared to the control group without intervention. In this study no consistent reduction of coronary risk could be shown (118). Another large scale study, the Multiple Risk Factor Intervention Trial (MRFIT) also did not show any significant improvement of the outcome of the intervention group compared to that of the control group (119). In this study, very intensive medical management was used as interventional strategy in order to treat hypercholesterolemia, hypertension and smoking. The control group was taken care of by the usual medical care system i. e. practicing physicians. The lack of a real difference between those two groups in terms of reduction of coronary risk can, however, be partly explained by the fact, that during the course of the trial the approach of the practicing physicians towards treating risk factors did significantly change. This led to a much better risk profile in the control group participants than had been expected.

6.2 Single risk factor trials: dietary and drug treatment of hyperlipidemia

The Lipid Research Clinics 'Coronary Primary Prevention Trial' (LRC-CPPT) was one of the trials specifically designed to definitely prove the importance of a reduction of LDL cholesterol for the prevention of coronary heart disease. The study was conducted double blind in a highly selected group of hypercholesterolemic patients with type II hyperlipoproteinemia . Over a total duration of between 7 and 10 years (mean 7.4 years) the participants were either treated with cholestyramine, a bile acid sequestrant resin, or with a placebo (120). Total cholesterol levels were reduced by an average of 8.5% as compared to the placebo group. This lowering of total cholesterol led to a decrease of coronary heart disease events. Fatal myocardial and nonfatal infarcts were reduced by 19% in the cholestyramine groups as compared to the placebo group. A more detailed analysis of the relationship between cholesterol lowering and risk reduction showed that the decrease of total cholesterol by 25% or of LDL cholesterol by 35% was associated with a risk reduction of 49% (121). The results of the LRC-CPPT were obtained in a relatively homogeneous population of patients with a relatively well defined disorder as the basis for coronary risk. It remains to be proven, however, that large scale intervention by diet or drug treatment is also efficacious in reducing coronary risk in the large group of patients suffering from an increase of predominantly triglyceride-rich lipoproteins. Using the most recent analytical tools, especially those directed towards apolipoprotein analysis, may provide the necessary means to identify atherogenic particles among the much more heterogeneous triglyceride-rich lipoprotein particle population. Using such analytical techniques it might be possible to show the benefit of intervention also in this group of hyperlipidemic patients.

References

1. McGee & Gordon, T. (1976) Section 31, DHEW Public No (NHI) *76*, 1083 (Pooling Project).
2. Intersociety Commission for Heart Disease Resources (Atherosclerosis Study Group and Epidemology Study Group) (1970) Circulation *42*, A54.
3. Carlson, L. A. & Böttiger, L. E. (1972) Lancet *1*, 865.
4. Keys, A. (1970) AHA Monograph No 29 (Suppl. 1) Circulation *41*, 1.
5. Wilhelmsen, L., Wedel, H. & Tibblin, G. (1973) Circulation *48*, 950.

6. Rosenman, R. H., Friedman, M., Jenkins, C. D., Straus, R., Wurm, M. & Kositchek, R. (1967) Amer. J. Cardiol. *19*, 771.
7. Doyle, J. T., Heslin, A. S., Hilleboe, H. E., Formel, P. F. & Korns, R. F. (1957) Amer. J. Public Health *47* (Suppl.), 25.
8. Chapman, J. M., Goerke, S., Dixon, W., Loveland, D. B. & Phillips, E. (1957) Amer. J. Public Health *47* (Suppl.), 33.
9. Stamler, J., Lindberg, H. A., Berkson, D. M., Schaffer, A., Miller, W. & Poindexter, A. (1960) J. Chronic Dis. *11*, 405.
10. Keys, A., Taylor, H. L., Blackburn, H., Brozek, J., Anderson, J. T. & Simonson, E. (1963) Circulation *28*, 381.
11. Kannel, W. B., Castelli, W. P. & Gordon, T. (1979) Ann. Intern. Med. *90*, 85.
12. Ducimetiere, P. (1985) Rev. Epidem. Santé Publique *33*, 331.
13. Keys, A. (1980) Seven countries. A multivariate analysis of death and coronary heart disease, Harvard University Press, Cambridge, Mass., London, p. 122.
14. Ducimetiere, P., Richard, J. L., Cambien, F., Rakotovao, R. & Claude, J. R. (1980) Lancet *1*, 1346.
15. Oliver, M. F. & Boyd, G. S. (1955) Brit. Heart. J. *17*, 299.
16. Gordon, T., Castelli, W. P., Hjortland, M. C., Kannel, W. B. & Dawber, T. R. (1977) Amer. J. Med. *62*, 707.
17. Glueck, C. J., Gartside, P., Fallat, R. W., Sielski, J. & Steiner, P. M. (1976) J. Lab. Clin. Med. *88*, 941.
18. Miller, N. E., Hamatt, F., Saltissi, S., Rao, S., Van Zeller, H., Coltart, J. & Lewis, B. (1981) Br. Med. J. *282*, 1741.
19. Goldstein, J. L., Brown, M. S. (1983) In: Metabolic basis of inherited disease (Stanbury, Wyngaarden, Fredrickson, Goldstein, Brown, eds.) 5th Ed., p. 672, McGraw Hill, New York.
20. Goldstein, J. L. & Brown, M. S. (1974) J. Biol. Chem. *249*, 5153.
21. Brown, M. S. & Goldstein, J. L. (1976) Science *181*, 150.
22. Tolleshaug, H., Goldstein, J. L., Schneider, W. J. et al. (1982) Cell *30*, 715.
23. Francke, U., Brown, M. S. & Goldstein, J. L. (1984) Proc. Natl. Acad. Sci. USA *81*, 2826.
24. Brown, M. S. & Goldstein, J. L. (1976) Cell *9*, 663.
25. Tolleshaug, H., Hobgood, K. K., Brown, M. S. et al. (1983) Cell *32*, 941.
26. Schneider, W. J., Brown, M. S. & Goldstein, J. L. (1983) Mol. Biol. Med. *1*, 355.
27. Shepherd J., Packard, C. J., Becker, S., Lawrie, T. D. V. and Morgan, H. G. (1980) N. Engl. J. Med. *302*, 1219.
28. Bilheimer, D. W., Stone, N. J. & Grundy, S. M. (1979) J. Clin. Invest. *64*, 524.
29. Simons, L. A., Reichl, D., Myant, N. B. & Mancini, M. (1975) Atherosclerosis *21*, 283.
30. Thompson, G. R. & Myant, N. B. (1976) Atherosclerosis *23*, 371.
31. Thompson, G. R., Spinks, T., Ranicar, A. & Myant, N. B. (1977) Clin. Sci. Mol. Med. *52*, 361.
32. Soutar, A. K., Myant, N. B. & Thompson, G. R. (1977) Atherosclerosis *28*, 247.
33. Goldstein, J. L. & Brown, M. S. (1977) Metabolism *26*, 1257.
34. Goldstein, J. L. & Brown, M. S. (1978) Johns Hopkins Med. *143*, 8.
35. Luc, G., Turpin, G. & De Gennes, J. L. (1986) Ann. Méd. Interne *137*, 155.
36. Kita, T., Brown, M. S., Bilheimer, D. W. et al. (1982) Proc. Natl. Acad. Sci. USA *79*, 5693.
37. Bilheimer, D. W., Goldstein, J. L., Grundy, S. M. et al. (1984) N. Engl. J. Med. *311*, 1658.
38. Mahley, R. W. & Angelin, B. O. (1984) Adv. Intern. Med. *29*, 385.
39. Rall, S. C., Weisgraber, K. H. & Mahley, R. W. (1982) J. Biol. Chem. *257*, 4171.
40. Gedde-Dahl, T., Olaisen, B. & Teisberg, P. (1984) Hum. Genet. *67*, 178.
41. Zannis, V. I., Just, P. W. & Breslow, J. L. (1981) Am. J. Hum. Genet. *33*, 11.
42. Gregg, R. E., Zech, L. A., Schaeffer, E. J. et al. (1981) Science *211*, 584.
43. Utermann, G., Hees, M. & Steinmetz, A. (1977) Nature *269*, 604.
44. Zannis, V. I. & Breslow, J. L. (1981) Biochemistry *20*, 1033.
45. Utermann, G., Jaeckle, M. & Menzel, H. J. (1975) FEBS Lett. *56*, 352.
46. Assmann, G., Brown, B. G. & Mahley, R. W. (1975) Biochemistry *14*, 3996.
47. Bersot, T. P., Mahley, R. W. & Brown, M. S. (1976) J. Biol. Chem. *251*, 2395.
48. Pitas, R. E., Innerarity, T. L., Arnold, K. S. et al. (1979) Proc. Natl. Acad. Sci. USA *76*, 2311.
49. Rall, S. C., Weisgraber, K. H. & Innerarity, T. L. (1982) Proc. Natl. Acad. Sci. USA *79*, 4696.
50. Rall, S. C., Weisgraber, K. H., Innerarity, T. L. et al. (1983) J. Clin. Invest. *72*, 1288.
51. Innerarity, T. L., Friedlander, E. J., Rall, S. C. et al. (1983) J. Biol. Chem. *258*, 12341.

52. Weisgraber, K. H., Innerarity, T. L., Harder K. J. et al. (1983) J. Biol. Chem. *258*, 12348.
53. Havel, R. J., Kotite, L., Kane, J. P. et al. (1983) J. Clin. Invest. *72*, 379.
54. Havekes, L. M., Gevers Leuven, J. A., Van Corven, E. et al. (1984) Eur. J. Clin. Invest. *14*, 7.
55. Gregg, R. E., Ghiselli, G. & Brewer, H. B. (1983) J. Clin. Endocrinol. Metab. *57*, 969.
56. Ghiselli, G., Schaeffer, E. J., Gascon, P. et al. (1981) Science *214*, 1239.
57. Hazzard, W. R., Warnick, G. R., Utermann, G. et al. (1981) Metabolism *30*, 79.
58. Hazzard, W. R., Miller, N. E., Albers, J. J. et al. (1981) Lancet *1*, 298.
59. Fainaru, M., Mahley, R. W., Hamilton, R. L. et al. (1982) J. Lipid Res. *23*, 702.
60. Kovanen, P. T., Brown, M. S. & Goldstein, J. L. (1979) J. Biol. Chem. *254*, 11367.
61. Thompson, G. R., Soutar, A. K., Spengel, F. A. et al. (1981) Proc. Natl. Acad. Sci. USA *78*, 2591.
62. Mahley, R. W., Hui, D. Y., Innerarity, T. L. et al. (1981) J. Clin. Invest. *68*, 1197.
63. Greenberg, B. K., Blackwelder, W. C. & Levy, R. I. (1977) Ann. Int. Med. *87*, 526.
64. Fallat, R. W. & Glueck, C. J. (1976) Atherosclerosis *23*, 41.
65. Lees, R. S., Wilson, D. E., Schoenfeld, G. & Fleet, S. (1966) In: Progress in Medical Genetics (Steinberg, A. G., Bearn, A. G., eds.) p. 429, McGraw-Hill, New York.
66. Farmer, R. G., Winkelmann, E. I., Brown, H. B. & Lewis, L. A. (1973) Amer. J. Med. *54*, 161.
67. Brunzell, J. D. & Bierman, E. L. (1982) Med. Clinics North America *66*, 455.
68. Nikkila, E. A. (1981) In: The Metabolic Basis of Inherited Disease (Stanbury, J. B., Wyngaarden, J. B., Frederickson, D. S., eds.), 5th Ed., McGraw-Hill, New York.
69. Kannel, W. B., Castelli, W. P., Gordon, T. & McNamara, P. M. (1971) Ann. Med. Interne *74*, 1.
70. Packard, C. J., Shepherd, J., Joerns, S., Gotto, R. M. & Taunton, O. D. (1980) Metabolism *29*, 213.
71. Grundy, S. M. & Vega, G. L. (1984) In: Latent Dyslipoproteinemias and Atherosclerosis (De Gennes, J. L., Polonovski, J., Paoletti, R., eds.) p. 165, Raven Press, New York.
72. Goldstein, J. L., Hazzard, W. R., Schrott, G., Bierman, E. L. & Motulsky, A. G. (1973) J. Clin. Invest. *52*, 1544.
73. Brunzell, J. D., Schrott, H. G., Motulsky, A. G. & Eierman, E. L. (1976) Metabolism *25*, 313.
74. Nikkila, E. A. & Aro, A. (1973) Lancet *1*, 954–958.
75. Hazzard, W. R., Goldstein, J. L., Schrott, H. G., Motulsky, A. G. & Birman, E. L. (1973) J. Clin. Invest. *52*, 1569.
76. De Gennes, J. L., Truffert, J. & Colas-Bellourt, J. F. (1975) Ann. Med. Interne *126*, 19.
77. Avogaro, P., Cazzolato, G., Bittolo-Bon, G., Rorai, E. & Pontoglio, E. (1982) In: Lipoproteins and coronary atherosclerosis (Noseda, G., GFragiacomo, C., Fumagelli, R., Paoletti, R., eds.) p. 123, Elsevier, Amsterdam.
78. De Backer, G., Rosseneu, M. & Deslypere, J. P. (1982) Atherosclerosis *42*, 197.
79. Riesen, W. D., Mordasini, R., Salzmann, C., Theler, A. & Gurtner, H. P. (1980) Atherosclerosis *37*, 157.
80. Whayne, T. F., Alaupovic, P., Curry, M. D., Lees, E. T., Anderson, P. S. & Schechter, E. (1981) Atherosclerosis *39*, 411.
81. Kladetzky, R. G., Assmann, G., Wagenbach, S., Tauchert, P. & Helb, H. D. (1980) Artery *7*, 191.
82. Desager, J. P., Rousseau, M., Riesen, W. F. & Harvengt, C. (1984) In: Latent Dyslipoproteinemias and Atherosclerosis (De Gennes, J. L., Polonovski, J., Padetti, R., eds.) p. 165, Raven Press, New York.
83. Douste-Blazy, Ph., Scheidegger, B., Camare, R., Marcel, Y. L., Marco, J. & Bernadet, P. (1984) In: Latent Dyslipoproteinemias and Atherosclerosis (De Gennes, J. L., Polonovsky, J., Paoletti, R., eds.) p. 179, Raven Press, New York.
84. Fruchart, J.-C., Bertrand, M., Parra, H., Gentilini, J. L., Boniface, B. & Boniface, M. (1982) Nouv. Presse Med. *11*, 3491.
85. Kupper, W., Hamm, C. W., Steinhagen-Thiessen, E., Daerr, W., Klose, G. & Greten, H. (1984) Circulation *70*, 129.
86. Snidermann, A., D., Shapiro, S., Marpole, D., Skinner, B., Teng, B. & Kwiterovich, P. O. (1980) Proc. Natl. Acad. Sci. USA *77*, 604.
87. Douste-Blazy, Ph., Marcel, Y. L., Cohen, L., Giroux, J. M. & Davignon, J. (1982) Ann. Intern. Med. *96*, 164.
88. Brown, M. S. & Goldstein, J. L. (1977) Trans. Ass. Amer. Physicians *90*, 91.
89. Goldstein, J. L. & Brown, M. S. (1974) J. Biol. Chem. *249*, 5163.

90. Brown, M. S. & Goldstein, J. L. (1984) Sci. Amer. *251*, 58.
91. Bilheimer, D. W., Goldstein, J. L., Grundy, S. M. & Brown, M. S. (1975) J. Clin. Invest. *56*, 1420.
92. Chait, A., Albers, J. J. & Brunzell, J. D. (1980) Eur. J. Clin. Invest. *10*, 17.
93. Snidermann, A., Wolfson, C., Teng, B., Franklin, F. A., Bachorik, P. S. & Kwiterovich, P. O. (1982) Ann. Int. Med. *97*, 833.
94. Janus, E. D., Nicholl, A. M., Turner, P. R., Magill, P. & Lewis, B. (1980) Eur. J. Clin. Invest. *10*, 161.
95. Schaefer, E. J. (NIH Publication, 1983), 54.
96. Schaefer, E. J., Heaton, W. H., Wetzel, M. G. & Brewer, H. B. (1982) Atheriosclerosis *2*, 16.
97. Norum, R. A., Lakier, J. B., Goldstein, S., Angel, A., Goldberg, R. B., Block, W. D., Noffze, D. K., Dolphin, P. J., Edelglass, M. D., Bogorad, D. D. & Alaupovic, P. (1982) N. Engl. J. Med. *306*, 1513.
98. Little, J. A., Cox, D., Breckenridge, W. C. & McGuire, V. M. (1980) In: Atherosclerosis V (Gotto, A. M., Smith, L. C., Allen, B., eds.) p. 671, Springer, Berlin, Heidelberg, New York.
99. Alaupovic, P. (1982) Ric. Clin. Lab. *12*, 3.
100. Riesen, W., Mordasini, R., Salzmann, C. & Gurtner, H. P. (1982) In: Lipoproteins and coronary atherosclerosis (Noseda, G., Fragiacomo, A., Fumagelli, R., Paoletti, R., eds.) p. 123, Elsevier, Amsterdam.
101. Davignon, J., Sing, C. F., Lussier-Cacan, S. & Bouthillier, D. (1984) In: Latent Dyslipoproteinemias and Atherosclerosis (De Gennes, J. L., Polonovski, J., Paoletti, R., eds.) p. 213, Raven Press, New York.
102. Utermann, G., Pruin, N. & Steinmetz, A. (1979) Clin. Genet. *15*, 63.
103. Utermann, G. (1983) In: Atherosclerosis VI (Schettler, G., Gotto, A. M., Middelhoff, G., Habenicht, A. J. R., Jurutka, K. R., eds.) pp. 611–615, Springer, Berlin, Heidelberg, New York.
104. Atmeh, R. F., Shepherd, J. & Packard, C. J. (1983) Biochim. Biophys. Acta *751*, 175.
105. Castro, G. R. & Fielding, C. J. (1984) J. Lipid Res. *25*, 53.
106. Cheung, M. C. & Albers, J. J. (1982) J. Lipid Res. *23*, 747.
107. Koren, E., McConathy, W. J. & Alaupovic, P. (1982) Biochem. *21*, 5347.
108. Alaupovic, P., McConathy, W. J. & Fesmire, J. C. NIH Publication, 1983, 27.
109. Koren, E., Solter, D., Knight, C. & Alaupovic, P. (1983) Circulation *68*, 111.
110. Fievet, C., Parra, H., Luyeye, I., Demarquilly, C., Fievet, P. Fruchart, J.-C., Bertrand, M., Lablanche, J. M., Drouin, P. & Gross, P. (1985) Monog. Atherosclerosis *13*, 134.
111. Puchois, P., Alaupovic, P. & Fruchart, J.-C. (1985) Ann. Biol. Clin. *43*, 831.
112. Puchois, P., Bertrand, M., Lablanche, J. M. & Fruchart, J.-C. (1985) Arteriosclerosis *5*, 513.
113. Curtiss, K. L. & Edgington, T. S. (1982) J. Biol. Chem. *257*, 15213.
114. Marcel, Y. L., Douste-Blazy, Ph. & Milne, R. W. (1984) In: Lippel, K., Workshop on antilipoprotein methods (NIH Publication) p. 414.
115. Fruchart, J.-C., Ghesquiere, J. C., Delpierre, C., Cachera, C. & Tartar, A. (1984) Clin. Chem. *30*, 992.
116. Patton, J. G., Badimon, J. J. & Mao, S. T. T. (1983) Clin. Chem. *29*, 1898.
117. Hjermann, I., Holme, I. & Byre, K. V. (1982) Lancet *2*, 1303.
118. World Health Organization (WHO) European Collaborative Group (1973) Eur. Heart J. *4*, 141.
119. Multiple Risk Factor Intervention Trial Research Group (1982) J. Amer. Med. Ass. *248*, 1465.
120. Lipid Research Clinics Program (1984) J. Amer. Med. Ass. *251*, 351.
121. Lipid Research Clinics Program (1984) J. Amer. Med. Ass. *251*, 365.

Subject Index

Concise Encyclopedia Biochemistry

Second Edition,
revised and expanded by *Thomas Scott* and *Mary Eagleson*

1988. 17 cm x 24 cm. 650 pages. Hardcover. ISBN 3 11 011625 1; 0-89925-457-8 (U.S.)

The only single work of its kind in English, the **Concise Encyclopedia of Biochemistry** provides a comprehensive, yet compact, source of biochemical data and information for the researcher, teacher, and student.

Following a five-year program of collecting and editing new material, as well as the revision of existing entries, the author-editors and the publishers are pleased to announce the new expanded Second Edition of this valuable reference work.

Major entries concerning the latest developments in DNA structure, synthesis, sequencing, binding proteins and methods, oncogenes, lymphokines and other newly discovered regulatory peptides, structural proteins, inositol phosphates, and protein kinases have been added. Graphic illustration has been given high priority, so that regulatory processes, transport, subcellular structures, etc. are abundantly and clearly illustrated.

The coverage of plant biochemistry has also been greatly expanded. Another new addition is a section on buffers which will be useful to anyone involved in laboratory work. Because of its comprehensiveness and multidisciplinary nature, we are sure that you will find it an indispensable reference tool.

Special features of this edition include:

- Approximately 4,500 entries
- Up-to-date, comprehensive
- Coverage of medical, animal, microbial, plant, and physical biochemistry, natural products, molecular biology, molecular genetics, and biotechnology
- Hundreds of illustrations, including structural formulas, schemes, and metabolic pathways
- Over 100 tables
- Modern terminology based on standard sources, e. g., IUB Enzyme Nomenclature
- Standard biochemical abbreviations
- Extensive cross references with synonyms provided
- Literature references are cited to aid the reader in locating original sources

Potential audience: biochemists, clinical biochemists, clinical chemists, medical researchers, clinicians, plant scientists, experimental biologists, lecturers and students of the life sciences.

W DE G **de Gruyter · Berlin · New York**

de Gruyter & Co., Genthiner Str. 13, D-1000 Berlin 30, Tel.: (0 30) 2 60 05-0, Telex 1 84 027, Telefax (0 30) 2 60 05-2 51
de Gruyter, Inc., 200 Saw Mill River Road, Hawthorne, N. Y. 10532, Tel. (914) 747-0110, Telex 64 66 77, Telefax (914) 7 47-13 26

Bei Fragen zur Produktsicherheit wenden Sie sich bitte an:
If you have any questions regarding product safety,
please contact:

Walter de Gruyter GmbH
Genthiner Straße 13
10785 Berlin
productsafety@degruyterbrill.com